Vector Measures

Mathematical
Surveys
and
Monographs

Volume 15

Vector Measures

J. Diestel
J. J. Uhl, Jr.

AMERICAN
MATHEMATICAL
SOCIETY

Providence, Rhode Island

AMS (MOS) subject classifications (1970). Primary 28A45; Secondary 28A15, 28A20, 46B10, 46B99, 46E15, 46E30, 46G05, 46G10, 47A65, 47B05, 47B10, 47B99, 52-00, 60G45.

Library of Congress Cataloging-in-Publication Data

Diestel, Joseph, 1943–
Vector measures.
(Mathematical surveys; no. 15)
Bibliography: p.
Includes indexes.
1. Vector-valued measures. 2. Banach spaces. 3. Linear operators. I. Uhl, John Jerry, 1940– joint author. II. Title. III. Series: American Mathematical Society. Mathematical surveys; no. 15.
QA312.D43 515′.73 77-9625
ISBN 0-8218-1515-6 (alk. paper)

Printed in the United States of America
Reprinted by the American Mathematical Society, 2025

The paper used in this book is acid-free and fall within the guidelines
established to ensure permanence and durability. ♾

13 12 11 10 9 8 7 30 29 28 27 26 25

FOREWORD

Much of the work on Banach spaces done in the 1930's resulted from investigating how much of real variable theory might be extended to functions taking values in such spaces. Members of E. H. Moore's school of general analysis at Chicago, including Graves and Hildebrandt, and functional analysts in Italy and Poland (Orlicz in particular) had already done pioneer work in convergence of functions, certain aspects of integration and differentiation, and the relationships between various convergence properties for series. In the 1930's Hildebrandt's group in Ann Arbor and Tamarkin's at Brown expanded the effort in the U.S.A., the strong Russian school developed, and the influence of the Polish group spread, via Banach's book, more deeply and widely. In developing integration and differentiation theory for functions defined on Euclidean space to a Banach space B in the period subsequent to Bochner's 1933 papers the important pioneer figures were Dunford and Gel'fand.

It was in the study of differentiation of functions on Euclidean figures that the role of the character of B emerged. Although some functions, such as Bochner integrals, were differentiable a.e. regardless of B, many were not, their differentiability depending on the characteristics of their range spaces; more precisely, it depended on what properties the function developed for its range set as a subset of B. (Clarkson invented uniformly convex spaces for the purpose of universal differentiation; reflexive spaces reappeared on the stage for the same purpose.) Moreover differentiation, aside from its intrinsic interest, was fundamental in efforts to represent linear operators by means of integrals, and when operations from spaces of functions whose domains were an abstract space were to be represented, differentiation had to be replaced by Radon-Nikodým theorems. Here Dunford led by proving the earliest R-N theorem (N. Dunford, *Integration and linear operations*, Trans. Amer. Math. Soc. **40** (1936), 474–494) and by giving the first proof of a R-N theorem, now well known, when B is a dual space (N. Dunford and B. J. Pettis, *Linear operations on summable functions*, Trans. Amer. Math. Soc. **47** (1940), 323–392; second proof). The study of Banach-space-valued functions waned in the 1940's, was revived and partly redirected by the deep work of Grothendieck, and generally relapsed again until late in the 1960's. Since then vigorous work by many here and in various parts of Europe and elsewhere has produced a flourishing body of results, a considerable amount of which has been organized and presented in the present volume in useful and no doubt fertile form. The notion of vector measures can be made central to a study of Banach-space-valued functions (series, integrals, differentiation, R-N theorems), to the representation and classification of linear operations between certain kinds of spaces, and the classification of Banach spaces. This is the view presented by the authors of this work, who display very effectively the interplay between properties of B and properties of vector measures taking their values in B, to the understanding of which they have themselves contributed substantially in recent years. Those who now or in the future work with Banach-space-valued functions or in the classification of geometric properties of Banach spaces, as well as those who have done so in the past, should be grateful to Professors Diestel and Uhl for their substantial contribution.

B. J. PETTIS

CONTENTS

INTRODUCTION

It seems to be well forgotten that many of the first ideas in geometry, basis theory and isomorphic theory of Banach spaces have vector measure-theoretic origins. Equally well forgotten is the fact that much of the early interest in weak and weak* compactness was motivated by vector measure-theoretic considerations.

In 1936, J. A. Clarkson introduced the notion of uniform convexity to prove that absolutely continuous functions on a Euclidean space with values in a uniformly convex Banach space are the integrals of their derivatives. At the same time, Clarkson used vector measure-theoretic ideas to prove that many familiar Banach spaces do not admit equivalent uniformly convex norms.

N. Dunford and A. P. Morse, in 1936, introduced the notion of a boundedly complete basis to prove that absolutely continuous functions on a Euclidean space with values in a Banach space with a boundedly complete basis are the integrals of their derivatives. Shortly thereafter Dunford was able to recognize the Dunford-Morse theorem and the Clarkson theorem as genuine Radon-Nikodým theorems for the Bochner integral. This was the first Radon-Nikodým theorem for vector measures on abstract measure spaces.

B. J. Pettis, in 1938, made his contribution to the Orlicz-Pettis theorem for the purpose of proving that weakly countably additive vector measures are norm countably additive.

In 1938, I. Gel'fand used vector measure-theoretic methods to prove that $L_1[0, 1]$ is not isomorphic to a dual of a Banach space.

In 1939, Pettis showed that the notions of weak and weak* compactness are intimately related to the problem of differentiating vector-valued functions on Euclidean space. Dunford and Pettis, in 1940, built on their earlier work to represent weakly compact operators on L_1 and the general operator from L_1 to a separable dual space by means of a Bochner integral. By means of their integral representation they were able to prove that L_1 has the property now known as the Dunford-Pettis property.

Then came the war. By the end of the war, the love affair between vector measure theory and Banach space theory had cooled. They began to drift down separate paths. Neither prospered. Much of Banach space theory became lost in the mazes of the theory of locally convex spaces. The work in vector measure theory became little more than formal generalizations of the scalar theory. Representation theory for operators on function spaces became the vogue. But all too often these representation theories gave no new information about the operators they represented. During the fifties and early sixties the theory of vector measures languished in sterility.

There were one or two bright spots. In the mid-fifties, A. Grothendieck used the then ignored vector measure theory of the late thirties and early forties to launch a monumental study of linear operators. The repercussions of Grothendieck's work are still being felt today. Also in the mid-fifties, Grothendieck and (independently) R. G. Bartle, N. Dunford and J. T. Schwartz studied operators on spaces of continuous functions and proved the first important theorems in the theory of vector measures in some fifteen years. But the unfortunate truth is that, aside from I. Kluvanek and a few others, no one followed their lead.

In the early sixties, largely through the pioneering work of A. Pełczyński and J. Lindenstrauss, Banach space theory came back to life and today has re-emerged as a deep and vigorous area of mathematical inquiry. Vector measure theory did not come around so quickly.

In the mid-sixties, N. Dinculeanu gave an intensive study of many of the theorems of vector measure theory that had been proven between 1950 and 1965. Dinculeanu's monograph was the catalytic agent that the theory of vector measures needed. Upon the appearance of Dinculeanu's book, interest in vector measures began to grow. It was not long before a number of mathematicians addressed themselves to the basic unsolved problems of vector measure theory. The study of the Radon-Nikodým theorem for the Bochner integral and the Orlicz-Pettis theorem served to re-establish the links between vector measures and the analytic, geometric and isomorphic theory of Banach spaces. Today the theory of vector measures stands as a hearty cousin and proud servant of the theory of Banach spaces. This survey is a report on how this has come about.

We endeavor to give a comprehensive survey of the theory of vector measures as we see it. It is our overriding desire to emphasize the fruitful (and we think exciting) interplay between properties of Banach spaces and measures taking values in Banach spaces. Thus the exposition of the relationships among vector measures, operators on L_1, operators on spaces of bounded measurable functions, topological structure of Banach spaces and geometric structure of Banach spaces is our unifying theme. We feel that any attempt to divorce vector measures from these latter areas would wallow in artificiality.

This survey is written for the student as well as the advanced mathematician. Much of it originated in lectures given by the authors at Kent State University and the University of Illinois. Other parts of the survey have grown from conversations with our colleagues in the classroom and other places where mathematicians gravitate to talk. We assume that the reader has some familiarity with basic Banach space theory as presented in Chapters II, V and VI of *Linear operators* by Dunford and Schwartz[1] and with basic measure theory as presented in Bartle's *Elements of integration* or Halmos's *Measure theory*. Other than this, this survey is self-contained.

[1]It may be noted that much of this survey is an outgrowth of Chapters IV and VI of Dunford and Schwartz.

The first chapter deals with countably additive and finitely additive vector measures. The basic behavior of countably additive measures is presented from the viewpoint of the fundamental work of Bartle, Dunford and Schwartz and Kluvanek. We base the theory of finitely additive vector measures on Rosenthal's lemma. With the help of this lemma, we examine the roles of the spaces c_0 and l_∞ in the theory of vector measures. Included here are the Vitali-Hahn-Saks-Nikodým theorem and the Nikodým boundedness theorem for finitely additive vector measures.

The second chapter, which is for the most part independent of Chapter I, is devoted to measurable functions with values in Banach spaces and the problem of integrating them. The Bochner integral receives most of the attention, but basic material on the Pettis, Dunford and Gel'fand integrals is found in this chapter.

The Radon-Nikodým theorem for the Bochner integral is the subject of Chapter III. We try to follow the genetic approach of treating the Radon-Nikodým theorem and L_1 operator theory as one unified theory. The analytic (i.e., topological) aspects of the Radon-Nikodým theorem are found here. The roles of compact operators on L_1, weakly compact operators on L_1, reflexive spaces, separable dual spaces and weakly compactly generated dual spaces in the Radon-Nikodým property are also discussed here.

Chapter IV continues with a potpourri of applications of the Radon-Nikodým theorem for the Bochner integral. The duals of the L_p-spaces of Bochner integrable functions are derived and weak compactness in the space of Bochner integrable functions is discussed. The relationship between differentiable vector-valued functions of a real variable and the Radon-Nikodým theorem is next. Then the relationship between the classical integral operators on L_p and the Bochner integral is surveyed. The chapter concludes with the Lewis Stegall theorem on complemented subspaces of L_1.

Martingales of Bochner integrable functions headline Chapter V. In addition to martingale convergence theorems, we observe a basic phenomenon in the theory of vector measures. Through a meld of the Radon-Nikodým theorem, elementary martingale theory, and geometry of Banach spaces, we see the Radon-Nikodým property transfer itself from an analytic property of Banach spaces to a geometric property of Banach spaces. This is the link between geometry and measure theory in Banach spaces.

Structural properties of operators on spaces of continuous functions $C(\Omega)$ are under scrutiny in Chapter VI. The basic work of Bartle, Dunford and Schwartz and Grothendieck is discussed from the viewpoint of Chapter I in the first part. The second part deals with absolutely summing and nuclear operators on $C(\Omega)$ and their relationship with the Radon-Nikodým theorem. Included here is a discussion of Pietsch integral operators on Banach spaces.

The seventh chapter builds on the martingale theory of Chapter V to give an exposition of the repercussions of the Radon-Nikodým theorem in the geometry of Banach spaces. Studied here are the relationships among the Radon-Nikodým theorem, the Kreĭn-Mil'man theorem, properties of strongly exposed points, and other extreme point phenomena. This chapter can be read directly after Chapter V.

Tensor products of Banach spaces and how the Radon-Nikodým theorem can be used within the theory of tensor products to study Banach spaces is the theme of Chapter VIII.

Chapter IX concludes the survey with a discussion of the Liapounoff convexity theorem and other geometric properties of the range of a vector measure.

At the end of each chapter is a section called "Notes and remarks." These sections, which are modeled after similiar sections in Dunford and Schwartz, attempt to discuss the original and subsequent versions of the results presented in the chapter in question. Sometimes they contain additional results, often with proofs, that could not be fitted into the main text. In each of these sections, there is an attempt to discuss additional results that bear on the theorems presented in the text. Sometimes these discussions contain terminology that is not defined in the text. Usually the terminology is standard. When in doubt, the reader should consult the appropriate reference.

We envision that this survey will be useful in a variety of ways. Those who want to study the Radon-Nikodým theorem and its relation to the topological and geometric structure of Banach spaces should read Chapters II, III, V and VII. Those who have an additional interest in applications of the Radon-Nikodým theorem may also want to look at Chapters IV, VI and VIII. Those who want to study measures of unbounded variation can read Chapter I, the first part of Chapter VI and Chapter IX. We have attempted to minimize the introduction of weighty terminology and notation. Thus it should be possible for someone who has not read the early chapter to be able to understand the content of a theorem in a late chapter with a minimum of frustration and page turning. We hope that this will make this survey useful for spot references.

The numbering of theorems is the same as in Dunford and Schwartz; thus Theorem V.2.6 is the sixth numbered item in the second section of the fifth chapter. Within the second section of the fifth chapter this theorem is referred to as Theorem 6; within the other sections of the fifth chapter this theorem is referred to as Theorem 2.6. Elsewhere it is referred to as V.2.6.

We hope our terminology is standard. To prevent any doubt let us fix some terminology. When we say that (Ω, Σ, μ) is a measure space, we mean that μ is an extended real-valued nonnegative countably additive measure defined on a σ-field Σ of subsets of a point set Ω. The triple (Ω, Σ, μ) is called a finite measure space if it is a measure space and $\mu(\Omega)$ is finite. A subset of a Banach space is called relatively norm (weakly) compact if its norm (weak) closure is norm (weakly) compact. A subset of a Banach space is called conditionally weakly compact if every sequence in it has a weakly Cauchy subsequence. If X and Y are Banach spaces, $\mathscr{L}(X, Y)$ stands for the space of bounded linear operators from X to Y; the space X contains a copy of Y if X has a subspace that is linearly homeomorphic to Y. There is one theorem that will be used from time to time and that may be unfamiliar to some readers. This is Stone's representation theorem which says that if $\mathscr{F}$ is a field of subsets of a point set Ω, then there is a compact Hausdorff space Ω_1, a field Σ_1 consisting of subsets of Ω_1 that are both closed and open, and there is a Boolean isomorphism between $\mathscr{F}$ and $\mathscr{F}_1$. The field $\mathscr{F}_1$ will be called the Stone representation algebra of $\mathscr{F}$.

Some readers may find some serious omissions in this survey. We deal with finite measure spaces only. Many of the theorems we present here have extensions to more general situations; some do not. When a theorem has an extension to more general situations, its extension is usually a routine extension. We feel that capturing this extra bit of generality is not worth the space and its inclusion would obscure the exposition in a mass of trivial details.

Although we treat in detail the representation of operators on L_1, L_∞ and $C(\Omega)$, the representation of the general operator on L_p for $1 < p < \infty$ is conspicuously absent. Our reason for this is that we do not know any applications of this representation theory. However we do study some important classes of operators on L_p.

A third omission is the integration and differentiation theory for functions that are not norm (strongly) measurable. We are quick to admit that an extensive theory exists for such functions. We know of very few honest applications of this theory. Additional omissions include measures with values in linear topological spaces other than Banach spaces, orthogonally scattered measures, vector-valued stochastic processes (other than martingales) and the lifting theory for vector-valued functions. A very serious omission is most of the material found in the monograph of Igor Kluvanek and Greg Knowles (*Vector measures and control systems*, North-Holland, Amsterdam, 1976). Those who desire more material on the range of a vector measure than found in Chapter IX or who want to study infinite dimensional control theory should consult this spendid volume.

During the preparation of this survey, we have been helped immeasurably by a number of our colleagues who have freely contributed their advice and criticism. A partial list of those to whom we owe our heartfelt thanks is: R. G. Bartle, W. J. Davis, M. M. Day, L. Dor, G. A. Edgar, B. T. Faires, T. Figiel, J. Hagler, R. E. Huff, J. A. Johnson, W. B. Johnson, N. J. Kalton, R. Kaufman, I. Kluvanek, G. Knowles, D. R. Lewis, H. B. Maynard, P. D. Morris, T. J. Morrison, R. E. Olson, N. T. Peck, A. Pełczyński, A. L. Peressini, B. J. Pettis, R. R. Phelps, H. P. Rosenthal, E. Saab, C. J. Seifert, T. W. Starbird, F. E. Sullivan, J. B. Turett and A. Vento.

We also owe a measure of gratitude to the editors of this series, especially R. G. Bartle, P. R. Halmos and M. Rosenlicht for wrestling with our sometimes unconventional style. We are much indebted to Carolyn Bloemker and Kathy Morrison for typing this survey. Their job was not easy. Finally we thank Linda Diestel for putting up with both of us.

As we progressed in the study of the history of the basic theorems of the theory of vector measures, we were not surprised by learning that most of them, in one way or another, have their origins in the fertile mind of one man, B. J. Pettis, who was kind enough to give us the benefit of his wisdom on many matters and to agree to write the foreword. To this mathematician and gentleman we dedicate our work.

KENT, OHIO — J. DIESTEL

URBANA, ILLINOIS — J. J. UHL, JR.

I. GENERAL VECTOR MEASURE THEORY

Grubby set-theoretic manipulations cannot be avoided in measure theory and most of them are found in this chapter. This is not all bad because they are at the base of a number of fundamental theorems of vector measure theory and Banach space theory. The first section introduces the notions of variation, semivariation, strong additivity (s-boundedness) and countable additivity. Also in this section is a brief look at integration with respect to a vector measure. Most of this section consists of straightforward manipulations of definitions.

§2 is a basic section which examines the essential properties of countably additive vector measures on σ-fields. Here the Bartle-Dunford-Schwartz theorems are found. §3 continues with an exposition of the Nikodým Boundedness Theorem.

Rosenthal's lemma forms the core of §4 which is one of the major sections of this book. In this section the interchange between the spaces c_0 and l_∞ and vector measure theory begins to emerge. From this perspective, the Orlicz-Pettis theorem, the Bessaga-Pełczyński c_0 theorem, and the Vitali-Hahn-Saks-Nikodým theorem are deduced very simply.

The last section begins with the Carathéodory-Hahn-Kluvanek Extension Theorem for vector measures. Then by Stone space arguments it is shown that strongly additive vector measures have almost all the properties of countably additive vector measures except countable additivity. The section concludes with derivations of the Yosida-Hewitt and Lebesgue decomposition theorems for vector measures.

1. Elementary properties of vector measures. This section deals with basic straightforward properties of vector measures. The familiar notions of variation and countable additivity are introduced together with the concepts of semivariation and strong additivity. Finally, an elementary integral is introduced to help establish a basic relationship between vector measures and operators on spaces of bounded measurable functions.

DEFINITION 1. A function F from a field $\mathscr{F}$ of subsets of a set Ω to a Banach space X is called a *finitely additive vector measure*, or simply a *vector measure*, if whenever E_1 and E_2 are disjoint members of $\mathscr{F}$ then $F(E_1 \cup E_2) = F(E_1) + F(E_2)$.

If, in addition, $F(\bigcup_{n=1}^\infty E_n) = \sum_{n=1}^\infty F(E_n)$ in the norm topology of X for all sequences (E_n) of pairwise disjoint members of $\mathscr{F}$ such that $\bigcup_{n=1}^{\infty} E_n \in \mathscr{F}$, then F is termed a *countably additive vector measure* or simply, F is countably additive.

EXAMPLE 2. *A finitely additive vector measure.* Let T: $L_\infty[0, 1] \to X$ be a continuous linear operator. For each Lebesgue measurable set $E \subseteq [0, 1]$, define $F(E)$ to be $T(\chi_E)$ (χ_E denotes the characteristic or indicator function of E). Then by the linearity of T, F is seen to be a finitely additive vector measure which may—even in the case that X is the real numbers—fail to be countably additive. The simplest such general example of a noncountably additive measure is provided by considering any Hahn-Banach extension to $L_\infty[0, 1]$ of a point mass functional on $C[0, 1]$.

EXAMPLE 3. *A countably additive vector measure.* Let T: $L_1[0, 1] \to X$ be a continuous linear operator. Again define $F(E) = T(\chi_E)$ for each Lebesgue measurable set $E \subseteq [0, 1]$. Then F is evidently finitely additive. Moreover, for each E, one has $\|F(E)\| \leqq \lambda(E)\|T\|$. Consequently, if $(E_n)_{n=1}^{\infty}$ is a sequence of disjoint Lebesgue measurable subsets of $[0, 1]$, then

$$\begin{aligned}\lim_m \left\| F\left(\bigcup_{n=1}^{\infty} E_n\right) - \sum_{n=1}^{m} F(E_n)\right\| &= \lim_m \left\| F\left(\bigcup_{n=m+1}^{\infty} E_n\right)\right\| \\ &\leqq \lim_m \lambda\left(\bigcup_{n=m+1}^{\infty} E_n\right)\|T\| = 0.\end{aligned}$$

All the measures produced via Example 3 have a property isolated by the first part of

DEFINITION 4. Let F: $\mathscr{F} \to X$ be a vector measure. The *variation of* F is the extended nonnegative function $|F|$ whose value on a set $E \in \mathscr{F}$ is given by

$$|F|(E) = \sup_\pi \sum_{A \in \pi} \|F(A)\|,$$

where the supremum is taken over all partitions π of E into a finite number of pairwise disjoint members of $\mathscr{F}$. If $|F|(\Omega) < \infty$, then F will be called a *measure of bounded variation.*

The *semivariation of* F is the extended nonnegative function $\|F\|$ whose value on a set $E \in \mathscr{F}$ is given by

$$\|F\|(E) = \sup\{|x^*F|(E): x^* \in X^*, \|x^*\| \leqq 1\},$$

where $|x^*F|$ is the variation of the real-valued measure x^*F. If $\|F\|(\Omega) < \infty$, then F will be called a *measure of bounded semivariation.*

Direct verifications show that the variation of F is a monotone finitely additive function on $\mathscr{F}$, while the semivariation of F is a monotone subadditive function on $\mathscr{F}$. Also it is easy to see that for each $E \in \mathscr{F}$ one has $\|F\|(E) \leqq |F|(E)$.

Now Examples 2 and 3 will be re-examined from the point of view of variations and semivariations.

EXAMPLE 5. *A measure of bounded variation.* Let F be a measure of the type discussed in Example 3. Since $\|F(E)\| \leqq \|T\|\lambda(E)$, it is plain that $|F|(E) \leqq \|T\|\lambda(E)$, so that F is of bounded variation.

EXAMPLE 6. *A measure of bounded semivariation but not of bounded variation.* Let Σ be the σ-field of Lebesgue measurable subsets of $[0, 1]$ and define F: $\Sigma \to L_\infty[0, 1]$, by $F(E) = \chi_E$. If $E \in \Sigma$ and $\lambda(E) > 0$, select a disjoint sequence (E_n) of subsets of E each with positive measure such that $\bigcup_n E_n = E$. Set $\pi_n = \{E_1, E_2, \cdots, E_{n-1}, \bigcup_{k=n}^{\infty} E_k\}$. Then for each n,

$$\sum_{A\in\pi_n} \|F(A)\| = \sum_{k=1}^{n-1} \|\chi_{E_k}\| + \|\chi_{\bigcup_{k=n}^{\infty} E_k} \bigcup_{k=n}^{\infty} E_k\| = n.$$

Accordingly $|F|(E)$ is infinite. The fact that this measure is of bounded semivariation is a consequence of

EXAMPLE 7. *Vector measures of bounded semivariation.* Let $T: L_\infty[0, 1] \to X$ be a continuous linear operator and for a Lebesgue measurable set $E \subseteq [0, 1]$ define $F(E) = T(\chi_E)$. If $x^* \in X^*$ and $\|x^*\| \leqq 1$ and π is a partition of $[0, 1]$ into Lebesgue measurable sets, then

$$\begin{aligned}\sum_{A\in\pi} |x^*F(A)| &= \sum_{A\in\pi} |x^*T\chi_A| = \sum_{A\in\pi} \operatorname{sgn} x^*T\chi_A x^*T\chi_A \\ &= x^*T\Big(\sum_{A\in\pi} (\operatorname{sgn} x^*T\chi_A)\chi_A\Big) \\ &\leqq \|x^*T\| \Big\|\sum_{A\in\pi} (\operatorname{sgn} x^*T\chi_A)\,\chi_A\Big\| \leqq \|T\|.\end{aligned}$$

Thus F is of bounded semivariation.

EXAMPLE 8. *A measure of unbounded semivariation.* Although little can be said of such measures it is worth noting that a vector measure (in fact, a real-valued measure) need not be of bounded semivariation. Indeed, if $\mathscr{F}$ is the field of subsets of $\boldsymbol{N}$, the positive integers, consisting of sets that are either finite or have finite complements, then the measure $F: \mathscr{F} \to \boldsymbol{R}$ defined by

$$\begin{aligned}F(E) &= \text{cardinality of } E, &&\text{if } E \text{ is finite,}\\ &= -\text{ cardinality of } \boldsymbol{N}\backslash E, &&\text{if } \boldsymbol{N}\backslash E \text{ is finite,}\end{aligned}$$

produces an example of real-valued measure with unbounded semivariation.

Of some use is the easily verified fact that if $F: \mathscr{F} \to X$ is a vector measure of bounded variation, then a nonnegative measure μ on $\mathscr{F}$ is the variation $|F|$ of F if and only if μ satisfies: (i) $|x^*F|(E) \leqq \mu(E)$ for all $E \in \mathscr{F}$ and all $x^* \in X^*$ with $\|x^*\| \leqq 1$, and (ii) if $\lambda: \mathscr{F} \to \boldsymbol{R}$ is any measure satisfying $|x^*F|(E) \leqq \lambda(E)$ for all $E \in \mathscr{F}$ and all $x^* \in X^*$ with $\|x^*\| \leqq 1$ then $\mu(E) \leqq \lambda(E)$ for all $E \in \Sigma$. In terms of the lattice structure of the space of set functions, $|F|$ is the least upper bound (if it exists) of the collection $\{|x^*F|: x^* \in X^* \text{ and } \|x^*\| \leqq 1\}$.

PROPOSITION 9. *A vector measure of bounded variation is countably additive if and only if its variation is also countably additive.*

PROOF. Suppose $F: \mathscr{F} \to X$ is of bounded variation. Since $\|F(E)\| \leqq |F|(E)$ for each $E \in \mathscr{F}$, it is plain that F is countably additive if $|F|$ is countably additive.

Conversely, suppose that $F: \mathscr{F} \to X$ is a countably additive vector measure of bounded variation. Let (E_n) be a sequence of pairwise disjoint members of $\mathscr{F}$ such that $\bigcup_n E_n \in \mathscr{F}$ and let π be a partition of $\bigcup_n E_n$ into pairwise disjoint members of $\mathscr{F}$. Then

$$\begin{aligned}\sum_{A\in\pi} \|F(A)\| &= \sum_{A\in\pi} \Big\|F\Big(A \cap \bigcup_n E_n\Big)\Big\| = \sum_{A\in\pi} \Big\|\sum_n F(A \cap E_n)\Big\| \leqq \sum_{A\in\pi} \sum_n \|F(A \cap E_n)\| \\ &= \sum_n \sum_{A\in\pi} \|F(A \cap E_n)\| \leqq \sum_n |F|(E_n).\end{aligned}$$

Since this holds for any partition π, the inequality $|F|(\bigcup_n E_n) \leqq \sum_n |F|(E_n)$ obtains.

But now recall that $|F|$ is finitely additive and monotone on F. Thus for each n

$$\sum_{k=1}^{n} |F|(E_k) = |F|\left(\bigcup_{k=1}^{n} E_k\right) \leqq |F|\left(\bigcup_{n} E_n\right).$$

This proves the reverse inequality $\sum_n |F|(E_n) \leqq |F|(\bigcup_n E_n)$ and shows that $|F|$ is countably additive on $\mathscr{F}$.

COROLLARY 10. *Let Σ be a σ-field generated by a subfield $\mathscr{F}$. If $F: \Sigma \to X$ is a countably additive vector measure of bounded variation and $F|_{\mathscr{F}}$ is the restriction of F to $\mathscr{F}$, then for each $E \in \mathscr{F}$, one has*

$$|F|_{\mathscr{F}}|(E) = |F|(E);$$

i.e., $|F|$ is the Carathéodory-Hahn extension of $|F|_{\mathscr{F}}|$ to Σ.

PROOF. Let μ be the countably additive Carathéodory-Hahn extension of $|F|_{\mathscr{F}}|$ to Σ. Then for each $E \in \mathscr{F}$ and for each $x^* \in X^*$ with $\|x^*\| \leqq 1$, one has

$$|x^*F|_{\mathscr{F}}|(E) \leqq \mu(E).$$

But for the same x^* and E,

$$|x^*F|_{\mathscr{F}}|(E) = |x^*F|(E).$$

Consequently one has $|x^*F|(E) \leqq \mu(E)$ for all $E \in \mathscr{F}$ and all $x^* \in X^*$ with $\|x^*\| \leqq 1$. It follows now from the facts that both $|x^*F|$ and μ are countably additive on Σ and the fact that $\mathscr{F}$ generates Σ that the inequality

$$|x^*F|(E) \leqq \mu(E)$$

holds for all $E \in \Sigma$ and all $x^* \in X^*$ with $\|x^*\| \leqq 1$. But then, as was remarked before Proposition 9, the inequality $|F|(E) \leqq \mu(E)$ holds for all $E \in \Sigma$.

On the other hand, it is plain that for any $E \in \mathscr{F}$ one has

$$|F|_{\mathscr{F}}|(E) \leqq |F|(E).$$

Hence $\mu(E) \leqq |F|(E)$ for all $E \in \mathscr{F}$ and hence, since $\mathscr{F}$ generates Σ, for all $E \in \Sigma$. Consequently $\mu = |F|$ and the proof is complete.

The next proposition presents two basic facts about the semivariation of a vector measure.

PROPOSITION 11. *Let $F: \mathscr{F} \to X$ be a vector measure. Then for $E \in \mathscr{F}$, one has*

(a) $$\|F\|(E) = \sup\left\{\left\|\sum_{A_n \in \pi} \varepsilon_n F(A_n)\right\|\right\}$$

where the supremum is taken over all partitions π of E into finitely many disjoint members of $\mathscr{F}$ and all finite collections $\{\varepsilon_n\}$ satisfying $|\varepsilon_k| \leqq 1$; and

(b) $$\sup\{\|F(H)\| : E \supseteq H \in \mathscr{F}\} \leqq \|F\|(E) \leqq 4 \sup\{\|F(H)\| : E \supseteq H \in \mathscr{F}\}.$$

Consequently a vector measure is of bounded semivariation on Ω if and only if its range is bounded in X.

PROOF. If $\pi = \{E_1, \cdots, E_m\}$ is a partition of E into pairwise disjoint members of $\mathscr{F}$ and $\varepsilon_1, \cdots, \varepsilon_m$ are scalars such that $|\varepsilon_1|, \cdots, |\varepsilon_m| \leqq 1$ then

$$\begin{aligned}\left\|\sum_{n=1}^{m}\varepsilon_n F(E_n)\right\| &= \sup\left\{\left|x^*\left(\sum_{n=1}^{m}\varepsilon_n F(E_n)\right)\right| : x^* \in X^*, \|x^*\| \leqq 1\right\}\\ &\leqq \sup\left\{\sum_{n=1}^{m}|\varepsilon_n x^* F(E_n)| : x^* \in X^*, \|x^*\| \leqq 1\right\}\\ &\leqq \sup\left\{\sum_{n=1}^{m}|x^* F(E_n)| : x^* \in X^*, \|x^*\| \leqq 1\right\}\\ &\leqq \|F\|(E).\end{aligned}$$

For the reverse inequality, let $x^* \in X^*$ with $\|x^*\| \leqq 1$ and suppose $\pi = \{E_1, \cdots, E_m\}$ is a partition of $E \in \mathscr{F}$ into pairwise disjoint members of $\mathscr{F}$. Then

$$\begin{aligned}\sum_{n=1}^{m}|x^*F(E_n)| &= \sum_{n=1}^{m}(\operatorname{sgn} x^*F(E_n))x^*F(E_n)\\ &= \left|x^*\left(\sum_{n=1}^{m}(\operatorname{sgn} x^*F(E_n))F(E_n)\right)\right|\\ &\leqq \left\|\sum_{n=1}^{m}(\operatorname{sgn} x^*F(E_n))F(E_n)\right\|.\end{aligned}$$

This proves (a).

To prove (b) note that for $E \in \mathscr{F}$, one has

$$\begin{aligned}\sup\{\|F(H)\| : E \supseteq H \in \mathscr{F}\} &= \sup\{\sup\{|x^*F(H)| : x^* \in X^*, \|x^*\| \leqq 1\} : E \supseteq H \in \mathscr{F}\}\\ &\leqq \|F\|(E).\end{aligned}$$

Also, if $\pi = \{E_1, \cdots, E_m\}$ is a partition of a member E of $\mathscr{F}$ into pairwise disjoint members of $\mathscr{F}$ and if $x^* \in X^*$ satisfies $\|x^*\| \leqq 1$, then (in case X is a *real* Banach space)

$$\sum_{E_n \in \pi}|x^*F(E_n)| = \sum_{n \in \pi^+} x^*F(E_n) - \sum_{n \in \pi^-} x^*F(E_n),$$

where $\pi^+ = \{n : 1 \leqq n \leqq m, x^*F(E_n) \geqq 0\}$ and $\pi^- = \{n : 1 \leqq n \leqq m, x^*F(E_n) < 0\}$,

$$\begin{aligned}&= x^*\left(\sum_{n \in \pi^+} F(E_n)\right) - x^*\left(\sum_{n \in \pi^-} F(E_n)\right)\\ &\leqq 2\sup\{\|F(H)\| : E \supseteq H \in \mathscr{F}\}\end{aligned}$$

as required. In case X is a complex Banach space, it is easy to see that a similar estimate holds if the number 2 is replaced by the number 4. Simply split x^*F into real and imaginary parts and apply the real case.

In view of Proposition 11(b) a vector measure of bounded semivariation will also be called a *bounded vector measure*.

As will be seen presently, it is easy to define the integral of a bounded measurable function with respect to a bounded vector measure. To this end, let $\mathscr{F}$ be a field of subsets of Ω and $F: \mathscr{F} \to X$ be a bounded vector measure. If f is a scalar-valued simple function on Ω, say $f = \sum_{i=1}^{n}\alpha_i\chi_{E_i}$ where α_i are nonzero scalars and $E_1, \cdots, E_n$ are pairwise disjoint members of $\mathscr{F}$, define $T_F(f) = \sum_{i=1}^{n}\alpha_i F(E_i)$. It is dreadfully boring to show that this formula defines a linear map T_F from the space of simple functions of the above form into X and we leave this as an exercise for masochists. Moreover, if f is as above and $\beta = \sup\{|f(\omega)| : \omega \in \Omega\}$, then

$$\begin{aligned}\|T_F(f)\| &= \left\|\sum_{i=1}^{n} \alpha_i F(E_i)\right\| \\ &= \beta\left\|\sum_{i=1}^{n} (\alpha_i/\beta)F(E_i)\right\| \leqq \beta\|F\|(\Omega)\end{aligned}$$

by Proposition 11(a). Thus, if the space of simple functions over $\mathscr{F}$ is given the supremum norm, T_F acts on this space as a continuous linear operator with $\|T_F\| \leqq \|F\|(\Omega)$. Another look at Proposition 11(a) and the above calculations shows that in fact $\|T_F\| = \|F\|(\Omega)$.

Next note that since T_F is continuous and linear from the simple functions modeled on $\mathscr{F}$ to X, T_F has a unique continous linear extension, still denoted by T_F, to $B(\mathscr{F})$, the space of all scalar-valued functions on Ω that are uniform limits of simple functions modeled on $\mathscr{F}$. (Note that in the case $\mathscr{F}$ is a σ-field, $B(\mathscr{F})$ is precisely the familiar space of bounded $\mathscr{F}$-measurable scalar-valued functions defined on Ω.)

This discussion allows us to make

Definition 12. Let $\mathscr{F}$ be a field of subsets of the set Ω and let $F: \mathscr{F} \to X$ be a bounded vector measure. For each $f \in B(\mathscr{F})$, $\int f\, dF$ is defined by

$$\int f\, dF = T_F(f)$$

where T_F is as above.

The general subject of integration with respect to a vector measure will be discussed later. For the present, this cheap integral as defined above has some convenient properties and uses. It is, of course, linear in f (and also in F) and satisfies

$$\left\|\int f\, dF\right\| \leqq \|f\|_\infty \|F\|(\Omega).$$

Moreover, if $x^* \in X^*$, then $x^* \int f\, dF = \int f\, dx^*F$ holds; indeed, for simple functions f this equality is trivial and density of simple functions in $B(\mathscr{F})$ proves the identity for all $f \in B(\mathscr{F})$. The following formality will allow various properties of vector measures to be translated into properties of linear operators and vice versa.

Theorem 13. *Let $\mathscr{F}(\Sigma)$ be a field (resp. σ-field) of subsets of the set Ω. Suppose μ is an extended real-valued nonnegative finitely additive measure on Σ. Then there is a one-to-one linear correspondence between $\mathscr{L}(B(\mathscr{F}); X)$ (resp. $\mathscr{L}(L_\infty(\mu); X)$) and the space of all bounded vector measures $F: \mathscr{F} \to X$ (resp. all bounded vector measures $F: \Sigma \to X$ that vanish on μ-null sets) defined by $F \leftrightarrow T_F$ if $T_F f = \int f\, dF$ for all $f \in B(\mathscr{F})$ (resp. $L_\infty(\mu)$). Moreover $\|T_F\| = \|F\|(\Omega)$.*

The proof is an easy combination of the observations and propositions preceding the statement of the theorem and is left as an exercise.

One obvious property of a countably additive vector measure F defined on a σ-field Σ (with values in X) is that if (E_n) is a sequence of pairwise disjoint members of Σ, then $\sum_n F(E_n)$ is an unconditionally convergent series (with norm limit $F(\bigcup_n E_n)$) in X. This property is shared by many noncountably additive vector measures. For instance, if $F: \Sigma \to \boldsymbol{R}$ is a nonnegative finitely additive measure, then for any sequence (E_n) of pairwise disjoint members of Σ we have

$$\sum_{n=1}^{k} F(E_n) = F\left(\bigcup_{n=1}^{k} E_n\right) \leqq F(\Omega),$$

so $\sum_n F(E_n) < \infty$. On the other hand, not all bounded vector measures have this property; in fact, if Σ is infinite then the map $F: \Sigma \to B(\Sigma)$ given by $F(A) = \chi_A$ is a bounded vector measure that lacks the above property. Because of its importance in the theory of vector measures this property will be isolated.

DEFINITION 14. Let $\mathscr{F}$ be a field of subsets of the set Ω and let $F: \mathscr{F} \to X$ be a vector measure. F is said to be *strongly additive* whenever given a sequence (E_n) of pairwise disjoint members of $\mathscr{F}$, the series $\sum_{n=1}^{\infty} F(E_n)$ converges in norm.

A family $\{F_\tau: \mathscr{F} \to X \mid \tau \in T\}$ of strongly additive vector measures is said to be *uniformly strongly additive* whenever for any sequence (E_n) of pairwise disjoint members of $\mathscr{F}$, then $\lim_n \| \sum_{m=n}^{\infty} F_\tau(E_m) \| = 0$ uniformly in $\tau \in T$.

Of course, *countably additive vector measures on sigma-fields are strongly additive.* It is important to realize that in the definition of strong additivity the convergence of the series $\sum_{n=1}^{\infty} F(E_n)$ is unconditional in norm (since every subseries also converges). It should also be noted that for families of countably additive measures on a σ-field the concept of uniform strong additivity is precisely the familiar concept of uniform countable additivity.

A wide but by no means exhaustive class of strongly additive vector measures is furnished by

PROPOSITION 15. *If $F: \mathscr{F} \to X$ is a vector measure of bounded variation, then F is strongly additive.*

PROOF. If (E_n) is a sequence of pairwise disjoint members of $\mathscr{F}$, then

$$\sum_{n=1}^{k} \|F(E_n)\| \leqq |F|\left(\bigcup_{n=1}^{k} E_n\right) \leqq |F|(\Omega).$$

Thus $\sum_n \|F(E_n)\| \leqq |F|(\Omega) < \infty$, and $\sum_{n=1} F(E_n)$ is an absolutely convergent, hence convergent, series in the Banach space X.

EXAMPLE 16. *A countably additive, hence strongly additive, vector measure on a sigma-field that is of unbounded variation over every nontrivial set.* Let $\Omega = [0, 1]$, Σ = Lebesgue measurable subsets of $[0, 1]$, λ = Lebesgue measure, $1 < p < \infty$, and $X = L_p[0, 1]$. Define $F: \Sigma \to L_p[0, 1]$ by $F(E) = \chi_E$. Then it is easily checked that if (E_n) is any sequence of pairwise disjoint Lebesgue measurable subsets of $[0, 1]$, then

$$\left\| F\left(\bigcup_{n=1}^{\infty} E_n\right) - \sum_{n=1}^{m} F(E_n) \right\|_p^p = \lambda\left(\bigcup_{n=m+1}^{\infty} E_n\right) \to 0$$

as $m \to \infty$. Thus F is countably additive on the σ-field Σ.

We now claim that if $E \subseteq [0, 1]$ is Lebesgue measurable and $\lambda(E) > 0$, then $|F|(E) = \infty$. To prove this fix a positive integer n and pick disjoint measurable subsets $E_1, E_2, \cdots, E_n$ of E such that $\lambda(E_i) = \lambda(E)/n$ for all $i = 1, \cdots, n$. Note that

$$\sum_{i=1}^{n} \|F(E_i)\| = \sum_{i=1}^{n} (\lambda(E)/n)^{1/p} = n^{(1-1/p)} \lambda(E)^{1/p}.$$

Plainly this means that $|F|(E) = \infty$.

There are many alternative and useful formulations of strong additivity. Most of them are consequences of the next result.

PROPOSITION 17. *Any one of the following statements about a collection $\{F_\tau : \tau \in T\}$ of X-valued measures defined on a field $\mathscr{F}$ implies all the others.*

(i) *The set $\{F_\tau : \tau \in T\}$ is uniformly strongly additive.*

(ii) *The set $\{x^*F_\tau : \tau \in T, x^* \in X^*, \|x^*\| \leqq 1\}$ is uniformly strongly additive.*

(iii) *If (E_n) is a sequence of pairwise disjoint members of $\mathscr{F}$, then* $\lim_n \|F_\tau(E_n)\| = 0$, *uniformly in $\tau \in T$.*

(iv) *If (E_n) is a sequence of pairwise disjoint members of $\mathscr{F}$, then* $\lim_n \|F_\tau\|(E_n) = 0$, *uniformly in $\tau \in T$.*

(v) *The set $\{|x^*F_\tau| : \tau \in T, x^* \in X^*, \|x^*\| \leqq 1\}$ is uniformly strongly additive.*

PROOF. That (i) implies (ii) and (ii) implies (iii) are obvious. To prove that (iii) implies (iv), suppose (iv) fails. Then there exists a $\delta > 0$ and a sequence (E_n) of pairwise disjoint members of $\mathscr{F}$ for which $\sup_{\tau\in T}\|F_\tau\|(E_n) \geqq 4\delta > 0$ holds for all n. By Proposition 11(b), for each n there is $H_n \in \mathscr{F}$ such that $H_n \subseteq E_n$ and $\sup_{\tau\in T}\|F_\tau\|(E_n) \leqq 4\sup_{\tau\in T}\|F_\tau(H_n)\|$. The sequence (H_n) consists of pairwise disjoint members of $\mathscr{F}$ which satisfy

$$\sup_{\tau\in T}\|F_\tau(H_n)\| \geqq \delta > 0$$

for each n. Thus (iii) fails to hold. This shows that (iii) implies (iv).

To prove that (iv) implies (v), suppose that $\{|x^*F_\tau| : \tau \in T, x^* \in X^*, \|x^*\| \leqq 1\}$ is not uniformly strongly additive. Then there exists a disjoint sequence (E_n) in $\mathscr{F}$ and a $\delta > 0$ such that for all m one has

$$\sup\left\{\sum_{n=m}^{\infty} |x^*F_\tau|(E_n) : \tau \in T,\ x^* \in X^*,\ \|x^*\| \leqq 1\right\} \geqq 2\delta > 0.$$

Thus there is an increasing sequence (m_j) of positive integers such that for all j

$$\sup\left\{\sum_{n=m_j+1}^{m_{j+1}} |x^*F_\tau|(E_n) : \tau \in T,\ x^* \in X^*,\ \|x^*\| \leqq 1\right\}$$

$$= \sup\left\{|x^*F_\tau|\left(\bigcup_{n=m_j+1}^{m_{j+1}} E_n\right) : \tau \in T,\ x^* \in X^*,\ \|x^*\| \leqq 1\right\} \geqq \delta > 0.$$

Therefore setting $H_j = \bigcup_{n=m_j+1}^{m_{j+1}} E_n$ produces a sequence (H_j) of pairwise disjoint members of $\mathscr{F}$ such that

$$\sup\{\|F_\tau\|(H_j) : \tau \in T\} = \sup\{|x^*F_\tau|(H_j) : \tau \in T, x^* \in X^*, \|x^*\| \leqq 1\} \geqq \delta > 0.$$

This denies (iv) and proves that (iv) implies (v). That (v) implies (i) is obvious.

The following corollary is the principal result of this section.

COROLLARY 18. *Any one of the following statements about a vector measure F defined on a field $\mathscr{F}$ implies all the others.*

(i) *F is strongly additive.*

(ii) $\{x^*F : x^* \in X^*, \|x^*\| \leqq 1\}$ *is uniformly strongly additive.*

(iii) *F is strongly bounded, i.e., if* (E_n) *is a sequence of pairwise disjoint members of* $\mathcal{F}$, *then* $\lim_n F(E_n) = 0$.

(iv) $\|F\|$ *is strongly bounded, i.e., if* (E_n) *is a sequence of pairwise disjoint members of* $\mathcal{F}$, *then* $\lim_n \|F\|(E_n) = 0$.

(v) $\{|x^*F| : x^* \in X^*, \|x^*\| \leqq 1\}$ *is uniformly strongly additive.*

(vi) $\lim_n F(E_n)$ *exists for every nondecreasing monotone sequence* (E_n) *of members of* $\mathcal{F}$.

(vii) $\lim_n F(E_n)$ *exists for every nonincreasing monotone sequence* (E_n) *of members of* $\mathcal{F}$.

PROOF. The equivalence of statements (i) through (v) is clear from Proposition 17. The equivalence of (vi) and (vii) follows from the identity $F(E) + F(\Omega\backslash E) = F(\Omega)$. To see that (i) implies (vi), let (E_n) be a nondecreasing sequence of members of $\mathcal{F}$. Then

$$\lim_n F(E_n) = F(E_1) + \lim_n \sum_{j=2}^{n} F(E_{j+1}\backslash E_j)$$

exists since the sequence $(E_{j+1}\backslash E_j)$ consists of disjoint members of $\mathcal{F}$. This proves that (i) implies (vi).

On the other hand, if (E_n) is a sequence of pairwise disjoint members of $\mathcal{F}$, then $\lim_n F(\bigcup_{k=1}^n E_k)$ exists by (vi). Thus

$$\lim_n F(E_n) = \lim_n \left[F\left(\bigcup_{k=1}^{n} E_k\right) - F\left(\bigcup_{k=1}^{n-1} E_k\right)\right] = 0.$$

This completes the proof.

Another basic fact about strongly additive vector measures is contained in

COROLLARY 19. *A strongly additive vector measure on a field is bounded.*

PROOF. Let $\mathcal{F}$ be a field of sets and $F: \mathcal{F} \to X$ be a strongly additive measure. If $\|F\|(\Omega) = +\infty$, choose $H_1 \in \mathcal{F}$ such that

$$\|F(H_1)\| \geqq 1 + 2\|F(\Omega)\|.$$

Then since $F(H_1) = F(\Omega) - F(\Omega\backslash H_1)$, it follows that

$$\|F(H_1)\| - \|F(\Omega)\| \leqq \|F(\Omega\backslash H_1)\|.$$

Thus $\|F(\Omega\backslash H_1)\| \geqq 1$. Now $\|F\|$ is subadditive on disjoint sets so either $\|F\|(H_1)$ or $\|F\|(\Omega\backslash H_1)$ is infinite. If $\|F\|(H_1) = \infty$, let $E_1 = H_1$; otherwise, let $E_1 = \Omega\backslash H_1$. In either case,

$$\|F\|(E_1) = \infty \quad \text{and} \quad \|F(E_1)\| \geqq 1.$$

Replacing Ω by E_1 in the above line of reasoning produces a member E_2 of $\mathcal{F}$ contained in E_1 such that

$$\|F\|(E_2) = \infty \quad \text{and} \quad \|F(E_2)\| \geqq 2.$$

Iterating this procedure yields a nonincreasing sequence (E_n) of members of $\mathcal{F}$ such that

$$\|F\|(E_n) = \infty \quad \text{and} \quad \|F(E_n)\| \geqq n.$$

Thus $\lim_n F(E_n)$ does not exist and an appeal to Corollary 18(vii) shows F is not strongly additive.

2. Countably additive vector measures. Countably additive vector measures on σ-fields inherit a good deal of their structure from the theory of uniformly countably additive families of scalar-valued measures. In this section, this fact will be exploited to show that a countably additive vector measure F on a σ-field takes its values in a weakly compact subset of its range space and that there exists a (finite) nonnegative real-valued countably additive measure μ such that $\lim_{\mu(E)\to 0} F(E) = 0$.

The first theorem shows that countably additive vector measures defined on a σ-field share a common property with their scalar counterparts.

THEOREM 1 (PETTIS). *Let Σ be a σ-field, $F: \Sigma \to X$ be a countably additive vector measure and μ be a finite nonnegative real-valued measure on Σ. Then F is μ-continuous, i.e.,*

$$\lim_{\mu(E)\to 0} F(E) = 0$$

if and only if F vanishes on sets of μ-measure zero.

PROOF. To prove the sufficiency, suppose F vanishes on sets of μ-measure zero, but $\overline{\lim}_{\mu(E)\to 0}\|F(E)\| > 0$. Then there exists an $\varepsilon > 0$ and a sequence (A_n) in Σ such that

$$\|F(A_n)\| \geqq \varepsilon \quad \text{and} \quad \mu(A_n) \leqq 2^{-n}$$

for all n. For each n select an $x_n^* \in X^*$ such that

$$\|x_n^*\| \leqq 1 \quad \text{and} \quad \|x_n^* F(A_n)\| \geqq \varepsilon/2.$$

Now since F is countably additive on Σ, the family $\{x_n^* F\}$ is uniformly strongly additive. By Proposition 1.17, the family $\{|x_n^* F|\}$ is also uniformly strongly additive. Now set $B_n = \bigcup_{j=n}^\infty A_j$. Evidently $\mu(\bigcap_{n=1}^\infty B_n) = 0$. Consequently F vanishes on every set $E \in \Sigma$ that is contained in $\bigcap_{n=1}^\infty B_n = B$. It follows that $|x_n^* F|(B) = 0$.

Let $E_1 = \Omega \backslash B_1$ and $E_{n+1} = B_n \backslash B_{n+1}$ for $n \geqq 1$. Then (E_n) constitutes a sequence of pairwise disjoint members of Σ for which

$$B_{m-1} \backslash B = \bigcup_{k=m}^\infty E_k.$$

Also, since $|x_n^* F|(B) = 0$ for all n, one has

$$\lim_m |x_n^* F|(B_{m-1}) = \lim_m |x_n^* F|\left(\bigcup_{k=m}^\infty E_k\right)$$

$$= \lim_m \sum_{k=m}^\infty |x_n^* F|(E_k) = 0$$

uniformly in n, by the uniform strong additivity of the family $\{|x_n^* F|\}$. But now

$$|x_{n-1}^* F|(B_{n-1}) \geqq |x_{n-1}^* F|(A_{n-1}) \geqq |x_{n-1}^* F(A_{n-1})| \geqq \varepsilon/2.$$

This contradicts the last calculation and proves the sufficiency; the converse is transparent.

The role of sigma-fields in Theorem 1 is crucial.

EXAMPLE 2. (*Theorem 1 fails for countably additive measures on fields.*) Let $\mathscr{F}$ be the field of subsets of the natural numbers that are finite or have finite complements. Define $\mu: \mathscr{F} \to \boldsymbol{R}$ by

$$\begin{aligned} \mu(E) &= \text{cardinality of } E, && \text{if } E \text{ is finite,} \\ &= -\text{cardinality of } \boldsymbol{N}\backslash E, && \text{if } \boldsymbol{N}\backslash E \text{ is finite.} \end{aligned}$$

As we observed in Example 1.8, μ is an unbounded real-valued measure. Define $\nu: \mathscr{F} \to [0, 1]$ by $\nu(E) = \sum_{n \in E} 2^{-n}$. Then ν is countably additive on $\mathscr{F}$. Clearly μ is not ν-continuous, i.e., $\lim_{\nu(E)\to 0} \mu(E) = 0$ is false. If μ were countably additive, we would have produced the advertised example. So we will make μ countably additive by changing the appearance of $\mathscr{F}$.

By Stone's Representation Theorem, there exists a compact, Hausdorff totally disconnected topological space $\tilde{\Omega}$ such that $\mathscr{F}$ is isomorphic (as a Boolean algebra) with the algebra $\tilde{\mathscr{F}}$ of clopen subsets of $\tilde{\Omega}$. Now make the crucial observation that if $(\tilde{A}_n)$ is a sequence of pairwise disjoint members of $\tilde{\mathscr{F}}$ such that $\bigcup_n \tilde{A}_n \in \tilde{\mathscr{F}}$, then for all but finitely many n we have $\tilde{A}_n = \varnothing$! In fact, $\bigcup_n \tilde{A}_n \in \tilde{\mathscr{F}}$ implies that $\bigcup_n \tilde{A}_n$ is a closed subset of $\tilde{\Omega}$, hence compact which, since $\{\tilde{A}_n\}$ is clearly an open cover of $\bigcup_n \tilde{A}_n$, establishes our claim. Therefore every finitely additive measure on $\tilde{\mathscr{F}}$ is countably additive by default. In particular, if we define $\tilde{\mu}, \tilde{\nu}: \tilde{\mathscr{F}} \to \boldsymbol{R}$ by

$$\tilde{\mu}(\tilde{E}) = \mu(E) \quad \text{and} \quad \tilde{\nu}(\tilde{E}) = \nu(E),$$

where E and $\tilde{E}$ are corresponding members of $\mathscr{F}$ and $\tilde{\mathscr{F}}$, then $\tilde{\mu}$ and $\tilde{\nu}$ are countably additive and $\tilde{\mu}$ vanishes on sets of $\tilde{\nu}$-measure zero but $\tilde{\mu}$ is not $\tilde{\nu}$-continuous.

We also remark that the measure $\tilde{\mu}$ is an example of an *unbounded countably additive measure defined on a field.*

We have already slipped into using the label "μ-continuous." The next definition formalizes this notion.

DEFINITION 3. Let $\mathscr{F}$ be a field of subsets of Ω, $F: \mathscr{F} \to X$ be a vector measure and μ be a (finite) nonnegative real-valued measure on $\mathscr{F}$. If $\lim_{\mu(E)\to 0} F(E) = 0$, then F is called *μ-continuous* and this is signified by $F \ll \mu$.

As a precautionary measure, it should be noted that writing $F \ll \mu$ is *not* the same as saying F vanishes on μ-null sets unless both F and μ are countably additive and defined on a σ-field.

When $F \ll \mu$, sometimes μ is called "a control measure for F." This terminology will not be used here. Sometimes, however, we will say F is continuous with respect to μ or F is absolutely continuous with respect to μ.

The consequences of the next theorem are the main results of this section.

THEOREM 4. *Let $\{F_\tau: \Sigma \to X \mid \tau \in T\}$ be a uniformly bounded family of countably additive vector measures on a σ-field Σ. The family $\{F_\tau : \tau \in T\}$ is uniformly countably* *additive* (= *uniformly strongly additive*) *if and only if there exists a nonnegative real-valued countably additive measure* μ *on* Σ *such that* $\{F_\tau : \tau \in T\}$ *is uniformly* μ*-continuous, i.e.,*

$$\lim_{\mu(E)\to 0} \|F_\tau(E)\| = 0$$

uniformly in $\tau \in T$.

PROOF. If $\{F_\tau : \tau \in T\}$ is uniformly μ-continuous for some nonnegative real-valued countably additive finite measure μ defined on Σ and (E_n) is a sequence of pairwise disjoint members of Σ, then

$$\lim_m \mu\Big(\bigcup_{n=m}^{\infty} E_n\Big) = 0;$$

so that

$$0 = \lim_m \sup_\tau \Big\| F_\tau\Big(\bigcup_{n=m}^{\infty} E_n\Big)\Big\| = \lim_m \sup_\tau \Big\| \sum_{n=m}^{\infty} F_\tau(E_n)\Big\|.$$

Thus $\{F_\tau : \tau \in T\}$ is uniformly countably additive.

For the converse, note that $\{F_\tau : \tau \in T\}$ is uniformly countably additive if and only if the family $\{x^*F_\tau : \tau \in T, x^* \in X^*, \|x^*\| \leqq 1\}$ is uniformly countably additive. Hence it suffices to prove the converse for scalar-valued countably additive measures. To this end, assume that $\{\mu_\tau : \tau \in T\}$ is a bounded family of uniformly countably additive scalar-valued measures defined on Σ.

First it will be shown that given $\varepsilon > 0$ there exists a finite family of indices $\{\tau_1, \cdots, \tau_n\} \subseteq T$, dependent upon ε, such that $\sup_{1\leqq i\leqq n} |\mu_{\tau_i}|(E) = 0$ implies $\sup_{\tau\in T} |\mu_\tau(E)| < \varepsilon$.

Suppose not and fix $\tau_1 \in T$. Then there exists $E_1 \in \Sigma$ and $\tau_2 \in T$ such that

$$|\mu_{\tau_1}|(E_1) = 0 \quad \text{yet} \quad |\mu_{\tau_2}(E_1)| \geqq \varepsilon.$$

Again, there exists $E_2 \in \Sigma$ and $\tau_3 \in T$ such that

$$|\mu_{\tau_1}|(E_2), \qquad |\mu_{\tau_2}|(E_2) = 0 \quad \text{yet} \quad |\mu_{\tau_3}(E_2)| \geqq \varepsilon.$$

Continuing this process produces a sequence (E_n) of members of Σ and a sequence (τ_n) of members of T such that

$$\sup_{1\leqq i\leqq n} |\mu_{\tau_i}|(E_n) = 0 \quad \text{yet} \quad |\mu_{\tau_{n+1}}(E_n)| \geqq \varepsilon$$

for each n. Let $H_n = \bigcup_{j=n}^{\infty} E_j$. Then (H_n) is a nonincreasing sequence in Σ. It follows from 1.17 that $\lim_n |\mu_{\tau_k}(H_n)|$ exists uniformly in k. Since, for each k, $\lim_n \mu_{\tau_k}(H_n) = 0$, then

$$\lim_n \sup_k |\mu_{\tau_k}(H_n)| = 0.$$

On the other hand, one has

$$\mu_{\tau_{n+1}}(H_n) = \mu_{\tau_{n+1}}(E_n) + \mu_{\tau_{n+1}}\Big(\bigcup_{j=n+1}^{\infty} E_j \backslash E_n\Big) = \mu_{\tau_{n+1}}(E_n),$$

since $|\mu_{\tau_{n+1}}|(E_j) = 0$ for $j \geqq n+1$. Thus

$$\overline{\lim_n} \sup_k |\mu_{\tau_{k+1}}(H_n)| \geqq \varepsilon,$$

which contradicts the fact that $\lim_n \sup_k |\mu_{\tau_k}(H_n)| = 0$.

Recapitulating, we have shown that given $\varepsilon > 0$ there exists a finite set of indices $\{\tau_1, \cdots, \tau_{n(\varepsilon)}\} \subseteq T$ such that

$$\sup_{1 \leq i \leq n(\varepsilon)} |\mu_{\tau_i}|(E) = 0 \quad \text{implies} \quad \sup_{\tau \in T} |\mu_\tau|(E) < \varepsilon.$$

To complete the proof, choose for each m a finite set $J_m = \{\tau_1^m, \cdots, \tau_{n(m)}^m\}$ of indices such that

$$\sup_{1 \leq j \leq n(m)} |\mu_{\tau_j^m}|(E) = 0 \quad \text{implies} \quad \sup_{\tau \in T} |\mu_\tau(E)| < \frac{1}{m}.$$

Let $\lambda_m \colon \Sigma \to [0, \infty)$ be defined by

$$\lambda_m(E) = \frac{1}{n(m)} \sum_{j=1}^{n(m)} |\mu_{\tau_j^m}|(E).$$

Then λ_m is a nonnegative real-valued countably additive measure such that

$$\lambda_m(E) = 0 \quad \text{implies} \quad \sup_{\tau \in T} |\mu_\tau(E)| < \frac{1}{m}.$$

Moreover, the sequence (λ_m) is uniformly bounded. Now let $\mu \colon \Sigma \to [0, \infty)$ be defined by

$$\mu(E) = \sum_m \lambda_m(E) 2^{-m} |\lambda_m|(\Omega)^{-1}.$$

Then μ is a nonnegative real-valued countably additive measure on Σ such that $\mu(E) = 0$ implies $|\mu_\tau(E)| = 0$ for all $\tau \in T$. To see that

$$\lim_{\mu(E) \to 0} \sup_\tau |\mu_\tau(E)| = 0,$$

define $F \colon \Sigma \to l_\infty(T)$ (where $l_\infty(T)$ denotes the Banach space of scalar-valued bounded functions defined on T equipped with the supremum norm) by

$$F(E)(\tau) = \mu_\tau(E).$$

By the uniform countable additivity of $\{\mu_\tau \colon \tau \in T\}$, it is readily seen that F is a countably additive vector measure. Moreover, $F(E) = 0$ whenever $\mu(E) = 0$. Hence by Theorem 1, F is μ-continuous, i.e.,

$$\lim_{\mu(E) \to 0} \|F(E)\|_{l_\infty(T)} = 0,$$

which is the desired result.

Some convenient and important corollaries follow.

COROLLARY 5. *Suppose $\{F_\tau \colon \tau \in T\}$ is a bounded uniformly countably additive family of X-valued measures defined on a σ-field Σ. If $\mu \colon \Sigma \to [0, \infty)$ is a countably additive measure and F_τ is μ-continuous for each $\tau \in T$ then*

$$\lim_{\mu(E) \to 0} \sup_{\tau \in T} \|F_\tau(E)\| = 0.$$

Moreover there is such a μ with

$$0 \leqq \mu(E) \leqq \sup_{\tau \in T} \|F_\tau\|(E)$$

for all $E \in \Sigma$. *Consequently* $\mu(E) \to 0$ *if and only if* $\sup_{\tau \in T}\|F_\tau\|(E) \to 0$.

PROOF. Only the second assertion needs to be proved and its proof is embedded in the proof of Theorem 4 (look at the construction of μ).

A specialization of Corollary 5 to a family consisting of one measure results in a theorem which is central to the theory of vector measures.

COROLLARY 6 (BARTLE-DUNFORD-SCHWARTZ). *Let F be a countably additive vector measure defined on a σ-field Σ. Then there exists a nonnegative real-valued countably additive measure μ on Σ such that $\mu(E) \to 0$ if and only if $\|F\|(E) \to 0$; in fact μ can be chosen so that $0 \leqq \mu(E) \leqq \|F\|(E)$ for all $E \in \Sigma$.*

Following immediately is a key structure theorem for countably additive vector measures on σ-fields.

COROLLARY 7 (BARTLE-DUNFORD-SCHWARTZ). *Let F be a countably additive vector measure defined on a σ-field Σ. Then F has a relatively weakly compact range.*

PROOF. Let $\mu: \Sigma \to [0, \infty)$ be a countably additive measure such that $F \ll \mu$. Define $T: L_\infty(\mu) \to X$ by $Tf = \int f\,dF$. Then, for each $x^* \in X^*$, one has

$$x^*Tf = \int f\,dx^*F = \int f \frac{dx^*F}{d\mu}\,d\mu,$$

where $dx^*F/d\mu = g_{x^*} \in L_1(\mu)$ is the Radon-Nikodým derivative of x^*F with respect to μ. If (f_α) is a net in $L_\infty(\mu)$ converging weak* to f_0 then, for each $x^* \in X^*$,

$$\lim_\alpha x^*Tf_\alpha = \lim_\alpha \int f_\alpha g_{x^*}\,d\mu = \int f_0 g_{x^*}\,d\mu = x^*Tf_0,$$

i.e., (Tf_α) converges weakly to Tf_0. Hence T is a weak*- to weak-continuous linear operator. It follows that T maps the weak*-compact set $\{f \in L_\infty(\mu) : \|f\|_\infty \leqq 1\}$ onto a weakly compact set $K \subseteq X$. But now

$$\{F(E): E \in \Sigma\} = \{T(\chi_E): E \in \Sigma\} \subseteq \{T(f): \|f\|_\infty \leqq 1\} \subseteq K,$$

and the proof is complete.

3. The Nikodym Boundedness Theorem. The subject of this short section is one of the truly impressive theorems of measure theory, the Nikodým Boundedness Theorem, which asserts that if a family of bounded measures on a σ-field is setwise bounded, then the family is uniformly bounded on the whole σ-field. "A striking improvement of the uniform boundedness principle" is how Dunford and Schwartz introduce this theorem, and they are right!

THEOREM 1 (NIKODÝM BOUNDEDNESS THOEREM). *Let Σ be a σ-field of subsets of Ω and let $\{F_\tau : \tau \in T\}$ be a family of X-valued bounded vector measures defined on Σ. If $\sup_\tau \|F_\tau(E)\| < \infty$ for each $E \in \Sigma$, then the family $\{F_\tau: \tau \in T\}$ is uniformly bounded, i.e.,*

$$\sup_{\tau \in T} \|F_\tau\|(\Omega) < \infty.$$

PROOF. By replacing the family $\{F_\tau : \tau \in T\}$ by the family $\{x^*F_\tau : \tau \in T, x^* \in X^*, \|x^*\| \leqq 1\}$ one can see quickly that the theorem need be proved only for a family of scalar-valued measures. Moreover to prove the theorem for families of scalar-valued measures it is plainly sufficient to prove the theorem for sequences of scalar-valued measures. Thus suppose (μ_n) is a sequence of scalar-valued finitely additive measures on Σ with the property that

$$\sup_n |\mu_n(E)| < \infty$$

for each $E \in \Sigma$.

Proceeding by contradiction, suppose $\sup_n \sup_{E\in\Sigma} |\mu_n(E)| = \infty$. First note that if $\rho > 0$, then there is a positive integer n and a partition $\{E, F\}$ of Ω into disjoint members of Σ such that $|\mu_n(E)|, |\mu_n(F)| > \rho$. In fact, choose n and E such that

$$|\mu_n(E)| > \sup_k |\mu_k(\Omega)| + \rho;$$

then

$$\begin{aligned} |\mu_n(F)| &= |\mu_n(\Omega \backslash E)| = |\mu_n(E) - \mu_n(\Omega)| \\ &\geqq |\mu_n(E)| - |\mu_n(\Omega)| > \rho. \end{aligned}$$

To begin an inductive process leading to the ultimate contradiction, let n_1 be the least positive integer such that there exists a partition (E_1, F_1) of Ω into disjoint members of Σ such that $|\mu_{n_1}(E_1)|, |\mu_{n_1}(F_1)| > 2$. At least one of $\sup_n \sup_{E\in\Sigma} |\mu_n(E \cap E_1)|$ and $\sup_n \sup_{E\in\Sigma} |\mu_n(E \cap F_1)|$ is infinite. If the former is infinite, set $S_1 = E_1$ and $T_1 = F_1$; otherwise, set $S_1 = F_1$ and $T_1 = E_1$. In either case, there is a least integer $n_2 > n_1$ such that there exists a partition (E_2, F_2) of S_1 into disjoint members of Σ such that

$$|\mu_{n_2}(E_2)|, |\mu_{n_2}(F_2)| > 3 + |\mu_{n_2}(T_1)|.$$

Now at least one of $\sup_n \sup_{E\in\Sigma} |\mu_n(E \cap E_2)|$ and $\sup_n \sup_{E\in\Sigma} |\mu_n(E \cap F_2)|$ is infinite. If the former, set $S_2 = E_2$ and $T_2 = F_2$; otherwise, set $S_2 = F_2$ and $T_2 = E_2$.

Continue in this fashion, obtaining a sequence (T_n) of pairwise disjoint members of Σ and a strictly increasing sequence (n_k) of positive integers such that, for each $k > 1$, we have

$$|\mu_{n_k}(T_k)| > \sum_{j=1}^{k-1} |\mu_{n_k}(T_j)| + k + 1.$$

Relabel (μ_{n_k}) by (μ_k).

Now partition $\boldsymbol{N}$ into infinitely many disjoint infinite subsets $N_1, N_2, \cdots, N_k, \cdots$. Since $|\mu_1|$ is additive, one finds that

$$\sum_{k=1}^{\infty} |\mu_1|\Big(\bigcup_{n\in N_k} T_n\Big) \leqq |\mu_1|\Big(\bigcup_{n=1}^{\infty} T_n\Big) \leqq |\mu_1|(\Omega).$$

It follows that there exists a subsequence (T_{k_i}) of $(T_k)_{k\geqq 2}$ such that $|\mu_1|((\bigcup_{i=1}^{\infty} T_{k_i}) < 1$.

Repeat this argument with $|\mu_1|$ replaced by $|\mu_{k_1}|$ and $(T_k)_{k\geqq 2}$ replaced by $(T_{k_i})_{i\geqq 2}$ to produce a subsequence $(T_{k_{i_j}})$ of $(T_{k_i})_{i\geqq 2}$ such that

$$|\mu_{k1}|\Big(\bigcup_{j=1}^{\infty} T_{k_{i_j}}\Big) < 1.$$

Continuing, repeat this argument with $|\mu_{k1}|$ replaced by $|\mu_{k_{i_1}}|$ and $(T_{k_i})_{i\geq 2}$ replaced by $(T_{k_{i_j}})_{j\geq 2}$. Iterate.

If T_{m_i} denotes the first member of the subsequence generated at the ith stage (so $m_1 = 1$, $m_2 = n_1$, $m_3 = n_{k_1}$, $m_4 = n_{k_{i_1}}, \cdots$), then it follows that

$$|\mu_{m_j}|\Big(\bigcup_{i=j+1}^{\infty} T_{m_i}\Big) < 1.$$

Finally set $D = \bigcup_{j=1}^{\infty} T_{m_j}$. Note that

$$\begin{aligned} |\mu_{m_j}(D)| &\geq |\mu_{m_j}(T_{m_j})| - \Big|\mu_{m_j}\Big(\bigcup_{i=1}^{j-1} T_{m_i}\Big)\Big| - \Big|\mu_{m_j}\Big(\bigcup_{i=j+1}^{\infty} T_{m_i}\Big)\Big| \\ &\geq |\mu_{m_j}(T_{m_j})| - \sum_{i=1}^{j-1} |\mu_{m_j}(T_{m_i})| - |\mu_{m_j}|\Big(\bigcup_{i=j+1}^{\infty} T_{m_i}\Big) \\ &\geq m_j + 1 - 1 = m_j. \end{aligned}$$

Hence $\sup_j |\mu_{m_j}(D)| = \infty$, a contradiction which completes the proof.

The next three corollaries should serve to illustrate the power of this specialized uniform boundedness principle for vector measures. The first corollary is indeed a "striking improvement of the uniform boundedness principle."

COROLLARY 2. *Let Σ be a σ-field of subsets of a set Ω. Suppose $\{T_\alpha : \alpha \in A\}$ is a collection of bounded linear operators from $B(\Sigma)$ to X such that $\sup_{\alpha\in A}\|T_\alpha \chi_E\| < \infty$ for each $E \in \Sigma$. Then*

$$\sup_{\alpha\in A}\|T_\alpha\| < \infty.$$

In case μ is a nonnegative extended real-valued countably additive measure defined on Σ, then the same statement is true with $B(\Sigma)$ replaced by $L_\infty(\mu)$.

The proof of Corollary 2 is immediate from the Nikodým Boundedness Theorem and Theorem 1.13.

The next corollary is rather surprising and very useful.

COROLLARY 3 (DIEUDONNÉ-GROTHENDIECK). *Let F be an X-valued function defined on the σ-field Σ and suppose that x^*F is bounded and finitely additive for each x^* belonging to some total subset Γ of X^*. Then F is a bounded vector measure.*

PROOF. The finite additivity of F is an obvious consequence of the totality of Γ. By virtue of Theorem 1, to prove F is bounded, it is enough to prove x^*F is bounded for each $x^* \in X^*$. To this end, let

$$M = \{x^* \in X^* : |x^*F|(\Omega) < \infty\}.$$

Now M is a linear subspace of X^* which contains the total set Γ; consequently M is a weak*-dense linear subspace of X^*. If it can be shown that $M_1 = \{x^* \in M : \|x^*\| \leq 1\}$ is weak*-closed, then an appeal to the Kreĭn-Smulian theorem will establish that M is weak*-closed in X^* and hence that $M = X^*$.

To show that M_1 is weak*-closed, first note that for each $E \in \Sigma$ one has

$$\sup_{x^*\in M_1} |x^*F(E)| \leqq \|F(E)\| < \infty.$$

By the Nikodým Boundedness Theorem, then we have

$$\sup_{x^*\in M_1} \sup_{E\in\Sigma} |x^*F(E)| = K < \infty.$$

Now let (x_α^*) be a net in M_1 such that $\lim_\alpha x_\alpha^* = x_0^*$ exists in the weak*-topology of X^*. Then $\|x_0^*\| \leqq 1$ and

$$|x_0^*F(E)| = \lim_\alpha |x_\alpha^*F(E)| \leqq K$$

for all $E \in \Sigma$. Thus $|x_0^*F|(\Omega) < \infty$ and $x_0^* \in M_1$. This completes the proof.

Another application of the Nikodým Boundedness Theorem is nothing less than spectacular.

COROLLARY 4 (SEEVER). *Let Σ be a σ-field of subsets of Ω. Let $T: X \to B(\Sigma)$ be a bounded linear operator whose range includes the set $\{\chi_A: A \in \Sigma\}$. Then $TX = B(\Sigma)$.*

PROOF. It is plain that TX is dense in $B(\Sigma)$; so the proof collapses to showing that T has a closed range. For this it suffices to show that $T^*: B(\Sigma)^* \to X^*$ has a closed range.

To this end, recall that (as a consequence of Theorem 1.13) $B(\Sigma)^*$ is the Banach space of bounded scalar-valued measures on Σ equipped with the variation norm. Note that if T^* does not have a closed range, then there exists a sequence (μ_n) in $B(\Sigma)^*$ such that

$$\|T^*\mu_n\| = 1 \quad \text{and} \quad \lim_n \|\mu_n\| = +\infty.$$

To see this, note that if T^* does not have a closed range, then there exists $x^* \in X^*$, $\|x^*\| = 1$, such that $x^* \notin T^*(B(\Sigma)^*)$, and there exists a sequence (μ_n) of members of $B(\Sigma)^*$ such that $\|T^*\mu_n\| = 1$ and

$$\lim_n \|x^* - T^*\mu_n\| = 0.$$

Now if (μ_n) is bounded in $B(\Sigma)^*$, then there is a weak*-limit point μ_0 of the sequence (μ_n). Since T^* is weak*-continuous, then $T^*\mu_0$ is a weak*-limit point of $T^*\mu_n$. But then $x^* = T^*\mu_0$. Thus (μ_n) is not bounded in $B(\Sigma)^*$. By passing to an appropriate subsequence we verify our claim, namely, if T^* has a nonclosed range, then there is a sequence (μ_n) of members of $B(\Sigma)^*$ with $\|T^*\mu_n\| = 1$ and $\lim_n |\mu_n|(\Omega) = \infty$.

To complete the proof, it will be shown that this is impossible. Indeed, if $E \in \Sigma$, there is an $x \in X$ with $Tx = \chi_E$. Thus

$$\sup_n |\mu_n(E)| = \sup_n |T^*\mu_n(x)| \leqq \|x\| < \infty.$$

An appeal to the Nikodým Boundedness Theorem reveals that

$$\sup_{E\in\Sigma} \sup_n |\mu_n(E)| < \infty.$$

This combined with Proposition 1.11(b) shows that

$$\sup_n |\mu_n|(\Omega) < \infty,$$

a contradiction which establishes the corollary.

EXAMPLE 5. *The Nikodým Boundedness Theorem fails for fields of sets.* Let $\mathscr{F}$ be the field of all subsets of the positive integers that are either finite or have finite complements. Let ε_n be the point-mass at n, i.e., $\varepsilon_n(E) = 1$ if $n \in E$ and $\varepsilon_n(E) = 0$ otherwise. Define

$$\begin{aligned} \mu_n(E) &= n(\varepsilon_{n+1}(E) - \varepsilon_n(E)) \quad \text{for } E \text{ finite,} \\ &= -n(\varepsilon_{n+1}(E) - \varepsilon_n(E)) \quad \text{for } \boldsymbol{N} \backslash E \text{ finite,} \end{aligned}$$

and note that $\sup_n |\mu_n(E)|$ is finite for all $E \in \mathscr{F}$. On the other hand, $\sup_n \sup_{E \in \mathscr{F}} |\mu_n(E)|$ is plainly infinite.

4. Rosenthal's lemma and the structure of a vector measure. This section is by far the most important section in Chapter I and is one of the most important sections of this book. For it is in this section that the interplay between vector measure theory and Banach space theory begins to merge in a fruitful way. In this section it will be seen that structural properties of the classical Banach spaces c_0 and l_∞ play a major role in structural properties of vector measures. Conversely, it will be seen in this section that properties of vector measures can be parlayed into structural properties of Banach spaces that contain or do not contain c_0 or l_∞. The basis for this section is a stunning lemma of Rosenthal which is beautiful for its elegance and simplicity and powerful because of its utility.

LEMMA 1 (ROSENTHAL'S LEMMA). *Let $\mathscr{F}$ be a field of subsets of the set Ω. Let (μ_n) be a uniformly bounded sequence of finitely additive scalar-valued measures defined on $\mathscr{F}$. Then, if (E_n) is a disjoint sequence of members of $\mathscr{F}$ and $\varepsilon > 0$, there is a subsequence (E_{n_j}) of (E_n) such that*

$$|\mu_{n_j}|\Big(\bigcup_{k \neq j; k \in \Delta} E_{n_k}\Big) < \varepsilon$$

for all finite subsets Δ of $\boldsymbol{N}$ and for all $j = 1, 2, \cdots$.

If in addition $\mathscr{F}$ is a σ-field, then the subsequence (E_{n_j}) may be chosen such that

$$|\mu_{n_j}|\Big(\bigcup_{k \neq j} E_{n_k}\Big) < \varepsilon$$

for all $j = 1, 2, \cdots$.

PROOF. The second statement will be established first; assume $\mathscr{F}$ is a σ-field and with no loss of generality assume $\sup_n |\mu_n|(\Omega) \leqq 1$. To begin an iterative process, partition the positive integers $\boldsymbol{N}$ into an infinite number of infinite (disjoint) subsets $M_1, M_2, \cdots$ with $\bigcup_p M_p = \boldsymbol{N}$. If for some p there is no $k \in M_p$ with $|\mu_k|(\bigcup_{j \neq k;\, j \in M_p} E_j) \geqq \varepsilon$, the goal is achieved by enumerating $M_p = n_1 < n_2 < \cdots$, since then $|\mu_{n_i}|(\bigcup_{j \neq i} E_{n_j}) < \varepsilon$ for all $i = 1, 2, \cdots$.

If for each p there is a $k_p \in M_p$ with $|\mu_{k_p}|(\bigcup_{j \neq k_p;\, j \in M_p} E_j) \geqq \varepsilon$, note that for each p, we have

$$|\mu_{k_p}|\Big(\bigcup_{q=1}^{\infty} E_{k_q}\Big) + |\mu_{k_p}|\Big(\bigcup_{n=1}^{\infty} E_n \setminus \bigcup_{n=1}^{\infty} E_{k_n}\Big) \leqq 1.$$

Since

$$\bigcup_{j \neq k_p;\, j \in M_p} E_j \subseteq \bigcup_{n=1}^{\infty} E_n \setminus \bigcup_{n=1}^{\infty} E_{k_n}$$

for all p, we have $|\mu_{k_p}|(\bigcup_{q=1}^{\infty} E_{k_q}) + \varepsilon \leqq 1$, for all p. Hence, for all p,

$$|\mu_{k_p}|\Big(\bigcup_{q=1}^{\infty} E_{k_q}\Big) \leqq 1 - \varepsilon.$$

Next, apply the same argument to (μ_{k_p}) and (E_{k_p}) by replacing (μ_n) by (μ_{k_n}) and (E_n) by (E_{k_n}). If the process does not stop, then there is a new subsequence (E_{n_i}) of (E_n) with

$$|\mu_{n_i}|\Big(\bigcup_{i=1}^{\infty} E_{n_i}\Big) \leqq 1 - 2\varepsilon.$$

It now becomes apparent that the process must come to a halt before n iterations, if n is the smallest positive integer such that $1 - n\varepsilon < 0$.

The first assertion is a direct consequence of the second assertion, the Hahn-Banach theorem, and some isomorphic formalities. Assume that $\mathscr{F}$ is a field. By virtue of Theorem 1.13, the map $f \to \int_\Omega f\, d\mu_n$ is a bounded linear functional on $B(\mathscr{F})$ with norm $|\mu_n|(\Omega)$. If Σ is the σ-field generated by $\mathscr{F}$, then $B(\mathscr{F})$ is a closed linear subspace of $B(\Sigma)$. By the Hahn-Banach theorem, the continuous linear functional $f \to \int_\Omega f\, d\mu_n$ on $B(\mathscr{F})$ has a norm preserving linear extension to $B(\Sigma)$. Another appeal to Theorem 1.13 produces a finitely additive scalar measure $\bar{\mu}_n$ on Σ with $|\bar{\mu}_n|(\Omega) = |\mu_n|(\Omega)$ and such that

$$\int_\Omega f\, d\mu_n = \int_\Omega f\, d\bar{\mu}_n$$

for $f \in B(\mathscr{F})$. In particular, if $E \in \mathscr{F}$, we have

$$\mu_n(E) = \int \chi_E\, d\mu_n = \int \chi_E\, d\bar{\mu}_n = \bar{\mu}_n(E).$$

Thus $\bar{\mu}_n$ is a finitely additive extension of μ_n to Σ. By the first part of the proof, for a fixed $\varepsilon > 0$, there is a subsequence $(\bar{\mu}_{n_i})$ such that

$$|\bar{\mu}_{n_i}|\Big(\bigcup_{j \neq i} E_{n_j}\Big) < \varepsilon$$

for $i = 1, 2, \cdots$. Consequently, if Δ is a finite subset of $\boldsymbol{N}$, then

$$\begin{aligned} |\mu_{n_i}|\Big(\bigcup_{j \neq i;\, j \in \Delta} E_{n_j}\Big) &\leqq |\bar{\mu}_{n_j}|\Big(\bigcup_{j \neq i;\, j \in \Delta} E_{n_j}\Big) \\ &\leqq |\bar{\mu}_{n_i}|\Big(\bigcup_{j \neq i} E_{n_j}\Big) < \varepsilon. \end{aligned}$$

This completes the proof.

Rosenthal's lemma has a speedy translation into a fundamental theorem about vector measures. The nth unit vector in c_0 is the vector that has a 1 in its nth slot and zeros elsewhere; we denote this vector by e_n.

THEOREM 2 (DIESTEL-FAIRES). *Let $\mathcal{F}$ be a field of subsets of the set Ω and $G: \mathcal{F} \to X$ be a bounded vector measure. If G is not strongly additive, then there is a topological isomorphism T: $c_0 \to X$ and a sequence (E_n) of disjoint members of $\mathcal{F}$ such that $T(e_n) = G(E_n)$. Consequently, $G(\mathcal{F})$ contains the image under T of all the $\{0, 1\}$-valued sequences in c_0.*

If in addition $\mathcal{F}$ is a σ-field, then the above statement remains true if the space c_0 is replaced by the space l_∞.

PROOF. Assume first that $\mathcal{F}$ is a field. If G is not strongly additive, then Corollary 1.18(iii) produces a sequence (E_n) of disjoint members of $\mathcal{F}$ such that

$$\limsup_n \|G(E_n)\| = \alpha > 0.$$

Discarding some of the E_n's and relabeling the others allows us to write

$$\|G(E_n)\| > \alpha/2 = \varepsilon$$

for all n. Now with the help of the Hahn-Banach theorem, select $x_n^* \in X^*$ such that $\|x_n^*\| = 1$ and $|x_n^* G(E_n)| > \varepsilon$ for all n. Since G is bounded, the sequence $(x_n^* G)$ is a uniformly bounded sequence. By Rosenthal's lemma, there is a subsequence $(x_{n_j}^*, G)$ with

$$|x_{n_j}^* G|\Big(\bigcup_{i \neq j;\, i \in \Delta} E_{n_i}\Big) < \varepsilon/2$$

for all i and all finite subsets Δ of N. Now relabel $x_{n_j}^*$ by x_j^* and E_{n_j} by E_i. Then

$$|x_j^* G|\Big(\bigcup_{i \neq j;\, i \in \Delta} E_i\Big) < \varepsilon/2 \quad \text{and} \quad |x_j^* G(E_j)| > \varepsilon$$

for all j and every finite subset Δ of N. At this point the bumps corresponding to the unit vector basis of c_0 begin to emerge. For a finitely nonzero sequence $(\alpha_i) \in c_0$, define

$$T((\alpha_i)) = \sum_{i=1}^{\infty} \alpha_i G(E_i).$$

The map T is plainly linear on the dense linear subspace of c_0 consisting of the finitely nonzero sequences. Moreover, if $(\alpha_i) \in c_0$ is finitely nonzero and $x^* \in X^*$, $\|x^*\| \leqq 1$, then

$$\begin{aligned} |x^* T((\alpha_i))| &\leqq \sum_{i=1}^{\infty} |\alpha_i| |x^* G(E_i)| \\ &\leqq \|(\alpha_i)\|_{c_0} \sum_{i=1}^{\infty} |x^* G(E_i)| \\ &\leqq \|(\alpha_i)\|_{c_0} \|G\|(\Omega). \end{aligned}$$

Hence T is bounded on a dense subspace of c_0 and has a bounded extension, still denoted by T, with norm no greater than $\|G\|(\Omega)$. From this it follows that if $(\alpha_i) \in c_0$, then $T((\alpha_i)) = \sum_{i=1}^{\infty} \alpha_i G(E_i)$. Now to see that T is a topological isomorphism, note that if $(\alpha_i) \in c_0$, then

$$
\begin{aligned}
|x_j^* T((\alpha_i))| &= \left|\sum_{i=1}^{\infty} \alpha_i x_j^* G(E_i)\right| \\
&\geqq |\alpha_j x_j^* G(E_j)| - \left|\lim_m \sum_{i\neq j; i=1}^{m} \alpha_i x_j^* G(E_i)\right| \\
&\geqq |\alpha_j|\varepsilon - \|(\alpha_i)\|_{c_0} \lim_m \sum_{i\neq j; i=1}^{m} |x_j^* G|(E_i) \\
&\geqq |\alpha_j|\varepsilon - \|(\alpha_i)\|_{c_0}\, \varepsilon/2
\end{aligned}
$$

by our choice of x_j^* and (E_j). Taking suprema over all the x_j^* yields

$$\|T((\alpha_i))\| \geqq \sup_n |x_n^* T((\alpha_i))| \geqq \frac{\varepsilon}{2} \|\alpha\|_{c_0}.$$

Finally, note that (E_n) was chosen to be a sequence of pairwise disjoint members of $\mathscr{F}$ and T was constructed so that $Te_n = G(E_n)$.

Moving to the case in which $\mathscr{F}$ is a σ-field, proceed as above to produce (with the help of Rosenthal's lemma in σ-field form) an $\varepsilon > 0$, a sequence (x_n^*) in the closed unit ball of X^* and a sequence (E_n) of pairwise disjoint members of $\mathscr{F}$ such that

$$|x_j^* G(E_j)| > \varepsilon \quad \text{and} \quad |x_j^* G|\Big(\bigcup_{i\neq j} E_i\Big) < \varepsilon/2.$$

If $(\alpha_i) \in l_\infty$ is a finitely-valued sequence, write

$$(\alpha_i) = \sum_{j=1}^{n} \beta_j \chi_{A_j}$$

where $A_1, \cdots, A_n$ are pairwise disjoint sets of positive integers. Define $T((\alpha_i))$ by

$$T((\alpha_i)) = \sum_{j=1}^{n} \beta_j G\Big(\bigcup_{k\in A_j} E_k\Big).$$

Then T is linear on the dense linear subspace of l_∞ consisting of finitely-valued sequences. Computations and estimates similar to those used above in the c_0 case show that T is bounded and is an isomorphism on the finitely-valued sequences in l_∞. Further it is evident that $G(\mathscr{F})$ contains the image under T of all $\{0, 1\}$-valued sequences.

As a direct consequence of Theorem 2 we see that the measure $G: 2^N \to l_\infty$ (where 2^N denotes the collection of all subsets of N, the natural numbers) defined by $G(E) = \chi_E$ is an archetypical representative of the class of nonstrongly additive vector measures on a σ-field.

Also, if $\mathscr{F}$ is the field of subsets of N consisting of the finite sets and their complements, then the measure $G: \mathscr{F} \to c_0$ defined by

$$
\begin{aligned}
G(E) &= \chi_E && \text{if } E \text{ is finite,} \\
&= -\chi_E && \text{if } N\backslash E \text{ is finite,}
\end{aligned}
$$

is an archetypical representative of the class of nonstrongly additive measures on a field.

The first corollary is a disguised version of the Orlicz-Pettis theorem.

COROLLARY 3. *Let $\mathscr{F}$ be a field of subsets of the set Ω. A bounded vector measure*

$G: \mathcal{F} \to X$ is strongly additive if and only if for every monotone nondecreasing sequence (E_n) in $\mathcal{F}$, then $\lim_n G(E_n)$ exists weakly in X.

PROOF. If G is strongly additive and (E_n) is a sequence in $\mathcal{F}$ with $E_n \subseteq E_{n+1}$ for all n, then Corollary 1.18(vi) guarantees that $\lim_n G(E_n)$ exists in norm.

For the converse, suppose G is not strongly additive. By Theorem 2 there is a topological isomorphism $T: c_0 \to X$ and a sequence (A_n) of disjoint members of $\mathcal{F}$ such that $Te_n = G(A_n)$ where e_n is the nth unit vector in c_0. Set $E_m = \bigcup_{n=1}^m A_n$; then $T(\sum_{n=1}^m e_n) = G(E_m)$. But now $\lim_m G(E_m)$ cannot exist weakly in X, since $\lim_m \sum_{n=1}^m e_n$ does not exist weakly in c_0.

An immediate consequence of Corollary 3 is the Orlicz-Pettis theorem.

COROLLARY 4 (ORLICZ-PETTIS). *Let $\sum_n x_n$ be a (formal) series in X such that every subseries of $\sum_n x_n$ is weakly convergent. Then $\sum_n x_n$ is unconditionally convergent in norm.*

Consequently a weakly countably additive vector measure on a sigma-field is (norm) countably additive.

PROOF. Let $\sum_n x_n$ be a series such that every subseries of $\sum_n x_n$ is weakly convergent. It follows quickly that $\sum_n |x^*x_n| < \infty$ for all $x^* \in X^*$. Define $T: X^* \to l_1$ by $Tx^* = (x^*x_n)$. The closed graph theorem shows that T is continuous; hence

$$\sup_{\|x^*\| \leq 1} \sum_n |x^*x_n| < \infty.$$

Now let $\mathcal{F}$ be the field of subsets of N consisting of finite sets and their complements. Define $G: \mathcal{F} \to X$ by

$$\begin{aligned} G(E) &= \sum_{n \in E} x_n, && \text{if } E \text{ is finite,} \\ &= -\sum_{n \notin E} x_n, && \text{if } N \backslash E \text{ is finite.} \end{aligned}$$

Evidently $G: \mathcal{F} \to X$ is finitely additive and, since $\sup_{\|x^*\| \leq 1} \sum_n |x^*x_n| < \infty$, the measure G is bounded. Now let (E_n) be a nondecreasing monotone sequence in $\mathcal{F}$. If (E_n) is not eventually constant, then (E_n) consists of finite sets. By hypothesis $\lim_n \sum_{m \in E_n} x_m = \lim_n G(E_n)$ exists weakly. Therefore G is strongly additive, by Corollary 3. Hence

$$\sum_{n \in N} G(\{n\}) = \sum_n x_n$$

converges unconditionally in norm.

Similarly, Theorem 2.1 quickly establishes a characterization of Banach spaces not containing c_0.

COROLLARY 5 (BESSAGA-PEŁCZYŃSKI). *The space X contains no copy of c_0 if and only if all series $\sum_n x_n$ in X, with $\sum_n |x^*x_n| < \infty$ for all $x^* \in X^*$, are unconditionally convergent in norm.*

PROOF. Suppose X contains no copy of c_0. Let $\sum_n x_n$ be a series in X with $\sum_n |x^*x_n| < \infty$ for all $x^* \in X^*$. Then, as in the proof of Corollary 4, one has $\sup_{\|x^*\| \leq 1} \sum_n |x^*x_n| < \infty$. Let $\mathcal{F}$ be as in the proof of Corollary 4 and define $G: \mathcal{F} \to X$ precisely as in the proof of Corollary 4. Then G is a bounded vector measure. Since X

contains no copy of c_0, Theorem 2 guarantees that G is strongly additive. Hence $\sum_n G(\{n\}) = \sum_n x_n$ is unconditionally convergent in norm.

The converse is transparent since c_0 contains an abundance of nonconvergent series $\sum_n x_n$ with $\sum_n |x^* x_n| < \infty$ for all $x^* \in X^*$.

The next corollary is an example of an application of vector measure theory to Banach space theory. It is a measure-theoretic result which makes no mention of measures.

COROLLARY 6 (BESSAGA-PEŁCZYŃSKI). *If X^* is the dual of a Banach space X and X^* contains no copy of l_∞, then X^* contains no copy of c_0.*

PROOF. Let $\sum_n x_n^*$ be a series in X^* such that $\sum_n |x^{**} x_n^*| < \infty$ for all $x^{**} \in X^{**}$. Evidently $\sum_{n \in E} x_n^*$ converges in the weak* topology of X^* for each subset E of $\boldsymbol{N}$. Define $G: 2^{\boldsymbol{N}} \to X^*$ by

$$G(E) = \text{weak}^* - \sum_{n \in E} x_n^*;$$

then G is a bounded vector measure. Since X^* contains no copy of l_∞, then Theorem 2 guarantees that G is strongly additive. Hence $\sum_n x_n^* = \sum_n G(\{n\})$ converges unconditionally in norm. An appeal to Corollary 5 concludes the proof.

The next corollary is a major strengthening of the Orlicz-Pettis theorem. It is a simple consequence of Theorem 2.

COROLLARY 7. *Let X contain no copy of l_∞ and let Γ be a total subset of X^*. If $\sum_n x_n$ is a (formal) series in X such that every subseries is Γ-convergent in the sense that for each subset A of $\boldsymbol{N}$ there exists $x_A \in X$ such that*

$$\sum_{n \in A} x^* x_n = x^* x_A$$

for all $x^* \in \Gamma$, then $\sum_n x_n$ is norm unconditionally convergent.

In particular, if X is a dual space containing no copy of l_∞, then a weak-countably additive X-valued measure on a σ-field is countably additive.*

PROOF. Define a vector measure $G: 2^{\boldsymbol{N}} \to X$ by $G(A) = x_A$ as above. By virtue of Corollary 3.3, the measure G is a bounded vector measure. Since X contains no copy of l_∞, Theorem 2 insures that G is strongly additive. Thus $\sum_n G(\{n\}) = \sum_n x_n$ converges unconditionally in norm.

The second assertion is an immediate consequence of the first assertion and the observation that if $G: \Sigma \to X^*$ is weak*-countably additive and (E_n) is a sequence of pairwise disjoint members of Σ, then $\sum_n G(E_n)$ is weak*-unconditionally convergent (to $G(\bigcup_n E_n)$), and hence norm unconditionally convergent (to $G(\bigcup_n E_n)$).

It should be noted that the Orlicz-Pettis theorem is an easy special case of this theorem. A glance at the statement of the Orlicz-Pettis theorem reveals that there is an implicit separability condition present in the statement of that theorem.

The next result, which is one of the most beautiful (and important) results of vector measure theory, is a direct consequence of the Nikodým Boundedness Theorem (Theorem 3.1) and Rosenthal's lemma.

THEOREM 8 (VITALI-HAHN-SAKS-NIKODÝM). *Let Σ be a σ-field of subsets of a set Ω and (F_n) be a sequence of strongly additive X-valued measures on Σ. If $\lim_n F_n(E)$ exists in X-norm for each $E \in \Sigma$, then the sequence (F_n) is uniformly strongly additive.*

PROOF. Since $\lim_n F_n(E)$ exists for each $E \in \Sigma$, an appeal to the Nikodým Boundedness Theorem 3.1 shows that the sequence (F_n) is uniformly bounded. Now, for the moment, assume that $\lim_n F_n(E) = 0$ for all $E \in \Sigma$. If (F_n) is not uniformly strongly additive, then there exists a sequence (x_n^*) in X^* with $\|x_n^*\| \leqq 1$ for all n such that the sequence of scalar measures $(x_n^* F_n)$ is not uniformly strongly additive. Moreover, since (x_n^*) is a bounded sequence, then $\lim_n x_n^* F_n(E) = 0$ for all $E \in \Sigma$. Define $G: \Sigma \to c_0$ by

$$G(E) = (x_n^* F_n(E))$$

for all $E \in \Sigma$. Then G is a c_0-valued bounded measure defined on the σ-field Σ. According to Theorem 2, the measure G is strongly additive. A glance at the definition of the norm in c_0 reveals that $(x_n^* F_n)$ is a uniformly strongly additive sequence, a contradiction.

Now release the assumption that $\lim_n F_n(E) = 0$ and assume merely that $\lim_n F_n(E)$ exists for all $E \in \Sigma$. If the sequence (F_n) is not uniformly strongly additive, then Proposition 1.17 produces a sequence (E_n) of pairwise disjoint members of Σ such that

$$\limsup_n \sup_m \|F_m(E_n)\| > 0.$$

By passing to appropriate subsequences and relabeling, one can arrange to have a $\delta > 0$ such that $\|F_n(E_n)\| > \delta$ for all n. Then, by making use of the fact that each F_n is strongly additive, one can pass to another subsequence and relabel to have

$$\|F_n(E_n)\| > \delta \quad \text{and} \quad \|F_n(E_{n+1})\| < \delta/2$$

for all n. Now set $G_n = F_{n+1} - F_n$. Since $\lim_n F_n(E)$ exists for all $E \in \Sigma$, one has $\lim_n G_n(E) = 0$ for all $E \in \Sigma$. Therefore the sequence (G_n) is uniformly strongly additive on Σ. On the other hand,

$$\begin{aligned}\|G_n(E_{n+1})\| &\geqq \|F_{n+1}(E_{n+1})\| - \|F_n(E_{n+1})\| \\ &> \delta - \delta/2 = \delta/2.\end{aligned}$$

Hence $\lim_n \sup_m \|G_m(E_n)\| > 0$ and (G_n) is not uniformly strongly additive. But according to the first part of the proof, (G_n) must be uniformly strongly additive because it tends setwise to 0. This contradiction completes the proof.

EXAMPLE 9. *Theorem 8 fails for measures on certain fields.* To see this let $\mathscr{F}$ be the field of subsets of N consisting of finite sets and their complements. Let $F: \mathscr{F} \to c_0$ be defined by

$$\begin{aligned} F(E) &= \chi_E \qquad \text{if } E \text{ is finite,} \\ &= -\chi_{N \setminus E} \quad \text{if } N \setminus E \text{ is finite.}\end{aligned}$$

The measure F is readily seen to be the setwise (norm) limit on $\mathscr{F}$ of the strongly additive vector measures $F_n: \mathscr{F} \to c_0$ given by $F_n(E) = F(E)\chi_{\{1,2,\cdots,n\}}$. But F is certainly not strongly additive.

COROLLARY 10 (VITALI-HAHN-SAKS). *Let Σ be a σ-field of subsets of Ω and (F_n) be a sequence of X-valued countably additive vector measures such that $\lim_n F_n(E) =$*

$F(E)$ exists in norm for each $E \in \Sigma$. If μ is a nonnegative real-valued (finite) countably additive measure such that $F_n \ll \mu$ for each n, then the sequence (F_n) is equi-μ-continuous in the sense that $\lim_{\mu(E)\to 0} F_n(E) = 0$ uniformly in n in N. Consequently $F \ll \mu$.

PROOF. By Theorem 8 the sequence (F_n) is uniformly strongly additive. Since each F_n is countably additive, the sequence (F_n) is uniformly countably additive. An appeal to Corollary 2.5 finishes the proof.

It is not difficult to show that the above corollary remains true when the measures F_n are assumed to be strongly additive on a σ-field and the measure μ is assumed to be only finitely additive. A painless presentation of this fact will appear in the next section. We close this section with the beautiful

THEOREM 11 (BACHELIS AND ROSENTHAL). *Let X be a Banach space not containing a copy of l_∞ and let $\{(x_n, x_n^*)\}$ be a biorthogonal sequence in X with $\{x_n^*\}$ a total subset of X^*. Suppose $x \in X$ satisfies the condition that for each sequence $\alpha = (\alpha_n)$ of 0's and 1's there is a $y_\alpha \in X$ for which $x_n^*(y_\alpha) = \alpha_n x_n^*(x)$. Then $\sum_n x_n^*(x)x_n$ converges unconditionally to x.*

PROOF. Let $S = \{\lambda \in l_\infty$: there is $y_\lambda \in X$ for which $x_n^*(y_\lambda) = \lambda_n x_n^*(x)$ for all $n\}$. If $\lambda \in S$, then the totality of $\{x_n^*\}$ insures the uniqueness of the corresponding y_λ. If we equip S with the norm

$$\|\lambda\| = \|\lambda\|_\infty + \|y_\lambda\|_X,$$

then S is a Banach space. Moreover the natural inclusion mapping $I: S \to l_\infty$ is a continuous one-to-one linear operator whose range contains all the sequences of 0's and 1's by hypothesis. By Corollary 4, the map I is an isomorphism of S and l_∞. The inverse of I, $R: l_\infty \to S$, is a continuous linear operator. The map $B: S \to X$ defined by $B(\lambda) = y_\lambda$ is seen to be a bounded linear operator by a simple closed graph argument.

Consider the bounded linear operator $BR: l_\infty \to X$. This operator is represented by a bounded vector measure $F: 2^N \to X$. Since X contains no copy of l_∞, F is strongly additive. Therefore $\sum_n F(\{n\})$ converges unconditionally. But $F(\{n\}) = x_n^*(x)x_n$; so $\sum_n x_n^*(x)x_n$ is an unconditionally convergent series. That x is its sum is an easy consequence of the totality of $\{x_n^*\}$.

5. The Carathéodory-Hahn-Kluvanek Extension Theorem and strongly additive vector measures. This section is devoted to the exposition of the principle that a strongly additive vector measure on a field has virtually every property of a countably additive vector measure on a σ-field, save countable additivity. For instance, it will be seen that a vector measure F is strongly additive if and only if its range is relatively weakly compact, and that F is strongly additive if and only if there is a finite nonnegative finitely additive measure μ with $F \ll \mu$. In addition Lebesgue and Yosida-Hewitt decomposition theorems will be established for strongly additive vector measures. Two basic theorems provide the foundation for these results. The first is Kluvanek's Extension Theorem which bootstraps the Carathéodory-Hahn extension procedure from the context of scalar-valued measures to vector measures. The second is Stone's representation of a Boolean algebra as the field of clopen subsets of some totally disconnected compact Hausdorff space.

Basic to Kluvanek's Extension Theorem is the following straightforward lemma which has its home in elementary measure theory.

LEMMA 1. *Let $\{F_\tau : \tau \in T\}$ be a family of countably additive vector measures on a σ-field Σ of subsets of Ω. If $\mathscr{F}$ is a subfield of Σ and Σ is the σ-field generated by $\mathscr{F}$, then $\{F_\tau : \tau \in T\}$ is uniformly countably additive on Σ if and only if the family of restrictions $\{F_\tau|_{\mathscr{F}} : \tau \in T\}$ is uniformly strongly additive on $\mathscr{F}$.*

PROOF. The necessity is obvious; to prove the sufficiency, we make use of Proposition 1.17 to reduce the question to a question about a family of real-valued non-negative (finite) countably additive measures $\{\mu_\tau : \tau \in T\}$. To prove the sufficiency for such a family it is evidently sufficient to prove it for a sequence (μ_n). To this end, suppose $\{\mu_n\}$ is a uniformly strongly additive family on $\mathscr{F}$ but $\{\mu_n\}$ is not uniformly countably additive on Σ. Then there exists a monotone decreasing sequence $(E_n) \subseteq \Sigma$ with

$$\lim_n E_n = \bigcap_n E_n = \varnothing,$$

such that $\lim_n \mu_m(E_n) = 0$ is not uniform in m. By passing to a subsequence and relabeling, if necessary, we may assume that $\mu_n(E_n) > 2\varepsilon$ for all n and some fixed $\varepsilon > 0$. The remainder of the proof is based on a careful look at the Carathéodory-Hahn extension procedure. With the help of this procedure, select a disjoint sequence (S_j) in $\mathscr{F}$ such that

$$E_1 \subseteq \bigcup_{j=1}^{\infty} S_j \quad \text{and} \quad \sum_{j=1}^{\infty} \mu_1(S_j) < \mu_1(E_1) + \varepsilon/2.$$

Next note that $E_m \subseteq E_1 \subseteq \bigcup_{j=1}^{\infty} S_j$ for all m. Consequently,

$$E_m = \Big(\bigcup_{j=1}^{p} S_j \cap E_m\Big) \cup \Big(\bigcup_{j=p+1}^{\infty} S_j \cap E_m\Big)$$

for all p and for all m. Since (μ_n) is uniformly strongly additive on $\mathscr{F}$, then there exists p_0 such that

$$\mu_n\Big(\bigcup_{j=p_0+1}^{\infty} S_j\Big) < \varepsilon/2$$

for all n. Set $B_1 = \bigcup_{j=1}^{p_0} S_j \in \mathscr{F}$. Then

$$\mu_n(E_m \cap B_1) > \mu_n(E_m) - \varepsilon/2$$

for all n and m, and $\mu_1(E_1) + \varepsilon/2 > \mu_1(B_1)$.

Now, by analogous reasoning, there is a $B_2 \in \mathscr{F}$ such that

$$\mu_n((E_m \cap B_1) \cap B_2) > \mu_n(E_m) - \varepsilon/2 - \varepsilon/4$$

for all $n \geqq 1$ and $m \geqq 2$ and such that $\mu_j(E_2) + \varepsilon/4 \geqq \mu_j(B_2)$ for $j = 1, 2$.

Repeating this argument over and over produces a sequence (B_n) in $\mathscr{F}$ such that

$$\mu_n(E_m \cap B_1 \cap \cdots \cap B_k) \geqq \mu_n(E_m) - \sum_{j=1}^{k} \varepsilon/2^j$$

for all $n \geqq 1$ and $m \geqq k$ and such that $\mu_j(E_k) + \varepsilon/2^k \geqq \mu_j(B_k)$ for all $1 \leqq j \leqq k$. Now let us see what this means. Since $E_k \subseteq E_{k-1}$ and $\bigcap_k E_k = \emptyset$, the last inequality insures that for each j, we have $\lim_k \mu_j(B_k) = 0$. Thus

$$\lim_k \mu_j(B_1 \cap \cdots \cap B_k) = 0$$

for each j. But $\{\mu_j\}$ is a uniformly strongly additive family on $\mathscr{F}$ and each $B_j \in \mathscr{F}$ so that

$$\lim_k \mu_j(B_1 \cap \cdots \cap B_k) = 0$$

uniformly in j. But then our construction gives

$$\begin{aligned} 0 &= \lim_k \sup_j \mu_j(B_1 \cap B_2 \cap \cdots \cap B_k) \\ &\geqq \lim_k \mu_k(E_k \cap B_1 \cap B_2 \cap \cdots \cap B_k) \\ &\geqq \overline{\lim_k}\, \mu_k(E_k) - \varepsilon \\ &\geqq 2\varepsilon - \varepsilon = \varepsilon > 0, \end{aligned}$$

a contradiction which completes the proof.

The following theorem is the basic result of this section.

THEOREM 2 (CARATHÉODORY-HAHN-KLUVANEK EXTENSION THEOREM). *Let $\mathscr{F}$ be a field of subsets of a set Ω and let Σ be the σ-field generated by $\mathscr{F}$. Any one of the following four statements about a bounded weakly countably additive vector measure $F: \mathscr{F} \to X$ implies all the others:*

(i) *F has a (necessarily unique) countably additive extension $\bar{F}: \Sigma \to X$.*

(ii) *There exists a nonnegative real-valued (finite) countably additive measure μ on $\mathscr{F}$ such that $F \ll \mu$.*

(iii) *F is strongly additive.*

(iv) *$F(\mathscr{F})$ is a relatively weakly compact subset of X.*

PROOF. Assume (i). Then $\bar{F}(\Sigma)$ is relatively weakly compact by Corollary 2.7. Since $F(\mathscr{F}) \subseteq \bar{F}(\Sigma)$, statement (iv) holds.

To prove that (iii) follows from (iv), let (E_n) be a monotone nondecreasing sequence in $\mathscr{F}$. Then $(F(E_n))$ is a sequence in a weakly compact set. Moreover $\lim_n x^*F(E_n)$ exists for all $x^* \in X^*$. Consequently $(F(E_n))$ is a weakly Cauchy sequence lying in a weakly compact set; hence $(F(E_n))$ is weakly convergent. An appeal to Corollary 4.3 shows that F is strongly additive.

To check that (iii) implies (ii), note that, on the basis of (iii), the family $\{x^*F: x^* \in X^*, \|x^*\| \leqq 1\}$ is a uniformly strongly additive family of countably additive (finite) real-valued measures on $\mathscr{F}$. By the Carathéodory-Hahn Extension Theorem, there exist unique countably additive real-valued measures $\overline{x^*F}$ on Σ such that $\overline{x^*F} = x^*F$ on $\mathscr{F}$ for all $x^* \in X^*$. By Lemma 1, the bounded family $\{\overline{x^*F}: x^* \in X^*, \|x^*\| \leqq 1\}$ is uniformly countably additive on Σ. Now, by virtue of Theorem 2.4, there exists a nonnegative real-valued countably additive measure μ on Σ such that

$$\lim_{\mu(E) \to 0} \overline{x^*F}(E) = 0$$

uniformly in $\|x^*\| \leqq 1$. In other words,

$$\lim_{\mu(E)\to 0;\, E\in\mathscr{F}} \|F(E)\| = \lim_{\mu(E)\to 0;\, E\in\mathscr{F}} \sup_{\|x^*\|\leqq 1} |x^*F(E)| = 0.$$

Therefore $F \ll \mu|_{\mathscr{F}}$. This shows that (iii) implies (ii).

Finally it will be shown that (ii) implies (i). Let μ be extended as a nonnegative real-valued countably additive measure to Σ. Consider the (pseudo-) metric space $\Sigma(\mu)$ consisting of members of Σ equipped with the pseudometric

$$\rho(E_1, E_2) = \mu(E_1 \,\Delta\, E_2)$$

where $E_1 \,\Delta\, E_2 = (E_1 \backslash E_2) \cup (E_2 \backslash E_1)$. First note that the subspace $\mathscr{F}(\mu)$ consisting of members of $\mathscr{F}$ is dense in $\Sigma(\mu)$. Second observe that since $F \ll \mu$ and

$$F(E_1) - F(E_2) = F(E_1 \backslash E_1 \cap E_2) - F(E_2 \backslash E_1 \cap E_2),$$

the function $F\colon \mathscr{F} \to X$ is uniformly continuous. Accordingly, F has a uniformly continuous extension $\bar{F}\colon\Sigma(\mu) \to X$. It is easily verified that $\bar{F}$ is finitely additive and μ-continuous when viewed as a vector measure on Σ. Hence $\bar{F}$ is a countably additive extension of F to Σ.

The uniqueness of $\bar{F}$ follows from the uniqueness of the Carathéodory-Hahn extension of x^*F from $\mathscr{F}$ to Σ for each $x^* \in X^*$. This completes the proof.

Even though Theorem 2 appears to deal exclusively with *countably* additive measures on fields, when teamed with the Stone Representation Theorem for Boolean algebras, Theorem 2 says a good deal about *finitely* additive vector measures as well.

It is the next result which allows us to use countably additive intuition for strongly additive vector measures.

COROLLARY 3. *Let $\mathscr{F}$ be a field of subsets of Ω. Any one of the following three statements about a bounded vector measure $F\colon \mathscr{F} \to X$ implies all the others.*

(i) *There exists a finitely additive nonnegative real-valued measure μ on $\mathscr{F}$ such that $F \ll \mu$.*

(ii) *F is strongly additive.*

(iii) *$F(\mathscr{F})$ is a relatively weakly compact subset of X.*

PROOF. Let $\tilde{\mathscr{F}}$ be the Stone representation field for $\mathscr{F}$, and let $i\colon \mathscr{F} \to \tilde{\mathscr{F}}$ be a Boolean isomorphism. Define $\tilde{F}\colon \tilde{\mathscr{F}} \to X$ by $\tilde{F}(i(E)) = F(E)$ for all $E \in \mathscr{F}$. Then, for each $x^* \in X^*$, the function $x^*\tilde{F}$ is a countably additive scalar measure on $\tilde{\mathscr{F}}$, i.e., $\tilde{F}$ is weakly countably additive on $\tilde{\mathscr{F}}$. Since properties (i), (ii) and (iii) are equivalent for $\tilde{F}$, it is transparent that they are equivalent for F.

Corollary 3 plays a central role in the philosophy of strongly additive vector measures. In effect, Corollary 3 and its proof say that whenever a strongly additive vector measure fails to be countably additive, this "defect" is not the fault of the measure. Instead the real villain is the domain of the measure. Indeed the proof of Corollary 3 shows that a strongly additive vector measure can be regarded as the restriction to a field of sets of a countably additive vector measure defined on some σ-field.

Proved in much the same way as Corollary 3 are the next two corollaries.

COROLLARY 4. *A bounded family $\{F_\tau\colon \tau \in T\}$ of X-valued strongly additive vector* *measures defined on a field $\mathscr{F}$ of subsets of Ω is uniformly strongly additive if and only if there exists a nonnegative real-valued finitely additive measure μ on $\mathscr{F}$ such that*

$$\lim_{\mu(E)\to 0} \|F_\tau(E)\| = 0$$

uniformly in $\tau \in T$.

PROOF. If such a μ exists, then the uniform strong additivity if the family $\{F_\tau : \tau \in T\}$ is immediate. On the other hand, if $\{F_\tau : \tau \in T\}$ is uniformly strongly additive, consider the family $\{\tilde{F}_\tau : \tau \in T\}$ (here the same notation as in the proof of Corollary 3 is employed); then $\{\tilde{F}_\tau : \tau \in T\}$ is uniformly strongly additive on $\tilde{\mathscr{F}}$ and each $\tilde{F}_\tau$ is weakly countably additive on $\tilde{\mathscr{F}}$. By Theorem 2, each $\tilde{F}_\tau$ has a unique countably additive extension to $\sigma(\tilde{\mathscr{F}})$, the σ-field generated by $\tilde{\mathscr{F}}$. If this extension is denoted by $\tilde{F}_\tau$ also, then Lemma 1 guarantees that the family $\{\tilde{F}_\tau : \tau \in T\}$ is uniformly countably additive on $\sigma(\tilde{\mathscr{F}})$. An application of Theorem 2.4 produces a countably additive nonnegative real-valued measure $\tilde{\mu}$ on $\sigma(\tilde{\mathscr{F}})$ such that

$$\lim_{\tilde{\mu}(E)\to 0} \|\tilde{F}_\tau(E)\| = 0$$

uniformly for $\tau \in T$. Define μ by $\mu(E) = \tilde{\mu}(iE)$ for $E \in \mathscr{F}$ and obtain $\lim_{\mu(E)\to 0} \|F_\tau(E)\| = 0$ uniformly in $\tau \in T$, as required.

A similar proof yields

COROLLARY 5. *Let $\{F_\tau : \tau \in T\}$ be a bounded uniformly strongly additive family of vector measures on a field $\mathscr{F}$. If μ is a finitely additive nonnegative real-valued measure on $\mathscr{F}$ such that $F_\tau \ll \mu$ for each $\tau \in T$, then*

$$\lim_{\mu(E)\to 0} \|F_\tau(E)\| = 0$$

uniformly in $\tau \in T$.

The next result follows immediately from Corollary 5 and the Vitali-Hahn-Saks-Nikodým Theorem 4.8.

COROLLARY 6 (VITALI-HAHN-SAKS THEOREM). *Let Σ be a σ-field of subsets of Ω and μ be a finitely additive nonneagtive real-valued measure on Σ. If (F_n) is a sequence of X-valued μ-continuous vector measures on Σ such that $\lim_n F_n(E)$ exists for each $E \in \Sigma$, then*

$$\lim_{\mu(E)\to 0} F_n(E) = 0$$

uniformly in n.

To state the next result we recall some common notation. Let $\mathscr{F}$ be a field of subsets of the set Ω and let X be a Banach space. Denote by $ba(\mathscr{F}, X)$ the linear space of bounded X-valued vector measures on $\mathscr{F}$. If we define the norm of $F \in ba(\mathscr{F}, X)$ to be

$$\|F\| = \sup\{\|F(A)\| : A \in \Sigma\},$$

then $ba(\mathscr{F}, X)$ is a Banach space.

Denote by $sa(\mathscr{F}, X)$ and $ca(\mathscr{F}, X)$ the strongly additive and countably additive

X-valued vector measures defined on $\mathcal{F}$, respectively. It is easily seen that both $sa(\mathcal{F}, X)$ and $ca(\mathcal{F}, X)$ are closed linear subspaces of $ba(\mathcal{F}, X)$.

Denote by $bva(\mathcal{F}, X)$ and $bvca(\mathcal{F}, X)$ the linear subspaces of $ba(\mathcal{F}, X)$ consisting of the vector measures of bounded variation and the countably additive vector measures of bounded variation defined on $\mathcal{F}$. If we equip $bva(\mathcal{F}, X)$ with the variation norm, then $bva(\mathcal{F}, X)$ is a Banach space and $bvca(\mathcal{F}, X)$ is a closed linear subspace.

With these formalities accomplished, we can state the following theorem which graphically portrays the relationship between strongly additive vector measures and countably additive vector measures. The proof is contained essentially in the proof of Corollary 3 and is therefore omitted.

THEOREM 7. *Let $\mathcal{F}$ be a field of subsets of the set Ω. Then there exists a totally disconnected compact Hausdorff space $\tilde{\Omega}$ and a Boolean isomorphism i from $\mathcal{F}$ onto the field $\tilde{\mathcal{F}}$ of clopen subsets of $\tilde{\Omega}$. If $\sigma(\tilde{\mathcal{F}})$ denotes the σ-field generated by $\tilde{\mathcal{F}}$, then there is an isometric isomorphism B of $sa(\mathcal{F}, X)$ onto $ca(\sigma(\tilde{\mathcal{F}}), X)$ determined by the correspondence*

$$(BF)(iE) = F(E),$$

for $F \in sa(\mathcal{F}, X)$ and $E \in \mathcal{F}$,

Moreover, B maps $bva(\mathcal{F}, X)$ onto $bvca(\sigma(\tilde{\mathcal{F}}), X)$ and acts as an isometric isomorphism with respect to the variation norm on these spaces.

We shall now use the formalities set up in Theorem 8 together with Corollary 3 to prove the Lebesgue and Yosida-Hewitt decomposition theorems for strongly additive vector measures. Recall that a bounded scalar measure μ on a field is *purely finitely additive* if the only countably additive measure λ on $\mathcal{F}$ satisfying $0 \leqq \lambda(E) \leqq |\mu|(E)$ for all $E \in \mathcal{F}$ is the measure $\lambda = 0$.

THEOREM 8 (YOSIDA-HEWITT). *Let $\mathcal{F}$ be a field of subsets of a set Ω and $F: \mathcal{F} \to X$ be a strongly additive vector measure. Then there exist unique strongly additive measures F_c and F_p on $\mathcal{F}$ to X such that*

(i) F_c is countably additive on $\mathcal{F}$;

(ii) x^*F_p is purely finitely additive on $\mathcal{F}$ for each $x^* \in X^*$; and

(iii) $F = F_c + F_p$.

If, in addition, F is of bounded variation, then so are F_c and F_p. Moreover, $|F|(E) = |F_c|(E) + |F_p|(E)$ for each $E \in \mathcal{F}$ and the measures $|F_c|$ and $|F_p|$ are mutually singular in the sense that for each $\varepsilon \geqq 0$ there exists a set $A \in \mathcal{F}$ such that

$$|F_c|(\Omega \backslash A) + |F_p|(A) \leqq \varepsilon.$$

PROOF. According to Corollary 3 there is a finitely additive measure $\mu: \mathcal{F} \to [0, \infty)$ such that $F \ll \mu$. Let $B: sa(\mathcal{F}, X) \to ca(\sigma(\tilde{\mathcal{F}}), X)$ and $D: ba(\mathcal{F}, \boldsymbol{R}) \to ca(\sigma(\tilde{\mathcal{F}}), \boldsymbol{R})$ be the isometric isomorphisms guaranteed us by Theorem 8. By the scalar-valued version of the Yosida-Hewitt decomposition theorem there exist nonnegative measures $\mu_c \in ca(\mathcal{F}, \boldsymbol{R})$ and $\mu_p \in ba(\mathcal{F}, \boldsymbol{R})$ such that $\mu = \mu_c + \mu_p$. Moreover, μ_c and μ_p are mutually singular measures. By a brief computation it follows that $D\mu_c$ and $D\mu_p$ are mutually singular countably additive measures on $\sigma(\tilde{\mathcal{F}})$. Hence there exist sets E_1 and $E_2 \in \sigma(\tilde{\mathcal{F}})$ with $E_1 = \tilde{\Omega} \backslash E_2$ such that

$$(D\mu_c)(E) = (D\mu_c)(E \cap E_1) \quad \text{and} \quad (D\mu_p)(E) = (D\mu_p)(E \cap E_2),$$

for all $E \in \sigma(\tilde{\mathscr{F}})$.

Define $F_c, F_p : \mathscr{F} \to X$ by

$$F_c(E) = (BF)(iE \cap E_1) \quad \text{and} \quad F_p(E) = (BF)(iE \cap E_2)$$

for $E \in \mathscr{F}$. Clearly $F_c \ll \mu_c$, $F_p \ll \mu_p$ and $F = F_c + F_p$. An appeal to Corollary 3 shows that F_c and F_p are strongly additive. Since $F_c \ll \mu_c$ and μ_c is countably additive, F_c is countably additive. Since $F_p \ll \mu_p$ and $x^*F_p \ll \mu_p$ for each $x^* \in X^*$, it follows that the countably additive part of x^*F_p is zero for each $x^* \in X^*$. Hence x^*F_p is purely finitely additive for each $x^* \in X^*$. The uniqueness of the decomposition follows from the uniqueness of the decomposition $x^*F = (x^*F)_c + (x^*F)_p$ for each $x^* \in X^*$.

The last statement about measures of bounded variation can be proved by replacing μ above with the variation measure $|F|$ and applying Corollary 1.10.

Proved in an entirely analogous way is the Lebesgue decomposition theorem for strongly additive vector measures.

THEOREM 9 (LEBESGUE DECOMPOSITION THEOREM). *Let $\mathscr{F}$ be a field of subsets of the set Ω and $F: \mathscr{F} \to X$ be a strongly additive vector measure. Let $\lambda : \mathscr{F} \to [0, \infty)$ be a finitely additive measure. Then there exist unique strongly additive vector measures F_c and F_s on $\mathscr{F}$ to X such that*

(i) *F_c is λ-continuous;*

(ii) *x^*F_s and λ are mutually singular for each $x^* \in X^*$; and*

(iii) *$F = F_c + F_s$.*

If, in addition, F is countably additive and λ is countably additive, then F_c and F_s are countably additive.

If F is of bounded variation, then F_c and F_s are of bounded variation, $|F|(E) = |F_c|(E) + |F_s|(E)$ for each $E \in \mathscr{F}$ and $|F_s|$ and λ are mutually singular.

6. Notes and remarks. The notion of a finitely additive measure dates back at least as far as the days of Jordan content. With the advent of the Lebesgue theory, the theory of Jordan content faded and seems to live today mainly in undergraduate text books. This is not the case with the theory of finitely additive measures. In fact, interest in finitely additive measures seems to be on the increase. This apparently is the case for two reasons. First, there are many situations in which the only natural measure is not countably additive. Second, finitely additive measures no longer exist under the shroud of prejudice and fear of the past. Indeed, it has now been realized that in most cases finitely additive measures are only slightly more troublesome than their aristocrats—the countably additive measures.

The resurgence of finitely additive measures can be traced back to Hildebrandt **[1934]** and Fichtenholtz and Kantorovich **[1935]** and their work on the representation of the dual of $L_\infty[0, 1]$. Special, but important, examples of finitely additive measures in the context of invariant means discovered by Banach **[1932]** and von Neumann **[1929]** highlighted the potential role to be played in analysis by finite additivity. The theory of finitely additive measures began to take on an air of maturity in the work of Bochner **[1939]**, **[1940]**, Bochner and Phillips **[1941]**, Yosida and Hewitt **[1952]** and Leader **[1953]**. After reading Leader **[1953]** one might be convinced that countable additivity is often more of a hindrance than a help. More recent developments can be found in Fefferman **[1967]**, **[1968]**, Darst and Green **[1968]**, Darst and DeBoth **[1971]** and, in the vector case, in Uhl **[1967]**.

The notion of semivariation was introduced by Gowurin **[1936]**. For measures with values in finite dimensional spaces, the notions of bounded semivariation and bounded variation coalesce. This is not the case in *any* infinite dimensional Banach space; this is an easy consequence of the Dvoretsky-Rogers theorem. All one need do is to take a series $\sum_n x_n$ that is unconditionally convergent but not absolutely convergent and define $F(E) = \sum_{n \in E} x_n$ for each $E \subseteq N$. This fact is, in turn, related to the following result of Thomas **[1974]**: If X is an infinite dimensional Banach space, Σ is the class of Borel sets in $[0, 1]$ and μ is Lebesgue measure, then there is a μ-continuous vector measure $F: \Sigma \to X$ such that $|F|(E) = \infty$ whenever $\mu(E) > 0$.

The situation in locally convex spaces is quite different. Grothendieck **[1955]** shows that a complete metrizable locally convex space E is nuclear if and only if all E-valued measures of bounded semivariation are of bounded variation.

A curious result in the opposite direction is that of Fischer and Schöler **[1976]** who prove that if $0 < p < 1$, the only nonatomic countably additive vector measure of bounded variation (with respect to the paranorm $\| \quad \|_p$) with values in l_p is the zero measure. Along the same line, Turpin **[1975b]** has constructed an example of a countably additive vector measure defined on a σ-field with values in a "highly exotic" nonlocally convex space whose range is not bounded. Turpin **[1975b]** then goes on to provide conditions that ensure the boundedness of countably additive measures. Other related results can be found in the work of Fischer and Schöler and Turpin.

Vector measures of bounded variation with values in Banach spaces are extensively studied in the monograph of Dinculeanu **[1967]** who concentrates mainly on the theory of integration with respect to such vector measures. In particular Proposition 1.9 is from Dinculeanu **[1967]**.

The relationships stated in Proposition 1.11 appear in Bartle, Dunford and Schwartz **[1955]**. The elementary integral of §1 also can be found in Dinculeanu **[1967]**. Theorem 1.13 was studied by Day **[1942]**; its spirit goes back to Fichtenholtz and Kantorovich **[1935]** and Hildebrandt **[1934]**.

The important concept of a strongly additive vector measure is by no means new. It was introduced by Rickart **[1943]** (who used the term "strongly bounded") as a simultaneous generalization of finitely additive vector measures of bounded variation and countably additive measures on σ-fields. Strongly additive measures took a long time to assume their place at the foundation of vector measure theory with the paper of Brooks and Jewett **[1970]** finally securing that place. No doubt the idea of strong additivity is the most important notion in this chapter. Corollary 1.19 is due to Rickart **[1943]** as is the example mentioned before Definition 1.14.

Absolute continuity of point functions was introduced by Vitali **[1905]** who established the fundamental fact that a real-valued function on $[0, 1]$ is absolutely continuous if and only if it is the integral of its derivative. Absolute continuity of set functions was studied by Radon **[1919]** and Nikodým **[1930]** to whom we owe the classical theorem that bears their names.

Theorem 2.1 is due to Pettis **[1938]**. Theorem 2.4 was first discovered by Doubrovsky **[1947b]**. The proof in the text is from Gould **[1965]**. Corollaries 2.6 and 2.7 are from the fundamental Bartle, Dunford and Schwartz **[1955]** paper.

Theorem 3.1, the Nikodým Boundedness Theorem, was first established for countably additive scalar measures on σ-fields by Nikodým **[1930a]**, **[1933]**. The familar Baire category proof of Nikodým's theorem (Dunford and Schwartz **[1958]**) is due to Saks **[1933]**. It was extended to bounded vector measures by Grothendieck **[1957]** and has been proved by a variety of authors since by many different techniques. Our proof comes from Darst **[1967]**, **[1970]**. For more reading about this basic theorem see Antosik **[1973a]**, Drewnowski **[1972]**, Landers and Rogge **[1971b]**, Mikusinski **[1970]**, **[1971]**, Rosenthal **[1970]** and Thomas **[1970a]**. Corollary 3.3 is an exercise in Grothendieck **[1957]**.

Do not be misled by Example 3.6 (which is due to R. E. Huff). According to Example 3.6, the Nikodým Boundedness Theorem does not extend to bounded measures on fields of sets. However there do exist abstract Boolean algebras that are not σ-complete for which the Nikodým Boundedness Theorem holds. See Faires **[1974a]**, **[1974b]**, Grothendieck **[1953]**, Seever **[1968]** and Wells **[1969]**.

Corollary 3.4 is due to Seever **[1968]** and, as is plain from the text, is instrumental in the proof of Theorem 4.11. A particularly striking generalization of the Seever theorem is due to Bennett and Kalton **[1973]**. Let us agree that a dense linear subspace S of a Banach space X is *surjective* if for any Banach space Y and any bounded linear operator $T:Y \to X$ the inclusion $S \subseteq T(Y)$ guarantees that $T(Y) = X$. Thus by Seever's Theorem 3.4, the simple functions are a surjective subspace of $B(\Sigma)$. Bennett and Kalton have characterized those subspaces among the dense linear subspaces of a Banach space that are surjective as the barrelled subspaces.

Rosenthal's Lemma 4.1 and Theorem 4.2 are variations of a more general theorem of Rosenthal **[1968]**, **[1970a]**. Rosenthal's lemma seems to be the ultimate sharpening of a classical theorem of Phillips **[1940]**.

PHILLIPS'S LEMMA. *Let (μ_n) be a sequence of bounded finitely additive scalar-valued measures defined on all subsets of the positive integers N. If for each set $E \subseteq N$ one has $\lim_n \mu_n(E) = 0$, then*

$$\lim_n \sum_{k=1}^{\infty} |\mu_n(\{k\})| = 0.$$

Phillips's lemma was originally proved to obtain a counterexample to a slightly erroneous statement of Gel'fand **[1938]** concerning compact sets in Banach spaces. But Phillips also found several new applications of the lemma including the proof that c_0 is not complemented in l_∞. Grothendieck **[1953]** gave several deep applications of Phillips's lemma, the most striking of which is the fact that for compact Hausdorff Stonean spaces Ω, weak* and weak sequential convergence are the same in $C(\Omega)^*$. Rosenthal **[1968]**, **[1970a]** has used Lemma 4.1 as the basis for a deep study of the behavior of operators on spaces of continuous functions. We shall say more about these facts in the notes and remarks section of Chapter VI.

Our proof of Rosenthal's lemma for measures on a σ-field is from Kupka **[1975]**. Kupka's proof is a stunningly elegant improvement of Rosenthal's original proof. For measures on fields, the proof is taken from Uhl **[1973b]**. The technique of using the Hahn-Banach theorem to extend finitely additive scalar measures from one field to a larger field is due to Pettis **[1938b]**. Our treatment of Rosenthal's lemma and its applications follows Uhl **[1973b]** who shows how to deduce Theorem 4.2, Corollaries 4.3—4.7 and Theorem 4.8 from Rosenthal's lemma.

Theorem 4.2 appears in this form in Diestel **[1973a]** and Diestel and Faires **[1974]**. Tumarkin **[1970]** and Labuda **[1976a]** have extended Theorem 4.2 to sequentially complete locally convex spaces while Drewnowski **[1976a]**, **[1976b]** and Kalton **[1975]** have removed the assumption of local convexity!

Orlicz **[1929]** and Pettis **[1938]** are responsible for Corollary 4.4. The Orlicz-Pettis theorem is now one of the basic tools of Banach space theory. First discovered by Orlicz in the case of weakly sequentially complete Banach spaces, then announced by Banach **[1932]** as having been proved by Orlicz for general Banach spaces, the Orlicz-Pettis theorem was first proved independently for general Banach spaces for consumption by non-Polish speaking peoples by Pettis **[1938]**. Pettis also made clear the intimate relationships that exists between the Orlicz-Pettis theorem and the theory of vector measures. The second assertion of Corollary 4.4 is due to him.

Only relatively recently has it been realized that the Orlicz-Pettis theorem is subject to considerable generalization. The first generalization of the Orlicz-Pettis theorem is due to Kalton who imposed a separability condition that is implicit in the statement of Corollary 4.4. Previously, Grothendieck **[1953]** had remarked that the locally convex version of the Orlicz-Pettis theorem is true. Kalton **[1971]** gave a deep analysis of the theorem in the duality free context of Abelian group-valued measures and proved a forerunner of Corollary 4.7 which is essentially due to him. Drewnowski **[1975]** contains an elegant treatment of Kalton's theorems which highlights the basic separability restrictions inherent in the Orlicz-Pettis theorem. Anderson and Christensen **[1973]** have shown that the validity of Orlicz-Pettis type theorems in a topological group is dependent only on the Borel σ-field generated by the topology of the group. Batt **[1969]** and Kalton **[1974a]** have established criteria for the unconditional convergence of series of operators. A by no means complete list of related work includes Christensen **[1971]**, Coste **[1971]**, Dierolf **[1976]**, Drewnowski **[1973a]**, **[1974a]**, **[1976a]**, Drewnowski and Labuda **[1973]**, Labuda **[1973a]**, **[1973b]**, **[1973c]**, **[1974]**, **[1975]**, **[1976a]**, **[1976b]**, MacArthur **[1967]**, Musial, Ryll-Nardzewski and Woyczynski **[1975]**, Orlicz **[1948]**, Ryll-Nardzewski and Woyczynski **[1975]**, Schwartz **[1969]**, Stiles **[1970]**, Thomas **[1968]**, and Woyczysnki **[1969]**.

Corollaries 4.5 and 4.6 are due to Bessaga and Pełczyński **[1958]**. Tumarkin **[1970]** extended Corollary 4.6 to sequentially complete locally convex spaces, a context in which Thomas **[1970]** studied integration with respect to vector measures in the Bourbaki style with the conclusion of Corollary 4.6 playing a central role in his development. The second assertion of Corollary 4.7 as well as its converse appears in Diestel and Faires **[1974]**.

The Vitali-Hahn-Saks-Nikodým theorem (Theorem 4.8 and Corollary 5.6) has a rich history that reflects the development of modern integration theory. Lebesgue **[1909]** proved that if μ is Lebesgue measure on an interval and (f_n) is a sequence in $L_1(\mu)$ such that $\lim_n \int_E f_n \, d\mu = 0$ for every measurable set E, then $(\int_{(\cdot)} f_n \, d\mu)$ is an equi-μ-continuous sequence. This was a dramatic improvement of an earlier result of Vitali **[1907]** who proved that if $(f_n)_{n=0}^{\infty}$ is a sequence in $L_1(\mu)$ and $\lim_n f_n = f_0$ is μ-measure then $\lim_n \int_E f_n \, d\mu = \int_E f_0 \, d\mu$ for all measurable sets E if and only if $(\int_{(\cdot)} f_n \, d\mu)$ is an equi-μ-continuous sequence. Hahn **[1922]** improved Lebesgue's theorem by removing the assumption that $\lim_n \int_E f_n \, d\mu = 0$ for all E and replacing it by the assumption that $\lim_n \int_E f_n \, d\mu$ exists for every measurable set E.

The next step was taken by Nikodým **[1931]**, **[1933]** who proved that if Σ is a σ-field of subsets of an abstract point set and (μ_n) is a sequence of countably additive finite scalar (signed) measures on Σ and $\lim_n \mu_n(E) = \mu(E)$ exists for all $E \in \Sigma$, then μ is countably additive. Shortly thereafter Saks **[1933]** gave the first Baire category proof of this fact. Saks also proved Corollary 4.10 for scalar measures. Until Saks **[1933]**, all the proofs proceeded by sliding hump arguments.

Curiously, it remained unnoticed for some time that if Σ is a σ-field and (μ_n) is a sequence of countably additive scalar measures that converges setwise, then (μ_n) is uniformly countably additive. The first explicit observations of this fact are due to Doubrovski **[1947a]** and Pettis **[1951]**. As Pettis notes, this fact is essentially imbedded in one of Saks's **[1933]** proofs. Pettis's paper is radically different from its predecessors; he attempts to understand the Vitali-Hahn-Saks theorem from the point of view of pointwise convergence of nets of continuous functions on second category subsets of topological spaces. In this regard, see Alexiewicz **[1950]** who also deals with vector measures.

The stunning fact that the Vitali-Hahn-Saks-Nikodým theorem (Theorem 4.8 and Corollary 5.6) holds for finitely additive scalar measures was discovered by Ando **[1961]**. As will be seen below, Baire category methods seem to be unsuitable in the finitely additive case and Ando was forced to return to the more primitive sliding hump arguments of the type originally used by Lebesgue, Hahn, and Nikodým. As our proof of Theorem 4.8 shows, the vector-valued case can be quickly reduced to the scalar case. Apparently not realizing this, Brooks and Jewett **[1970]** proved Theorem 4.8 and Corollary 5.6 by a direct argument. Variations of Theorem 4.8 were also obtained by Darst **[1970a]** and Seever **[1968]**.

There are two approaches to proving theorems of the Vitali-Hahn-Saks type. The first approach, followed in Dunford and Schwartz **[1958]**, is due to M. Fréchet and O. Nikodým. It is a topological approach that is based on the fact that if Σ is a σ-field of sets and μ is a nonnegative real-valued countably additive measure defined on Σ, then $\{\chi_E: E \in \Sigma\}$ is a closed subset of $L_1(\mu)$ and is thus a complete metric space. The best known proof of the Vitali-Hahn-Saks theorem is due to Saks **[1933]** and is based on the validity of the Baire category theorem in this space. For finitely additive measures μ on σ-fields, this approach does not seem to be useful: Let Σ be the power set of the positive integers, let μ_1 be a $\{0, 1\}$-valued purely finitely additive measure on Σ and let μ_2 be defined on Σ by $\mu_2(E) = \sum_{n \in E} 2^{-n}$. Let $\mu = \mu_1 + \mu_2$ and note that μ never assumes the value 1. Next note that if $E_n = \{n, n+1, \cdots\}$, then χ_{E_n} is a Cauchy sequence in $L_1(\mu)$ and $\lim_n \mu(E_n) = 1$. It follows that $\{\chi_E: E \in \Sigma\}$ is not a complete subset of $L_1(\mu)$. This brings up an interesting question: If μ is a finite nonnegative finitely additive measure on a σ-field Σ, is the subset $\{\chi_E: E \in \Sigma\}$ of $L_1(\mu)$ of second category in itself?

For this reason, we are forced to use the second approach which relies on sliding hump arguments. Rosenthal's lemma is natural here because it produces a hump with no place to slide. In addition, the sliding hump arguments seem to have the advantage of giving more precise information about vector measures than the elegant topological methods of Fréchet and Nikodým.

In the Fréchet-Nikodým approach one considers a vector measure defined on a Boolean algebra and introduces a uniform structure on the Boolean algebra via the vector measure in such a way that the vector measure is uniformly continuous. By way of illustration suppose Σ is a σ-field of subsets of a set Ω, X is a Banach space and $F: \Sigma \to X$ is a bounded vector measure. For $A, B \in \Sigma$ define the pseudo-distance $\rho_F(A, B) = \|F\|(A \Delta B)$, where $A \Delta B$ denotes the symmetric difference of A and B, i.e., $A \Delta B = (A \backslash B) \cup (B \backslash A)$. By identifying the points of Σ that are ρ_F-distance zero apart we obtain a metric space $\Sigma(F)$. If F is countably additive, the metric space $\Sigma(F)$ is complete and F is uniformly continuous on this metric space. Moreover, if F and G are two vector measures defined on Σ, then G is F-continuous only if G is continuous on $\Sigma(F)$. In this way, uniform absolute continuity of measures translates into equicontinuity of families of continuous functions. See Drewnowski **[1972a]**, **[1972b]**, **[1973b]**, **[1974b]**, **[1976c]** for a masterful study of Fréchet-Nikodým topologies and their applications.

As is the case with the Nikodým Boundedness Theorem, the Vitali-Hahn-Saks theorem is true for measures on certain fields of sets that are not σ-fields (and for certain Boolean algebras that are not σ-complete). An algebraic characterization of those Boolean algebras (or dually a topological classification of their Stone spaces) for which the Nikodým Boundedness Theorem or the Vitali-Hahn-Saks theorem holds is unknown. Also unknown is the precise abstract relationship between the Nikodým Boundedness Theorem and the Vitali-Hahn-Saks theorem. It is known (see Seever **[1968]** or Faires **[1974b]**. **[1976]** that the Nikodým Boundedness Theorem is true for a Boolean algebra $\mathscr{F}$ whenever the Vitali-Hahn-Saks theorem holds for finitely additive bounded real-valued measures on $\mathscr{F}$. One broad class of Boolean algebras for which both theorems are valid is the class of algebras with the "interpolation property": Given sequences (a_n) and (b_n) with $a_m \leqq b_n$ for all m, n there exists a c with $a_n \leqq c \leqq b_n$ for all n; see Bade and Curtis **[1960]**, Seever **[1968]** and Faires **[1974a]**, **[1974b]**, **[1976]**. More recently, Dashiell **[1976]** has added another class of examples of Boolean algebras for which the Vitali-Hahn-Saks theorem is valid; the class considered by Dashiell is closely related to the problem of isomorphic classification of Banach spaces of bounded Baire functions. Incidentally, both the Boolean algebras with the interpolation property and those considered by Dashiell share another feature with σ-fields of sets: The weakly countably additive measures on these Boolean algebras are norm countably additive; see Faires **[1976]**. In this connection it seems to be unknown for which Boolean algebras $\mathscr{F}$ are countably additive real-valued measures on $\mathscr{F}$ necessarily bounded.

For an early and intriguing counterexample to the Vitali-Hahn-Saks theorem for real-valued measures on fields of sets, see Dunford **[1936b]**.

Theorem 4.11 is due to Bachelis and Rosenthal **[1971]**; a separable version of this theorem was proved by Davis, Dean and Singer **[1971]**.

The extension theorem. As we have seen in this chapter, the extension theorem for countably additive measures is a basic tool which unites the theory of countably additive and the theory of strongly additive vector measures. The antecedents of Theorem 5.2 go back to Carathéodory **[1927]** and Hahn **[1932]**. The extension theorem for vector measures as it appears here is a major contribution of I. Kluvanek to the theory of vector measures. Kluvanek **[1961]**, **[1966]** proved most of Theorem 5.2. When one recalls that a theory of strongly additive measures was not available at the time, one must be very impressed by Kluvanek's results. Here is a sample from Kluvanek **[1966]**.

Let X be a locally convex topological vector space and X^* its dual space. Let $\mathscr{R}$ be a ring of sets, $\mathscr{T}$ be the δ-ring and $\mathscr{S}$ the σ-ring generated by $\mathscr{R}$. Let G be a weak measure on $\mathscr{R}$ with values in X.

THEOREM. A. *Each of the following conditions is both necessary and sufficient for the existence of a measure $\bar{G}$ on $\mathscr{T}$ with values in X satisfying $\bar{G}(E) = G(E)$ for $E \in \mathscr{R}$.*

(i) *If (E_n) is a decreasing sequence of sets in $\mathscr{R}$, then $(G(E_n))$ is weakly convergent to an element of X.*

(ii) *If (E_n) is an increasing sequence of sets in $\mathscr{R}$ and if there is $F \in \mathscr{R}$ with $E_n \subseteq F$ for all n, then $(G(E_n))$ is weakly convergent to an element of X.*

(iii) *If (E_n) is a sequence of pairwise disjoint members of $\mathscr{R}$ and there is $F \in \mathscr{R}$ with $E_n \subseteq F$ for all n, then the series $\sum_{n=1}^{\infty} G(E_n)$ is weakly convergent to a sum in X.*

B. *Each of the following conditions* (iv), (v) *is both necessary and sufficient for the existence of a measure G' on $\mathscr{S}$ with values in X satisfying $G'(E) = G(E)$, $E \in \mathscr{R}$.*

(iv) *For every increasing sequence (E_n) of sets in $\mathscr{R}$, the sequence $(G(E_n))$ is weakly convergent to an element in X.*

(v) *For every sequence (E_n) of pairwise disjoint members of $\mathscr{R}$, the series $\sum_{n=1}^{\infty} G(E_n)$ is weakly convergent to a sum in X.*

COROLLARY 1. *If for each $E \in \mathscr{R}$, the set $\{G(A): A \subseteq E\}$ is relatively weakly sequentially compact in X, then the weak measure G can be extended to a measure on $\mathscr{T}$ with values in X.*

COROLLARY 2. *If the set $\{G(E): E \in \mathscr{R}\}$ is relatively weakly sequentially compact in X, then the weak measure G can be extended to a measure on $\mathscr{S}$ with values in X.*

COROLLARY 3. *If x^*G has finite variation for each $x^* \in X^*$ and X is a sequentially complete space which does not contain a copy of c_0, then G can be extended to a measure on $\mathscr{T}$ keeping values in X.*

In the text, we approach the extension problem by taking a vector measure G and stepping back to the associated family of scalar measures $\{x^*G: \|x^*\| \leqq 1\}$. Kluvanek's methods are strictly vector measure-theoretic. For a complete survey of the extension theorem, consult Kluvanek **[1973]**.

The argument that we use to prove Corollary 5.3 is sometimes called a "Stone space argument". The argument is formalized in Theorem 5.7 which is due to Stone **[1937]**. It has long been used in the theory of finitely additive scalar measures as a device to reduce the finitely additive case to the countably additive case as is done in the text. In the context of vector measures, it seems first to have been used by Uhl **[1967]**. Approximately two-thirds of the proof of Corollary 5.3 is taken directly from Uhl **[1971a]**, the other third was proved by Hoffman-Jorgensen **[1971]** and Brooks **[1971]**.

The Stone space arguments we give in §5 serve well to illustrate the principle that a strongly additive vector measure fails to be countably additive merely because of a deficiency in its domain and through no fault of its own. A strongly additive vector measure is eager to become countably additive if given the chance. Save countable additivity itself, there is no property of countably additive vector measures that is not shared by strongly additive vector measures.

The close relationships between strong additivity and countable additivity has been graphically accentuated by Drewnowski [**1973b**].

THEOREM (DREWNOWSKI). *Let X be a Banach space and F be a finitely additive X-valued vector measure defined on a σ-field Σ. The measure F is strongly additive if and only if for each sequence (E_n) of pairwise disjoint members of Σ there is a subsequence (A_m) of (E_n) such that F is countably additive on the σ-field generated by (A_m).*

PROOF. The sufficiency is a quick consequence of Corollary 1.18.

For the converse, suppose F is strongly additive and select, with the help of Corollary 5.3, a finite nonnegative finitely additive measure μ such that $F \ll \mu$. It is evidently sufficient to prove that if (E_n) is a pairwise disjoint sequence in Σ, then (E_n) has a subsequence (A_m) such that μ is countably additive on the σ-field generated by (A_m). We can assume that $\mu(\Sigma) \subseteq [0, 1]$.

To this end, let A and B be infinite subsets of $\boldsymbol{N}$ such that $A \cup B = \boldsymbol{N}$ and $A \cap B = \emptyset$. Then, either (a) $\sum_{n \in A} \mu(E_n) \leqq \frac{1}{2}$, or (b) $\sum_{n \in B} \mu(E_n) \leqq \frac{1}{2}$. If (a) obtains, let $\boldsymbol{N}_1 = A$; otherwise let $\boldsymbol{N}_1 = B$. Let $n_1 = \inf \boldsymbol{N}_1$ and partition $\boldsymbol{N}_1 \backslash \{n_1\}$ into disjoint infinite subsets A_1 and B_1 as above. Either (a_1) $\sum_{n \in A_1} \mu(E_n) \leqq \frac{1}{4}$, or ($\mathrm{b}_1$) $\sum_{n \in B_1} \mu(E_n) \leqq \frac{1}{4}$. If ($\mathrm{a}_1$) obtains, let $\boldsymbol{N}_2 = A_1$; otherwise let $\boldsymbol{N}_2 = B_1$ and let $n_2 = \inf \boldsymbol{N}_2$. Continue this process and then note that μ is countably additive on the σ-field generated by $\{E_{n_k}\}$. (We are indebted to H. P. Lotz for this proof.)

The following corollary is due to Diestel [**1973e**] and Drewnowski [**1973b**].

COROLLARY (DIESTEL, DREWNOWSKI). *Let Σ be a σ-field of subsets of Ω and let (F_n) be a sequence of strongly additive vector measures defined on Σ. If $\lim_n F_n(E) = F(E)$ exists weakly for each $E \in \Sigma$, then F is a strongly additive vector measure on Σ.*

PROOF. If F is not strongly additive, then there is an $\varepsilon > 0$ and a disjoint sequence (E_n) of sets belonging to Σ such that $\|F(E_n)\| > \varepsilon$ for all n. By a diagonal procedure, select a subsequence (A_m) of (E_n) such that each measure F_n is countably additive on the σ-field generated by (A_m). By the Vitali-Hahn-Saks-Nikodým theorem for countably additive scalar measures, x^*F is countably additive on this σ-field for each $x^* \in X^*$. By the Orlicz-Pettis theorem, F is countably additive on this σ-field. This contradiction completes the proof.

Huff [**1973b**] has used Drewnowski's theorem as a starting point for much of the theory of strongly additive measures including the Vitali-Hahn-Saks-Nikodým theorem.

The Vitali-Hahn-Saks-Nikodým theorem has already been discussed. The Vitali-Hahn-Saks theorem as stated above was first isolated by Brooks and Jewett [**1970**] and can be deduced immediately from its scalar-valued counterpart, which is due to Ando [**1961**]. It extends easily and without injury to the context of locally convex spaces. Faires [**1976**] has extended it to group-valued measures on Boolean algebras with the interpolation property.

The Yosida-Hewitt decomposition theorem for scalar measures was (naturally) proved by Yosida and Hewitt **[1952]**. It can also be found in Dunford and Schwartz **[1958]**. Its extension to the vector-valued case is due to Uhl **[1971a]** who obtained it from the scalar case by the Stone space argument in the text. A direct proof that includes both the vector case and the scalar case was given by Huff **[1973]**. Huff's argument comes from ergodic theory. Drewnowski **[1973b]** and Traynor **[1972b]** have obtained Yosida-Hewitt theorems for group-valued measures. Bilyeu and Lewis **[1976]** have looked at this theorem from the point of view of James orthogonality.

The Lebesgue decomposition theorem (5.9) for strongly additive measures is due to Rickart, who introduced the notion of strong additivity with precisely this result in mind. Wallowing in a state of ignorance, Uhl **[1971a]** rediscovered this theorem and Uhl's theorem was merged into Rickart's theorem as a consequence of Hoffman-Jorgensen **[1971]** and Brooks **[1971]**. Luckily for Uhl, his proof was easier then Rickart's proof and it is Uhl's proof that we give in the text. The work of Brooks **[1969c]**, Darst **[1962a]**, **[1962b]**, **[1963]**, **[1970b]**, Drewnowski **[1973b]**, **[1974b]**, Orlicz **[1968]** and Traynor **[1972a]** is of interest in connection with the Lebesgue decomposition theorem. Brooks **[1971b]** has given a particularly elegant derivation of the Lebesgue decomposition theorem for scalar-valued measures.

The other classical decomposition theorems of scalar measure theory, the Hahn decomposition and the Jordan decomposition theorems, have been investigated for vector-valued measures. Prerequisite to the study of these results is the assumption of an order-theoretic structure in the vector space. In case X is a Banach lattice of dimension greater than one the Hahn decomposition theorem fails to hold as simple examples show. If X is an order complete Banach lattice then an additive measure F with values in X admits a decomposition into the difference of positive X-valued measures if and only if the range of F is order-bounded; see Faires and Morrison **[1976]**.

Wright **[1968]**, **[1969a]**, **[1969b]**, **[1970]**, **[1971]** has made an extensive study of measures with values in order complete vector lattices where countable additivity refers to the topology of order-convergence. Sion **[1969]**, **[1973]** has conducted an investigation of measures with values in semigroups; the condition of strong additivity arises naturally in Sion's work.

II. INTEGRATION

This chapter deals with the definitions and basic properties of integrals of vector-valued functions with respect to scalar measures, and integrals of scalar-valued functions with respect to vector measures. We will not be striving for overwhelming generality here; rather, we will examine integrals that have enough structure to be worthwhile for the purpose of concrete applications. Thus most of the chapter is devoted to the Bochner integral (Dunford's first integral), the Pettis integral (Dunford's second integral) and the Dunford integral. At the end of the chapter we shall look at a version of the Bartle integral which we have already met in the first chapter. The basis for this material is a finite measure space (Ω, Σ, μ) and a Banach space X.

1. Measurable functions. Two forms of measurability—strong and weak measurability—form the core in this section. Most of the work will focus on strong measurability for, as we shall see in later chapters, the quality of measurability is directly proportional to the quality of applications.

DEFINITION 1. A function $f: \Omega \to X$ is called *simple* if there exist $x_1, x_2, \cdots, x_n \in X$ and $E_1, E_2, \cdots, E_n \in \Sigma$ such that $f = \sum_{i=1}^n x_i \chi_{E_i}$, where $\chi_{E_i}(\omega) = 1$ if $\omega \in E_i$ and $\chi_{E_i}(\omega) = 0$ if $\omega \notin E_i$. A function $f: \Omega \to X$ is called *μ-measurable* if there exists a sequence of simple functions (f_n) with $\lim_n \|f_n - f\| = 0$ μ-almost everywhere. A function $f: \Omega \to X$ is called *weakly μ-measurable* if for each $x^* \in X^*$ the numerical function x^*f is μ-measurable. More generally, if $\Gamma \subseteq X^*$ and x^*f is measurable for each $x^* \in \Gamma$, then f is called *Γ-measurable.* If $f: \Omega \to X^*$ is X-measurable (when X is identified with its image under the natural imbedding of X into X^{**}), then f is called *weak*-measurable.*

In the literature, the terms *strong measurability* and *scalar measurability* are often used to describe μ-measurability and weak μ-measurability, respectively. Sometimes reference to the measure μ will be suppressed when there is no chance of ambiguity.

The usual facts regarding the stability of measurable functions under sums, scalar multiples and pointwise (almost everywhere) limits hold. Replacing absolute values by norms throughout the usual proof of Egoroff's theorem generalizes that result to the vector-valued case. This is useful in the proof of the next theorem which is basic to the study of measurable functions.

THEOREM 2 (PETTIS'S MEASURABILITY THEOREM) *A function $f\colon \Omega \to X$ is μ-measurable if and only if*

(i) *f is μ-essentially separably valued, i.e., there exists $E \in \Sigma$ with $\mu(E) = 0$ and such that $f(\Omega \backslash E)$ is a (norm) separable subset of X, and*

(ii) *f is weakly μ-measurable.*

PROOF. Let $f\colon \Omega \to X$ be μ-measurable. Egoroff's theorem produces a sequence (f_n) of simple functions with $\lim_n \|f_n - f\| = 0$ μ-almost uniformly. Thus, for each positive integer n, there is a set $E_n \in \Sigma$ such that $\mu(E_n) < 1/n$ and $\lim_n f_n = f$ uniformly on $\Omega \backslash E_n$. Since each f_n has a finite dimensional bounded range, it follows that $f(\Omega \backslash E_n)$ is totally bounded and therefore separable. Accordingly

$$f\Big(\bigcup_{n=1}^{\infty} (\Omega \backslash E_n)\Big) = \bigcup_{n=1}^{\infty} f(\Omega \backslash E_n)$$

is separable. Moreover $\Omega \backslash \bigcup_{n=1}^{\infty} (\Omega \backslash E_n) = \bigcap_{n=1}^{\infty} E_n$ is a set of μ-measure zero since $\mu(E_n) < 1/n$ for each n. This proves the necessity of (i).

To prove the necessity of (ii), note that $f_n(\omega) \to f(\omega)$ for almost all $\omega \in \Omega$ guarantees that for $x^* \in X^*$, we have $x^*(f_n(\omega)) \to x^*(f(\omega))$ for almost all $\omega \in \Omega$. Since each f_n is simple, then $x^* f_n$ is also simple. Therefore $x^* f$ is measurable for each $x^* \in X^*$. This proves the necessity of (ii).

To prove the converse, let $E \in \Sigma$ be chosen such that $\mu(E) = 0$ and $f(\Omega \backslash E)$ is separable. Let $\{x_n\}$ be a countable dense subset of $f(\Omega \backslash E)$. With the help of the Hahn-Banach theorem, choose a sequence (x_n^*) of members of X^* such that $x_n^*(x_n) = \|x_n\|$ and $\|x_n^*\| = 1$. A moment's reflection reveals that $\|f(\omega)\| = \sup_n |x_n^*(f(\omega))|$ for each $\omega \in \Omega \backslash E$. Therefore the function $\|f(\cdot)\|$ is μ-measurable. By the same argument, the functions g_n defined by $g_n(\cdot) = \|f(\cdot) - x_n\|$ for each n, are all μ-measurable. Now let $\varepsilon > 0$. Write $E_n = \{\omega \in \Omega\colon g_n(\omega) < \varepsilon\}$. If μ is complete, each E_n belongs to Σ. In any case, for each n there is a set $B_n \in \Sigma$ with $\mu(B_n \,\Delta\, E_n) = 0$. Define $g\colon \Omega \to X$ by

$$\begin{aligned} g(\omega) &= x_n \quad \text{if } \omega \in B_n \Big\backslash \bigcup_{m<n} B_m, \\ &= 0 \quad \text{otherwise.} \end{aligned}$$

Then $\|g - f\| < \varepsilon$ μ-almost everywhere. Therefore f can be μ-essentially uniformly approximated by a countably valued function. Taking $\varepsilon = 1/n$ and letting n run through the positive integers produces a sequence (g_n') of countably valued functions with $\|g_n' - f\| < 1/n$ a.e. for each n. Since each g_n has the form $g_n' = \sum_{m=1}^{\infty} x_{n,m} \chi_{E_{n,m}}$ with $E_{n,i} \cap E_{n,j} = \emptyset$ for $i \neq j$, and $E_{n,m} \in \Sigma$ and since μ is a finite measure, it is a simple matter to clip off the g_n's in such a way as to define a sequence (f_n) of simple functions converging μ-almost everywhere to f.

The above proof yields somewhat stronger results than those advertised in the statement of Theorem 2.

COROLLARY 3. *A function $f\colon \Omega \to X$ is μ-measurable if and only if f is the μ-almost everywhere uniform limit of a sequence of countably valued μ-measurable functions.*

Also immediate from the proof is

COROLLARY 4. *A μ-essentially separably valued function $f\colon \Omega \to X$ is μ-measurable* *if there exists a norming set $\Gamma \subseteq X^*$ such that the numerical function x^*f is μ-measurable for each $x^* \in \Gamma$. (Recall $\Gamma \subseteq X^*$ is norming if*

$$\|x\| = \sup\{|x^*x| / \|x^*\| : x^* \in \Gamma\}$$

for each $x \in X$.)

EXAMPLE 5. *A weakly measurable function that is not measurable.* Let $\{e_t : t \in [0, 1]\}$ be an orthonormal basis for the nonseparable Hilbert space $l_2([0, 1])$. Define f: $[0, 1] \to l_2([0, 1])$ by $f(t) = e_t$. If μ is Lebesgue measure on [0, 1], then the Riesz Representation Theorem reveals that $x^*f = 0$ μ-almost everywhere for each $x^* \in l_2([0, 1])^* = l_2([0, 1])$. Therefore f is weakly Lebesgue measurable. On the other hand, if $E \subseteq [0, 1]$, then $f([0, 1]\backslash E)$ is separable if and only if $[0, 1]\backslash E$ is countable. Therefore f is not essentially separably valued.

EXAMPLE 6 (SIERPINSKI). *A weak*-measurable function that is not weakly measurable.* Let (r_n) be the sequence of Rademacher functions on [0, 1], i.e., for $t \in [0, 1]$, $r_n(t) = \text{sign}(\sin(2^n\pi t))$. Define f: $[0, 1] \to l_\infty$ by $f(t) = ((r_1(t) + 1)/2, (r_2(t) + 1)/2, \cdots)$. First, it is obvious that f is not almost separably valued with respect to Lebesgue measure on [0, 1] since $\|f(t) - f(t')\|_\infty = 1$ for nondyadic rational members t and t' of [0, 1] with $t \neq t'$. More interesting is the fact that f is not weakly measurable. An outline of this argument will be given.

Let β be a $\{0, 1\}$-valued, nonzero purely finitely additive measure on $\mathscr{P}(N)$, the power set of the positive integers. Make note of the following facts:

(a) If $t = \sum_{n=1}^\infty c_n 2^{-n}$ is any nondyadic number in [0, 1], then $1 - t = \sum_{n=1}^\infty (1 - c_n)2^{-n}$ and $f(1 - t) = (1, 1, \cdots) - f(t)$.

(b) Integration with respect to β over N in the sense of I.1.12 defines a bounded linear functional on l_∞.

(c) Since $\beta(E) = 0$ for every finite set $E \subseteq N$, it follows that for each dyadic rational d in [0, 1]

$$\int_N f(t)\, d\beta = \int_N f(t+d)\, d\beta$$

whenever $t, t + d \in [0, 1]$.

Set $\phi(t) = \int_N f(t)\, d\beta$. Then either $\phi(t)$ is 0 or 1. In fact, $\phi(t) = 0$ if $\beta(\{n: r_n(t) + 1 = 2\}) = 0$, and $\phi(t) = 1$, if $\beta(\{n: r_n(t) + 1 = 2\}) = 1$. Now if ϕ is Lebesgue measurable, then a glance at (c) reveals that ϕ has a dense set of periods. Consequently ϕ is a constant k almost everywhere with respect to Lebesgue measure on [0, 1]. Making use of fact (a) above, one sees that

$$\begin{aligned}\phi(1 - t) &= \int_N f(1 - t)\, d\beta = \int_N [(1, 1, \cdots) - f(t)]\, d\beta \\ &= \beta(N) - \int_N f(t)\, d\beta = 1 - \phi(t)\end{aligned}$$

for every nondyadic $t \in [0, 1]$. Hence $k = 1 - k$ and $k = \frac{1}{2}$. But this contradicts the fact that ϕ assumes only the values 0 and 1. Thus ϕ is not Lebesgue measurable and f is not weakly measurable.

It is plain that f is weak*-measurable.

The next example is a relative of Example 6 but with strikingly different properties.

EXAMPLE 7 (HAGLER). *A nontrivially weakly measurable function.* Let (A_n) be a sequence of subintervals of $[0, 1]$ such that: (i) $A_1 = [0, 1]$; (ii) each A_n is a nonempty subinterval of $[0, 1]$; (iii) $\lim_n \mu(A_n) = 0$ where μ is Lebesgue measure; (iv) $A_n = A_{2n} \cup A_{2n+1}$ for all n, and (v) $A_m \cap A_j = \emptyset$ for each pair m and j with $m \neq j$ and $2^i \leqq m, j \leqq 2^{i+1} - 1$ (bisect the interval and keep bisecting). Define $f\colon [0, 1] \to l_\infty$ by $f(t) = (\chi_{A_n}(t))$ for $t \in [0, 1]$. Let $\phi \in l_\infty^*$, and let ϕ_1 be the countably additive part of ϕ and ϕ_2 be the purely finitely additive part of ϕ. (Here ϕ_1 is the countably additive measure on $\mathscr{P}(N)$ given by $\phi_1(E) = \sum_{n\in E} \phi(\{n\})$ for $E \subseteq N$.) Now

$$\begin{aligned}\phi(f(t)) &= \phi_1(f(t)) + \phi_2(f(t)) \\ &= \sum_{n=1}^{\infty} [\chi_{A_n}(t)\phi\{n\} + \phi_2(f(t))].\end{aligned}$$

To show f is weakly (Lebesgue) measurable, it is evidently enough to show $\phi_2(f(\cdot))$ is measurable, and this will be established if it can be shown that $\phi_2(f(\cdot))$ is countably nonzero. To prove this, it is enough to show that

$$\sum_{[t\in 0,1]} \phi_2(f(t)) \leqq \|\phi_2\| \qquad (\leqq \|\phi\|).$$

To this end, let $t_1, \cdots, t_k$ be distinct points in $[0, 1]$ and write $B_i = \{j : f(t_i)(j) = 1\}$, $i = 1, 2, \cdots, k$. Thus $f(t_i) = \chi_{B_i}$. By the "tree property" ($A_{2n} \cup A_{2n+1} = A_n$) of the sequence (A_n) and the fact that distinct t_i's lie in distinct A_j's, there is an m such that the sets $B_i \cap \{m, m + 1, \cdots\}$, $i = 1, 2, \cdots, k$, are pairwise disjoint. Hence

$$\|\phi_2\| = |\phi_2| \geqq \sum_{i=1}^{k} |\phi_2(B_i \cap \{m, m + 1, \cdots\})|,$$

where $|\phi_2|$ is the total variation of ϕ_2. But ϕ_2 vanishes on finite sets and the above inequality takes the form

$$\|\phi_2\| \geqq \sum_{i=1}^{k} |\phi_2(B_i)| = \sum_{i=1}^{k} |\phi_2(f(t_i))|,$$

and this proves that $\phi f(\cdot)$ is measurable.

2. The Bochner integral. This section is devoted to an examination of the Bochner integral. Some know it as the "Dunford and Schwartz integral" and some old timers know it as Dunford's first integral. It is a straightforward abstraction of the Lebesgue integral. Indeed, some have said that the Bochner integral is only the Lebesgue integral with absolute value signs replaced by norm signs. We shall see that often this is the case, and sometimes it is a totally ignorant appraisal of the Bochner integral. In fact, as we shall see later, the failure of the Radon-Nikodým theorem for the Bochner integral lies at the base of some of the most intriguing results in the theory of vector measures and the structure theory of Banach spaces.

DEFINITION 1. A μ-measurable function $f\colon \Omega \to X$ is called *Bochner integrable* if there exists a sequence of simple functions (f_n) such that

$$\lim_n \int_\Omega \|f_n - f\| \, d\mu = 0.$$

In this case, $\int_E f \, d\mu$ is defined for each $E \in \Sigma$ by

$$\int_E f\,d\mu = \lim_n \int_E f_n\,d\mu,$$

where $\int_E f_n\,d\mu$ is defined in the obvious way.

In the interest of the sanity of the readers and the authors, the verification of the facts that the above limits exists and is independent of the defining sequence (f_n) will be omitted.

A concise characterization of Bochner integrable functions is given next.

THEOREM 2. *A μ-measurable function $f\colon \Omega \to X$ is Bochner integrable if and only if $\int_\Omega \|f\|\,d\mu < \infty$.*

PROOF. If f is Bochner integrable, let (f_n) be a defining sequence of simple functions for $\int_{(\cdot)} f\,d\mu$. Then

$$\int_\Omega \|f\|\,d\mu \leqq \int_\Omega \|f - f_n\|\,d\mu + \int_\Omega \|f_n\|\,d\mu < \infty$$

for sufficiently large n.

Conversely, suppose f (and consequently $\|f\|$) is μ-measurable and $\int_\Omega \|f\|\,d\mu < \infty$. With the help of Corollary 1.3, choose a sequence of countably valued measurable functions (f_n) such that $\|f - f_n\| \leqq 1/n$ for each positive integer n. Since $\|f_n\| \leqq \|f\| + 1/n$ μ-a.e. and μ is finite, then $\int_\Omega \|f_n\|\,d\mu < \infty$. For each positive integer n, write

$$f_n = \sum_{m=1}^{\infty} x_{n,m}\,\chi_{E_{n,m}},$$

where $E_{n,i} \cap E_{n,j} = \emptyset$ for $i \neq j$, $E_{n,m} \in \Sigma$, $x_{n,m} \in X$. For each n, choose p_n so large that

$$\int_{\bigcup_{m=p_n+1}^{\infty} E_{n,m}} \|f_n\|\,d\mu < \mu(\Omega)/n$$

and set $g_n = \sum_{m=1}^{p_n} x_{n,m}\,\chi_{E_{n,m}}$. Then each g_n is a simple function and

$$\begin{aligned}\int_\Omega \|f - g_n\|\,d\mu &\leqq \int_\Omega \|f - f_n\|\,d\mu + \int_\Omega \|f_n - g_n\|\,d\mu \\ &\leqq \mu(\Omega)/n + \mu(\Omega)/n = 2\mu(\Omega)/n.\end{aligned}$$

Therefore f is Bochner integrable, as was to be proven.

THEOREM 3 (DOMINATED CONVERGENCE THEOREM). *Let (Ω, Σ, μ) be a finite measure space and (f_n) be a sequence of Bochner integrable X-valued functions on Ω. If $\lim_n f_n = f$ in μ-measure, (i.e., $\lim_n \mu\{\omega \in \Omega\colon \|f_n - f\| \geqq \varepsilon\} = 0$ for every $\varepsilon > 0$) and if there exists a real-valued Lebesgue integrable function g on Ω with $\|f_n\| \leqq g$ μ-almost everywhere, then f is Bochner integrable and $\lim_n \int_E f_n\,d\mu = \int_E f\,d\mu$ for each $E \in \Sigma$. In fact, $\lim_n \int_\Omega \|f - f_n\|\,d\mu = 0$.*

PROOF. Just apply the scalar Dominated Convergence Theorem to $\|f - f_n\|$ with dominating function $2g$.

Further elementary facts about the Bochner integral are collected next.

THEOREM 4. *If f is a μ-Bochner integrable function, then*
(i) $\lim_{\mu(E)\to 0} \int_E f\, d\mu = 0$;
(ii) $\left\|\int_E f\, d\mu\right\| \leqq \int_E \|f\|\, d\mu$, *for all* $E \in \Sigma$;
(iii) *if* (E_n) *is a sequence of pairwise disjoint members of* Σ *and* $E = \bigcup_{n=1}^{\infty} E_n$, *then*

$$\int_E f\, d\mu = \sum_{n=1}^{\infty} \int_{E_n} f\, d\mu,$$

where the sum on the right is absolutely convergent;
(iv) *if* $F(E) = \int_E f\, d\mu$, *then* F *is of bounded variation and*

$$|F|(E) = \int_E \|f\|\, d\mu \quad \textit{for all } E \in \Sigma.$$

PROOF. (i) Since $\lim_{\mu(E)\to 0} \int_E \|f\|\, d\mu = 0$ for $f \in L_1(\mu)$, statement (i) follows from (ii). To prove (ii), note that the triangle inequality establishes (ii) for simple functions. For the general case, pass to the appropriate limit.

(iii) First note that the series $\sum_{n=1}^{\infty} \int_{E_n} f\, d\mu$ is dominated term-by-term by the convergent series of nonnegative numbers $\sum_{n=1}^{\infty} \int_{E_n} \|f\|\, d\mu\ (\leqq \int_\Omega \|f\|\, d\mu < \infty)$. Therefore $\sum_{n=1}^{\infty} \int_{E_n} f\, d\mu$ is absolutely convergent. To check its limit note that

$$\left\|\int_{\bigcup_{n=1}^{\infty} E_n} f\, d\mu - \sum_{n=1}^{m} \int_{E_n} f\, d\mu\right\| = \left\|\int_{\bigcup_{n=m+1}^{\infty} E_n} f\, d\mu\right\|,$$

by the obvious finite additivity of the Bochner integral. Moreover $\lim_m \mu(\bigcup_{n=m+1}^{\infty} E_n) = 0$. An appeal to (a) reveals that $\lim_m \|\int_{\bigcup_{n=m+1}^{\infty} E_n} f\, d\mu\| = 0$ and, consequently,

$$\int_{\bigcup_{n=1}^{\infty} E_n} f\, d\mu = \sum_{n=1}^{\infty} \int_{E_n} f\, d\mu,$$

as required.

(iv) To prove (iv), note that if π is a partition of a set $E \in \Sigma$, then

$$\sum_{A\in\pi} \|F(A)\| = \sum_{A\in\pi} \left\|\int_A f\, d\mu\right\|$$
$$\leqq \sum_{A\in\pi} \int_A \|f\|\, d\mu = \int_E \|f\|\, d\mu.$$

Hence $|F|(E) \leqq \int_E \|f\|\, d\mu$ and F is of bounded variation by Theorem 2.

To prove the reverse inequality, let $\varepsilon > 0$ and select a sequence (f_n) of simple functions such that

$$\lim_n \int_\Omega \|f - f_n\|\, d\mu = 0.$$

Fix n_0 such that $\int_\Omega \|f - f_{n_0}\|\, d\mu < \varepsilon$ and choose a partition π' of E such that

$$\sum_{A\in\pi'} \left\|\int_A f_{n_0}\, d\mu\right\| = \int_E \|f_{n_0}\|\, d\mu.$$

Next choose a partition π of E refining π' such that

$$|F|(E) - \sum_{B\in\pi} \left\|\int_B f\, d\mu\right\| < \varepsilon.$$

One still has

$$\int_E \|f_{n_0}\|\, d\mu = \sum_{B\in\pi} \left\|\int_B f_{n_0}\, d\mu\right\|.$$

Moreover

$$\sum_{B\in\pi} \left|\left\|\int_B f\, d\mu\right\| - \left\|\int_B f_{n_0}\, d\mu\right\|\right| \leqq \int_E \|f - f_{n_0}\|\, d\mu < \varepsilon.$$

Hence one has

$$\left||F|(E) - \int_E \|f_{n_0}\|\, d\mu\right| = \left||F|(E) - \sum_{B\in\pi} \int_B \|f_{n_0}\|\, d\mu\right| < 2\,\varepsilon.$$

Since this holds for all sufficiently large n_0 we infer from the above that

$$|F|(E) = \lim_n \int_E \|f_n\|\, d\mu = \int_E \|f\|\, d\mu,$$

as required.

COROLLARY 5. *If f and g are Bochner integrable and $\int_E f\, d\mu = \int_E g\, d\mu$ for each $E \in \Sigma$, then $f = g$ μ-almost everywhere.*

PROOF. Set $F(E) = \int_E (f - g)\, d\mu$. Then $F(E) = 0$ for each $E \in \Sigma$. Therefore $|F|(E) = 0$ for each $E \in \Sigma$. But then $0 = |F|(\Omega) = \int_\Omega \|f - g\|\, d\mu$ and so $\|f - g\| = 0$ μ-almost everywhere. This can happen only if $f = g$ μ-almost everywhere.

The next theorem exhibits a strong property of Bochner integration that has no analogue in the theory of Lebesgue integration. For bounded operators, its proof is a simple exercise.

THEOREM 6 (HILLE). *Let T be a closed linear operator defined inside X and having values in a Banach space Y. If f and Tf are Bochner integrable with respect to μ, then*

$$T\left(\int_E f\, d\mu\right) = \int_E Tf\, d\mu$$

for all $E \in \Sigma$.

PROOF. Let $\varepsilon > 0$ and select a function $\sum_{n=1}^\infty x_n \chi_{E_n} = h_\varepsilon$, where (E_n) is a sequence of pairwise disjoint members of Σ, $E_n \subseteq E$ and $x_n \in X$, such that

$$\sup\{\|f(\omega) - h_\varepsilon(\omega)\| : \omega \in E\backslash N_1\} < \varepsilon/2$$

for some μ-null set N_1. Also we can find a function g_ε of the form $g_\varepsilon = \sum_{n,m=1}^\infty y_{n,m} \chi_{E_{n,m}}$, where $(E_{n,m})$ is a sequence of pairwise disjoint members of Σ, $\bigcup_{m=1}^\infty E_{n,m} = E_n$, $y_{n,m} \in Y$, such that

$$\sup\{\|Tf(\omega) - g_\varepsilon(\omega)\| : \omega \in E\backslash N_2\} < \varepsilon/2$$

for some μ-null set N_2. For each pair (n, m) of positive integers, pick $\omega_{n,m} \in E_{n,m}$ arbitrarily. Write $\phi = \sum_{n,m} f(\omega_{n,m}) \chi_{E_{n,m}}$. Then it follows that

$$\|f(\omega) - \phi(\omega)\| < \varepsilon \quad \text{for } \omega \notin N_1$$

and

$$\|Tf(\omega) - T\phi(\omega)\| < \varepsilon \quad \text{for } \omega \notin N_2.$$

Moreover one has

$$\int_\Omega \|f - \phi\| \, d\mu < \varepsilon\mu(\Omega) \quad \text{and} \quad \int_\Omega \|Tf - T\phi\| \, d\mu < \varepsilon\mu(\Omega).$$

Also

$$\int_E \phi \, d\mu = \lim_{k,j} \sum_{n=1}^{k} \sum_{m=1}^{j} f(\omega_{n,m})\mu(E_{n,m})$$

and

$$\int_E T\phi \, d\mu = \lim_{k,j} \sum_{n=1}^{k} \sum_{m=1}^{j} Tf(\omega_{n,m})\mu(E_{n,m}).$$

Since T is closed, it follows that $\int_E \phi \, d\mu$ belongs to the domain of T and that $T(\int_E \phi \, d\mu) = \int_E T\phi \, d\mu$. Now replace ε with a sequence $\varepsilon_i \to 0$ and replace ϕ with the corresponding sequence ϕ_i. Then one has

$$\int_E \phi_i \, d\mu \to \int_E f \, d\mu, \quad \text{and} \quad \int_E T\phi_i \, d\mu \to \int_E Tf \, d\mu.$$

Since $T(\int_E \phi_i \, d\mu) = \int_E T\phi_i d\mu$ and T is closed, it follows that $T(\int_E f \, d\mu) = \int_E Tf \, d\mu$.

It is instructive to see how many ways Theorem 6 can be used to prove "differentiation under the integral sign" theorems. More important for us is a straightforward application of Theorem 6.

COROLLARY 7. *Let f and g be μ-measurable. If for each $x^* \in X^*$, $x^*f = x^*g$ μ-almost everywhere, then $f = g$ μ-almost everywhere.*

PROOF. Select a sequence (E_n) in Σ such that $E_n \subseteq E_{n+1}$, $\bigcup_{n=1}^\infty E_n = \Omega$ and f and g are both bounded on each E_n. Fix n. Since f and g are both bounded on E_n, the Bochner integrals

$$\int_E f\chi_{E_n} \, d\mu \quad \text{and} \quad \int_E g\chi_{E_n} \, d\mu$$

exist for all $E \in \Sigma$. Since for each $x^* \in X^*$, we have $x^*f = x^*g$ μ-almost everywhere, these integrals are equal by Theorem 6. An appeal to Corollary 5 establishes that $f\chi_{E_n} = g\chi_{E_n}$ μ-almost everywhere for each n. Consequently $f = g$ almost everywhere.

Another straightforward application of Theorem 6 is a version of the mean value theorem for the Bochner integral.

COROLLARY 8. *Let f be Bochner integrable with respect to μ. Then for each $E \in \Sigma$ with $\mu(E) > 0$ one has*

$$\frac{1}{\mu(E)} \int_E f \, d\mu \in \overline{\text{co}}(f(E)).$$

PROOF. Proceeding by contradiction, suppose there is a set $E \in \Sigma$ of positive

μ-measure such that $(\mu(E))^{-1} \int_E f\, d\mu \notin \overline{\text{co}}(f(E))$. With the aid of the geometric version of the Hahn-Banach theorem, select $x^* \in X^*$ and real α such that

$$x^*\left((\mu(E))^{-1} \int_E f\, d\mu\right) < \alpha \leqq x^* f(\omega)$$

for all $\omega \in E$ (the obvious variations can be made in the case of complex scalars). Then, by Theorem 6, one has

$$(\mu(E))^{-1} \int_E x^* f\, d\mu < \alpha \leqq x^* f(\omega)$$

for all $\omega \in E$. Integrating over E yields

$$\int_E x^* f\, d\mu < \alpha\mu(E) \leqq \int_E x^* f\, d\mu,$$

a blatant contradiction.

The next result shows that indefinite Bochner integrals share a most important property with indefinite Lebesgue integrals.

THEOREM 9. *Let f be Bochner integrable on* $[0, 1]$ *with respect to Lebesgue measure. Then for almost all* $s \in [0, 1]$ *one has*

$$\lim_{h\to 0} \frac{1}{h} \int_s^{s+h} \|f(t) - f(s)\|\, dt = 0.$$

Consequently, for almost all $s \in [0, 1]$ *one has*

$$\lim_{h\to 0} \frac{1}{h} \int_s^{s+h} f(t)\, dt = f(s).$$

PROOF. Since $\|h^{-1} \int_s^{s+h} f(t)\, dt - f(s)\| \leqq h^{-1} \int_s^{s+h} \|f(t) - f(s)\|\, dt$, the second assertion follows from the first statement. To prove the first statement, assume without loss of generality that f is separably valued. Let $\{x_n\}$ be a countable dense subset of $f([0, 1])$. By the Lebesgue differentiation theorem, one has

$$\lim_{h\to 0} \frac{1}{h} \int_s^{s+h} \|f(t) - x_n\|\, dt = \|f(s) - x_n\|$$

for almost all $s \in [0, 1]$ and for all n. For any s such that this holds for all n, one obtains

$$\begin{aligned} \limsup_{h\to 0} \frac{1}{h} \int_s^{s+h} \|f(t) - f(s)\|\, dt & \\ &\leqq \limsup_{h\to 0} \left(\frac{1}{h} \int_s^{s+h} \|f(t) - x_n\|\, dt\right) + \|x_n - f(s)\| \\ &= 2\|f(s) - x_n\|, \end{aligned}$$

for all n. If $\varepsilon > 0$ is given, a choice of n such that $\|f(s) - x_n\| < \varepsilon/2$ completes the proof of the theorem.

Next we shall take a rather cavalier look at the Lebesgue-Bochner spaces. If $1 \leqq p < \infty$, the symbol $L_p(\Omega, \Sigma, \mu, X)$ $(= L_p(\mu, X))$ will stand for all (equivalence classes of) μ-Bochner integrable functions $f: \Omega \to X$ such that

$$\|f\|_p = \left(\int_\Omega \|f\|_X^p d\mu\right)^{1/p} < \infty.$$

Normed by the functional $\|\cdot\|_p$ defined above $L_p(\mu, X)$ becomes a Banach space, a fact whose proof is the same as the real-valued case. The symbol $L_\infty(\Omega, \Sigma, \mu, X)$ $(= L_\infty(\mu, X))$ will stand for all (equivalence classes of) essentially bounded μ-Bochner integrable functions $f : \Omega \to X$. Normed by the functional $\|\cdot\|_\infty$ defined for $f \in L_\infty(\mu, X)$ by

$$\|f\|_\infty = \text{ess sup}\|f\|_X,$$

$L_\infty(\mu, X)$ becomes a Banach space. The symbol $L_p(\mu)$ $(1 \leqq p \leqq \infty)$ will always mean $L_p(\mu, X)$ for X = scalars.

One of the most interesting aspects of the theory of the Bochner integral centers about the following question: When does a vector measure arise as an indefinite Bochner integral? Let us briefly examine the situation: If (Ω, Σ, μ) is a finite measure space and $F : \Sigma \to X$ is a vector measure of the form

$$F(E) = \int_E f\, d\mu$$

for some Bochner integrable f, then Theorem 4 guarantees that F is countably additive, μ-continuous and of bounded variation. Conversely, if $F : \Sigma \to X$ is any countably additive μ-continuous measure of bounded variation with a finite dimensional range, then the classical Radon-Nikodým theorem produces a Bochner integrable function f for which $F(E) = \int_E f\, d\mu$. For the general Bochner integral, this is no longer necessarily true.

EXAMPLE 10. *A countably additive c_0-valued vector measure of bounded variation that has no Radon-Nikodým derivative.* Let μ be Lebesgue measure on [0, 1]. For a measurable set $E \subseteq [0, 1]$, write

$$\lambda_n(E) = \int_E \sin(2^n\pi t)\, dt$$

and let

$$F(E) = (\lambda_1(E), \lambda_2(E), \cdots, \lambda_n(E), \cdots).$$

The Riemann-Lebesgue lemma guarantees that the finitely additive measure F defined above is c_0-valued. Moreover

$$\|F(E)\|_{c_0} \leqq \sup_n \int_E |\sin(2^n\pi t)|\, dt \leqq \mu(E)$$

for each measurable set $E \subseteq [0, 1]$. It follows that F is countably additive, μ-continuous and of bounded variation. Now suppose there is a Bochner integrable $f : [0, 1] \to c_0$ such that $F(E) = \int_E f\, d\mu$ for each measurable $E \subseteq [0, 1]$. Write $f = (f_1, f_2, \cdots, f_n, \cdots)$. Since the coordinate functionals on c_0 are all continuous linear functionals, each f_n is measurable and the equality $\lambda_n(E) = \int_E f_n\, d\mu$ holds for each E and each n. Hence $f_n(t) = \sin(2^n\pi t)$ for almost all $t \in [0, 1]$. Consider

$$E_n = \{t \in [0, 1] : f_n(t) \geqq 1/\sqrt{2}\}.$$

Then $\mu(E_n) = \frac{1}{4}$ for each n. Moreover $\mu(\overline{\lim}_j(E_j)) \geqq \overline{\lim}_j \mu(E_j) \geqq \frac{1}{4}$. Hence $\mu(\{t \in [0, 1]: f(t) \in c_0\}) \leqq 3/4$, a contradiction.

The failure of the Radon-Nikodým thereom for the Bochner integral is not to be intepreted as a negative aspect of the Bochner integral. Indeed, the failure of a general Radon-Nikodým theorem for the Bochner integral in special cases has powerful repercussions in operator theory, the geometry of Banach spaces, duality theory for Banach spaces, vector-valued probability theory and integration theory itself. Much of the later part of this monograph is devoted to the enjoyment and the exposition of these repercussions.

Closing this section are two fundamental theorems of Banach space theory. It is not always recognized that both of them are simple consequences of properties of the Bochner integral.

THEOREM 11 (KREĬN-SMULIAN). *The closed convex hull of a weakly compact subset of a Banach space is weakly compact.*

PROOF. Let W be a weakly compact set in a Banach space X. To show that the closed convex hull of W is weakly compact, it suffices by the Eberlien-Smulian theorem to show that the convex hull of W is relatively weakly sequentially compact. Since any sequence in the convex hull of W is in a separable subspace of X, it follows from the Hahn-Banach theorem that W itself may be assumed to be norm separable.

Thus suppose W is a norm separable weakly compact set in X and let g be the identity function on W. Evidently g is separably valued and x^*g is continuous on W equipped with the weak topology for all $x^* \in X^*$. From the Pettis Measurability Theorem 1.2, it follows that g is μ-measurable for every regular measure μ defined on the (weak) Borel sets of W.

Now W is a compact Hausdorff space in its weak topology. Thus for $\mu \in C(W)^*$, the Bochner integral $\int_W g d\mu$ exists since g is μ-measurable and bounded. Define $T: C(W)^* \to X$ by $T(\mu) = \int_W g d\mu$ for $\mu \in C(W)^*$. Then if (μ_α) is a net in $C(W)^*$ that converges to $\mu \in C(W)^*$ in the weak*-topology and $x^* \in X^*$, then

$$\lim_\alpha x^*T(\mu_\alpha) = \lim_\alpha x^* \int_W g d\mu_\alpha$$

$$= \lim_\alpha \int_W x^* g d\mu_\alpha = x^*T(\mu)$$

since $x^*g \in C(W)$ for every $x^* \in X^*$. Hence T is continuous for the weak*-topology of $C(W)^*$ and weak topology of X; accordingly T is a weakly compact operator. Thus if S^* is the closed unit ball of $C(W)^*$, then $T(S^*)$ is a weakly compact and convex subset of X. Moreover the point mass measures on W are mapped onto W by T. Hence $W \subseteq T(S^*)$ and the closed convex hull of W is a subset of the weakly compact set $T(S^*)$. This completes the proof.

THEOREM 12 (MAZUR). *The closed convex hull of a norm compact subset of a Banach space is norm compact.*

PROOF. The proof is a simple streamlining of the proof of Theorem 11. This time let W be a compact set in a Banach space X. Then W is separable and the identity function g on W is a continuous function on W. Equip W with its norm topology and define $T:C(W)^* \to X$ by $T(\mu) = \int_W g\, d\mu$, for $\mu \in C(W)^*$. Since g has a totally bounded range, the proof of Theorem 1.2 shows that there is a sequence (g_n) of (Borel) measurable simple functions on W such that $\lim_n g_n = g$ uniformly on W. Define $T_n:C(W)^* \to X$ by $T_n(\mu) = \int_W g_n d\mu$ for $\mu \in C(W)^*$. Then each T_n has a finite dimensional range and

$$\|(T - T_n)(\mu)\| \leqq \sup_{x \in W} \|g_n(x) - g(x)\| \, \|\mu\|.$$

Thus T is a compact operator on $C(W)^*$. The rest of the proof proceeds as the proof of Theorem 11 with some obvious changes.

3. The Pettis integral. A theory of integration similar to the Bochner integral is impossible for functions that are only weakly measurable. Furthermore, it is impossible to use the Bochner integral theory directly to integrate a function f if $\|f\|$ is not integrable. Nevertheless, there are rather simple methods available to integrate some such functions and, as a small part of Pettis's contribution to functional analysis shows, these simple methods have some unexpectedly strong properties which will be presently investigated. The following lemma provides the basis for this section.

LEMMA 1 (DUNFORD). *Suppose f is a weakly μ-measurable function on Ω and $x^*f \in L_1(\mu)$ for each $x^* \in X^*$. Then for each $E \in \Sigma$ there exists $x_E^{**} \in X^{**}$ satisfying*

$$x_E^{**}(x^*) = \int_E x^*(f)\, d\mu$$

for all $x^* \in X^*$.

PROOF. Let $E \in \Sigma$ and define $T: X^* \to L_1(\mu)$ by $T(x^*) = x^*(f\chi_E)$. Note that T is closed. Indeed, if $\lim_n x_n^* = x^*$ and $\lim_n T(x_n^*) = g$ exists in $L_1(\mu)$, then some subsequence $(x_{n_j}^*(f\chi_E) = T(x_{n_j}^*))$ tends μ-almost everywhere to g. But $\lim_n x_n^*(f\chi_E) = x^*(f\chi_E)$ everywhere. Hence $x^*f = g$ μ-almost everywhere and T is a closed linear operator. An appeal to Banach's closed graph theorem shows that T is continuous. Hence

$$\|x^*(f)\|_1 \leqq \|T(x^*)\| \leqq \|T\|\,\|x^*\|.$$

Since the operation of integrating over E is a continuous linear functional of norm at most 1, it follows that

$$\left|\int_E x^*f\, d\mu\right| \leqq \|T\|\,\|x^*\|.$$

Hence the mapping $x^* \to \int_E x^*f\, d\mu$ defines a continuous linear functional on X^* and, as such, defines a member x_E^{**} of X^{**}.

With the help of Lemma 1 the Pettis integral and the Dunford integral can be defined very simply.

DEFINITION 2. If f is a weakly μ-measurable X-valued function on Ω such that $x^*f \in L_1(\mu)$ for all $x^* \in X^*$, then f is called *Dunford integrable* The *Dunford integral of f over $E \in \Sigma$* is defined by the element x_E^{**} of X^{**} such that

$$x_E^{**}(x^*) = \int_E x^* f \, d\mu$$

for all $x^* \in X^*$, and we write $x_E^{**} = (D) - \int_E f \, d\mu$.

In the case that $(D) - \int_E f \in X$ for each $E \in \Sigma$, then f is called *Pettis integrable* and we write $(P) - \int_E f \, d\mu$ instead of $(D) - \int_E f \, d\mu$ to denote the *Pettis integral of f over $E \in \Sigma$.*

By the same closed graph argument it is possible to show that if $f: \Omega \to X^*$ is a function such that $xf \in L_1(\mu)$ for all x in X, then for each set $E \in \Sigma$ there is a vector x_E^* in X^* such that

$$x_E^*(x) = \int_E xf \, d\mu \quad \text{for all } x \in X.$$

The element x_E^* is called the *Gel'fand* (or weak*-) *integral* of f over E.

Naturally the Dunford and Pettis integrals coincide when X is reflexive. When X is not reflexive, this may not be the case.

EXAMPLE 3. *A Dunford integrable function that is not Pettis integrable.* Define f: $[0, 1] \to c_0$ by

$$f(t) = (\chi_{(0,1]}(t), 2\chi_{(0,1/2]}(t), \cdots, n\chi_{(0,1/n]}(t), \cdots)$$

for $t \in [0, 1]$. If $x^* \in c_0^* = l_1$ and $x^* = (\alpha_1, \alpha_2, \cdots, \alpha_n, \cdots)$ then $x^*f = \sum_{n=1}^{\infty} \alpha_n n \chi_{(0,1/n]}$, a function which is certainly Lebesgue integrable. However, if μ is Lebesgue measure, then

$$(D) - \int_{(0,1]} f \, d\mu = (1, 1, \cdots, 1, \cdots) \in l_\infty \backslash c_0,$$

since if $x^* = (\alpha_1, \alpha_2, \cdots, \alpha_n, \cdots) \in l_1$, then

$$\int_{(0,1]} x^* f \, d\mu = \sum_{n=1}^{\infty} \alpha_n$$

and the mapping $(\alpha_1, \alpha_2, \cdots, \alpha_n, \cdots) \to \sum_{n=1}^{\infty} \alpha_n$ is the linear functional on l_1 corresponding to $(1, 1, \cdots, 1, \cdots) \in l_\infty \backslash c_0$.

EXAMPLE 4. *The Dunford integral may fail to be countably additive.* Let $f: [0, 1] \to c_0$ be the function defined in Example 3. It is a simple matter to check that

$$\left\| (D) - \int_{(0,1/n]} f \, d\mu \right\|_{l_\infty} = 1$$

for each n. This makes it impossible for $(D) - \int_{(\cdot)} f \, d\mu$ to be countably additive on the Lebesgue measurable sets. It is also clear that $(D) - \int_{(\cdot)} f \, d\mu$ is not μ-continuous.

Taking another look at the above example, one can see quickly that $(D) - \int_{(\cdot)} f \, d\mu$ is countably additive in the weak*-topology of l_∞. In view of the remarks following I.1.14, $(D) - \int_{(\cdot)} f \, d\mu$ is not even strongly additive—such a situation is impossible for Pettis integrable functions.

THEOREM 5 (PETTIS). *If f is Pettis integrable, then $(P) - \int_{(\cdot)} f \, d\mu$ is a countably additive μ-continuous vector measure on Σ.*

PROOF. If (E_n) is a sequence of disjoint members of Σ, then

$$x^*\Big((P) - \int_{\bigcup_{n=1}^{\infty} E_n} f\, d\mu\Big) = \int_{\bigcup_{n=1}^{\infty} E_n} x^* f\, d\mu$$
$$= \sum_{n=1}^{\infty} \int_{E_n} x^* f\, d\mu$$
$$= \sum_{n=1}^{\infty} x^*\Big((P) - \int_{E_n} f\, d\mu\Big).$$

Thus $(P) - \int_{(\cdot)} f\, d\mu$ is weakly countably additive. Since the same argument applies to any subsequence of (E_n), an appeal to I.4.4 shows that $(P) - \int_{(\cdot)} f\, d\mu$ is norm countably additive. Since it is plain that $(P) - \int_E f\, d\mu = 0$ whenever $\mu(E) = 0$, one sees that $(P) - \int_{(\cdot)} f\, d\mu$ is μ-continuous by I.2.1.

COROLLARY 6. *A μ-measurable Dunford integrable function f is Pettis integrable if and only if $(D) - \int_{(\cdot)} f\, d\mu$ is strongly additive.*

Consequently, a μ-measurable Dunford integrable function f is Pettis integrable if and only if $(D) - \int_{(\cdot)} f\, d\mu$ is countably additive.

PROOF. Since $(D) - \int_{(\cdot)} f\, d\mu$ is countably additive in the weak*-topology of X^{**}, the remarks after I.1.14 insure that $(D) - \int_{(\cdot)} f\, d\mu$ is countably additive if $(D) - \int_{(\cdot)} f\, d\mu$ is strongly additive. Thus the second statement follows from the first statement.

To prove the first statement, note that the indefinite Pettis integral is countably additive and hence is strongly additive. For the converse, choose an increasing sequence (E_n) in Σ such that $\bigcup_{n=1}^{\infty} E_n = \Omega$ and each $f\chi_{E_n}$ is bounded. If $E \in \Sigma$, then the Bochner integral $\int_{E \cap E_n} f\, d\mu$ exists for each n. In addition, Theorem 2.6 guarantees that the Dunford, Pettis and Bochner integrals of f over $E \cap E_n$ coincide. If $(D) - \int_{(\cdot)} f\, d\mu$ is strongly additive, then $\lim_n (D) - \int_{E \cap E_n} f\, d\mu$ exists in X by I.1.18. On the other hand,

$$\lim_n \int_{E \cap E_n} x^* f\, d\mu = \int_E x^* f\, d\mu$$

for all $x^* \in X^*$. Hence $(D) - \int_E f\, d\mu = \lim_n \int_{E \cap E_n} f\, d\mu \in X$, for all $E \in \Sigma$, and f is Pettis integrable.

The next result is closely related to Corollary 6 and indicates that Example 3 is archetypical.

THEOREM 7. *If (Ω, Σ, μ) is a finite measure space and X contains no copy of c_0, then every μ-measurable Dunford integrable function $f: \Omega \to X$ is Pettis integrable.*

PROOF. Suppose f is μ-measurable and $\int_\Omega |x^* f|\, d\mu < \infty$ for each $x^* \in X^*$. According to Corollary 1.3 there is a countably valued μ-measurable $g: \Omega \to X$ such that $\operatorname{ess\,sup} \|f - g\| \leqq 1$. Hence $f - g$ is Bochner integrable and therefore Pettis integrable. Now write

$$g = \sum_{n=1}^{\infty} x_n \chi_{E_n},$$

where $x_n \in X$ and $(E_n) \subseteq \Sigma$ with $E_n \cap E_m = \emptyset$, for $n \neq m$. If $x^* \in X^*$, then

$|x^*f| \geqq ||x^*(f-g)| - |x^*g||$. Since $\int_\Omega |x^*f|\, d\mu < \infty$ and $\int_\Omega |x^*(f-g)|\, d\mu < \infty$, it follows that for any $E \in \Sigma$, we have

$$\sum_{n=1}^{\infty} |x^*(x_n)|\, \mu(E_n \cap E) = \int_E |x^*g|\, d\mu < \infty$$

for all $x^* \in X^*$. According to the Bessaga-Pełczyński characterization of Banach spaces not containing a copy of c_0 (I.4.5), the series $\sum_{n=1}^{\infty} x_n \mu(E \cap E_n)$ is unconditionally norm convergent. Evidently $\sum_{n=1}^{\infty} x_n \mu(E \cap E_n) = (P) - \int_E g\, d\mu$. Hence $g, f - g$ and $f = (f - g) + g$ are Pettis integrable.

Two incidental comments on Theorem 7 are in order. First, the proof of Theorem 7 reveals another feature of the Pettis integral that contrasts with the Bochner integral. By telescoping, one can always write a measurable function $f: \Omega \to X$ as a sum $\sum_{n=1}^{\infty} x_n \chi_{E_n}$ with $x_n \in X$ and $E_n \in \Sigma$ (with (E_n) not necessarily disjoint). It is a quick exercise to see that f is Pettis integrable if and only if such a sum exists such that $\sum_{n=1}^{\infty} x_n \mu(E_n \cap E)$ is *unconditionally* convergent for each $E \in \Sigma$. Also f is Bochner integrable if and only if such a sum exists such that $\sum_{n=1}^{\infty} x_n \mu(E_n \cap E)$ is *absolutely* convergent for each $E \in \Sigma$.

THEOREM 8. *Let $f: \Omega \to X$ be Dunford integrable with respect to μ. Define T: $L_\infty(\mu) \to X^{**}$ by $T(g) = (D) - \int_\Omega gf\, d\mu$. Then*

(a) *T is bounded;*

(b) *if T is weakly compact and f is measurable, then f is Pettis integrable, and if f is Pettis integrable, then T is weakly compact;*

(c) *T is compact iff f is Bochner integrable.*

PROOF. The simple proof of (a) is omitted. Directing our attention to (b), suppose T is weakly compact and f is measurable. Set

$$F(E) = (D) - \int_E f\, d\mu, \quad \text{for } E \in \Sigma.$$

Then $F(\Sigma) \subseteq \{T(g): \|g\| \leqq 1\}$; consequently F has relatively weakly compact range. Combining this fact with I.5.3 shows that F is strongly additive. Since F is weak*-countably additive the measure F must be countably additive. Hence f is Pettis integrable by Corollary 6. For the converse, suppose f is Pettis integrable and write

$$F(E) = (P) - \int_E f\, d\mu, \qquad E \in \Sigma.$$

By Theorem 5 the measure F is countably additive. By I.2.7 the set $F(\Sigma)$ is relatively weakly compact. To show T is weakly compact, it is obviously sufficient to show the set $\{\int_\Omega gf\, d\mu$: $g = \sum_{i=1}^{n} \alpha_i \chi_{E_i}$, $0 \leqq \alpha_i \leqq 1$, $E_i \cap E_j = \emptyset$, $i \neq j$, $E_i \in \Sigma$, $n \in N\}$ is contained in the closed convex hull of $F(\Sigma)$. To this end, suppose $g = \sum_{i=1}^{n} \alpha_i \chi_{E_i}$, where $E_1, \cdots, E_n$ are pairwise disjoint members of Σ and $0 \leqq \alpha_1 \leqq \alpha_2 \leqq \cdots \leqq \alpha_n \leqq 1$. Then

$$\int_\Omega gf\, d\mu = \alpha_1 \int_{\cup_{i=1}^{n} E_i} f\, d\mu + \sum_{j=2}^{n} (\alpha_j - \alpha_{j-1}) \int_{\cup_{i=j}^{n} E_i} f\, d\mu \in \overline{\text{co}}(F(\Sigma)),$$

since $\alpha_1 + \sum_{i=2}^{n} (\alpha_j - \alpha_{j-1}) = \alpha_n \leqq 1$.

To prove (c), select a sequence (f_n) of simple functions such that

$$\lim_n \int_\Omega \|f - f_n\| \, d\mu = 0.$$

Define operators $T_n: L_\infty(\mu) \to X$ by $T_n(g) = \int_\Omega g f_n \, d\mu$, for $g \in L_\infty(\mu)$. Since each f_n has a finite range, each T_n is a finite rank continuous linear operator. Moreover, if $g \in L_\infty(\mu)$, then

$$\|(T - T_n)(g)\| \leqq \int_\Omega |g| \, \|f - f_n\| \, d\mu \leqq \|g\|_\infty \int_\Omega \|f - f_n\| \, d\mu.$$

From this it follows immediately that T, as the uniform operator limit of compact operators, is compact.

Two comments are in order: A considerable strengthening of (c) is true. The full truth is that T is nuclear if and only if f is Bochner integrable. For more on this consult Chapter 6. Second, the proof of Theorem 8 shows that an equivalent formulation of Theorem 8 is

COROLLARY 9. *Let $f: \Omega \to X$ be Dunford integrable with respect to μ. Define $F: \Sigma \to X^{**}$ by $F(E) = (D) - \int_E f \, d\mu$. Then*

(a) *F has bounded range.*

(b) *If $F(\Sigma)$ is relatively weakly compact and f is measurable, then f is Pettis integrable, and if f is Pettis integrable, then $F(\Sigma)$ is relatively weakly compact.*

(c) *F has a relatively compact range iff f is Bochner integrable.*

4. An elementary version of the Bartle integral. There is a substantial theory of integration of vector-valued functions with respect to vector-valued measures. The mature version of this theory, sometimes called the Bartle integral, is becoming increasingly important in the theory of operators on spaces of vector-valued functions. Since the theory of spaces of vector-valued functions is not occupying a central role in this book, we shall limit ourselves to a rather cursory look at the Bartle integral and refer the reader to the literature (see notes and remarks) for more study.

The theory we present here is the elementary integral of I.1.12. The main fact we would like to emphasize is

THEOREM 1 (BARTLE'S BOUNDED CONVERGENCE THEOREM). *Let (Ω, Σ) be a measurable space and X be a Banach space. Suppose $G: \Sigma \to X$ is a countably additive vector measure and (f_n) is a uniformly bounded sequence in $B(\Sigma)$. If $\lim_n f_n = f$ pointwise, then*

$$\lim_n \int_\Omega f_n \, dG = \int_\Omega f \, dG.$$

PROOF. Select with the help of I.2.6 a nonnegative finite countably additive measure μ on Σ such that $G \ll \mu$. Then by Egoroff's theorem we see that $\lim_n f_n = f$ almost uniformly. Suppose $\|f_n\|_\infty \leqq k$ for all n. Note that, for each $E \in \Sigma$, we have

$$\left\| \int_E f_n \, dG \right\| \leqq k \|G\|(E),$$

a fact which is obvious if f_n is simple and, therefore, simple if f_n is not simple. Finally, if $\varepsilon > 0$, select $\delta > 0$ such that $\|G\|(E) < \varepsilon/2k$ whenever $\mu(E) < \delta$. Choose $E_0 \in \Sigma$ such that $\mu(E_0) < \delta$ and $\lim_n f_n = f$ uniformly on $\Omega \backslash E_0$. Then we have

$$\overline{\lim_n} \left\| \int_\Omega f\, dG - \int_\Omega f_n\, dG \right\| \leqq \overline{\lim_n} \left\| \int_{\Omega \backslash E_0} (f - f_n)\, dG \right\| + \overline{\lim_n} \left\| \int_{E_0} (f - f_n)\, dG \right\|$$
$$\leqq 0 + 2k \|G\|(E_0) < \varepsilon.$$

Thus $\lim_n \int_\Omega f_n\, dG = \int_\Omega f\, dG$.

5. Notes and remarks. Most of this chapter is devoted to properties of the Bochner integral for (strongly) measurable functions. In the literature there is no shortage of integrals for functions that are not necessarily measurable. Generally integrals for functions that are not measurable are plagued with one common defect; they seem to have no applications outside their own contexts. Thanks to measurability, the Bochner integral has enough solid structure to have substantive applications that reverberate throughout vector measure theory, operator theory and geometry of Banach spaces. In fact a good deal of this survey is devoted to the enjoyment of applications of the Bochner integral.

At this point we must hasten to call attention to the fact that the general Pettis integral, now approaching the age of forty, has some unexpectedly strong properties. In fact, the Orlicz-Pettis theorem owes its origin (at least its American origin) to the Pettis integral. Presently the Pettis integral has very few applications. But our prediction is that when (and if) the general Pettis integral is understood it will pay off in deep applications.

Bochner integration and measurability. The Bochner integral can be traced back to two origins. It first appeared in Bochner **[1932]** and later it appeared in Dunford **[1935]**. It has also been called Dunford's first integral. For countably additive nonnegative measures, the integral found in Dunford and Schwartz **[1958]** coincides with the Bochner integral. Those interested in integrating with respect to finitely additive measures should consult Dunford and Schwartz **[1958]** who, as pointed out by Bartle **[1966]**, have "exploded" the prevailing feeling "that countable additivity is a necessary ingredient of a 'decent' integration theory." Space limitations forced us to confine ourselves to the context of countably additive finite measures.

Most of the basic properties of Bochner integration are forced on it by the classical Lebesgue integration and the definition of measurability. Thus, from our point of view, the most basic theorem of this chapter is the Pettis Measurability Theorem 1.2 which can be found in Pettis **[1938a]**. It is impossible to integrate a function until that function is measurable but once the function can be integrated good applications follow. For a hint of what we mean, look at the proof of the Kreĭn-Smulian Theorem 2.11. The heart of its proof is the Pettis Measurability Theorem. The Pettis Measurability Theorem will be used repeatedly throughout this survey. Be on the lookout for it.

Example 1.6 is due to Sierpinski **[1938]**. Example 1.7 was custom made for us in 1973 at Murphy's Pub, Champaign, Illinois by James Hagler.

Theorem 2.6 is from Hille and Phillips **[1957]**. An alternate approach to the Bochner integral has been suggested by Bogdanowicz **[1965a]**.

Kreĭn-Smulian theorem and Mazur's theorem. Theorems 2.11 and 2.12 are standard basic theorems of Banach space theory. The Kreĭn-Smulian theorem rightfully belongs to the theory of Bochner integration as our proof from Dunford and Schwartz **[1958]** shows. We cannot seriously argue that our proof of Mazur's Theorem 2.12 is the most natural proof. This theorem can be proved easily without the help of the Bochner integral.

The Pettis integral. We have already made some general comments on the Pettis integral. Its basic properties were established by Pettis **[1938a]** and no one else has been able to uncover additional information about its basic structure. The Pettis integral was studied by Dunford **[1936]** and correctly should be called "Dunford's second integral". Theorem 3.7 was observed by Diestel **[1973a]**.

The Dunford and Gel'fand integrals. The integral that we call the "Dunford integral" traces its life back to Dunford **[1937]**. History seems to indicate that the integral that we call the "Gel'fand integral" was studied first by Gel'fand **[1936]**.

The Bartle integral. Bartle **[1956]** launched a theory of integration that includes most of the known integration procedures that have any claim to quality. His integral specializes to include the Bochner integral but does not include the general Pettis integral. Possibly workers in the theory of vector measures would be better off if they attempted to use the Bartle integral rather than inventing their own. The Bartle integral has proved useful for representing operators on spaces of vector-valued functions (see notes and remarks to Chapter VI) but this representation theory, in our opinion, has not achieved the maturity to warrant inclusion in this survey. In the future, we expect to see properties of the Bartle integral exploited more often than they have been in the past.

History of vector integration. We cannot give a survey of the history of vector integration any better than those found in Bartle **[1956]** and Hildebrandt **[1953]**.

III. ANALYTIC RADON-NIKODÝM THEOREMS AND OPERATORS ON $L_1(\mu)$

If (Ω, Σ, μ) is a finite measure space, then two basic theorems of measure theory, the Riesz Representation Theorem and the Radon-Nikodým theorem, guarantee that $L_1(\mu)^* = L_\infty(\mu)$ and that if λ is a finite μ-continuous scalar-valued measure on Σ, then there exists $f \in L_1(\mu)$ such that $\lambda(E) = \int_E f\, d\mu$ for all $E \in \Sigma$. Each of these theorems can be derived from the other and it is not surprising that their vector-valued extensions are related in the most intimate of ways.

RIESZ REPRESENTATION THEOREM. *If X is a Banach space and T: $L_1(\mu) \to X$ is a continuous linear operator then there exists $g \in L_\infty(\mu, X)$ such that $Tf = \int_\Omega fg\, d\mu$ for all $f \in L_1(\mu)$.*

RADON-NIKODÝM THEOREM. *If G: $\Sigma \to X$ is a μ-continuous vector measure of bounded variation, then there exists a Bochner integrable g ($\in L_1(\mu, X)$) such that $G(E) = \int_E g\, d\mu$ for all $E \in \Sigma$.*

This chapter is devoted to the study of both these statements and the interplay between them. We shall see in §1 that the connection between these statements is basically purely formal and that if X is a fixed Banach space, then the Riesz Representation Theorem describes all operators T as above if and only if the Radon-Nikodým theorem describes all measures G as above. In §2, the interchange between the Riesz Representation Theorem and the Radon-Nikodým theorem will be exploited. After showing that the Riesz Representation Theorem is true for compact and weakly compact operators on $L_1(\mu)$, we shall deduce several Radon-Nikodým theorems for vector measures. Finally, in §3, the crucial role of separable dual spaces in Radon-Nikodým theory is examined.

As the reader progresses through this chapter, he may note that the presentation is not the most efficient possible. In this case he will be correct, for at times we have proceeded on less than efficient routes because these routes convey more "intuitions" than quicker ones.

1. The Radon-Nikodým theorem and Riesz representable operators on $L_1(\mu)$. The main purpose of this section is to examine the essentially formal relationship between the Riesz Representation Theorem for operators on $L_1(\mu)$ and the Radon-

Nikodým theorem for the Bochner integral. After making this relationship precise, the section continues with a look at the classical Dunford-Morse Radon-Nikodým theorem which guarantees that the Radon-Nikodým theorem holds for measures with values in a Banach space with a boundedly complete Schauder basis. The section ends with a look at the curious role of l_1 in spaces with the Radon-Nikodým property.

EXAMPLE 1. *The failure of the Radon-Nikodým theorem for a c_0-valued measure.* Let $\Omega = [0, 1]$ and μ be Lebesgue measure on Σ, the σ-field of Lebesgue measurable subsets of [0, 1]. Define a measure $G: \Sigma \to c_0$ by

$$G(E) = \left(\int_E \sin(2^n \pi t)\, d\mu(t)\right).$$

According to the Riemann-Lebesgue lemma, G has its values in c_0. Since $|\sin t| \leqq 1$ for all real t, one has

$$\|G(E)\| = \sup_n \left|\int_E \sin(2^n \pi t)\, d\mu(t)\right| \leqq \mu(E).$$

Hence G is countably additive, μ-continuous and is of bounded variation. Suppose there exists a Bochner integrable $g: \Omega \to c_0$ for which $G(E) = \int_E g\, d\mu$ for all $E \in \Sigma$. If $g = (g_n)$ and l_n is evaluation at the nth coordinate, then for each $E \in \Sigma$,

$$l_n G(E) = \int_E l_n g\, d\mu = \int_E g_n\, d\mu.$$

It follows that $g_n(t) = \sin(2^n \pi t)$ for almost all $t \in [0, 1]$. Consequently, $g(t) = (\sin(2^n \pi t))$ for almost all $t \in [0, 1]$. However, if we set $E_n = \{t \in [0, 1]: |\sin(2^n \pi t)| \geqq 1/\sqrt{2}\}$ then $\mu(E_n) = \frac{1}{4}$ for all n. Thus

$$\mu\left(\overline{\lim_n}\, E_n\right) \geqq \overline{\lim_n}\, \mu(E_n) \geqq \frac{1}{4}.$$

Hence $\mu(\{t \in [0, 1]: g(t) \in c_0\}) \leqq \frac{3}{4}$. This crude argument shows g is not almost everywhere c_0-valued and shows that G has no Radon-Nikodým derivative with respect to μ.

EXAMPLE 1′. *Failure of the Riesz Representation Theorem for an operator $T: L_1[0, 1] \to c_0$.* Let (Ω, Σ, μ) be as in Example 1. Define $T: L_1(\mu) \to c_0$ by

$$Tf = \left(\int_{[0,1]} f(t) \sin(2^n \pi t)\, d\mu(t)\right).$$

According to the Riemann-Lebesgue lemma, T is a c_0-valued linear operator on $L_1(\mu)$. In addition, one has

$$\begin{aligned} \|Tf\| &= \sup_n \left|\int_{[0,1]} f(t) \sin(2^n \pi t)\, d\mu(t)\right| \\ &\leqq \sup_n \int_{[0,1]} |f(t) \sin(2^n \pi t)|\, d\mu(t) \\ &\leqq \int_{[0,1]} |f(t)|\, d\mu(t) = \|f\|_1. \end{aligned}$$

Therefore T is bounded.

Now suppose there exists $g \in L_\infty(\mu, c_0)$ such that

$$Tf = \int_\Omega fg \, d\mu \quad \text{for all } f \in L_1(\mu).$$

Then if G is the vector measure of Example 1 and $E \in \Sigma$, it follows that $G(E) = T(\chi_E) = \int_E g \, d\mu$. According to Example 1, such a g does not exist.

EXAMPLE 2. *An $L_1(\mu)$-valued measure with no Radon-Nikodým derivative.* Suppose (Ω, Σ, μ) is a finite measure space without atoms. Define $G: \Sigma \to L_1(\mu)$ by $G(E) = \chi_E$. Then G is countably additive, μ-continuous and is of bounded variation; in fact, $|G|(E) = \mu(E)$ for each $E \in \Sigma$. Suppose that there exists a Bochner integrable $g: \Omega \to L_1(\mu)$ such that

$$G(E) = \int_E g \, d\mu \quad \text{for all } E \in \Sigma.$$

Define an operator $T: L_\infty(\mu) \to L_1(\mu)$ by $Tf = \int_\Omega fg \, d\mu$ for $f \in L_\infty(\mu)$. According to II.3.8, T is a compact linear operator. Consequently $\{T(\chi_E): E \in \Sigma\}$ is a relatively compact set in $L_1(\mu)$. On the other hand, $\{T(\chi_E): E \in \Sigma\}$ contains a sequence (χ_{E_n}) such that $\mu(E_n) = \mu(\Omega)/2$ and, for $m \neq n$, $\mu(E_n \,\Delta\, E_m) = \mu(\Omega)/4$. Thus $\|\chi_{E_n} - \chi_{E_m}\|_1 = \mu(E_n \,\Delta\, E_m) = \mu(\Omega)/4$ and T is not compact. This is a contradiction.

EXAMPLE 2′. *Failure of the Riesz Representation Theorem for the identity operator on $L_1(\mu)$.* Let (Ω, Σ, μ) be a finite measure space without atoms. Let T be the identity operator on $L_1(\mu)$. If the Riesz Representation Theorem holds for this operator, then there exists an essentially bounded μ-measurable $g: \Omega \to L_1(\mu)$ such that

$$f = Tf = \int_\Omega fg \, d\mu$$

for all $f \in L_1(\mu)$. In particular, we have $\chi_E = \int_E g \, d\mu$ for all $E \in \Sigma$. A glance at Example 2 shows that g is the Radon-Nikodým derivative of the measure G defined in Example 2. This is impossible.

The relationship between Examples 1 and 2 and the relationship between Examples 1′ and 2′ are no accidents. The next few theorems of this section are devoted to making this relationship precise. Throughout this section, (Ω, Σ, μ) is a finite measure space and X is a Banach space.

The following definitions establish the terminology of this section.

DEFINITION 3. A Banach space X has the *Radon-Nikodým property with respect to* (Ω, Σ, μ) if for each μ-continuous vector measure $G: \Sigma \to X$ of bounded variation there exists $g \in L_1(\mu, X)$ such that $G(E) = \int_E g \, d\mu$ for all $E \in \Sigma$.

A Banach space X has the *Radon-Nikodým property* if X has the Radon-Nikodým property with respect to every finite measure space.

A bounded linear operator $T: L_1(\mu) \to X$ is *Riesz representable* (or simply *representable*) if there exists $g \in L_\infty(\mu, X)$ such that

$$Tf = \int_\Omega fg \, d\mu \quad \text{for all } f \in L_1(\mu).$$

According to Examples 1 and 2, the space c_0 does not have the Radon-Nikodým property and $L_1(\mu)$ does not have the Radon-Nikodým property when μ has no atoms. (If μ is not purely atomic, then Ω contains a subset Ω_0 such that $\mu|_{\Omega_0}$ has no atoms. By modifying Example 1 by defining $G(E) = \chi_{E \cap \Omega_0}$, we obtain an example of an $L_1(\mu)$-valued measure without a derivative. Therefore, whenever μ is not purely atomic, the space $L_1(\mu)$ does not have the Radon-Nikodým property.) On the other hand, if (Ω, Σ, μ) is purely atomic, then every Banach space has the Radon-Nikodým property with respect to (Ω, Σ, μ). To see this, suppose (E_n) is a sequence of disjoint atoms of Σ with the properties that $\bigcup_n E_n = \Omega$ and $\mu(E_n) > 0$. If $G: \Sigma \to X$ is a μ-continuous vector measure of bounded variation, define $g: \Omega \to X$ by

$$g = \sum_{n=1}^{\infty} \frac{G(E_n)}{\mu(E_n)} \chi_{E_n}.$$

A routine computation shows that

$$\int_{\Omega} \|g\| \, d\mu = \sum_{n=1}^{\infty} \|G(E_n)\| < \infty \quad \text{and} \quad G(E) = \int_E g \, d\mu$$

for all $E \in \Sigma$.

The fundamental connection between representable operators on $L_1(\mu)$ and vector measures with Radon-Nikodým derivatives is contained in the following straightforward lemma:

LEMMA 4. *Let $T: L_1(\mu) \to X$ be a bounded linear operator. For $E \in \Sigma$, define $G(E)$ by*

$$G(E) = T(\chi_E).$$

Then T is representable if and only if there exists $g \in L_1(\mu, X)$ such that

$$G(E) = \int_E g \, d\mu$$

for all $E \in \Sigma$. In this case, the function $g \in L_\infty(\mu, X)$ and

$$T(f) = \int_{\Omega} fg \, d\mu$$

for all $f \in L_1(\mu)$. Moreover $\|g\|_\infty = \|T\|$.

PROOF. If T is representable, then there exists $g \in L_\infty(\mu, X)$ such that $T(f) = \int_\Omega fg \, d\mu$ for all $f \in L_1(\mu)$. Thus, if $E \in \Sigma$, then

$$G(E) = T(\chi_E) = \int_E g \, d\mu.$$

This proves the necessity.

For the converse, let $G(E) = T(\chi_E) = \int_E g \, d\mu$ for some $g \in L_1(\mu, X)$ and all $E \in \Sigma$. Since for $E \in \Sigma$ one has

$$\|G(E)\| = \|T(\chi_E)\| \leqq \|T\| \, \|\chi_E\|_1 = \|T\| \mu(E),$$

it follows that the variation $|G|$ of G satisfies

$$|G|(E) \leqq \|T\| \mu(E)$$

for all $E \in \Sigma$. Since for each $E \in \Sigma$ one has $|G|(E) = \int_E \|g\| \, d\mu$, it follows immediately that $\|g\| \leqq \|T\|$ almost everywhere. Hence $g \in L_\infty(\mu, X)$.

To prove that $\|g\|_\infty = \|T\|$, note that if $f \in L_1(\mu)$, then

$$\|T(f)\| = \left\|\int_\Omega fg\,d\mu\right\| \leqq \int_\Omega |f|\,\|g\|_\infty\,d\mu = \|g\|_\infty \|f\|_1.$$

Hence $\|T\| \leqq \|g\|_\infty$ and the equality $\|T\| = \|g\|_\infty$ follows.

The next theorem cements the connection between the Radon-Nikodým theorem and the Riesz Representation Theorem.

THEOREM 5. *Let X be a Banach space and (Ω, Σ, μ) be a finite measure space. Then X has the Radon-Nikodým property with respect to (Ω, Σ, μ) if and only if each $T \in \mathscr{L}(L_1(\mu); X)$ is representable.*

PROOF. Suppose X has the Radon-Nikodým property with respect to (Ω, Σ, μ). Let $T: L_1(\mu) \to X$ be a continuous linear operator. Define $G: \Sigma \to X$ by $G(E) = T(\chi_E)$. Since $\|G(E)\| \leqq \|T\|\mu(E)$, it follows that G is countably additive, μ-continuous and is of bounded variation. Since X has the Radon-Nikodým property with respect to (Ω, Σ, μ), there is $g \in L_1(\mu, X)$ such that $G(E) = \int_E g\,d\mu$ for all $E \in \Sigma$. An appeal to Lemma 4 finishes the proof of the necessity.

For the converse, suppose every member of $\mathscr{L}(L_1(\mu); X)$ is representable. Let $G: \Sigma \to X$ be a μ-continuous vector measure of bounded variation. Since G is countably additive, so is $|G|$ by virtue of Proposition I.1.9. Since $|G|$ vanishes on μ-null sets, the measure $|G|$ is μ-continuous. According to the Hahn Decomposition Theorem for scalar measures, there exists a sequence (E_n) of disjoint members of Σ such that $\Omega = \bigcup_{n=1}^\infty E_n$ and with the property that

$$(n-1)\mu(E) \leqq |G|(E) \leqq n\mu(E)$$

for any member E of Σ contained in E_n for $n = 1, 2, \cdots$. (Alternatively, choose an everywhere finite nonnegative function $h \in L_1(\mu)$ such that $|G|(E) = \int_E h\,d\mu$ for $E \in \Sigma$ and write $E_n = \{\omega \in \Omega: n-1 \leqq h(\omega) < n\}$, $n = 1, 2, \cdots$.) Fix n and for a simple function $f = \sum_{i=1}^p \alpha_i \chi_{A_i}$, $A_i \in \Sigma$, $A_i \cap A_j = \emptyset$ for $i \neq j$, define an operator

$$T_n(f) = \sum_{i=1}^p \alpha_i G(E_n \cap A_i) = \int_{E_n} f\,dG.$$

Then one has

$$\begin{aligned}\|T_n(f)\| &= \left\|\sum_{i=1}^p \alpha_i G(A_i \cap E_n)\right\| \\ &\leqq \sum_{i=1}^p |\alpha_i|\,|G|(A_i \cap E_n) \\ &\leqq \sum_{i=1}^p |\alpha_i|\, n\mu(A_i \cap E_n) = n\|f\|_1.\end{aligned}$$

It follows that T_n extends to a continuous linear operator from $L_1(\mu)$ to X. Since every member of $\mathscr{L}(L_1(\mu); X)$ is representable, there exists $g_n \in L_\infty(\mu, X)$ such that

$$T_n(f) = \int_\Omega fg_n\,d\mu.$$

Moreover, if $E \in \Sigma$, then

$$G(E \cap E_n) = T_n(\chi_E) = \int_E g_n\,d\mu.$$

Doing this for each n produces a sequence (g_n) in $L_\infty(\mu, X)$ such that $G(E \cap E_n) = \int_E g_n \, d\mu$ for all $E \in \Sigma$. Define $g : \Omega \to X$ by $g(\omega) = g_n(\omega)$ for $\omega \in E_n$. Since G is countably additive, we have

$$G(E) = \lim_m G\left(E \cap \left(\bigcup_{n=1}^m E_n\right)\right) = \lim_m \int_{E \cap (\bigcup_{n=1}^m E_n)} g \, d\mu.$$

Since G is of bounded variation, we have

$$\int_{\bigcup_{n=1}^m E_n} \|g\| \, d\mu \leqq |G|(\Omega)$$

and so $\|g\| \in L_1(\mu)$ by the Monotone Convergence Theorem. An appeal to the Dominated Convergence Theorem with dominating function $\|g\|$ yields the equalities

$$G(E) = \lim_m \int_{E \cap (\bigcup_{n=1}^m E_n)} g \, d\mu = \int_E g \, d\mu.$$

Hence X has the Radon-Nikodym property with respect to (Ω, Σ, μ).

An easy example of a space with the Radon-Nikodým property is the space l_1. Indeed, if $G: \Sigma \to l_1$ is a μ-continuous vector measure of bounded variation, then one can apply the scalar Radon-Nikodým theorem to each coordinate measure to produce a Radon-Nikodým derivative of G with respect to μ. The fact that the coordinate derivatives add up to a derivative for G depends on a special feature of the norm in l_1: if (α_n) is a scalar sequence and $\sup_n \| \sum_{k=1}^n \alpha_k e_k \| < \infty$, where e_n is the kth unit vector in l_1, then $\sum_{k=1}^n \alpha_k e_k$ converges in l_1. Such a phenomenon occurs in other Banach spaces with Schauder bases, namely, those Banach spaces with boundedly complete Schauder bases. A Schauder basis (x_n) of X is called *boundedly complete* if for each scalar sequence (α_n) such that

$$\sup_n \left\| \sum_{k=1}^n \alpha_k x_k \right\| < \infty,$$

then $\sum_{n=1}^\infty \alpha_n x_n$ converges.

THEOREM 6 (DUNFORD). *If X has a boundedly complete Schauder basis (x_n), then X has the Radon-Nikodým property.*

PROOF. Denote by (x_n^*) the sequence of coefficient functionals of the basis (x_n), so each $x \in X$ has the form $x = \sum_{n=1}^\infty x_n^*(x) x_n$. We start by making the norm "monotone" with respect to (x_n). Define a new norm $|||\cdot|||$ on X by writing for $x \in X$

$$|||x||| = \sup_n \left\| \sum_{k=1}^n x_k^*(x) x_k \right\|.$$

The new norm $|||\cdot|||$ is equivalent to the old norm and for any sequence (α_n) of scalars, one has

$$\left|\left|\left| \sum_{k=1}^n \alpha_k x_k \right|\right|\right| \leqq \left|\left|\left| \sum_{k=1}^{n+m} \alpha_k x_k \right|\right|\right|$$

for all positive integers m and n.

Now let $G: \Sigma \to X$ be a μ-continuous vector measure of bounded variation. It follows directly from Theorem II.2.6 that the Radon-Nikodým property is invariant under linear homeomorphisms. Thus without loss of generality, we shall show G has a Bochner integrable Radon-Nikodým derivative under the new norm $|||\cdot|||$ on X. For each n and $E \in \Sigma$, let $\lambda_n(E) = x_n^*(G(E))$. Since each $x_n^* \in X^*$, the measure λ_n is a scalar-valued μ-continuous finite measure on Σ. By the (scalar) Radon-Nikodým theorem there exists a measurable g_n on Ω such that

$$\lambda_n(E) = \int_E g_n \, d\mu$$

for all $E \in \Sigma$. Also, by the monotonicity of $|||\cdot|||$, the inequalities

$$\left|\left|\left| \sum_{k=1}^{n} g_k(\omega) x_k \right|\right|\right| \leqq \left|\left|\left| \sum_{k=1}^{n+m} g_k(\omega) x_k \right|\right|\right|$$

and

$$\left|\left|\left| \sum_{k=1}^{n} \int_E g_k \, d\mu \, x_k \right|\right|\right| \leqq \left|\left|\left| \sum_{k=1}^{n+m} \int_E g_k \, d\mu \, x_k \right|\right|\right| \leqq |||G(E)|||$$

and

$$\int_E \left|\left|\left| \sum_{k=1}^{n} g_k x_k \right|\right|\right| d\mu \leqq |G|(E)$$

obtain for all positive integers m, n, all $\omega \in \Omega$ and all $E \in \Sigma$, with the last inequality following from the penultimate inequality and Theorem II.2.4. The first and third inequalities together with the Monotone Convergence Theorem show that

$$\lim_n \left|\left|\left| \sum_{k=1}^{n} g_k(\omega) x_k \right|\right|\right|$$

exists for almost all $\omega \in \Omega$. Since (x_n) is a boundedly complete basis, this means

$$\lim_n \sum_{k=1}^{n} g_k(\cdot) x_k = g$$

exists almost everywhere. Consequently g is measurable and, by Fatou's lemma and the third inequality, is Bochner integrable. Finally, from the fact that

$$\left|\left|\left| \sum_{k=1}^{n} g_k(\cdot) x_k \right|\right|\right| \leqq |||g(\cdot)|||$$

almost everywhere and the Dominated Convergence Theorem, it follows that for each $E \in \Sigma$

$$\begin{aligned} G(E) &= \lim_n \sum_{k=1}^{n} \lambda_k(E) x_k \\ &= \lim_n \sum_{k=1}^{n} \int_E g_k \, d\mu \, x_k = \int_E g \, d\mu; \end{aligned}$$

this completes the proof.

COROLLARY 7. *Neither $L_1(\mu)$ (μ nonatomic) nor c_0 has a boundedly complete Schauder basis.*

PROOF. Examples 1 and 2.

In spite of the fact that a class of spaces strictly including l_1 has the Radon-Nikodým property, the space l_1 plays a curious role in the study of the Radon-Nikodým property.

THEOREM 8 (LEWIS-STEGALL). *A Banach space X has the Radon-Nikodým property with respect to (Ω, Σ, μ) if and only if every bounded linear operator $T: L_1(\mu) \to X$ admits a factorization $T = LS$.*

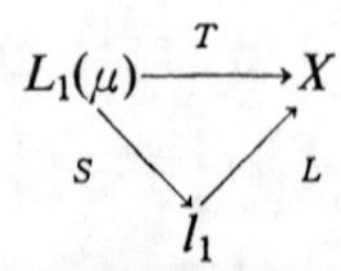

where $L: l_1 \to X$ and $S: L_1(\mu) \to l_1$ are continuous linear operators.

In this case, for each $\varepsilon > 0$, L, S can be chosen such that $\|S\| \leqq \|T\| + \varepsilon$ and $\|L\| \leqq 1$.

PROOF. Suppose $T: L_1(\mu) \to X$ admits such a factorization. Since l_1 has the Radon-Nikodým property, there exists a Bochner μ-integrable $g: \Omega \to l_1$ such that for all $f \in L_1(\mu)$ one has $S(f) = \int_\Omega fg \, d\mu$. This fact combined with Theorem II.2.6 shows

$$T(f) = LS(f) = \int_\Omega fL(g) \, d\mu$$

for all $f \in L_1(\mu)$. Hence T is Riesz representable. The proof of the sufficiency is concluded by an appeal to Theorem 5.

For the converse, suppose X has the Radon-Nikodým property with respect to (Ω, Σ, μ). Let $T: L_1(\mu) \to X$ be a bounded linear operator. According to Theorem 5, there exists $g \in L_\infty(\mu, X)$ such that $T(f) = \int_\Omega fg \, d\mu$ for all $f \in L_1(\mu)$. Let $\varepsilon > 0$. By Corollary II.1.3, there exists a sequence (f_n) of countably valued μ-measurable functions such that $\|g - f_n\|_\infty < \varepsilon 2^{-n-1}$. Writing $g_1 = f_1$ and $g_n = f_n - f_{n-1}$ for all $n \geqq 2$ one has

$$\left\| g - \sum_{m=1}^{n} g_m \right\|_\infty < \varepsilon 2^{-n-1}$$

for all n. Now write for each n, $g_n = \sum_{k=1}^\infty x_{n,k} \chi_{E_{n,k}}$ where $(E_{n,k})_{k=1}^\infty$ is a sequence of disjoint members of Σ and $\|x_{n,k}\| < \varepsilon 2^{-n}$ for all $n \geqq 2$. Define $S: L_1(\mu) \to l_1(N \times N)$ by

$$S(f)(n, k) = \|x_{n,k}\| \int_{E_{n,k}} f \, d\mu$$

for $f \in L_1(\mu)$. Then one has

$$\begin{aligned} \|S(f)\| &\leqq \sum_{n=1}^\infty \sum_{k=1}^\infty \|x_{n,k}\| \left| \int_{E_{n,k}} f \, d\mu \right| \\ &\leqq \sum_{k=1}^\infty \|x_{1,k}\| \int_{E_{1,k}} |f| \, d\mu + \sum_{n=2}^\infty \sum_{k=1}^\infty \|x_{n,k}\| \int_{E_{n,k}} |f| \, d\mu. \end{aligned}$$

Now since $\|g - g_1\|_\infty < \varepsilon/2$ and $\|g\|_\infty = \|T\|$, it follows that $\|x_{1,k}\| < \|T\| + \varepsilon/2$ for all k. On the other hand, $\|x_{n,k}\| < \varepsilon 2^{-n}$ for all $n \geqq 2$, so that

$$\begin{aligned}\|S(f)\| &\leqq (\|T\| + \varepsilon/2)\sum_{k=1}^{\infty}\int_{E_{1,k}}|f|\,d\mu + \sum_{n=2}^{\infty}\sum_{k=1}^{\infty}\varepsilon 2^{-n}\int_{E_{n,k}}|f|\,d\mu \\ &\leqq (\|T\| + \varepsilon)\|f\|_1.\end{aligned}$$

Therefore $\|S\| \leqq (\|T\| + \varepsilon)$.

Next, define $L: l_1(N \times N) \to X$ by

$$L((\alpha_{n,k})) = \sum_{n=1}^{\infty}\sum_{k=1}^{\infty}\alpha_{n,k}\frac{x_{n,k}}{\|x_{n,k}\|},$$

with the usual proviso that $0/0 = 0$. It is clear that $\|L\| \leqq 1$.

Finally, note that if $f \in L_1(\mu)$, then

$$\begin{aligned}LS(f) &= \sum_{n=1}^{\infty}\sum_{k=1}^{\infty}\|x_{n,k}\|\int_{E_{n,k}} f\,d\mu\,\frac{x_{n,k}}{\|x_{n,k}\|} \\ &= \sum_{n=1}^{\infty}\int_{\Omega} fg_n\,d\mu = \int_{\Omega} fg\,d\mu = T(f),\end{aligned}$$

by the Dominated Convergence Theorem.

In a certain sense, Theorem 8 seems to emphasize the role of l_1 in the theory of the Radon-Nikodým property a bit too much. Upon first glance, one is tempted to say that on the basis of Theorem 8, every space with the Radon-Nikodým property contains a copy of l_1. But on the basis of Theorem 6, one finds that l_2 has the Radon-Nikodým property, and l_2 certainly contains no copy of l_1. The role of l_1 in Radon-Nikodým theory is important and that importance derives largely through Theorem 8; the reader will see evidence of the applicability of Theorem 8 in various forms throughout Chapter 6 and in the notes and remarks section of Chapter 8.

2. Representable operators, weak compactness and Radon-Nikodým theorems. The roles of compactness and weak compactness in the theory of Radon-Nikodým derivatives of vector measures form the core of this section. Briefly, the plan is to prove that compact and weakly compact operators on $L_1(\mu)$ are representable and with the help of an exhaustion lemma, to parlay these facts into two Radon-Nikodým theorems for the Bochner and Pettis integrals. In the course of the work, some basic properties of weakly compact operators on $L_1(\mu)$ arise. For instance, the facts that weakly compact operators on $L_1(\mu)$ have separable ranges and map weakly compact sets onto norm compact sets come about in a very natural way. Throughout this section, (Ω, Σ, μ) is a finite measure space and X is a Banach space with dual X^*.

LEMMA 1. *For each partition π of Ω (into a finite set of disjoint members of Σ) define the linear operator $E_\pi: L_1(\mu, X) \to L_1(\mu, X)$ by*

$$E_\pi(f) = \sum_{A \in \pi} \frac{\int_A f \, d\mu}{\mu(A)} \chi_A$$

(observing the convention $0/0 = 0$*) for all* $f \in L_1(\mu, X)$*. Then* E_π *is a contraction on* $L_1(\mu, X)$ *which maps* $L_\infty(\mu, X)$ *into* $L_\infty(\mu, X)$ *in a contractive manner. Moreover, if the partitions are directed by refinement, then*

$$\lim_\pi \|E_\pi(f) - f\|_1 = 0 \quad \textit{for all } f \in L_1(\mu, X)$$

and

$$\lim_\pi \|E_\pi f - f\|_\infty = 0 \quad \textit{for all } f \in L_\infty(\mu, X)$$

with relatively norm compact ranges.

PROOF. If $f \in L_1(\mu, X)$, then

$$\|E_\pi(f)\|_1 = \left\| \sum_{A \in \pi} \frac{\int_A f \, d\mu}{\mu(A)} \chi_A \right\|_1 = \sum_{A \in \pi} \left\| \int_A f \, d\mu \right\| \leq \int_\Omega \|f\| \, d\mu = \|f\|_1.$$

Also, if f is a simple function, then the net $(E_\pi(f))$ is eventually constant. Hence $E_\pi(f) \to f$ for all f in a dense linear subset of $L_1(\mu, X)$. Since $\|E_\pi\| \leq 1$ for all π, it follows that $\lim_\pi E_\pi(f) = f$ in $L_1(\mu, X)$-norm for all $f \in L_1(\mu, X)$. This proves the assertions dealing with the action of E_π as an operator on $L_1(\mu, X)$.

For the $L_\infty(\mu, X)$ case, note that

$$\|E_\pi(f)\|_\infty = \max \left\{ \frac{\|\int_A f \, d\mu\|}{\mu(A)} : A \in \pi \right\}.$$

But $\|\int_A f \, d\mu\| \leq \|f\|_\infty \mu(A)$ for all $A \in \Sigma$. Hence $\|E_\pi(f)\|_\infty \leq \|f\|_\infty$ for each $f \in L_\infty(\mu, X)$. The proof that $\lim_\pi E_\pi(f) = f$ in $L_\infty(\mu, X)$-norm provided f has an essentially relatively compact range is now based on the fact that the subspace of $L_\infty(\mu, X)$ spanned by the simple functions is dense in the (closed linear) subspace of $L_\infty(\mu, X)$ consisting of functions whose ranges are essentially relatively compact subsets of X.

We are now ready to investigate representable operators on $L_1(\mu)$ and measures with Radon-Nikodým derivatives. It is transparent that if X is the scalars, then every $T \in \mathscr{L}(L_1(\mu); X)$ is representable. Consequently, if X is any Banach space and $T \in \mathscr{L}(L_1(\mu); X)$ is a finite rank operator, then T is representable. This suggests the following fact.

THEOREM 2 (REPRESENTATION OF COMPACT OPERATORS ON $L_1(\mu)$). *Every compact member of* $\mathscr{L}(L_1(\mu); X)$ *is representable. In fact, if* $K_\infty(\mu, X)$ *is the subspace of* $L_\infty(\mu, X)$ *consisting of members of* $L_\infty(\mu, X)$ *whose ranges are essentially relatively compact then the correspondence* $T \leftrightarrow g$ *given by*

$$T(f) = \int_\Omega fg \, d\mu \quad \textit{for} \quad f \in L_1(\mu)$$

establishes an isometric isomorphism between the space of compact members of $\mathscr{L}(L_1(\mu); X)$ *and* $K_\infty(\mu, X)$.

PROOF. Let $T \in \mathscr{L}(L_1(\mu); X)$ be compact. We shall first show that $\lim_\pi \|TE_\pi - T\| = 0$. For this, note that if $f \in L_1(\mu)$ and $g \in L_\infty(\mu)$, then

$$\int_\Omega E_\pi(f)g\,d\mu = \sum_{A\in\pi} \frac{\int_A f\,d\mu \int_A g\,d\mu}{\mu(A)} = \int_\Omega f E_\pi(g)\,d\mu.$$

From this it follows that E_π is "selfadjoint." Hence the adjoint of TE_π is $E_\pi T^*$. Now by Schauder's theorem $T^*: X^* \to L_\infty(\mu)$ is compact. Since $\lim_\pi E_\pi(f) = f$ for each $f \in L_\infty(\mu)$ and $\|E_\pi\| \leqq 1$, we see that $\lim_\pi E_\pi(f) = f$ uniformly on compact subsets of $L_\infty(\mu)$. Hence

$$\lim_\pi (E_\pi T^*)(x^*) = T^*(x^*)$$

uniformly for $\|x^*\| \leqq 1$. Therefore $\lim_\pi E_\pi T^* = T^*$ in the operator norm. Since $E_\pi T^* = (TE_\pi)^*$, it follows that $\lim_\pi TE_\pi = T$ in operator norm. In particular, (TE_π) is a Cauchy net in operator norm.

Now towards finding a Bochner integrable kernel for T, for a partition π, write

$$g_\pi = \sum_{A\in\pi} \frac{T(\chi_A)}{\mu(A)} \chi_A.$$

A quick computation establishes that $TE_\pi(f) = \int_\Omega f g_\pi\,d\mu$. Hence if π_1 and π_2 are partitions, then

$$(TE_{\pi_1} - TE_{\pi_2})(f) = \int_\Omega f(g_{\pi_1} - g_{\pi_2})\,d\mu.$$

Since $\lim_{\pi_1,\pi_2} \|TE_{\pi_1} - TE_{\pi_2}\| = 0$, an appeal to Lemma 1.4 shows that

$$\lim_{\pi_1,\pi_2} \|g_{\pi_1} - g_{\pi_2}\|_\infty = \lim_{\pi_1,\pi_2} \|TE_{\pi_1} - TE_{\pi_2}\| = 0.$$

It follows that there exists $g \in L_\infty(\mu, X)$ such that $\lim_\pi \|g_\pi - g\|_\infty = 0$. Since each g_π has finite range, the function g has relatively compact range and so $g \in K_\infty(\mu, X)$. Also, for a fixed $f \in L_1(\mu)$, the Dominated Convergence Theorem with dominating function $|f|\ \|T\|$ justifies the equalities

$$T(f) = \lim_\pi TE_\pi(f) = \lim_\pi \int_\Omega f g_\pi\,d\mu = \int_\Omega fg\,d\mu.$$

This proves the representability of T by a member of $K_\infty(\mu, X)$.

Conversely, suppose $g \in K_\infty(\mu, X)$. Define $T: L_1(\mu) \to X$ by $T(f) = \int_\Omega fg\,d\mu$ for $f \in L_1(\mu)$. By Lemma 1.4, $\|T\| = \|g\|_\infty$. Moreover, since g has an essentially totally bounded range, it is not difficult to show that for each $\varepsilon > 0$, there is a simple function $g_\varepsilon \in K_\infty(\mu, X)$ such that $\|g_\varepsilon - g\|_\infty < \varepsilon$. Define $T_\varepsilon: L_1(\mu) \to X$ by $T_\varepsilon(f) = \int_\Omega f g_\varepsilon\,d\mu$ for $f \in L_1(\mu)$. Then T_ε has a finite dimensional range and $\|T - T_\varepsilon\| = \|g - g_\varepsilon\|_\infty < \varepsilon$. Hence, as the operator limit of finite rank continuous operators, T is compact. This completes the proof.

Here is a well-known corollary of the above proof:

COROLLARY 3. *Every compact linear operator $T: L_1(\mu) \to X$ is the limit in operator norm of a sequence of finite rank continuous linear operators. In fact, $\lim_\pi \|TE_\pi - T\| = 0$.*

One interesting feature of the kernel g of Theorem 2 is that, off a set of measure zero, g takes values in the closure of the set $\{T(\chi_A)/\mu(A): \mu(A) > 0\}$. In other words, g takes values in the closure of $\{T(f): \|f\|_1 = 1\}$.

With the help of the following technical lemma, we shall translate Theorem 2 into a Radon-Nikodým theorem for the Bochner integral.

LEMMA 4 (EXHAUSTION LEMMA). *Let $G: \Sigma \to X$ be a vector measure. Suppose* P *is a property of G such that*

(a) *G has* P *on every μ-null set;*

(b) *if G has property* P *on $E \in \Sigma$, then G has property* P *on every $A \in \Sigma$ contained in E;*

(c) *if G has property* P *on E_1 and E_2 (both members of Σ), then G has property* P *on $E_1 \cup E_2$; and*

(d) *every set $A \in \Sigma$ of positive μ-measure contains a set $B \in \Sigma$ of positive μ-measure such that G has property* P *on B.*

Then there exists a sequence (A_n) of disjoint members of Σ such that $\Omega = \bigcup_{n=1}^{\infty} A_n$ and such that G has property P *on each A_n.*

PROOF. Let $\mathscr{A} = \{E \in \Sigma: G$ has property P on $E\}$ and let $c = \sup\{\mu(A): A \in \mathscr{A}\}$. Choose a sequence (B_n) from $\mathscr{A}$ such that $\lim_n \mu(B_n) = c$. Let $E_n = \bigcup_{k=1}^{n} B_k$. Then each $E_n \in \mathscr{A}$ and $\lim_n \mu(E_n) = c$. Moreover, $E_n \subseteq E_{n+1}$ for each n. Now, if $\mu(\Omega \backslash \bigcup_{n=1}^{\infty} E_n) > 0$, then (d) insures the existence of $A \in \mathscr{A}$ with $\mu(A) > 0$ such that $A \subseteq \Omega \backslash \bigcup_{n=1}^{\infty} E_n$. But $(A \cup E_n)$ is a sequence in $\mathscr{A}$ with

$$\lim_n \mu(A \cup E_n) = \lim_n \mu(A) + \mu(E_n) = \mu(A) + c > c.$$

This contradicts the definition of c. Thus $A_0 = \Omega \backslash \bigcup_{n=1}^{\infty} E_n$ has μ-measure zero. Set $A_1 = E_1$, $A_n = E_2 \backslash E_1, \cdots$, $A_n = E_n \backslash E_{n-1}, \cdots$. Then $(A_n)_{n=0}^{\infty}$ is the desired sequence.

The following corollary specializes the Exhaustion Lemma to a form useful for proving Radon-Nikodým theorems.

COROLLARY 5. *Let $G: \Sigma \to X$ be a μ-continuous vector measure. If for each $E_1 \in \Sigma$ with $\mu(E_1) > 0$, there exists $E_2 \in \Sigma$ with $E_2 \subseteq E_1$ and $\mu(E_2) > 0$ and a Bochner integrable $h\ (= h_{E_1})$ such that*

$$G(E) = \int_E h\, d\mu$$

for all $E \in \Sigma$ with $E \subseteq E_2$, then there is a μ-measurable Pettis integrable function g such that

$$G(E) = \text{Pettis-}\int_E g\, d\mu$$

for all $E \in \Sigma$.

If G is of bounded variation, then g is Bochner integrable and $G(E) =$ (Bochner)-$\int_E g\, d\mu$ for all $E \in \Sigma$.

PROOF. By direct application of the Exhaustion Lemma, there exists a sequence (A_n) of pairwise disjoint members of Σ such that $\bigcup_{n=1}^{\infty} A_n = \Omega$ and a sequence (h_n) of Bochner integrable functions on Ω such that

$$G(E \cap A_n) = \int_{E \cap A_n} h_n \, d\mu$$

for each $E \in \Sigma$ and all n. Define $g\colon \Omega \to X$ by $g(\omega) = h_n(\omega)$ if $\omega \in A_n$; clearly g is μ-measurable. Moreover,

$$G\left(E \cap \left(\bigcup_{n=1}^{m} A_n\right)\right) = \int_E g\chi_{\bigcup_{n=1}^m A_n} \, d\mu$$

for each $E \in \Sigma$ and all m. Consequently

$$G(E) = \lim_m \int_E g\chi_{\bigcup_{n=1}^m A_n} \, d\mu$$

for each $E \in \Sigma$. But if $x^* \in X^*$, then the variation

$$|x^*G|(\Omega) \geqq \lim_m \int_\Omega |x^*g| \chi_{\bigcup_{n=1}^m A_n} \, d\mu.$$

Hence by the Monotone Convergence Theorem, $x^*g \in L_1(\mu)$ for each $x^* \in X^*$. Thus if $E \in \Sigma$ and $x^* \in X^*$, then

$$x^*(G(E)) = \lim_m \int_E x^*(g)\chi_{\bigcup_{n=1}^m A_n} \, d\mu = \int_E x^*g \, d\mu$$

by the Dominated Convergence Theorem. Therefore g is Pettis integrable and Pettis-$\int_E g \, d\mu = G(E)$ for each $E \in \Sigma$.

To prove the second assertion, suppose the variation $|G|(\Omega)$ is finite. Then

$$\int \|g\| \chi_{\bigcup_{n=1}^m A_n} \, d\mu \leqq |G|(\Omega)$$

for all m. Again by the Monotone Convergence Theorem, $\|g\| \in L_1(\mu)$. Hence g is Bochner integrable. Since the Bochner and Pettis integrals coincide whenever they coexist, we have $G(E) =$ Bochner-$\int_E g \, d\mu$ for each $E \in \Sigma$.

Now we are in a position to translate the representation of compact operators on $L_1(\mu)$ into a rather crude Radon-Nikodým theorem.

THEOREM 6 (JUNIOR GRADE RADON-NIKODÝM THEOREM). *Let $G\colon \Sigma \to X$ be a μ-continuous vector measure. If for each $E_1 \in \Sigma$ with $\mu(E_1) > 0$ there exists $E_2 \in \Sigma$ with $E_2 \subseteq E_1$ and $\mu(E_2) > 0$ such that $\{G(E)/\mu(E)\colon E \in \Sigma,\ E \subseteq E_2,\ \mu(E) > 0\}$ is relatively norm compact, then there exists a μ-measurable Pettis integrable $g\colon \Omega \to X$ such that*

$$G(E) = \textit{Pettis-}\int_E g \, d\mu$$

for each $E \in \Sigma$.

If G is of bounded variation, then g is also Bochner integrable and $G(E) =$ Bochner-$\int_E g \, d\mu$ for each $E \in \Sigma$.

PROOF. According to Corollary 5, this theorem will be proved if for each $E_1 \in \Sigma$ with $\mu(E_1) > 0$ we can find $E_2 \in \Sigma$, with $E_2 \subseteq E_1$ and $\mu(E_2) > 0$ and a Bochner integrable g such that $G(E) = \int_E g \, d\mu$ for all $E \in \Sigma$ with $E \subseteq E_2$. To this end, let $E_1 \in \Sigma$ with $\mu(E_1) > 0$. Select $E_2 \subseteq E_1$ with $E_2 \in \Sigma$ and $\mu(E_2) > 0$ such that

$$K = \{G(E)/\mu(E): E \in \Sigma, E \subseteq E_2, \mu(E) > 0\}$$

is relatively norm compact. By Mazur's theorem (II.2.12) M, the absolutely closed convex hull of K, is norm compact. Now define an operator T on the simple functions in $L_1(\mu)$ by

$$T(f) = \sum_{i=1}^{n} \alpha_i \, G(A_i \cap E_2),$$

where $f = \sum_{i=1}^{n} \alpha_i \chi_{A_i}$, $A_i \in \Sigma$ and $A_i \cap A_j = \emptyset$ for $i \neq j$. Note that

$$T(f) = \sum_{i=1}^{n} \alpha_i \mu(A_i \cap E_2) \frac{G(A_i \cap E_2)}{\mu(A_i \cap E_2)}$$

$(0/0 = 0)$. But if $\|f\|_1 \leqq 1$, then

$$\sum_{i=1}^{n} |\alpha_i \mu(A_i \cap E_2)| \leqq \sum_{i=1}^{n} |\alpha_i| \mu(A_i) = \|f\|_1 \leqq 1.$$

Hence $T(f) \in M$ for $\|f\|_1 \leqq 1$. Thus T has a compact linear extension, still denoted by T, to all of $L_1(\mu)$. According to Theorem 2, there exists a $g \in L_\infty(\mu, X)$ such that $T(f) = \int_\Omega fg \, d\mu$ for all $f \in L_1(\mu)$. In particular, if $E \in \Sigma$ is contained in E_2, then

$$G(E) = T(\chi_E) = \int_E g \, d\mu$$

as required.

Unfortunately, the hypothesis of Theorem 6 is unduly restrictive. Probably the most important characteristic of Theorem 6 is the technique of proof—prove a representation theorem for a class of operators on $L_1(\mu)$ and translate to a Radon-Nikodým theorem. This technique will be used again.

Theorem 6 does have at least one major positive attribute; its converse is true.

THEOREM 7. *Suppose $g: \Omega \to X$ is μ-measurable.*

(a) *If g is Dunford integrable and $G(E) =$ Dunford-$\int_E g \, d\mu$ for $E \in \Sigma$, then $|G|$ is a σ-finite measure and for each $\varepsilon > 0$ there exists E_ε with $\mu(\Omega \backslash E_\varepsilon) < \varepsilon$ such that $\{G(E)/\mu(E): E \in \Sigma, E \subseteq E_\varepsilon, \mu(E) > 0\}$ is relatively norm compact.*

(b) *If G is as above and g is Pettis integrable, then G is also μ-continuous (and therefore countably additive).*

(c) *If G is as above and g is Bochner integrable, then $|G|$ is also finite.*

PROOF. (a) Suppose $G(E) =$ Dunford-$\int_E g \, d\mu$. For each n, let A_n be the set $\{\omega \in \Omega: \|g(\omega)\| \leqq n\}$. Select $E_n \in \Sigma$ such that $\chi_{E_n} = \chi_{A_n}$ μ-a.e. Then $|G|$ is finite on each E_n. Since $E_n \uparrow \Omega$ μ-a.e. and $|G|$ vanishes on μ-null sets, $|G|$ is σ-finite.

To prove the second part of (a), select n_0, such that $\mu(\Omega \backslash E_{n_0}) < \varepsilon/2$. Let (g_n) be a sequence of simple X-valued functions converging μ-a.e. to g. By Egoroff's theorem, there is a set $A \in \Sigma$ with $A \subseteq E_{n_0}$ and $\mu(E_{n_0} \backslash A) < \varepsilon/2$ such that (g_n) converges to g uniformly on A. Since the g_n's are simple functions and the convergence of the sequence (g_n) is uniform on A, $g\chi_A \in K_\infty(\mu, X)$. If $T(f) = \int_A fg \, d\mu$ for $f \in L_1(\mu)$, then T is compact by Theorem 2. Moreover

$$\left\{\frac{G(E)}{\mu(E)}: E \in \Sigma,\ E \subseteq A,\ \mu(E) > 0\right\} = \left\{\frac{T(\chi_E)}{\mu(E)}: E \in \Sigma,\ E \subseteq A,\ \mu(E) > 0\right\}$$
$$\subseteq T \quad (\text{unit ball of } L_1(\mu)).$$

Since the last set is relatively norm compact, (a) is proved.

In spite of the fact that Theorems 6 and 7 give a necessary and sufficient condition for a vector measure to arise as an indefinite Bochner integral, they are by no means the end of the Radon-Nikodým story. As we shall see shortly, the hypothesis of Theorem 6 can be weakened considerably. On the other hand, there is no reason to try to weaken the necessary condition of Theorem 7.

EXAMPLE 8 (LEWIS). *A $C[0, 1]$-valued measure with no Radon-Nikodým derivative.* Let $\Omega = [0, 1]$ and μ be Lebesgue measure on Σ, the σ-field of Lebesgue measurable sets. Define $G: \Sigma \to C[0, 1]$ by $G(E)(t) = \mu(E \cap [0, t])$ for $E \in \Sigma$ and $t \in [0, 1]$. Evidently $\|G(E)\| = \mu(E)$. Therefore G is μ-continuous and of bounded variation. Select any Lebesgue measurable set $E_1 \subseteq [0, 1]$ with $\mu(E_1) > 0$. Fix m and choose $0 = t_0 < t_1 < t_2 < \cdots < t_{m-1} < t_m = 1$ such that

$$\mu([t_{n-1}, t_n) \cap E_1) = \mu(E_1)/m$$

for $n = 1, 2, \cdots, m$. Suppose $1 \leqq i < j \leqq m$. Consider for $t \in [0, 1]$

$$\left\| \frac{G([t_{i-1}, t_i) \cap E_1)(t)}{\mu([t_{i-1}, t_i) \cap E_1)} - \frac{G([t_{j-1}, t_j) \cap E_1)(t)}{\mu([t_{j-1}, t_j) \cap E_1)} \right\|.$$

At $t = t_i$, this quantity is equal to 1. Consequently, $\{G(E)/\mu(E): E \in \Sigma,\ E \subseteq E_1,\ \mu(E) > 0\}$ contains m elements of distance 1 apart. It follows that $\{G(E)/\mu(E): E \in \Sigma, E \subseteq E_1, \mu(E) > 0\}$ is not totally bounded. Hence G has no Radon-Nikodým derivative. It should be noted that the method used in Example 1.2 does not work for this example because the operator $\int_\Omega \cdot\, dG$ from $L_\infty(\mu)$ to $C[0, 1]$ is, in this case, a compact operator.

Incidentally, this example shows that the operator $T: L_1[0, 1] \to C[0, 1]$ defined by

$$(Tf)(t) = \int_{[0,t]} f\, d\mu$$

(μ = Lebesgue measure) is not representable. It is a consequence of the next group of results that if this operator is followed by the natural inclusion of $C[0, 1]$ into any $L_p[0, 1]$ $(1 \leqq p < \infty)$, then the resulting operator on $L_1[0, 1]$ is representable. (The alert reader should have no problem in proving this directly.)

The next result is a fundamental theorem of the theory of vector measures.

LEMMA 9 (DUNFORD-PETTIS). *A weakly compact linear operator $T: L_1(\mu) \to X$ whose range is separable is representable. In fact, there exists $g \in L_\infty(\mu, X)$ with an essentially relatively weakly compact range such that*

$$T(f) = \int_\Omega fg\, d\mu$$

for all $f \in L_1(\mu)$.

PROOF. For each partition π define

$$g_\pi = \sum_{A\in\pi} \frac{T(\chi_A)}{\mu(A)} \chi_A \qquad (0/0 = 0).$$

Then there is a norm separable weakly compact set K such that $g_\pi(\Omega) \subseteq T$ (unit ball of $L_1(\mu)) \subseteq K$ for all partitions π. Now there is no loss of generality in assuming that X is separable. Consequently we may and do assume that X^* contains a countable norming set (x_n^*). For each n, pick $g_n \in L_\infty(\mu)$ such that

$$x_n^* T(f) = \int_\Omega f g_n \, d\mu$$

for all $f \in L_1(\mu)$. A quick computation establishes that

$$\int_\Omega f x_n^*(g_\pi) \, d\mu = \int_\Omega E_\pi(f) g_n \, d\mu = \int_\Omega f E_\pi(g_n) \, d\mu$$

for all $f \in L_1(\mu)$ and all partitions π. Thus $x_n^* g_\pi = E_\pi(g_n)$ for all partitions π and all n. According to Lemma 1, this means that

$$\lim_\pi \|x_n^* g_\pi - g_n\|_\infty = \lim_\pi \|E_\pi g_n - g_n\| = 0$$

for each n. It follows that there exists a sequence (π_n) of partitions and a μ-null set P such that for each n

$$\lim_m x_n^* g_{\pi_m}(\omega) = g_n(\omega)$$

uniformly in $\omega \in \Omega \backslash P$. Next, for each $\omega \in \Omega$ define $g(\omega)$ to be an arbitrary weak cluster point of the sequence $(g_{\pi_n}(\omega))$. Then g is a separably valued bounded function taking its values in the weakly compact set K. Moreover, since for each n, one has $\lim_m x_n^* g_{\pi_m} = g_n$ uniformly on $\Omega \backslash P$, it follows that $\lim_m x_n^* g_{\pi_m} = x_n^* g$ almost everywhere for each n. Hence $x_n^* g$ is measurable for all n. This fact combined with the facts that (x_n^*) is a norming sequence and g is separably valued shows that g is μ-measurable by Theorem II.1.2. Since g is bounded, and since we have

$$x_n^* T(f) = \int_\Omega f g_n \, d\mu = \int_\Omega f x_n^* g \, d\mu = x_n^* \int_\Omega f g \, d\mu$$

for all $f \in L_1(\mu)$ and all n, we see that $T(f) = \int_\Omega fg \, d\mu$ for all $f \in L_1(\mu)$, as required.

In all honesty, we should note that Theorem 2 could have been proven the same way as Lemma 9. In fact, Lemma 9 completely subsumes Theorem 2. We have included the separate proof of Theorem 2 mainly for reasons of taste; the proof of Theorem 2 is pleasing and instructive as is the proof of Lemma 9. For more on this, see the notes and remarks section.

The following definition and lemma will allow the separability condition in Lemma 9 to be scrapped.

DEFINITION 10. A subset K of $L_1(\mu)$ is called *uniformly integrable* if

$$\lim_{\mu(E)\to 0} \int_E |f| \, d\mu = 0$$

uniformly in $f \in K$.

LEMMA 11 (DUNFORD-PETTIS). *A representable operator $T: L_1(\mu) \to X$ maps bounded uniformly integrable sets into norm compact sets.*

PROOF. Let $T\colon L_1(\mu) \to X$ have the form $T(f) = \int_\Omega fg\,d\mu, f \in L_1(\mu)$, for some fixed $g \in L_\infty(\mu, X)$. Let $K \subseteq L_1(\mu)$ be a bounded uniformly integrable set and choose a sequence (g_n) of simple functions converging almost everywhere to g. Since each g_n is a simple function, it follows easily that $g\chi_E$ has a relatively norm compact range whenever $E \in \Sigma$ is such that $\lim_n g_n = g$ uniformly on E. Now with the help of Egoroff's theorem, choose a set $E_1 \in \Sigma$ such that $\lim_n g_n = g$ uniformly on E_1 and such that $\mu(\Omega \backslash E_1)$ is so small that

$$\int_{\Omega \backslash E_1} |f|\,d\mu < \varepsilon/(\|T\| + 1)$$

for all $f \in K$. Since $g\chi_{E_1}$ has a relatively compact range, the operator $f \to \int_{E_1} fg\,d\mu$ is compact and therefore $\{\int_{E_1} fg\,d\mu : f \in K\}$ is relatively norm compact. Also one has

$$\left\| \int_{\Omega \backslash E_1} fg\,d\mu \right\| \leqq \frac{\|T\|\varepsilon}{\|T\| + 1} < \varepsilon$$

for $f \in K$. It follows easily that

$$\{T(f) : f \in K\} = \left\{ \int_{E_1} fg\,d\mu + \int_{\Omega \backslash E_1} fg\,d\mu : f \in K \right\}$$

is totally bounded by 2ε-balls. Hence $T(K)$ is relatively compact.

The converse to Lemma 11 is false. Consider a bounded linear operator $T\colon L_1(\mu) \to c_0$. Then, as in Example 1′, there is a sequence (g_n) in $L_\infty(\mu)$ such that

$$T(f) = \left(\int_\Omega fg_n\,d\mu \right) \quad \text{for } f \in L_1(\mu).$$

It is easy to see that T is representable if and only if $\lim_n g_n = 0$ almost everywhere. It is only slightly less easy to see that T maps bounded uniformly integrable sets into norm compact sets if and only if $\lim_n g_n = 0$ in measure.

The next theorem, which is one of the main results of this section (indeed of the whole theory of Radon-Nikodým differentiation of vector measures), is now proved by a bit of "boot-strapping".

THEOREM 12 (DUNFORD-PETTIS-PHILLIPS). *A weakly compact linear operator on $L_1(\mu)$ has a norm-separable range. Consequently, every weakly compact operator on $L_1(\mu)$ is representable.*

In fact, a linear operator $T\colon L_1(\mu) \to X$ is weakly compact if and only if there exists a $g \in L_\infty(\mu, X)$ with an essentially relatively weakly compact range such that $T(f) = \int_\Omega fg\,d\mu$ for all $f \in L_1(\mu)$.

PROOF. To prove the first two statements, it is enough to prove that the range of a weakly compact operator on $L_1(\mu)$ is separable and then appeal to Lemma 9. To this end, let $T\colon L_1(\mu) \to X$ be a weakly compact operator. To prove $T(L_1(\mu))$ is separable, it suffices to prove that $\{T(\chi_E) : E \in \Sigma\}$ is separable. To prove this, it is obviously enough to prove $\{T(\chi_E) : E \in \Sigma\}$ is relatively compact. For this, consider a sequence $(T(\chi_{E_n}))$ and let Σ_1 be the σ-field generated by (E_n). Since Σ_1 is countably generated, the subspace $L_1(\Sigma_1, \mu)$ of $L_1(\mu)$ consisting of those members of $L_1(\mu)$

that are Σ_1-measurable is a separable closed linear subspace of $L_1(\mu)$. Hence the restriction T_1 of T to $L_1(\Sigma_1, \mu)$ is a weakly compact operator whose range is separable. According to Lemma 9, the operator T_1 is representable. But $\{\chi_{E_n}\}$ is a bounded uniformly integrable subset of $L_1(\Sigma_1, \mu)$. Hence $\{T_1(\chi_{E_n})\}$ is a relatively norm compact subset of X by Lemma 11. But $T(\chi_{E_n}) = T_1(\chi_{E_n})$ which, since the latter has a norm convergent subsequence, shows that $\{T(\chi_{E_n})\}$ is relatively norm compact. This proves the first two statements.

To prove the last assertion, note that if $T: L_1(\mu) \to X$ is a weakly compact operator, then the kernel g constructed in the proof of Lemma 9 has its range in the weak closure of $\{T(\chi_A)/\mu(A): \mu(A) > 0\}$, a weakly compact set. On the other hand, suppose $T(f) = \int_\Omega fg \, d\mu$, for all $f \in L_1(\mu)$ and some $g \in L_\infty(\mu, X)$ with an essentially relatively weakly compact range. By Corollary II.2.8, for each $A \in \Sigma$ with $\mu(A) > 0$, one has $T(\chi_A)/\mu(A) \in \overline{\text{co}}(g(\Omega))$, which is weakly compact by the Kreĭn-Smulian Theorem II.2.11. Consequently, the absolute closed convex hull of $\{T(\chi_A)/\mu(A): A \in \Sigma, \mu(A) > 0\}$ is weakly compact. But this set is nothing but the norm closure of $\{T(f): \|f\|_1 \leqq 1\}$ (see the proof of Theorem 6). Hence T is weakly compact.

Theorem 12 has a wealth of corollaries.

COROLLARY 13 (PHILLIPS). *Reflexive Banach spaces have the Radon-Nikodým property.*

PROOF. This is an immediate consequence of Theorem 12 and Theorem 1.4.

COROLLARY 14 (DUNFORD-PETTIS). *A weakly compact operator defined on $L_1(\mu)$ maps weakly compact sets into norm compact sets.*

PROOF. This is an immediate consequence of Theorem 12, Lemma 11 and the fact that the relatively weakly compact sets in $L_1(\mu)$ are precisely the bounded uniformly integrable sets. This last fact is isolated in

THEOREM 15 (DUNFORD). *A subset of $L_1(\mu)$ is relatively weakly compact if and only if it is bounded and uniformly integrable.*

PROOF. Let $K \subseteq L_1(\mu)$ be relatively weakly compact. Then K is bounded and if (f_n) is a sequence in K, then (f_n) has a weakly convergent subsequence by Eberlein's theorem. Hence there is a subsequence (f_{n_j}) such that

$$\lim_j \int_E f_{n_j} \, d\mu$$

exists for all $E \in \Sigma$. By the Vitali-Hahn-Saks theorem (Corollary I.4.10), (f_{n_j}) is uniformly integrable. Hence every sequence in K has a uniformly integrable subsequence. It follows immediately that K is uniformly integrable.

For the converse, suppose K is bounded and uniformly integrable. Let (f_n) be a sequence in K. Then there is a countable field $\mathcal{F}$ such that f_n is measurable relative to the σ-field, Σ_1, generated by $\mathcal{F}$. By a diagonal procedure, select a subsequence (f_{n_j}) such that $\lim \int_E f_{n_j} \, d\mu = F(E)$ exists for all $E \in \mathcal{F}$. Also, since K is uniformly integrable, it follows that F is μ-continuous. Thus there exists $f \in L_1(\Sigma_1, \mu)$ such that

$$\lim_j \int_E f_{n_j}\, d\mu = \int_E f\, d\mu$$

for each $E \in \Sigma_1$. From this point, it is a simple argument to verify that

$$\lim_j \int_\Omega f_{n_j} g\, d\mu = \int_\Omega fg\, d\mu$$

for each $g \in L_\infty(\Sigma_1, \mu)$. Hence $f_{n_j} \to f$ weakly in $L_1(\Sigma_1, \mu)$. But $L_1(\Sigma_1, \mu)$ is a closed linear subspace of $L_1(\mu)$. Hence $f_{n_j} \to f$ weakly in $L_1(\mu)$, and K is relatively weakly compact.

COROLLARY 16. *No infinite dimensional reflexive subspace of $L_1(\mu)$ is complemented in $L_1(\mu)$.*

PROOF. Let P be a continuous linear projection on $L_1(\mu)$. If the range of P is reflexive, then P maps the unit ball of $L^1(\mu)$ into a relatively weakly compact set.

By Corollary 14, P^2(unit ball of $L_1(\mu)$) = P(unit ball of $L_1(\mu)$) is compact. Hence the range of P is finite dimensional.

Embedded in the proof of Corollary 16 is the proof of

COROLLARY 17 (DUNFORD-PETTIS). *Weakly compact operators from $L_1(\mu)$ to $L_1(\mu)$ have compact squares.*

The following basic theorem is the improvement of Theorem 6 that we promised earlier.

THEOREM 18 (UTILITY GRADE RADON-NIKODÝM THEOREM). *Let $G: \Sigma \to X$ be a μ-continuous vector measure. If for each $E_1 \in \Sigma$ with $\mu(E_1) > 0$ there exists $E_2 \in \Sigma$ with $E_2 \subseteq E_1$ and $\mu(E_2) > 0$ such that $\{G(E)/\mu(E): E \in \Sigma, E \subseteq E_2, \mu(E) > 0\}$ is relatively weakly compact, then there exists a μ-measurable Pettis integrable g: $\Omega \to X$ such that*

$$G(E) = \textit{Pettis-}\int_E g\, d\mu$$

for all $E \in \Sigma$. If G is of bounded variation, g is also Bochner integrable and

$$G(E) = \textit{Bochner-}\int_E g\, d\mu$$

for all $E \in \Sigma$.

PROOF. Proceed as in the proof of Theorem 6, replacing the word "compact" by "weakly compact", using the Kreĭn-Smulian Theorem II.2.11 in place of Mazur's theorem and using Theorem 12 in place of Theorem 2.

EXAMPLE 19. *An $L_1[0, 1]$-valued measure with a Radon-Nikodým derivative.* Let $\Omega = [0, 1]$ and μ be Lebesgue measure on Σ, the class of Lebesgue measurable sets. Define $G: \Sigma \to L_1(\mu)$ by

$$G(E)(t) = \mu(E \cap [0, t))$$

for $t \in [0, 1]$. Then there exists a μ-measurable Bochner integrable g: $[0, 1] \to L_1[0, 1]$ such that

$$G(E) = \int_E g\, d\mu$$

for all $E \in \Sigma$. To verify this, note that $\|G(E)\| \leqq \mu(E)$ for each $E \in \Sigma$. Thus G is of bounded variation and is μ-continuous. Moreover, if $E \in \Sigma$ and $\mu(E) > 0$, then $0 \leqq G(E)/\mu(E) \leqq 1$. Thus $\{G(E)/\mu(E): E \in \Sigma\}$ is uniformly integrable and is therefore contained in a weakly compact set by Theorem 15. An appeal to Theorem 18 establishes the existence of the advertised μ-measurable g. (The alert reader should be able to write down the explicit form of g.)

Two comments are in order regarding this example. First, note that this example is different from Example 8 only insofar as the range of G is $L_1[0, 1]$ above, while in Example 8, the range is $C([0, 1])$. This example also shows via Lemma 1.4 that if $T: L_1[0, 1] \to L_1[0, 1]$ is defined by $T(f)(t) = \int_{[0,t]} f(s)\, d\mu(s)$ for each $t \in [0, 1]$, then T is representable. Another way to see this is to define $T_1: L_1[0, 1] \to C([0, 1])$ by

$$T_1(f)(t) = \int_{[0,t]} f(s)\, d\mu(s)$$

and let $T_2: C([0, 1]) \to L_1[0, 1]$ be the natural inclusion of $C([0, 1])$ into $L_1[0, 1]$. Since T_2 is clearly weakly compact (Theorem 15) so too is $T = T_2 T_1$. Therefore T is representable by Theorem 12.

This line of reasoning can be generalized at no expense.

COROLLARY 20. *A weakly compact linear operator composes with countably additive vector measures of bounded variation to yield countably additive vector measures of bounded variation that have Bochner integrable Radon-Nikodým derivatives with respect to their variations. Specifically, let $G: \Sigma \to X$ be a μ-continuous vector measure of bounded variation. If $T: X \to Y$ is a weakly compact linear operator, then there exists $g \in L_1(\mu, Y)$ such that*

$$T(G(E)) = \int_E g\, d\mu$$

for all $E \in \Sigma$.

PROOF. Define $F: \Sigma \to Y$ by $F(E) = T(G(E))$, for $E \in \Sigma$. Easy computations show that F is a μ-continuous measure of bounded variation. Moreover, if $E_1 \in \Sigma$ and $\mu(E_1) > 0$, then there is a positive integer N and a set E_2 in Σ with $E_2 \subseteq E_1$ and $\mu(E_2) > 0$ such that

$$|G|(E \cap E_2) \leqq N\mu(E \cap E_2)$$

for all $E \in \Sigma$. Indeed, let ϕ be the Radon-Nikodým derivative of $|G|$ with respect to μ. Then, for some n, $\mu(\{\omega \in \Omega: \phi(\omega) \leqq n\}) > 0$. Let N be any such n. Then the set $\{G(E)/\mu(E): E \in \Sigma, E \subseteq E_2, \mu(E) > 0\}$ is bounded in X. Thus the set $\{F(E)/\mu(E): E \in \Sigma, E \subseteq E_2, \mu(E) > 0\}$ is relatively weakly compact in Y. Apply Theorem 18.

The last result of this section is a proposition whose proof is left as an exercise.

PROPOSITION 21. *Let $T: L_1(\mu) \to X$ be a continuous linear operator. For $E \in \Sigma$, define T_E by*

$$T_E(f) = T(f\chi_E), \quad \textit{for } f \in L_1(\mu).$$

Any one of the following statements about T implies all the others:

(i) *The operator T is representable.*

(ii) *For each* $\varepsilon > 0$, *there exists* $E_\varepsilon \in \Sigma$ *with* $\mu(\Omega \backslash E_\varepsilon) < \varepsilon$ *such that* T_{E_ε} *is compact.*

(iii) *For each* $\varepsilon > 0$, *there exists* $E_\varepsilon \in \Sigma$ *with* $\mu(\Omega \backslash E_\varepsilon) < \varepsilon$ *such that* T_{E_ε} *is weakly compact.*

(iv) *For each* $E_1 \in \Sigma$ *with* $\mu(E_1) > 0$, *there is a set* $E_2 \in \Sigma$ *with* $E_2 \subseteq E_1$ *and* $\mu(E_2) > 0$ *such that* T_{E_2} *is compact.*

(v) *For each* $E_1 \in \Sigma$ *with* $\mu(E_1) > 0$, *there is a set* $E_2 \in \Sigma$ *with* $E_2 \subseteq E_1$ *and* $\mu(E_2) > 0$ *such* that T_{E_2} *is weakly compact.*

The heavy use of weakly compact operators on $L_1(\mu)$ to establish Radon-Nikodým theorems in this section may leave the false impression that only the weakly compact operators on $L_1(\mu)$ are representable. This can be corrected quickly.

EXAMPLE 22. *A representable operator on* $L_1[0, 1]$ *that is not weakly compact.* Let $T: L_1[0, 1] \to l_1$ be a quotient map, i.e., T is linear, continuous and onto. By Theorems 1.5 and 1.6, T is representable. Of course, T is not weakly compact since l_1 is not reflective.

3. Separable dual spaces and the Radon-Nikodým property. This section is devoted to an exposition of the connection between separable dual spaces and the Radon-Nikodým property. In this section we shall see that separable dual spaces have the Radon-Nikodým property and that a Banach space has the Radon-Nikodým property if each of its separable subspaces has this property. Moreover, we shall see that for a dual space to have the Radon-Nikodým property it suffices that every separable subspace of the predual have a separable dual. Several consequences of these results will be noted. As usual, X is a Banach space and (Ω, Σ, μ) is a finite measure space.

THEOREM 1 (DUNFORD-PETTIS). *Separable dual spaces have the Radon-Nikodým property.*

It is possible to proceed along lines similar to the proof of Lemma 2.9. In fact, if X is a space with separable dual space X^* and $T: L_1(\mu) \to X^*$, simply replace X in Lemma 2.9 by X^* and choose the countable norming set from X. Then define $g(\omega)$ to be a weak*-cluster point of the sequence $(g_{\pi_n}(\omega))$. The reader should have no problems furnishing the details of such an approach. However, Dunford and Pettis supplied another proof that is also exciting. Since the result plays such a central role in the theory and applications of vector measures, their proof will be presented.

PROOF. For sake of simplicity assume X is a real Banach space. Let X be a Banach space with separable dual X^* and suppose $F: \Sigma \to X^*$ is a countably additive vector measure of bounded variation, $|F|$. We will show that there exists a $|F|$-essentially bounded, $|F|$-measurable function $f: \Omega \to X^*$ such that

$$F(E) = \int_E f \, d|F|$$

for all $E \in \Sigma$. From this it readily follows that if F is μ-continuous, then F has a Bochner integrable μ-measurable Radon-Nikodým derivative with respect to μ, namely, $fd|F|/d\mu$.

Let $x \in X$ and consider $F_x(A) = F(A)(x)$ for $A \in \Sigma$. Clearly F_x is a countably additive scalar-valued measure satisfying

$$|F_x(A)| \leqq \|F(A)\| \; \|x\| \leqq \|x\| \; |F|(A)$$

for each $A \in \Sigma$ and each $x \in X$. Thus there exists $g_x \in L_\infty(|F|)$ such that $\|g_x\| \leqq \|x\|$ and

$$F_x(A) = \int_A g_x(\omega) \, d|F|(\omega)$$

for each $A \in \Sigma$.

Since X^* is separable, so is X. Let D be a countable dense subset of X. Suppose $q_1, \cdots, q_n$ are rational numbers and $x_1, \cdots, x_n \in D$. Consider $x = \sum_{i=1}^n q_i x_i$. Then

(a) there exists an $|F|$-null set $N_x^{(1)}$ such that $\omega \notin N_x^{(1)}$ implies $|g_x(\omega)| \leqq \|x\|$, and

(b) for $A \in \Sigma$, one has

$$\begin{aligned}\int_A g_x(\omega) \, d|F|(\omega) &= F(A)(x) = F(A)\Big(\sum_{i=1}^n q_i x_i\Big) \\ &= \sum_{i=1}^n q_i F(A)(x_i) = \sum_{i=1}^n q_i \int_A g_{x_i}(\omega) \, d|F|(\omega) \\ &= \int_A \sum_{i=1}^n q_i g_{x_i}(\omega) \, d|F|(\omega).\end{aligned}$$

Therefore there exists an $|F|$-null set $N_x^{(2)}$ such that for $\omega \notin N_x^{(2)}$, one has

$$g_{\sum_{i=1}^n q_i x_i}(\omega) = \sum_{i=1}^n q_i g_{x_i}(\omega).$$

Now the collection of rational linear combinations of members of D is a countable collection. Consequently, if N is the union of all the sets $N_x^{(1)}$ and $N_x^{(2)}$, where x ranges over the rational linear combinations of members of D, then N is $|F|$-null.

Moreover, if $\omega \in \Omega \backslash N$, then one has

$$\Big|\sum_{i=1}^n q_i g_{x_i}(\omega)\Big| = |g_{\sum_{i=1}^n q_i x_i}(\omega)| \leqq \Big\|\sum_{i=1}^n q_i x_i\Big\|$$

for any rational numbers $q_1, \cdots, q_n$ and $x_1, \cdots, x_n \in D$. It follows easily from this that if $\alpha_1, \cdots, \alpha_n$ are any scalars then for each $\omega \in \Omega \backslash N$ one obtains the inequality

$$\Big|\sum_{i=1}^n \alpha_i g_{x_i}(\omega)\Big| \leqq \Big\|\sum_{i=1}^n \alpha_i x_i\Big\|$$

for any $x_1, \cdots, x_n \in D$. It follows that, for each $\omega \in \Omega \backslash N$ there is $g(\omega) \in X^*$ such that $\|g(\omega)\| \leqq 1$ and

$$g(\omega)(y) = g_y(\omega)$$

for any $y \in D$. Letting $g(\omega) = 0$ for $\omega \in N$, we find that $g(\cdot)y \in L_1(|F|)$ for each $y \in D$ and that

$$F(A)(y) = \int_A g(\omega)(y) \, d|F|(\omega)$$

for each $y \in D$ and each $A \in \Sigma$.

Now fix $x \in X$ and pick a sequence (y_n) in D that converges to x. Then one has $\lim_n g(\omega)(y_n) = g(\omega)(x)$ for $\omega \in \Omega\backslash N$. Since for $\omega \in \Omega\backslash N$ one has

$$|g(\omega)(y_n)| \leqq \sup_n \|y_n\| \, \|g(\omega)\|,$$

the Dominated Convergence Theorem ensures $g(\cdot)(x) \in L_1(|F|)$ and

$$\begin{aligned}\int_A g(\omega)(x)\, d|F|(\omega) &= \lim_n \int_A g(\omega)(y_n)\, d|F|(\omega)\\ &= \lim_n F(A)(y_n) = F(A)(x).\end{aligned}$$

Thus $g\colon \Omega \to X^*$ is the weak-star derivative of F with respect to $|F|$. Since X^* is separable, g is separably valued and weak*-measurable. Moreover, X norms X^* so by II.1.4, the function g is $|F|$-measurable. Since g is $|F|$-essentially bounded, it is Bochner integrable. Of course, the indefinite Bochner integral of g must be F since the indefinite Gel'fand integral of g is F. This completes the proof.

With the help of the next result, Theorem 1 can be parlayed into a more general theorem (see Corollary 5).

THEOREM 2. *A Banach space has the Radon-Nikodým property if each of its closed separable linear subspaces has this property.*

If a Banach space has the Radon-Nikodým property, so does each of its closed linear subspaces.

PROOF. The proof of the first statement consists mainly of a re-examination of the proof of Theorem 2.12. In fact, as in the proof of Theorem 2.12, consider $\{T(\chi_E)\colon E \in \Sigma\}$. For a sequence (E_n) in Σ, let Σ_1 be the sub-σ-field of Σ generated by the set $\{E_n\}$. Then $L_1(\Sigma_1, \mu)$ is a separable closed linear subspace of $L_1(\mu)$. Let T_1 be the restriction of T to $L_1(\Sigma_1, \mu)$. Then T_1 is a continuous linear operator from $L_1(\Sigma_1, \mu)$ to $\overline{T_1(L_1(\Sigma_1, \mu))}$, a separable closed linear subspace of X. By hypothesis and Theorem 1.5, the operator T_1 is representable. By Lemma 11, the operator T_1 takes bounded uniformly integrable subsets of $L_1(\Sigma_1, \mu)$ into norm compact sets in X. One such set is $\{\chi_{E_n}\}$; thus, $\{T(\chi_{E_n})\}$ is a relatively norm compact set in X, i.e., $(T\chi_{E_n})$ has a norm convergent subsequence. It follows that $\{T(\chi_E)\colon E \in \Sigma\}$ is relatively norm compact and so T has separable range. Another appeal to Theorem 1.5 finishes the proof of the first assertion.

The proof of the second statement is simple. Suppose X has the Radon-Nikodým property and Y is a closed linear subspace of X. Let $G\colon \Sigma \to Y$ be a μ-continuous vector measure of bounded variation. Since X has the Radon-Nikodým property, there exists $g \in L_1(\mu, X)$ such that $G(E) = \int_E g\, d\mu$ for each $E \in \Sigma$. From Lemma 2.1, it follows that there exists a sequence (π_n) of partitions such that

$$\lim_n E_{\pi_n} g = g \quad \text{a.e.}$$

But for each n, one has

$$E_{\pi_n}(g) = \sum_{A \in \pi_n} \frac{\int_A g\, d\mu}{\mu(A)} \chi_A = \sum_{A \in \pi_n} \frac{G(A)}{\mu(A)} \chi_A.$$

Hence $E_{\pi_n}(g)$ is Y-valued and g is almost everywhere Y-valued. Hence there exists $h \in L_1(\mu, Y)$ such that $G(E) = \int_E h \, d\mu$ for all $E \in \Sigma$; thus Y has the Radon-Nikodým property.

A simple corollary to the proof of Theorem 2 is

COROLLARY 3. *A representable operator on $L_1(\mu)$ has a separable range. All operators on $L_1(\mu)$ to X are representable, if and only if each operator on $L_1(\mu)$ into a separable subspace of X is representable.*

Since every separable closed linear subspace of a reflexive Banach space is a separable dual space, Corollary 2.13 can be recast as a corollary of Theorem 2.

COROLLARY 4 (PHILLIPS). *Reflexive Banach spaces have the Radon-Nikodým property.*

A generalization (though slight) of this line of reasoning produces a quick and simple way of recognizing some Banach spaces with the Radon-Nikodým property.

COROLLARY 5 (UHL). *If every separable closed linear subspace of X is isomorphic to a subspace of a separable dual space, then X has the Radon-Nikodým property.*

A frequently useful consequence of Corollary 5 is

COROLLARY 6 (UHL). *If every separable subspace of X has a separable dual, then X^* possesses the Radon-Nikodým property.*

PROOF. Let M be a norm separable closed linear subspace of X^* and let $\{x_m^*\}$ be a countable dense subset of M. Select sequences $(x_{m,n})_{n=1}^{\infty}$ in X such that $\|x_{m,n}\| = 1$ and

$$|x_m^*(x_{m,n})| \geqq (1 - 1/n)\|x_m^*\|.$$

Let Y be the closed linear span of $\{x_{m,n}\colon m, n = 1, 2, \cdots\}$. Evidently Y is a separable subspace of X and so, by hypothesis, Y^* is separable. Define $T\colon M \to Y^*$ by $(Tx^*)(y) = x^*(y)$ for $x^* \in M$ and $y \in Y$. Clearly $\|T\| \leqq 1$. On the other hand, one has

$$\|Tx_m^*\| \geqq \sup_n |x_m^*(x_{m,n})| = \|x_m^*\|.$$

Thus T is a linear isometry of M into Y^*. Since Y^* is separable, the Dunford-Pettis theorem (Theorem 1) implies Y^* has the Radon-Nikodým property. An appeal to the last corollary shows M has the Radon-Nikodým property. Since M is an arbitrary separable closed linear subspace of X^*, Theorem 2 ensures that X^* has the Radon-Nikodým property.

It must be emphasized that the converse to Corollary 6 is true. The proof of this important theorem will be found in Chapter 7. *It is not known whether the converse of Corollary 5 is true.*

We conclude this section with a few simple applications.

Recall that a Banach space is *weakly compactly generated* whenever it is the closed linear span of one of its weakly compact subsets.

COROLLARY 7 (KUO). *If X is a Banach space whose dual X^* is a subspace of a weakly compactly generated Banach space Y, then X^* has the Radon-Nikodým property.*

PROOF. Let X_0 be a separable subspace of X. Then there is a bounded linear operator T from l_1 onto X_0. Since T is onto X_0, T^* maps X_0^* isomorphically into l_∞. Now note that X_0^* is a quotient space of X^*, i.e., there is a bounded linear operator Q from X^* onto X_0^*. Next, let I be the inclusion embedding of X^* into Y. Here is the situation:

$$\begin{array}{ccccc} X^* & \xrightarrow{Q} & X_0^* & \xrightarrow{T^*} & l_\infty \\ {\scriptstyle I}\downarrow & & & & \\ Y & & & & \end{array}$$

By the Hahn-Banach theorem (applied coordinate by coordinate) there is a bounded linear operator S: $Y \to l_\infty$ such that $T^*Q = SI$. But now since weakly compact subsets of l_∞ are norm separable (because they are weak*-metrizable and the weak and weak*-topologies agree on them), the space $S(Y)$ is norm separable. It follows that $T^*(X_0^*)$ is norm separable. Since T^* is an isomorphism, this means X_0^* is separable. Thus X^* has the Radon-Nikodým property by Corollary 6.

That weakly compactly generated dual spaces do not characterize duals with the Radon-Nikodým property is seen in the next corollary to Theorem 2.

COROLLARY 8. *For any set Γ, $l_1(\Gamma)$ has the Radon-Nikodým property.*

PROOF. If S is a separable closed linear subspace of $l_1(\Gamma)$, then there exists a countable subset Γ_0 of Γ such that $x(\gamma) = 0$ for any $\gamma \notin \Gamma_0$ and any $x \in S$. Thus S is a subspace of $l_1(\Gamma_0)$, a space isometric to l_1. Appeal to Theorem 2.

The next two corollaries are usually proved by methods unrelated to Radon-Nikodým arguments. Of course parts of both corollaries can be made to work for any Banach space that lacks the Radon-Nikodým property.

COROLLARY 9 (GEL'FAND-PEŁCZYŃSKI). *If μ is not purely atomic, the space $L_1(\mu)$ is not a copy of a subspace of a weakly compactly generated dual space.*

Consequently unless μ (μ finite!) is purely atomic, the space $L_1(\mu)$ is not isomorphic to a dual space.

PROOF. Since $L_1(\mu)$ lacks the Radon-Nikodým property, the first statement is obvious. The second statement follows directly from the first since the relatively weakly compact set $\{\chi_E: E \in \Sigma\}$ generates $L_1(\mu)$.

COROLLARY 10 (BESSAGA-PEŁCZYŃSKI). *The space c_0 is not a copy of a subspace of a separable dual space.*

In fact, if Γ is any infinite set $c_0(\Gamma)$ is not a copy of a subspace of a weakly compactly generated dual space.

4. Notes and Remarks. It is in this chapter that the fundamental pre-eminence of the Bochner integral over weaker integrals emerges once and for all. As we have seen in this chapter, the representation of a linear operator on $L_1(\mu)$ by means of a

Bochner integral provides very strong structural information about the operator under consideration. This is not the case for certain other integral representation theories for operators on $L_1(\mu)$. Let us look at the situation.

It is easy to prove that if (Ω, Σ, μ) is a finite measure space and X is a Banach space then the space of all vector measures $G: \Sigma \to X$ such that

$$\sup_{E \in \Sigma} \|G(E)\|/\mu(E) = \|G\|_\infty < \infty$$

is isometric to $\mathscr{L}(L_1(\mu); X)$ under the correspondence $T(f) = \int_\Omega f\,dG, f \in L_1(\mu)$. Unfortunately this representation theory stands as nothing more than a mere formality. Alone it gives no information about operators on $L_1(\mu)$ and it gives no information about vector measures.

There is a representation theorem for operators on $L_1(\mu)$ that is much deeper than the above representation and of comparable depth to the Bochner integral representation theory outlined in this chapter. This is the representation theory based on liftings (see Dinculeanu **[1967]**, Dinculeanu and Uhl **[1973]** and Ionescu Tulcea **[1969]**). With the help of the lifting theorem, it is possible to prove that if (Ω, Σ, μ) is a finite measure space and $T: L_1(\mu) \to X^*$ is a bounded linear operator, then there exists a function $g: \Omega \to X^*$ that is weak*-measurable and such that for each $x \in X$ and $f \in L_1(\mu)$, one has $(Tf)(x) = \int_\Omega f(\omega)g(\omega)(x)\,d\mu(\omega)$. On the surface, this is a vast generalization of Theorem 3.1 which it includes. Unfortunately, this generalization is mostly an esthetic generalization because the measurability properties of the kernel g are not, in general, strong enough to exhibit structural properties of the operator under representation. Thus this generalization has not yet proved to be a useful tool for the study of the operators it represents. The problem is that the generalization represents so many operators that it fails to provide a good representation for well-behaved operators. For instance, if X is not separable and X^* is weakly compactly generated, Corollary 3.7 guarantees that T is Riesz representable. This is well nigh impossible to deduce from the very general representation above. The trouble seems to center around the fact that the lifting theorem has not proved useful in the study of the Radon-Nikodým property. For a hint of the difficulties encountered in this regard, see Dinculeanu and Uhl **[1973]** and A. Ionescu Tulcea **[1974]**.

There is very little in this chapter that is not implicit in the fundamental papers of Dunford-Pettis **[1940]** and Phillips **[1940]**. For some reason, these papers were largely forgotten (even by workers in the theory of vector measures) until the late sixties when they again appeared out of the mist still in good working order. Aside from Grothendieck and a few others, no one paid attention to these seemingly special results. Now that it has been realized that there is still plenty of mystery in Banach spaces themselves, the importance of these papers is well accepted.

The Riesz Representation Theorem and the Radon-Nikodým theorem. Nowadays the relationship between these theorems is considered by many to be nothing more than the formality of translating a set of definitions from one context to another. Although this view is probably the correct view today, matters were not always so simple. It was none other than Dunford **[1936a]** who realized this connection and used it to great advantage. In this paper Dunford represented the general linear operator from $L_1[0, 1]$ to $L_p[0, 1]$ $(1 < p < \infty)$ and proved the beautiful Theorem

1.6. Earlier in 1936, Dunford and Morse **[1936]** introduced the notion of a boundedly complete basis to prove that if X has a boundedly complete basis, then an "additive function $F(R)$ which is defined for elementary figures R contained in a fixed figure R_0 in Euclidean space of n dimensions and has its values in the Banach space X, is of bounded variation, then it has a derivative $F'(P)$ for almost all points P in R_0. $F'(P)$ is summable [in the sense of Bochner] and if $F(R)$ is absolutely continuous, then for every elementary figure R in R_0 we have $F(R) = \int_R F'(P)\,dP$."

Today we recognize this as a Radon-Nikodým theorem for arbitrary finite measure spaces (see IV.3). In 1936, this was not so obvious. But Dunford **[1936a]** realized this and in so doing proved Theorem 1.6. Along the same lines, Clarkson **[1936]** introduced the notion of uniform convexity and proved that if "$F(R)$ (is) an additive function of elementary figures, defined for the figures within a fixed figure R_0 (in Euclidean n-space) and (assumes) values in a uniformly convex Banach space B (and) F is of bounded variation in R_0, then F is differentiable almost everywhere in R_0." Dunford **[1936a]** improved this by showing that uniformly convex spaces have the Radon-Nikodým property.

It is a pleasure to say that the concepts of boundedly complete basis and uniform convexity owe their origins to Radon-Nikodým considerations. It is sad to say that the Dunford-Morse **[1936]** and Dunford **[1936a]** papers remain the only substantial contribution to the problem of relating properties of bases to the Radon-Nikodým property. In this connection, an unsolved problem of Pełczyński arises: If X has the Radon-Nikodým property, does X have a subspace with a boundedly complete basis? This question is of considerable theoretic importance as we shall see later in this section.

Lemma 1.4 is due to Dunford **[1936a]**. Theorem 1.8 was formulated implicitly by D.R. Lewis and Stegall **[1973]** and explicitly by Rosenthal **[1975]**.

Compact and weakly compact operators on $L_1(\mu)$. We have already seen that important Radon-Nikodým theorems often sprang from theorems dealing with differentiability of additive functions on "figures" in Euclidean n-space. This is the case for the Radon-Nikodým theorems studied in §2. Here the seminal paper is Pettis **[1939a]**. Pettis's purpose was to find tests for "the strong differentiability of an individual function having its values in an unrestricted (and perhaps *unsatisfactory*) (our italics) space," and then using these tests to determine "whether or not a given condition on (a Banach space) X is $\cdots$ strong enough to insure the differentiability a.e. of every additive function of bounded variation (on figures in Euclidean n-space with range in X) $\cdots$. In each proof the essential idea is to show that if X satisfies the particular condition under consideration, then X is weakly compact in one generalized sense or another." The immediate predecessors of the fundamental Lemma 2.9, Corollary 2.13 and Theorem 3.1 are easily spotted in Pettis **[1939a]**. Furthermore, the idea of proving a Radon-Nikodým theorem for an individual vector measure without regard to the range space finds its predecessor in Pettis **[1939a]**.

Pettis's **[1939a]** theorems were quickly translated into genuine Radon-Nikodým theorems the following year by Dunford and Pettis **[1940]**. Their work and the work of Phillips **[1940]** form the core of §2.

Theorem 2.2 traces its life back to Dunford and Pettis **[1940]** and has antecedents in the works of Dunford **[1936]**, **[1938]**, Gel'fand **[1938]** and Pettis **[1939]**. In the text we have treated this as a prototype for what follows. The truth is that the very important Lemma 2.9 and Theorem 3.1 can be deduced cleanly from Theorem 2.2. This fact is essentially in Dunford and Schwartz **[1958]** and can be found explicitly in Rieffel **[1968]**. To prove Lemma 2.9 from Theorem 2.2 suppose X is separable and $K \subseteq X$ is a symmetric convex weakly compact set. Suppose $T: L_1(\mu) \to X$ maps the unit ball of $L_1(\mu)$ into K. Let (x_n^*) be a sequence in X^* with $\|x_n^*\| = 1$ for all n and such that $\|x\| = \sup_n |x_n^*(x)|$. Define $|||\cdot|||$ on X by

$$|||x||| = \sum_{n=1}^{\infty} 2^{-n} |x_n^*(x)|.$$

Since K is weakly compact it is compact in the $|||\cdot|||$-topology of X. Hence the $|||\cdot|||$-topology agrees with the weak topology of X on the set K. Since K is compact in the $|||\cdot|||$-topology, Theorem 2.2 produces a function $g: \Omega \to K$ that is measurable for the $|||\cdot|||$-topology on X such that $T(f) = \int_\Omega fg \, d\mu$ for all $f \in L_1(\mu)$. But now the Pettis Measurability Theorem guarantees that g is also measurable with respect to the original norm on X. Theorem 3.1 can be proved in a similar fashion.

The technique of exhaustion is folklore, but the form that it appears in Lemma 2.4 can be found in Maynard **[1970]**. Theorem 2.6 appears first in this form in Rieffel **[1968]**. Rieffel's proof is considerably more complicated than our operator-theoretic proof. His proof does have the advantage of requiring no operator theory and thus establishes Theorem 2.6 as a purely measure-theoretic phenomenon. Our proof is adapted from Moedomo and Uhl **[1971]**.

Theorem 2.7 is due to Rieffel **[1968]** whose proof is a bit too complicated. This is an easy consequence of Egoroff's theorem and can be proved without operator theory by use of Egoroff's theorem and Corollary II.2.8. The proof in the text is from Moedomo and Uhl **[1971]**. It must be remarked here that all of the theorems of this chapter were either derived thirty-five years ago or *could have been derived* thirty-five years ago. We find it curious that Theorem 2.7 was such a late bloomer. Example 2.8 is from D. R. Lewis **[1972a]**.

The fundamental Lemma 2.9 is right from the classic Dunford and Pettis **[1940]** though our proof is a somewhat dandified version of theirs. This theorem began to surface in Pettis [**1939a**, Theorem 3.1] and the proof we give is very close to Pettis's proof. Lemma 2.11 and its companion, Corollary 2.17, are probably the most famous results in Dunford and Pettis **[1940]**. The remarks following the proof of Lemma 2.11 were communicated to us by A. Pełczyński. Theorem 2.12 is one of the cornerstones of the theory of vector measures; its complete proof was first given by Phillips **[1940a]**. Our proof is from Moedomo and Uhl **[1971]** and has the flavor of the proof of Phillips [**1940a**, Theorem 5.5]. For some reason Corollary 2.13 is often attributed to Phillips **[1943]**; it appears first in Phillips **[1940a]**. Theorem 2.15 is due to Dunford **[1939]** and has been an important tool in the study of $L_1(\mu)$ since its discovery. Theorem 2.18 is a slight generalization of a theorem of Metivier **[1967]** which, in turn, is very close to a theorem of Phillips **[1943]**; again our proof follows Moedomo and Uhl **[1971]**.

Separable dual spaces. The fundamental Theorem 3.1 is due to Dunford and Pettis **[1940]**. Its very close ancestors can be found in Pettis **[1939a]** and Gel'fand **[1938]**. The important Lemma 2.9 can be deduced elegantly from Theorem 3.1 as follows. If X is separable and $T: L_1(\mu) \to X$ is weakly compact the factorization theorem of Davis, Figiel, Johnson and Pełczyński [**1974**] (VIII.4.8) produces a separable reflexive Banach space R and operators $S_1: L_1(\mu) \to R$, $S_2: R \to X$ such that $T = S_2S_1$. Since R is a separable dual space, S_1 is representable by Theorem 3.1. Hence $T = S_2S_1$ is also representable by II.2.6.

Theorem 3.2 is often attributed to Uhl [**1972a**] though like many other results related to vector measures it can be gleaned from Grothendieck [**1955a**]. Corollaries 3.5 and 3.6 are due to Uhl [**1972a**]. Their proofs are easy but most spaces with the Radon-Nikodým property have this property as an easy consequence of either Corollary 3.5 or 3.6.

Motivated by this, Uhl [**1972a**] asked whether a separable Banach space X with the Radon-Nikodým property is isomorphic to a subspace of separable dual space. Stegall [**1975**] showed (see VII.2.6) that if X is a dual space, the answer is yes. Since there are Banach spaces with the Radon-Nikodým property that are not dual spaces (see Lindenstrauss [**1964c**] for an example of a subspace of l_1 that is not a dual space), the question remains unsolved.

There is a relationship between this question and the Dunford Theorem 1.6. According to Davis, Figiel, Johnson and Pełczyński [**1974**], a separable dual space is a copy of a subspace of a Banach space with a boundedly complete basis. Also Johnson and Rosenthal [**1972**] have shown that every infinite dimensional subspace of a separable dual space has an infinite dimensional subspace with boundedly complete basis. This motivates the question of Pełczyński mentioned earlier: Does an infinite dimensional Banach space with the Radon-Nikodým property have an infinite dimensional subspace with a boundedly complete basis? A negative answer to Pełczyński's question is a negative answer to Uhl's question and a positive answer to Uhl's question provides a similar response to Pełczyński's.

Two related questions: If each subspace of X with a Schauder basis has the Radon-Nikodým property, need X have the Radon-Nikodým property, and what Banach spaces with a Schauder basis have the Radon-Nikodým property, i.e., characterize the class of Schauder bases that span spaces with the Radon-Nikodým property.

Corollary 3.7 is due to Kuo [**1974**]; the elegant proof given in the text is due to Peter Morris who asks whether a second dual space with the Radon-Nikodým property is weakly compactly generated. W. B. Johnson and C. Stegall observed earlier that weakly compactly generated duals have the Radon-Nikodým property. Here is a quick proof of this fact (which of course follows from Corollary 3.7): First note that if X^* is weakly compactly generated then each quotient of X^* is also weakly compactly generated. Thus if Y is a separable subspace of X, then Y^* is also weakly compactly generated. If $K \subseteq Y^*$ is a weakly compact set, then the weak* and weak topologies agree on K. Since Y is separable, it follows that K is a compact metric space in the weak* topology. Therefore K is weak* separable and hence weakly separable. This fact combined with the Hahn-Banach theorem shows that K is norm separable. It follows that Y^* is separable for each separable subspace Y of X. An appeal of Corollary 3.6 completes the proof.

Corollary 3.9 is due (in the separable case) to Gel'fand [**1938**]. It is the source of a number of interesting papers: Hagler [**1973**], Hagler and Stegall [**1973**], Pełczyński [**1961**], [**1968a**] and Stegall [**1973**]. Corollary 3.10 is due in spirit to Orlicz [**1929**] and was implicitly in Gel'fand [**1938**]. Bessaga and Pełczyński [**1958**] have derived a more incisive theorem: *If X contains a copy of c_0, then l_1 is complemented in X. Accordingly, if X^* contains c_0, then X^* contains a weak* closed copy of l_∞ that has a weak* closed complement.* Further, Rosenthal **[1970]** showed that if $c_0(\Gamma)$ is in a dual space then so is $l_\infty(\Gamma)$.

Weakly measurable functions that are equivalent to measurable functions. Example II.1.5 exhibits a function f: $[0, 1] \to l_2[0,1]$ that is not measurable but such that $x^*f = 0$ almost everywhere for all $x^* \in l_2[0, 1]^*$. The following unpublished theorem of D. R. Lewis shows that this is a very special case of a general phenomenon.

THEOREM (D. R. LEWIS). *Let (Ω, Σ, μ) be a finite measure space and X be a weakly compactly generated Banach space. If f: $\Omega \to X$ is a bounded weakly measurable function then there exists a bounded measurable $g: \Omega \to X$ such that, for each $x^* \in X^*$, $x^*f = x^*g$ μ-almost everywhere (the exceptional set may depend on x^*).*

Consequently f is Pettis integrable.

PROOF. The basis for the proof is the following fact about weakly compactly generated spaces proved by Amir and Lindenstrauss **[1968]**. If X_0 is a separable subspace of X and Y_0 is a separable subspace of X^*, then there is a projection $P: X \to X$ whose range is separable such that $X_0 \subseteq P(X)$ and $Y_0 \subseteq P^*(X^*)$.

Let $f: \Omega \to X$ be a bounded weakly measurable function. We shall show first that $\{x^*f: x^* \in X^*, \|x^*\| \leqq 1\}$ is relatively compact in $L_1(\mu)$. If not, then there exists a sequence (x_n^*) in the unit ball of X^* and a $\delta > 0$ such that

$$\int_\Omega |x_n^* f - x_m^* f| \, d\mu \geqq \delta$$

for $m \neq n$. Select a projection P with a separable range such that $P^*(x_n^*) = x_n^*$ for all n. Then we have

$$(*) \qquad \int_\Omega |x_n^* P(f) - x_m^* P(f)| \, d\mu > \delta$$

for $m \neq n$. But since P has a separable range, the function $P(f)$ is measurable by the Pettis Measurability Theorem II.1.2. According to Theorem II.3.8, the operator $S: L_\infty(\mu) \to X$ defined by $S(g) = \int_\Omega P(f) g \, d\mu$ for $g \in L_\infty(\mu)$ is compact. The operator $x^* \to x^*P(f)$ is the adjoint of S and is therefore compact. This contradicts $(*)$ and proves that the set

$$\{x^*f : x^* \in X^*, \|x^*\| \leqq 1\}$$

is relatively compact in $L_1(\mu)$.

Next select a sequence (x_n^*) in the unit ball of X^* such that the sequence $(x_n^* f)$ is $L_1(\mu)$-dense in the set

$$\{x^*f: x^* \in X^*, \|x^*\| \leqq 1\}.$$

Choose a projection P with a separable range such that $P^*(x_n^*) = x_n^*$ for all n. As above, the function $P(f)$ is measurable. By Egoroff's theorem, there is, for each $\delta > 0$, a set $A \in \Sigma$ with $\mu(\Omega \backslash A) < \delta$ such that $P(f)\chi_A$ can be uniformly approximated by simple functions. Fix $\delta > 0$ and choose such a set A. It follows quickly that the sequence $(x_n^* f \chi_A)$ $(= (x_n^* P(f)\, \chi_A))$ is relatively compact in $L_\infty(\mu)$. Since this sequence is $L_\infty(\mu)$-dense in the set

$$\{x^* f \chi_A : x^* \in X^*,\ \|x^*\| \leqq 1\},$$

and this sequence is $L_\infty(\mu)$-relatively compact, it follows that the set $\{x^* f \chi_A : x^* \in X^*,\ \|x^*\| \leqq 1\}$ is relatively compact in $L_\infty(\mu)$.

Now define $T: X^* \to L_\infty(\mu)$ by $T(x^*) = x^* f \chi_A$ for $x^* \in X^*$. It is clear that T is a compact operator and that, as an operator on $L_1(\mu)$, T^*: $L_1(\mu) \to X^{**}$ is a weak*-sequentially continuous compact operator. Since X is weakly compactly generated, this means that $T^*(L_1(\mu)) \subseteq X$ (see Diestel [**1975**, p. 148]). Since T^*: $L_1(\mu) \to X$ is compact, Theorem 2.2 produces a Bochner integrable function g: $A \to X$ such that $T^*(h) = \int_A hg\, d\mu$ for all $h \in L_1(\mu)$. But if $x^* \in X^*$, then $T^{**}(x^*) = x^* g$. It follows immediately that if $x^* \in X^*$ then $x^* f = x^* g$ almost everywhere on A. Since $\mu(\Omega \backslash A) < \delta$ this completes the proof.

This proof was adapted from Stegall [**1976**].

For some time it was thought that the conclusion of the above theorem might hold in any space containing no copy of l_∞. This turned out to be false; Lindenstrauss and Stegall [**1975**] have constructed a weakly measurable function with values in the James Tree space (JT) (for more on JT, see the notes and remarks section of Chapter VII) that is not equivalent to any measurable function. Thus the search continues for a characterization of those Banach spaces for which the conclusion of the above theorem holds.

One final remark: If X is a dual space with the Radon-Nikodým property (in particular if X is a weakly compactly generated dual space), then the theorem collapses into a triviality. To see why, let X be a dual space with the Radon-Nikodým property, (Ω, Σ, μ) be a finite measure space and f: $\Omega \to X$ be a bounded weakly measurable function. Then f is Dunford integrable. If P: X^{**} $(= Y^{***}) \to X$ $(= Y^*)$ is the restriction projection, define F: $\Sigma \to X$ by

$$F(E) = P\Big(\text{Dunford-}\int_E f\, d\mu\Big), \qquad E \in \Sigma.$$

It is easily seen that $\|F(E)\| \leqq K\mu(E)$ for some constant $K > 0$ and all $E \in \Sigma$. Since X has the Radon Nikodým property, there is a measurable function $g: \Omega \to X$ such that

$$F(E) = \text{Bochner-}\int_E g\, d\mu \quad \text{for all } E \in \Sigma.$$

This is the desired function.

Similarly, it is possible to use the Gel'fand integral instead of the Dunford integral to prove that if Y is a Banach space whose dual Y^* has the Radon-Nikodým property, then, for every bounded weak*-measurable function f with values in Y^*, there is a bounded measurable function g with values in Y^* such that $yf = yg$ almost everywhere for each y in Y.

Weakly compactly generated spaces. According to I.5.3, the range of a strongly additive vector measure is relatively weakly compact. Thus, in a sense, much of the theory of Banach-space-valued measures finds itself concerned with weakly compactly generated Banach spaces. These spaces have been under intensive study for the past decade. Now a coherent theory of these spaces, based on fundamental work of Lindenstrauss and others, is emerging. For our purposes, Lindenstrauss **[1972]**. Day **[1973]** and Diestel **[1975]** are sufficient references.

Kuo's Theorem 3.7 is a genuine strengthening of the Johnson-Stegall observation mentioned earlier. Indeed Rosenthal **[1974b]** has exhibited a nonweakly compactly generated dual space X that is a subspace of a certain $L_1(\mu)$ space for a suitable finite measure μ. By Theorem 3.7, X has the Radon-Nikodým property and by Rosenthal's example not all subspaces of weakly compactly generated spaces are weakly compactly generated. Of course the weakly compactly generated space which contains X is $L_1(\mu)$ and $L_1(\mu)^*$ fails the Radon-Nikodým property. In some way the Radon-Nikodým property for X^* may be related to whether all the subspaces of a weakly compactly generated space X are also weakly compactly generated.

For example, Johnson and Lindenstrauss **[1974]** have proven that if both X and X^* are weakly compactly generated then each subspace of X is weakly compactly generated. Further, Friedland **[1976]** and John and Zizler **[1974a]** have shown that if X is weakly compactly generated and admits a Fréchet differentiable norm, then every subspace of X is weakly compactly generated. On the other hand, Restrepo **[1964]** has proven that if Y is separable and has a Fréchet differentiable norm then Y^* is separable; thus, if X has a Fréchet differentiable norm X^* has the Radon-Nikodým property. Therefore both of these facts support the possibility that, if X is a weakly compactly generated Banach space and X^* has the Radon-Nikodým property, then every subspace of X is weakly compactly generated.

Operators from $L_1[0, 1]$ to $L_1[0, 1]$. Hilbert called an operator from one Banach space to another "completely continuous" if it maps weakly convergent sequences into norm convergent sequences. According to Lemma 2.11 and Theorem 2.15 every representable operator on $L_1(\mu)$ is completely continuous. There was some feeling that this may characterize the representable operators from $L_1(\mu)$ to itself. Costé **[1973]** showed that this intuition was incredibly naive by constructing a convolution type operator on an $L_1(\mu)$ space that is completely continuous and not representable. The first part of this section is devoted to convolution operators on $L_1(\mu)$ and Costé's example. After this we shall look at a convolution operator defined by Rosenthal **[1976]** that has some special properties. Finally, a summary of some important theorems of P. Enflo and T. Starbird on the action of operators from $L_1[0, 1]$ to $L_1[0, 1]$ will be given.

1. *Convolution operators.* Let G be a compact Abelian group with Haar measure μ. Let λ be a regular Borel measure on G and for each $y \in G$ define $(f * \lambda)(y)$ by $\int_G f(y - x)\, d\lambda(x)$. It is easily seen that the mapping $f \to f * \lambda$ defines a bounded linear operator from $L_1(\mu)$ to $L_1(\mu)$. We shall call this operator convolution with respect to λ.

THEOREM (COSTÉ **[1973]**). *Let μ be Haar measure on a compact Abelian group G. Convolution with respect to a regular Borel measure λ on G is a representable operator from $L_1(\mu)$ to $L_1(\mu)$ if and only if $\lambda \ll \mu$.*

PROOF. If $\lambda \ll \mu$ then there exists $g \in L_1(\mu)$ such that

$$(Tf)(y) = \int_G f(y-x)\,d\lambda(x) = \int_G f(y-x)g(x)\,d\mu(x) = \int_G f(x)g(y-x)\,d\mu(x)$$

for μ-almost all $y \in G$. Define $\phi \in L_\infty(\mu, L_1(\mu))$ by $\phi(x)(\cdot) = g(\cdot - x)$, $x \in G$. Then ϕ is Bochner integrable and $Tf = \int f\phi\,d\mu$ for all $f \in L_1(\mu)$. Thus T is representable.

For the converse, select $g \in L_\infty(\mu, L_1(\mu))$ such that $T(f) = \int_G f(x)g(x)d\mu(x)$ for all $f \in L_1(\mu)$. It is possible to consider g as a function of two variables and write for all $f \in L_1(\mu)$

$$Tf(y) = \int_G f(x)g(x,y)\,d\mu(x)$$

for μ-almost all $y \in G$. Now define $\tilde{g}$ on $G \times G$ by $\tilde{g}(x,y) = g(y-x, y)$ and define $h \in L_1(\mu)$ by $h(x) = \int_G \tilde{g}(x,y)\,d\mu(y)$. If we can show that $h(x) = \tilde{g}(x,y)$ $(\mu \times \mu)$-almost everywhere, then we will have proven that

$$h(y-x) = \tilde{g}(y-x, y) = g(x,y)$$

$(\mu \times \mu)$-almost everywhere. This will show that

$$\begin{aligned}(Tf)(y) &= \int_G f(x)h(y-x)\,d\mu(x) = \int_G f(y-x)h(x)\,d\mu(x)\\ &= \int_G f(y-x)\,d\lambda(x)\end{aligned}$$

for μ-almost all y. This will be enough to show that h is the Radon-Nikodým derivative of λ with respect to μ and will complete the proof.

Now let us see why $h(x) = \tilde{g}(x,y)$ $(\mu \times \mu)$-almost everywhere. For each $f \in L_1(\mu)$ and $a \in G$, let $f_a(x) = f(x-a)$. Also for each $a \in G$, let $g_a(x,y) = g(x+a, y+a)$. Then $g_a \in L_\infty(\mu, L_1(\mu))$. Let $T_a(f) = \int_G fg_a\,d\mu$ for $f \in L_1(\mu)$. Since T is a convolution operator, we see that $T(f_a) = (T(f))_a$ for all $a \in G$. On the other hand, one has

$$(T_a f)(y) = \int_G f(x)g(x+a, y+a)\,d\mu(x) = \int_G f(x-a)g(x, y+a)\,d\mu(x)$$

for μ-almost all $y \in G$. This means that we have $T_a(f) = T(f_a)_{-a} = ((T(f))_a)_{-a} = T(f)$ for all $f \in L_1(\mu)$. It follows quickly that $g_a = g$ $(\mu \times \mu)$-almost everywhere.

Next, as above write $\tilde{g}(x,y) = g(y-x, y)$ and $\tilde{g}_a(x,y) = g_a(y-x, y)$. Evidently $\tilde{g}_a = \tilde{g}$ $(\mu \times \mu)$-almost everywhere and equally evident is the identity $\tilde{g}_a(x,y) = \tilde{g}(x, y+a)$. Accordingly, $\tilde{g}(x, y+a) = \tilde{g}(x,y)$ $(\mu \times \mu)$-almost everywhere.

Now let h be defined as above and ϕ and χ belong to $L_\infty(\mu)$. Define $\Phi(y) = \int_G \tilde{g}(x,y)\phi(x)\,d\mu(x)$. Since $\tilde{g}(x, y+a) = \tilde{g}(x,y)$ $(\mu \times \mu)$-almost everywhere, the Fubini theorem helps us deduce that $\Phi_a = \Phi$ μ-almost everywhere. This means that Φ is almost everywhere equal to a constant. To see this, let $\theta \in L_\infty(\mu)$. For μ-almost all $x \in G$, one has

$$\int_G \Phi(y)\theta_x(y)\,d\mu(y) = \int_G \Phi_{-x}(y)\theta(y)\,d\mu(y) = \int_G \Phi(y)\theta(y)\,d\mu(y)$$

since $\Phi_a = \Phi$ μ-almost everywhere. Integrate both sides with respect to $d\mu(x)$, remembering that $\mu(G) = 1$, to obtain

$$\int_G \Phi(y)\theta(y)\,d\mu(y) = \int_G\int_G \Phi(y)\theta_x(y)\,d\mu(y)\,d\mu(x) = \int_G \Phi(y)\,d\mu(y)\int_G \theta(x)\,d\mu(x).$$

Hence Φ is constant μ-almost everywhere; thus $\Phi(y) = \int_G \Phi(x)\,d\mu(x)$ for μ-almost all $y \in G$.

Finally, use Fubini's theorem to verify the equalities

$$\begin{aligned}\int_G\int_G \bar{g}(x, y)\, \phi(x)\chi(y)\, d\mu(x)\, d\mu(y) &= \int_G \chi(y)\, \Phi(y)\, d\mu(y)\\ &= \int_G \chi(y)\, d\mu(y) \int_G \Phi(y)\, d\mu(y)\\ &= \int_G \chi(y)\, d\mu(y) \cdot \int_G\int_G \bar{g}(x, y)\phi(x)\, d\mu(x)\, d\mu(y)\\ &= \int_G \chi(y)\, d\mu(y) \int_G \phi(x)h(x)\, d\mu(x)\\ &= \int_G\int_G h(x)\phi(x)\, \chi(y)\, d\mu(x)\, d\mu(y).\end{aligned}$$

Since ϕ and χ are arbitrary in $L_\infty(\mu)$, this means $\bar{g}(x, y) = h(x)$ $(\mu \times \mu)$-almost everywhere; this completes the proof.

2. *Convolution operators on the Cantor group and completely continuous nonrepresentable operators.* Now we are in a good position to generate many examples of nonrepresentable completely continuous operators from $L_1[0, 1]$ to itself. Let G be the Cantor group $\prod_{n=1}^\infty \{-1, 1\}$. If (α_n) is a sequence with $0 \leqq \alpha_n \leqq 1$ define $\lambda_n(\{1\}) = \alpha_n$, $\lambda_n(\{-1\}) = 1 - \alpha_n$ and let $\lambda = \prod_{n=1}^\infty \lambda_n$ be the infinite product of the λ_n's on G. If $\alpha_n = \frac{1}{2}$ for all n, the resulting product measure is Haar measure which we denote by μ.

The main fact about product measures on G is a special case of a crisp theorem of Kakutani [**1948**].

THEOREM (KAKUTANI). *If there exists β with $0 < \beta < \frac{1}{2}$ such that $\beta < \alpha_n < 1 - \beta$ for all n then $\lambda = \prod_{n=1}^\infty \lambda_n$ and μ are equivalent or λ and μ are mutually singular according as $\sum_{n=1}^\infty (2\alpha_n - 1)^2$ is convergent or divergent.*

Now for a more comfortable environment, let us transfer to $L_1[0, 1]$. Note that the dyadic subintervals of $[0, 1]$ correspond naturally to "dyadic" compact subsets of G; e.g., $[0, \frac{1}{2}] \leftrightarrow \{-1\} \times \prod_{n=2}^\infty\{-1, 1\}$, $[\frac{1}{2}, 1] \leftrightarrow \{+1\} \times \prod_{n=2}^\infty\{-1, 1\}$, $[\frac{1}{2}, \frac{3}{4}] \leftrightarrow \{+1\} \times \{-1\} \times \prod_{n=3}^\infty \{-1, 1\}$, etc. This correspondence defines a linear isometry $C\colon L_1(\mu) \to L_1[0, 1]$ that also establishes an isometry between $L_2(\mu)$ and $L_2[0, 1]$. Now if $\lambda = \prod_{n=1}^\infty \lambda_n$ as above and $S_\lambda\colon L_1(\mu) \to L_1(\mu)$ is defined by $S_\lambda(f) = f * \lambda$, $f \in L_1(\mu)$, then $T_\lambda = C^{-1}S_\lambda C\colon L_1[0, 1] \to L_1[0, 1]$ is representable if and only if S_λ is representable. Let us examine the action of T_λ. A moment's reflection shows that if r is the nth Rademacher function (the Rademacher functions correspond to coordinate functionals), then $T_\lambda(r_n) = (2\alpha_n - 1)r_n$ and that

$$T_\lambda\Big(\prod_{n\in F} r_n\Big) = \prod_{n\in F} (2\alpha_n - 1) \prod_{n\in F} r_n$$

for every finite set F of positive integers. Now recall that the system of Walsh functions, i.e., all finite products of Rademacher functions, is a complete orthonormal sequence in $L_2[0, 1]$. Therefore, as an operator from $L_2[0, 1]$ to $L_2[0, 1]$, T_λ is a diagonal operator that is compact (from $L_2[0, 1]$ to $L_2[0, 1]$) if and only if $\lim_n(2\alpha_n - 1) = 0$.

THEOREM. *If* $\lim_n 2\alpha_n = 1$ *but* $\sum_{n=1}^{\infty}(2\alpha_n - 1)^2 = \infty$ *then* T_λ *is a completely continuous nonrepresentable operator on* $L_1[0, 1]$.

PROOF. From Kakutani's theorem, we find that λ is not μ-continuous on G. Thus S_λ is not representable by Costé's theorem. Therefore T_λ is not representable.

On the other hand, T_λ: $L_2[0, 1] \to L_2[0, 1]$ is a compact operator. It follows that T_λ maps bounded subsets of L_∞ into compact subsets of L_1. Now if $K \subseteq L_1[0, 1]$ is weakly compact, then K is bounded and uniformly integrable by III.2.15. Thus if $E_{f,n} = \{t: |f(t)| \geqq n\}$, then $\lim_n \|f\chi_{E_{f,n}}\|_1 = 0$ uniformly in $f \in K$. Note that, for each n, $\{T_\lambda(f\chi_{\Omega \setminus E_{f,n}}): f \in K\}$ is relatively compact in $L_1[0, 1]$. It follows easily that $T_\lambda(K)$ is totally bounded. Hence T_λ is completely continuous.

As those familiar with harmonic analysis have probably realized by now the phenomenon above is subject to considerable generalization. Indeed as Costé **[1973]** details, Salem **[1942]** has shown that if G is any compact group with Haar measure μ then there is a regular positive Borel measure λ on G that is μ-singular and whose Fourier transform vanishes at infinity. (For more on this see Varopoulis **[1966]**.) Define T: $L_1(\mu) \to L_1(\mu)$ by $Tf = f * \lambda$. Then T is not representable but since the Fourier transform of λ tends to zero at infinity, the Plancherel theorem guarantees that T: $L_2(\mu) \to L_2(\mu)$ is a compact operator. Hence as above T: $L_1(\mu) \to L_1(\mu)$ is completely continuous.

Returning to the discussion preceding the above paragraph, we note the following

THEOREM. *Any one of the following statements about* T_λ *implies all the others.*

(1) T_λ *maps* $L_1[0, 1]$ *into* $L_2[0, 1]$ *in a continuous way.*

(2) T_λ: $L_1[0, 1] \to L_1[0, 1]$ *is representable.*

(3) T_λ: $L_2[0, 1] \to L_2[0, 1]$ *is a Hilbert-Schmidt operator.*

PROOF. Since $L_2[0, 1]$ has the Radon-Nikodým property (1) obviously implies (2). If T_λ: $L_1[0, 1] \to L_1[0, 1]$ is representable, then S_λ is representable and $\lambda \ll \mu$ by Costé's theorem. Hence $\sum_{n=1}^{\infty}(2\alpha_n - 1)^2 < \infty$ by Kakutani's theorem. Now since $\sum_{n=1}^{\infty}(2\alpha_n - 1)^2$ is finite, the infinite product $\prod_{n=1}^{\infty}(1 + (2\alpha_n - 1)^2)$ is convergent. It follows that T_λ: $L_2[0, 1] \to L_2[0, 1]$ is a diagonal operator whose eigenvalues are square summable. Thus T_λ: $L_2[0, 1] \to L_2[0, 1]$ is a Hilbert-Schmidt operator and (2) implies (3).

To see that (3) implies (1) let (w_n) be the sequence of Walsh functions and let $T_\lambda(w_n) = \lambda_n w_n$ (the λ_n's are finite products of the numbers $(2\alpha_n - 1)$). Since T_λ is a Hilbert-Schmidt operator, we have $\sum_n \lambda_n^2 < \infty$. Now if $f \in L_2[0, 1]$, then

$$T_\lambda(f) = \sum_{n=1}^{\infty}\left(\int_0^1 f(t)w_n(t)\,dt \cdot \lambda_n\right) w_n.$$

Hence

$$\|T_\lambda(f)\|_2^2 = \sum_{n=1}^{\infty} \lambda_n^2 \left(\int_0^1 f(t)w_n(t)\,dt\right)^2 \leqq \sum_{n=1}^{\infty} \lambda_n^2 \|f\|_1^2$$

by the Hölder inequality and the fact that $\|w_n\|_\infty = 1$ for each n. Since $L_2[0, 1]$ is dense in $L_1[0, 1]$, this proves that (3) implies (1).

Three questions arise:

(a) The implication (3) implies (1) in the above theorem is easily generalized from the context of Walsh functions to complete orthonormal $L_\infty[0, 1]$-bounded sequences in $L_2[0, 1]$. For diagonal operators on $L_2[0, 1]$, how far can the implication (2) implies (3) be generalized?

(b) Is there a Banach space X without the Radon-Nikodým property such that $T: L_1[0, 1] \to X$ is representable if and only if it is completely continuous? This question is unresolved. Hagler **[1976]** has exhibited a Banach space X such that each operator $T: L_1[0, 1] \to X^*$ is completely continuous yet X^* does not have the Radon-Nikodým property. For more on Hagler's example we refer the reader to the notes and remarks of Chapter VII.

(c) Does a noncompletely continuous operator from $L_1[0, 1]$ to $L_1[0, 1]$ act as a topological isomorphism on a subspace of $L_1[0, 1]$ that is a copy of $L_1[0, 1]$? Adding fuel to the hopes that this question might have a positive answer, Rosenthal **[1976]** proved that a noncompletely continuous operator from $L_1[0, 1]$ to $L_1[0, 1]$ preserves a copy of Hilbert space in $L_1[0, 1]$. But then Rosenthal **[1976]** turned around and dashed the hopes of the optimists by defining a convolution operator on $L_1(\mu)$ where μ is Haar measure on the Cantor group that preserves no copy of $L_1(\mu)$. His example is built by setting $\lambda_n(\{1\}) = 2/3$ and $\lambda_n(\{-1\}) = 1/3$ and then building the product measure $\lambda = \Pi_{n=1}^{\infty} \lambda_n$ on the Cantor group as above. Convolution with respect to this measure is not a completely continuous operator since in the notation above $S_\lambda(r_n) = r_n/3$ for every Rademacher function r_n. Rosenthal **[1976]** goes on to show that S_λ preserves no copy of $L_1[0, 1]$; this is not easy. Aside from some of his own machinery, he makes crucial use of the following basic theorem of P. Enflo and T. Starbird.

3. *Enflo operators.* Following the lead of Starbird **[1976]**, let us call an operator $T: L_1[0, 1] \to L_1[0, 1]$ an *Enflo operator* if there is a subspace Y of $L_1[0, 1]$ that is a copy of $L_1[0, 1]$ and such that T acts as an isomorphism on Y.

THEOREM (ENFLO AND STARBIRD). (a) *Let μ be Lebesgue measure on* $[0, 1]$ *and let $T: L_1[0, 1] \to L_1[0, 1]$ be an Enflo operator. Then there is a measurable set E with $\mu(E) > 0$ and a sub-σ-algebra $\mathfrak{A}$ of the measurable subsets of E such that μ is nonatomic on $\mathfrak{A}$ and T acts as an isomorphism on $L_1(\mu|_{\mathfrak{A}})$.*

Consequently T acts as an isomorphism on an isometric copy of $L_1[0, 1]$ *that is complemented in* $L_1[0, 1]$.

(b) *Conversely, suppose $T: L_1[0, 1] \to L_1[0, 1]$ is a bounded linear operator. Then T is an Enflo operator if there exists a $\delta > 0$ and a sequence of partitions Π_n of* $[0, 1]$ *such that*

(i) *Π_{n+1} refines Π_n,*

(ii) $\lim_n \max_{E \in \Pi_n} \mu(E) = 0$, *and*

(iii) $\int_0^1 \max_{E \in \Pi_n} |T(\chi_E)|\, d\mu \geqq \delta.$

The hypotheses of (b) arise naturally in the following two natural and important situations. In each the sequence (Π_n) is a subsequence of the sequence of simplest dyadic partitions of $[0, 1]$.

COROLLARY 1 (ENFLO). *If $T_1, T_2: L_1[0, 1] \to L_1[0, 1]$ are bounded linear operators* *whose sum is the identity then the hypotheses of* (b) *hold with* $\delta = \frac{1}{2}$ *for either* T_1 *or* T_2.

Consequently $L_1[0, 1]$ is primary in the sense that if $L_1[0, 1]$ is the direct sum of two of its subspaces then one of these subspaces is $L_1[0, 1]$ by Pełczyński's **[1960]** decomposition method.

COROLLARY 2. *Every copy of* $L_1[0, 1]$ *in* $L_1[0, 1]$ *contains a complemented copy of* $L_1[0,1]$.

The proof uses Dor **[1975b]** to show that the hypotheses of (b) hold. Plug into (a) to get the complemented subspace.

The Radon-Nikodým property in Banach lattices. The Radon-Nikodým property assumes a particularly orderly form in dual Banach lattices. Working mainly with Corollary 3.7, Lotz **[1976]** has established the following result.

THEOREM. *Any of the following statements about a Banach lattice X implies the others.*

(a) X^* *has the Radon-Nikodým property.*

(b) X *contains no copy of* l_1.

(c) *No closed sublattice of* X^* *is lattice isomorphic to* c_0 *or* $L_1[0, 1]$.

Consequently, it can be seen from Lotz **[1974]** that if X is a Banach lattice, then X^{**} has the Radon-Nikodým property if and only if both X and X^* have the Radon-Nikodým property, in which case X is reflexive.

In addition Lotz has shown, among other things, that *if X is a Banach lattice whose dual has a weak order unit and Y is a subspace of X then* Y^* *has the Radon-Nikodým property if and only if* Y^* *is weakly compactly generated.* Lotz also asks whether a Banach lattice with the Radon-Nikodým property is a dual space.

The Radon-Nikodým property in spaces of operators. Slightly modifying the proof of Theorem 3.1, Diestel and Morrison **[1976]** have shown that *if* X^* *and Y are separable and both have the Radon-Nikodým property and every operator from X to Y is compact, then* $\mathscr{L}(X; Y)$ *has the Radon-Nikodým property.* A consequence of this is the fact that the space of unconditionally convergent series in a space with the Radon-Nikodým property also has the Radon-Nikodým property. The separability hypothesis can be dropped at the cost of increasing the complexity of the proof. The question as to whether the condition that each operator be compact is necessary is open.

Diestel and Faires **[1974]** and Diestel and Uhl **[1976]** have considered the Radon-Nikodým property for the space of nuclear operators; their observations motivate the question: If X^* and Y have the Radon-Nikodým property, then need the space of nuclear operators from X to Y have the Radon-Nikodým property? The answer is affirmative when Y is a dual space with the approximation property.

The Radon-Nikodým theorem for finitely additive vector measures. Let μ be a nonnegative finitely additive real-valued measure on a field $\mathscr{F}$ of subsets of Ω. If $F: \mathscr{F} \to X$ is a finitely additive vector measure of bounded variation and $\lim_{\mu(E)\to 0} F(E) = 0$, then, even in the case that X is one dimensional, there may be no μ-integrable function f such that $F(E) = \int_E f\, d\mu$ for all $E \in \mathscr{F}$. (The integral here is that found in Dunford and Schwartz **[1958]**.) Recognizing this, Bochner **[1939]** has shown that if X is the scalars, then for each $\varepsilon > 0$ there is a simple function f_ε such that the indefinite integral of f_ε with respect to μ approximates F in variation within ε, i.e., F has approximate Radon-Nikodým derivatives. (See also Bochner and Phillips **[1941]**, Leader **[1953]**, and Fefferman **[1967]**, **[1968]**.) Uhl **[1967]** showed that if X has the Radon-Nikodým property the same thing is true. For arbitrary X, Uhl **[1969b]** characterized those finitely additive vector measures that have approximate Radon-Nikodým derivatives.

Not satisfied with mere approximate Radon-Nikodým derivatives, Maynard **[1970]** went on to characterize those finitely additive vector measures that have *exact* Radon-Nikodým derivatives. This is a stunning achievement.

The Radon-Nikodým property in Fréchet spaces. D. R. Lewis **[1970]** and Thomas **[1974]** showed that in a complete metrizable locally convex space (Fréchet space) E the Radon-Nikodým theorem holds for any E-valued measure on a σ-field (without necessity of any hypothesis on finite variation) if and only if E is a nuclear space. In case the hypothesis of metrizability is dropped, Chi **[1973]** has followed the direction of Lewis and Thomas to extend their results. See also the notes and remarks of Chapter VII. See also Saab **[1976]**.

Differentiation of a vector measure with respect to an operator-valued measure. Let X and Y be Banach spaces and Σ be a σ-field of subsets of Ω. Suppose $G\colon \Sigma \to \mathscr{L}(X; Y)$ is a vector measure that is countably additive for the strong operator topology. Maynard **[1972]** has characterized those vector measures $F\colon \Sigma \to Y$ such that there exists a function $f\colon \Omega \to X$ such that $F(E) = \int_E f(\omega)\, dG(\omega)$ for each $E \in \Sigma$. The integral here is that of Dobrakov **[1970a]**, **[1970b]**. Maynard's theorem is a cousin of Theorem 2.2 which it includes.

Differentiation of one vector measure with respect to another vector measure. Aside from Rao **[1967]** and Bogdanowicz and Kritt **[1967]** little has been done in this area. Here is a small contribution. Suppose X and Y are Banach spaces and Σ is a σ-field of subsets of Ω. Let $F\colon \Sigma \to X$ and $G\colon \Sigma \to Y$ be vector measures, suppose G is of bounded variation and Y has the Radon-Nikodým property. If $G(E) = 0$ whenever $\|F\|(E) = 0$ then there is a measurable operator-valued function $g\colon \Omega \to \mathscr{L}(X; Y)$ such that $G(E) = \int_E g\, dF$ for all $E \in \Sigma$. To prove this select, by IX.2.2, a vector $x^* \in X^*$ such that $F \ll |x^*F|$; then $G \ll |x^*F|$. Since Y has the Radon-Nikodým property, there is a measurable function $\phi\colon \Omega \to Y$ such that $G(E) = \int_E \phi\, d|x^*F|$ for all $E \in \Sigma$. Select $E_1 \in \Sigma$ such that $|x^*F|(E) = x^*F(E \cap E_1)$. Let $\theta = x^*(\chi_{E_1} - \chi_{\Omega \setminus E_1})$. Then one has $G(E) = \int_E \phi\theta\, dF$ for all $E \in \Sigma$.

This does not go far enough since it is impossible to derive from it the identity $G(E) = \int_E I\, dG$ where I is the identity operator and G is a vector measure.

IV. APPLICATIONS OF ANALYTIC RADON-NIKODÝM THEOREMS

This chapter is a smorgasbord of applications of some of the Radon-Nikodým theorems established in the last chapter. The first section deals with the isolation of the dual of $L_p(\mu, X)$ for $1 \leqq p < \infty$. We shall learn that the statement $L_p(\mu, X)^* = L_q(\mu, X^*)$ $(p^{-1}+q^{-1} = 1)$ is in reality a statement about the Radon-Nikodým property for X^*. The second section deals with the problem of finding the weakly compact subsets of $L_1(\mu, X)$. The third section continues with a brief discussion about the relationship between absolutely continuous vector-valued functions of a real variable and the Radon-Nikodým theorem.

In the fourth section, we shall look at operators on $L_p(\mu)$ that are defined by means of Pettis and Bochner integrals. The last section is devoted to the Lewis-Stegall theorem which says that a complemented subspace of $L_1[0, 1]$ that has the Radon-Nikodým property is necessarily a copy of l_1. For the most part, these sections are independent of each other.

1. The dual of $L_p(\mu, X)$. Let be a Banach space and (Ω, Σ, μ) be a finite measure space. As we noted in II.2, if $1 \leqq p < \infty$, then $L_p(\mu, X)$ stands for the space of all (equivalence classes of) X-valued Bochner integrable functions f defined on Ω with $\int_\Omega \|f\|^p d\mu < \infty$. The norm $\|\cdot\|_p$ is defined by

$$\|f\|_p = \left(\int_\Omega \|f\|^p \, d\mu\right)^{1/p}, \qquad f \in L_p(\mu, X).$$

Routine computations show that $L_p(\mu, X)$ is a Banach space under $\|\cdot\|_p$. In addition, simple functions are dense in $L_p(\mu, X)$ for $1 \leqq p < \infty$. For $p = \infty$, the symbol $L_\infty(\mu, X)$ stands for the space of all (equivalence classes of) X-valued Bochner integrable functions defined on Ω that are essentially bounded, i. e., such that

$$\|f\|_\infty = \text{ess sup}\{\|f(\omega)\| : \omega \in \Omega\} < \infty.$$

This space is also a Banach space under the norm $\|\cdot\|_\infty$ and the countably valued functions in $L_\infty(\mu, X)$ are dense in $L_\infty(\mu, X)$.

For $1 \leqq p < \infty$, it is not difficult to recognize $L_q(\mu, X^*)$ $(1/q + 1/q = 1)$ isometrically as a subspace of $L_p(\mu, X)^*$. Let us take a look at this.

Let $g \in L_q(\mu, X^*)$ and let (g_n) be a sequence of simple functions in $L_q(\mu, X^*)$

converging to g a.e. Suppose $f \in L_p(\mu, X)$ and define $\langle f, g\rangle$ on Ω by $\langle f, g\rangle(\omega) = g(\omega)(f(\omega))$ for $\omega \in \Omega$. Certainly $\langle f, g_n\rangle$ is measurable for each n, and it is only slightly less evident that $\lim_n \langle f, g_n\rangle = \langle f, g\rangle$ a.e. Consequently, $\langle f, g\rangle$ is measurable. Moreover one has

$$\int_\Omega |\langle f, g\rangle| \, d\mu \leqq \int_\Omega \|f\| \, \|g\| \, d\mu \leqq \|f\|_p \, \|g\|_q$$

by the Hölder inequality. Therefore the functional $l(\cdot) = \int (\cdot) g \, d\mu$ is a member of $L_p(\mu, X)^*$ whose norm is not greater than $\|g\|_q$. To prove the reverse inequality ($\|l\| \geqq \|g\|_q$), we shall "boot-strap" from the scalar case. Let $\varepsilon > 0$ and suppose first that $g = \sum_{i=1}^\infty x_i^* \chi_{E_i}$, where (x_i^*) is a sequence in X^* and (E_i) is a countable partition of Ω by members of Σ with $\mu(E_i) > 0$ for all i. Choose $h \geqq 0$ in $L_p(\mu)$ such that $0 < \|h\|_p \leqq 1$ and such that

$$\|g\|_q - \varepsilon/2 < \int_\Omega \|g\| h \, d\mu.$$

Next choose $x_i \in X$ with $\|x_i\| = 1$ such that

$$\|x_i^*\| - \varepsilon/2\|h\|_1 < x_i^*(x_i)$$

and define $f \in L_p(\mu, X)$ by $f = \sum_{i=1}^\infty x_i h \chi_{E_i}$. Then we have $\|f\|_p = \|h\|_p \leqq 1$, and we have

$$\begin{aligned} \int \langle f, g\rangle \, d\mu &= \int_\Omega h \sum_{i=1}^\infty x_i^*(x_i) \chi_{E_i} \, d\mu \\ &\geqq \int_\Omega h \sum_{i=1}^\infty \left(\|x_i^*\| - \frac{\varepsilon}{2\|h\|_1}\right) \chi_{E_i} \, d\mu \\ &\geqq \int_\Omega h \, \|g\| \, d\mu - \frac{\varepsilon}{2} \frac{\int_\Omega h \, d\mu}{\|h\|_1} \\ &\geqq \|g\|_q - \varepsilon/2 - \varepsilon/2. \end{aligned}$$

Hence $\|l\| = \|g\|_q$ whenever $g \in L_q(\mu, X^*)$ is countably valued. For the general case, let $g \in L_q(\mu, X^*)$ and choose a sequence (g_n) of countably valued functions in $L_q(\mu, X^*)$ such that $\lim_n \|g_n - g\|_q = 0$. If $l_n(\cdot) = \int_\Omega \langle \cdot, g_n\rangle \, d\mu$ and $l(\cdot) = \int_\Omega \langle \cdot, g\rangle \, d\mu$, then we already know that $\|l_n\| = \|g_n\|_q$ and that $0 \leqq \|l_n - l\| \leqq \|g_n - g\|_q \to 0$. Hence we have $\|l\| = \lim_n \|l_n\| = \lim_n \|g_n\|_q = \|g\|_q$. Recapitulating, we have just seen that the mapping that takes g to $\int_\Omega \langle \cdot, g\rangle \, d\mu$ for $g \in L_q(\mu, X^*)$ is an isometric isomorphism carrying $L_q(\mu, X^*)$ onto a subspace of $L_p(\mu, X)^*$. It is quite easy to give examples of situations when $L_q(\mu, X^*)$ is a *proper* subspace of $L_p(\mu, X)^*$. We say this is easy because of the following theorem.

THEOREM 1. *Let (Ω, Σ, μ) be a finite measure space, $1 \leqq p < \infty$, and X be a Banach space. Then $L_p(\mu, X)^* = L_q(\mu, X^*)$ where $p^{-1} + q^{-1} = 1$, if and only if X^* has the Radon-Nikodým property with respect to μ.*

PROOF. So far we have proven that $L_q(\mu, X^*)$ is always contained isometrically as a subset of $L_p(\mu, X)^*$ for $1 \leqq p < \infty$. Now assume X^* has the Radon-Nikodým property. For $l \in L_p(\mu, X)^*$ define $G: \Sigma \to X^*$ by

$$G(E)(x) = l(x\chi_E) \quad \text{for } E \in \Sigma.$$

Since $\|l(x\chi_E)\| \leqq \|l\|\|x\chi_E\|_p = \|l\|\|x\|\|\chi_E\|_p$, it follows that G has its values in X^* and is countably additive. To see that $|G|(\Omega) < \infty$, let $\{E_1,\cdots,E_n\}$ be a partition and $x_1,\cdots, x_n$ be in the closed unit ball of X. Then one has

$$\begin{aligned}\left|\sum_{i=1}^n G(E_i)(x_i)\right| &= \left|l\Big(\sum_{i=1}^n x_i\chi_{E_i}\Big)\right| \\ &\leqq \|l\|\left\|\sum_{i=1}^n x_i\chi_{E_i}\right\|_p \\ &\leqq \|l\|\left\|\sum_{i=1}^n \chi_{E_i}\right\|_p \quad \text{since } \|x_i\| \leqq 1 \\ &= \|l\|\mu(\Omega)^{1/p}.\end{aligned}$$

Taking appropriate suprema shows that G is of bounded variation. Since X^* has the Radon-Nikodým property, there exists a Bochner integrable $g\colon \Omega \to X^*$ such that $G(E) = \int_E g\, d\mu$ for all $E \in \Sigma$. Plainly, if $f \in L_p(\mu, X)$ is a simple function, then $l(f) = \int_\Omega \langle f, g\rangle\, d\mu$.

Select an expanding sequence (E_n) in Σ such that $\bigcup_n E_n = \Omega$ and such that g is bounded on each E_n. Fix n_0 and note that $\int_{E_{n_0}} \langle \cdot, g\rangle\, d\mu$ is a bounded linear functional on $L_p(\mu, X)$ which agrees with l on all simple functions supported on E_{n_0}. It follows that

$$l(f\chi_{E_{n_0}}) = \int \langle f, g\chi_{E_{n_0}}\rangle\, d\mu$$

for all $f \in L_p(\mu, X)$. Further, since $g\chi_{E_{n_0}}$ is bounded, one has $g\chi_{E_{n_0}} \in L_q(\mu, X^*)$ and $\|g\chi_{E_{n_0}}\|_q \leqq \|l\|$. Since this last inequality obtains for each n_0, the Monotone Convergence Theorem guarantees that $g \in L_q(\mu, X^*)$. Armed with this knowledge and the Hölder inequality, one easily sees that

$$l(f) = \lim_n \int_\Omega \langle f\chi_{E_n}, g\rangle\, d\mu = \int_\Omega \langle f, g\rangle\, d\mu, \quad \text{for all } f \in L_p(\mu, X).$$

Hence $L_p(\mu, X)^*$ coincides with $L_q(\mu, X^*)$. This proves the sufficiency.

For the necessity, suppose $L_p(\mu, X)^* = L_q(\mu, X)^*$ and let $G\colon \Sigma \to X^*$ be a μ-continuous vector measure of bounded variation. We shall show that if $E_0 \in \Sigma$ has positive μ-measure, then G has a Bochner integrable Radon-Nikodým derivative on a set $B \in \Sigma$, $B \subseteq E_0$ with $\mu(B) > 0$. An appeal to the Exhaustion Lemma III.2.5 will then complete the proof. Thus let $E_0 \in \Sigma$ have positive μ-measure. Applying the Hahn Decomposition Theorem to the scalar measures $|G|$ and $k\mu$ for a large enough positive integer k produces a subset B of E_0, $B \in \Sigma$, $\mu(B) > 0$ such that $|G|(E) \leqq k\mu(E)$ for all $E \in \Sigma$ with $E \subseteq B$. Define for a simple function $f = \sum_{i=1}^n x_i\chi_{E_i}$, where $x_i \in X$, $E_i \in \Sigma$, and $E_i \cap E_j = \emptyset$ for $i \neq j$,

$$l(f) = \sum_{i=1}^n G(E_i \cap B)(x_i).$$

Then

$$\begin{aligned}|l(f)| &= \left|\sum_{i=1}^n G(E_i \cap B)(x_i)\right| = \left|\sum_{i=1}^n \frac{G(E_i \cap B)}{\mu(E_i \cap B)}(\mu(E_i \cap B)x_i)\right| \\ &\leqq \sum_{i=1}^n k\|\mu(E_i \cap B)x_i\| \leqq k\|f\|_1 \leqq k\mu(\Omega)^{1/q}\|f\|_p.\end{aligned}$$

Since l is evidently linear on the simple functions in $L_p(\mu, X)$, this shows that l is continuous on the simple functions in $L_p(\mu, X)$ and therefore has a bounded linear extension to all of $L_p(\mu, X)$. By hypothesis, there is $g \in L_q(\mu, X^*)$ such that

$$l(f) = \int_\Omega \langle f, g\rangle \, d\mu, \quad \text{for all } f \in L_p(\mu, X).$$

But one has $G(E \cap B)(x) = l(x\chi_E) = \int_E \langle x, g\rangle \, d\mu$ for all $x \in X$ and $E \in \Sigma$. Since each $g \in L_q(\mu, X^*)$ is Bochner integrable, we see that

$$G(E \cap B)(x) = \left(\int_E g \, d\mu\right)(x), \quad \text{for all } x \in X \text{ and } E \in \Sigma.$$

Consequently, $G(E \cap B) = \int_E g \, d\mu$, for all $E \in \Sigma$. This completes the proof.

Here are some elementary, yet basic, corollaries.

COROLLARY 2 (PHILLIPS). *Let X be a Banach space and (Ω, Σ, μ) be a finite measure space. For $1 \leqq p \leqq \infty$, the space $L_p(\mu, X)$ is reflexive if and only if both $L_p(\mu)$ and X are reflexive.*

PROOF. The sufficiency follows directly from Theorem 1 and the fact that reflexive spaces have the Radon-Nikodým property. For the necessity, note that the set $\{x\chi_\Omega : x \in X\}$ is a closed linear subspace of $L_p(\mu, X)$ which is isometric to X; while if $x_0 \in X$ has norm one, then $\{fx_0 : f \in L_p(\mu)\}$ is a closed linear subspace of $L_p(\mu, X)$ isometric to $L_p(\mu)$.

COROLLARY 3. *Let (Ω, Σ, μ) be a finite measure space and let $1 < p < \infty$. If every separable subspace of X has a separable dual, then $L_p(\mu, X^*)$ has the Radon-Nikodým property.*

PROOF. We shall see that if $q^{-1} + p^{-1} = 1$, then every separable subspace of $L_q(\mu, X)$ has a separable dual. Once this is shown an appeal to Corollary III.3.6 will prove that $L_q(\mu, X)^*$ has the Radon-Nikodým property. But then by Corollary III.3.6, X^* also has the Radon-Nikodým property. Hence $L_q(\mu, X)^* = L_p(\mu, X^*)$ also has the Radon-Nikodým property.

To see that every separable subspace of $L_q(\mu, X)$ has a separable dual, let M be a separable subspace of $L_q(\mu, X)$. Then there exists a countably generated σ-field $B \subseteq \Sigma$ and a separable subspace Y of X such that $M \subseteq L_q(\mu|_B, Y)$. Since Y^* is separable, Y^* has the Radon-Nikodým property. By Theorem 1, $L_q(\mu|_B, Y)^* = L_p(\mu|_B, Y^*)$. Since Y^* is separable and B is countably generated $L_p(\mu|_B, Y^*)$ is separable. Hence M^*, as a quotient of the separable space $L_p(\mu|_B, Y^*)$, is separable. Hence every separable subspace of $L_q(\mu, X)$ has a separable dual, as advertised.

The last corollary establishes nothing new. It is a very special case of III.2.13.

COROLLARY 4 (VON NEUMANN). *Hilbert spaces have the Radon-Nikodým property.*

PROOF. Let H be a Hilbert space and (Ω, Σ, μ) be a finite measure space. Then $L_2(\mu, H)$ is obviously a Hilbert space and self-dual; thus $L_2(\mu, H)^* = L_2(\mu, H^*)$. An appeal to Theorem 1 shows that $H^* = H$ has the Radon-Nikodým property.

As trivial as this approach may seem, this method of proving that a Hilbert space has the Radon-Nikodým property is nothing but a rather pedantic version of von Neumann's original proof of this fact.

2. Weakly compact subsets of $L_1(\mu, X)$. A classical theorem of Dunford (III.2.15) isolates the relatively weakly compact subsets of $L_1(\mu)$ as the bounded uniformly integrable subsets. If X is reflexive, the original proof for the $L_1(\mu)$ case extends with only notational changes to show that if X is reflexive, the relatively weakly compact subsets of $L_1(\mu, X)$ are precisely the bounded uniformly integrable sets. The facts to be presented here about weakly compact subsets of $L_1(\mu, X)$ will be proved as little more than technical extensions of Dunford's original proof.

Recall that a subset K of $L_1(\mu, X)$ is uniformly integrable if

$$\lim_{\mu(E)\to 0} \int_E \|f\| \, d\mu = 0$$

uniformly in $f \in K$.

THEOREM 1 (DUNFORD). *Let (Ω, Σ, μ) be a finite measure space and X be a Banach space such that both X and X^* have the Radon-Nikodým property. A subset K of $L_1(\mu, X)$ is relatively weakly compact if*

(i) *K is bounded,*

(ii) *K is uniformly integrable, and*

(iii) *for each $E \in \Sigma$, the set $\{\int_E f \, d\mu : f \in K\}$ is relatively weakly compact.*

PROOF. Let $K \subseteq L_1(\mu, X)$ satisfy (i), (ii) and (iii) and let (f_n) be a sequence in K. Let $\mathscr{F} \subseteq \Sigma$ be a countable field such that each f_n is measurable with respect to $\sigma(\mathscr{F}) = \Sigma_1 = \sigma$-field generated by $\mathscr{F}$. Let (E_k) be an enumeration of $\mathscr{F}$. An easy Cantor diagonalization applied to condition (iii) produces a subsequence (g_m) of (f_n) such that

$$\text{weak limit}_m \int_{E_k} g_m \, d\mu$$

exists for each $k = 1, 2, \cdots$. In other words, weak $\text{limit}_m \int_E g_m \, d\mu$ exists for each $E \in \mathscr{F}$. An appeal to (ii) reveals that the sequence $(\int_E g_m \, d\mu)$ is weakly Cauchy for all $E \in \sigma(\mathscr{F}) = \Sigma_1$. Invoking (iii) allows a vector measure $G\colon \Sigma_1 \to X$ to be determined by

$$G(E) = \text{weak limit}_m \int_E g_m \, d\mu, \qquad E \in \Sigma_1.$$

Since (ii) guarantees for $x^* \in X^*$ that

$$\lim_{\mu(E)\to 0} \int_E x^* g_m \, d\mu = 0$$

uniformly in m, it follows that $\lim_{\mu(E)\to 0} x^*G(E) = 0$ for each $x^* \in X^*$. Thus $G\colon \Sigma_1 \to X$ is weakly countably additive and is therefore norm countably additive by the Orlicz-Pettis theorem (Corollary I.4.4). Since G obviously vanishes on sets in Σ_1 of μ-measure zero, G is μ-continuous on Σ_1.

Next, it will be established that G is of bounded variation. Since $\lim_n x_n = x$ weakly in X implies $\|x\| \leqq \liminf_n \|x_n\|$, it follows that

$$\|G(E)\| \leqq \liminf_m \left\| \int_E g_m \, d\mu \right\|$$

for all $E \in \Sigma_1$. Thus if $\pi \subseteq \Sigma_1$ is a partition, then one has

$$\begin{aligned}\sum_{E\in\pi}\|G(E)\| &\leqq \sum_{E\in\pi}\liminf_m\left\|\int_E g_m\,d\mu\right\|\\ &\leqq \liminf_m \sum_{E\in\pi}\left\|\int_E g_m\,d\mu\right\|\\ &\leqq \sup_m \sum_{E\in\pi}\int_E \|g_m\|\,d\mu\\ &= \sup_m \|g_m\|_1 \leqq \sup_{f\in K}\|f\|_1 < \infty\end{aligned}$$

by (i). Hence $|G|(\Omega) < \infty$. Since X has the Radon-Nikodým property, there is $g \in L_1(\Sigma_1, \mu, X)$ such that $G(E) = \int_E g\,d\mu$ for all $E \in \Sigma_1$.

To complete the proof, it will be shown that $\lim_m g_m = g$ weakly in $L_1(\Sigma_1, \mu, X)$ and therefore weakly in $L_1(\mu, X)$. An appeal to Eberlein's theorem will finish the story.

First, note that the definition of g implies $\lim_m \int_E g_m\,d\mu = \int_E g\,d\mu$ weakly for each $E \in \Sigma_1$. Consequently, we have

$$\lim_m \int_\Omega \langle g_m, f\rangle\,d\mu = \int_\Omega \langle g, f\rangle\,d\mu$$

for all simple functions $f \in L_\infty(\Sigma_1, \mu, X^*)$. If $f \in L_\infty(\Sigma_1, \mu, X^*)$ is arbitrary and $\varepsilon > 0$ is given, then choose, with the aid of Egoroff's theorem and (ii), an $A_0 \in \Sigma_1$ such that $f\chi_{A_0}$ has a precompact range and such that for all m one has

$$\int_{\Omega\setminus A_0}\|g\|\,d\mu,\quad \int_{\Omega\setminus A_0}\|g_m\|\,d\mu \leqq \varepsilon/(\|f\|_\infty + 1).$$

Since the range of $f\chi_{A_0}$ is relatively compact, there is a simple function $h \in L_\infty(\Sigma_1, \mu, X^*)$ such that $\|f\chi_{A_0} - h\|_\infty < \varepsilon/p$ where p is so large that $p > \int_\Omega \|g\|\,d\mu$ and $p > \sup_{g\in K}\|g\|_1$. Then we have

$$\begin{aligned}&\left|\int_\Omega \langle g_m, f\rangle\,d\mu - \int_\Omega \langle g, f\rangle\,d\mu\right|\\ &\leqq \left|\int_{A_0}\langle g_m, f\rangle\,d\mu - \int_{A_0}\langle g, f\rangle\,d\mu\right| + 2\varepsilon\\ &\leqq \left|\int_{A_0}\langle g_m, f - h\rangle\,d\mu\right| + \left|\int_{A_0}\langle g_m - g, h\rangle\,d\mu\right|\\ &\quad + \left|\int_{A_0}\langle g, h - f\rangle\,d\mu\right| + 2\varepsilon\\ &\leqq \varepsilon + \left|\int_{A_0}\langle g_m - g, h\rangle\,d\mu\right| + \varepsilon + 2\varepsilon\\ &= 4\varepsilon + \left|\int_{A_0}\langle g_m - g, h\rangle\,d\mu\right|.\end{aligned}$$

Since h is simple, we obtain $\lim_m \int_{A_0}\langle g_m - g, h\rangle\,d\mu = 0$. Thus the above estimates imply that

$$\lim_m \int_\Omega \langle g_m, f\rangle\,d\mu = \int_\Omega \langle g, f\rangle\,d\mu$$

for each $f \in L_\infty(\Sigma_1, \mu, X^*)$. Since X^* has the Radon-Nikodým property, Theorem 1.1 guarantees that $\lim_m g_m = g$ weakly in $L_1(\Sigma_1, \mu, X)$. This completes the proof.

Next, some examples and theorems will be collected to indicate that the hypotheses of the above theorem are sensitive to radical change.

EXAMPLE 2 (BATT). *The condition that X^* has the Radon-Nikodým property cannot be removed from the statement of Theorem* 1. To see this, let $\Omega = [0, 1]$ and μ be Lebesgue measure. Let (r_n) be the sequence of Rademacher functions, i.e., if $0 \leqq t \leqq 1$,

$$r_n(t) = \operatorname{sgn}(\sin(2^n \pi t))$$

where $\operatorname{sgn} t = t/|t|$ for $t \neq 0$ and $\operatorname{sgn} 0 = 1$. Define a sequence (f_n) in $L_1(\mu, l_1)$ as follows: Let the nth entry in $f_n(t)$ be $r_n(t)$ and let all others be zero. Then $\|f_n(t)\|_{l_1} = 1$, $\|f_n\|_{L_1(\mu,l_1)} = 1$ and $\lim_n \|\int_E f_n \, d\mu\|_{l_1} = \lim_n |\int_E r_n \, d\mu| = 0$ for all Borel sets E. Hence $K = \{f_n\}$ satisfies (i), (ii) and (iii) of the statement of Theorem 1. Moreover, if $g \in L_\infty(\mu, l_\infty)$ is simple, then $\lim_n \int_\Omega \langle f_n, g\rangle \, d\mu = 0$ since $\lim_n \int_\Omega r_n \chi_E \, d\mu = 0$ for all Borel sets E. Since the simple functions in $L_\infty(\mu, l^\infty)$ separate points in $L_1(\mu, l_1)$, it follows that a weakly convergent subsequence of (f_n) must have weak limit 0. Now for $f = (\phi_1, \phi_2, \ldots) \in L_1(\mu, l_1)$ define

$$l(f) = \int \sum_{n=1}^{\infty} \phi_n r_n \, d\mu.$$

Then l is linear and

$$|l(f)| \leqq \int_\Omega \sum_{n=1}^{\infty} |\phi_n r_n| \, d\mu = \int_\Omega \sum_{n=1}^{\infty} |\phi_n| \, d\mu = \|f\|_{L_1(\mu,l_1)}.$$

Thus $l \in L_1(\mu, l_1)^*$. But $l(f_n) = \int_\Omega r_n^2 \, d\mu = 1$ for all n. Hence no subsequence of (f_n) can converge to zero weakly, and (f_n) has no weakly convergent subsequence. In view of Eberlein's theorem, $\{f_n\}$ is not relatively weakly compact.

When X lacks the Radon-Nikodým property the roof caves in on Theorem 1.

THEOREM 3. *Let (Ω, Σ, μ) be a finite measure space. If X lacks the Radon-Nikodým property with respect to μ, then there is a bounded uniformly integrable set $K \subseteq L_1(\mu, X)$ such that $\{\int_E f \, d\mu : E \in \Sigma, f \in K\}$ is relatively weakly compact in X but such that K is not relatively weakly compact in $L_1(\mu, X)$.*

PROOF. Since X lacks the Radon-Nikodým property with respect to μ, there is a μ-continuous vector measure $G: \Sigma \to X$ of bounded variation that does not have a Bochner integrable Radon-Nikodým derivative with respect to μ. Moreover, we may assume by the discussion of III.1 in Chapter III that $\|G(E)\| \leqq \mu(E)$ for all $E \in \Sigma$. Now let K be the set $\{g_\pi = \sum_{E \in \pi} G(E)\chi_E/\mu(E) : \pi$ is a Σ-partition of $\Omega\}$ (the usual convention $0/0 = 0$ is in effect). Then K is easily seen to be a bounded uniformly integrable subset of $L_1(\mu, X)$. Moreover, for any partition π and any $F \in \Sigma$,

$$\int_F g_\pi \, d\mu = \sum_{E \in \pi} G(E) \frac{\mu(E \cap F)}{\mu(E)}.$$

The right-hand sum is a sum of the form $\sum_{i=1}^n \alpha_i \, G(A_i)$ where $0 \leqq \alpha_1 \leqq \alpha_2 \leqq \cdots \leqq \alpha_n \leqq 1$, $A_i \in \Sigma$ and $A_i \cap A_j = \emptyset$ for $i \neq j$. By the summation by parts technique used in the proof of II.3.8, one sees that this sum is in the convex hull of $G(\Sigma)$. Since

G is μ-continuous, we see that $G(\Sigma)$ is relatively weakly compact by I.2.7. Thus, by the Kreĭn-Smulian theorem, the set co $G(\Sigma)$ is also relatively weakly compact. Hence $\{\int_E g_\pi \, d\mu : E \in \Sigma; \pi$ a partition$\}$ lies in the weakly compact set $\overline{\text{co}}\, G(\Sigma)$.

Therefore the set $\{\int_E f \, d\mu : f \in K\}$ is relatively weakly compact for each $E \in \Sigma$ and K obeys (iii) of the statement of Theorem 1. To verify that K is not relatively weakly compact, note that if the partitions are partially ordered by refinement $(g_\pi : \pi$ is a partition$)$ is a net. It is simply checked that any weakly convergent subnet of (g_π) must converge to a Radon-Nikodým derivative of G with respect to μ; this contradicts the choice of G.

Even in the absence of the Radon-Nikodým property, the converse to Theorem 1 holds. To see this, note that a relatively weakly compact set is always bounded and that since $f \to \int_E f \, d\mu$ is a bounded linear operator from $L_1(\mu, X)$ to X, it follows that if $K \subseteq L_1(\mu, X)$ is relatively weakly compact, then so is $\{\int_E f \, d\mu : f \in K\}$ for each $E \in \Sigma$. To verify the necessity of the uniform integrability hypothesis, we shall prove

THEOREM 4. *Let K be a bounded subset of $L_1(\mu, X)$. If K is not uniformly integrable, then there is a sequence (f_n) in K and, $\alpha, \beta > 0$ such that*

$$\alpha \sum_{n=1}^{\infty} |\gamma_n| \leqq \left\| \sum_{n=1}^{\infty} \gamma_n f_n \right\|_1 \leqq \beta \sum_{n=1}^{\infty} |\gamma_n|$$

for all $(\gamma_n) \in l_1$.

Consequently, a conditionally weakly compact subset of $L_1(\mu, X)$ is uniformly integrable.

PROOF. Suppose $\lim_{\mu(E) \to 0} \int_E \|f\| \, d\mu = 0$ is not uniform in $f \in K$. According to Theorem I.2.4, the measures $\int_{(\cdot)} \|f\| \, d\mu$ are not uniformly countably additive on Σ. Consequently there is a sequence (f_n) in K, a disjoint sequence (E_n) in Σ and a $\delta > 0$ such that $\int_{E_n} \|f_n\| \, d\mu > \delta$ for all n. By Rosenthal's lemma (Lemma I.3.1) we can and do arrange matters so that

$$\int_{E_n} \|f_n\| \, d\mu > \delta, \text{ and } \int_{\bigcup_{j \neq n} E_j} \|f_n\| d\mu < \delta/2.$$

Select $\beta > 0$ such that $\|f\|_1 \leqq \beta$ for all $f \in K$. Then, if $(\gamma_n) \in l_1$, then

$$\left\| \sum_{n=1}^{\infty} \gamma_n f_n \right\|_1 \leqq \beta \sum_{n=1}^{\infty} |\gamma_n|,$$

by the triangle inequality. On the other hand, for $(\gamma_n) \in l_1$, one has

$$\begin{aligned} \left\| \sum_{n=1}^{\infty} \gamma_n f_n \right\|_1 &\geqq \left\| \sum_{n=1}^{\infty} \gamma_n f_n \chi_{\bigcup_m E_m} \right\|_1 \\ &\geqq \sum_{n=1}^{\infty} \int_{E_n} \|\gamma_n f_n\| \, d\mu - \left\| \sum_{n=1}^{\infty} \gamma_n f_n \chi_{\bigcup_{m \neq n} E_m} \right\|_1 \\ &\geqq \delta \sum_{n=1}^{\infty} |\gamma_n| - (\delta/2) \sum_{n=1}^{\infty} |\gamma_n| \\ &= (\delta/2) \sum_{n=1}^{\infty} |\gamma_n|. \end{aligned}$$

Taking $\alpha = \delta/2$ completes the proof of the first assertion.

To prove the second statement, note that the first statement implies that K contains a copy of the unit vector basis of l_1, a wildly nonrelatively conditionally compact set. This completes the proof.

If X is reflexive, a restatement of Theorem 4 says that a nonreflexive subspace of $L_1(\mu, X)$ contains a copy of l_1.

Dunford's classical proof that a relatively weakly compact subset of $L_1(\mu)$ is uniformly integrable is based on the Vitali-Hahn-Saks theorem (Corollary I.4.10) and does not give as much information about the role of l_1 as the proof via Rosenthal's lemma as given above. From the point of view of Chapter I, this is not unexpected since, in Chapter I, the Vitali-Hahn-Saks theorem follows from Rosenthal's lemma. On the other hand, the original Dunford argument is so beautiful that to omit it would be a sin against nature. Here is the argument: (Keep in mind that the original Dunford result concerned itself with weakly sequentially compact sets in $L_1(\mu)$ so the use of Eberlein's theorem below is not really anything but a modern affectation.)

Suppose $K \subseteq L_1(\mu, X)$ is bounded and not uniformly integrable. As above, there is a $\delta > 0$, a sequence (E_n) of disjoint members of Σ and a sequence (f_n) in K such that $\int_{E_n} \|f_n\| \, d\mu > \delta$ for all n. If K is relatively weakly compact then we can assume that (f_n) is weakly convergent to some $f \in L_1(\mu, X)$. Now for each n, choose $g_n \in L_\infty(\mu, X^*)$ such that $\|g_n\|_\infty \leqq 1$, g_n vanishes off E_n and $\int_{E_n} \langle f_n, g_n\rangle \, d\mu > \delta$. Set $g = \sum_{n=1}^{\infty} g_n \chi_{E_n}$. Then we have $\|g\|_\infty \leqq 1$ and $\lim_n \int_E \langle f_n, g\rangle \, d\mu = \int_E \langle f, g\rangle \, d\mu$ for all $E \in \Sigma$. Invoking the Vitali-Hahn-Saks theorem, one sees that

$$\lim_m \int_{E_m} \langle f_n, g\rangle \, d\mu = 0$$

uniformly in n. But $\int_{E_n} \langle f_n, g\rangle \, d\mu = \int_{E_n} \langle f_n, g_n\rangle \, d\mu > \delta$; a contradiction which finishes the proof.

Theorem 4 has a beguiling consequence when it is applied to l_1. Define for $E \subseteq \mathscr{P}(N)$, $\mu(E) = \sum_{n \in E} 2^{-n}$. Then l_1 and $L_1(\mu)$ are isometric in an obvious way. Combining Theorem 1 and Theorem 4 with a moment of reflection reveals that the relatively norm compact subsets of l_1 are precisely the relatively weakly compact subsets of l_1 and a bounded sequence in l_1 is weakly convergent if and only if it is norm convergent. Thus the "Schur property" of l_1 may be viewed as a consequence of the Vitali-Hahn-Saks theorem or as a consequence of Rosenthal's lemma.

The following result "boot-straps" the $L_1(\mu, X)$ weak compactness results up to the space $bvca(\Sigma, X)$. Its proof is based on the fact that isometric isomorphisms preserve weak compactness and that a weakly compact set in $bvca(\Sigma, X)$ finds itself in an $L_1(\mu, X)$ space.

THEOREM 5 (BARTLE-DUNFORD-SCHWARTZ). *Let Σ be a σ-field of subsets of Ω. Suppose X is a Banach space such that both X and X^* have the Radon-Nikodým property. A subset K of $bvca(\Sigma, X)$ is relatively weakly compact if and only if*

(i) *K is bounded,*

(ii) *there is a nonnegative countably additive finite scalar measure μ on Σ such that $\lim_{\mu(E)\to 0} |G|(E) = 0$ uniformly in $G \in K$, and*

(iii) *for each $E \in \Sigma$, the set $\{G(E): G \in K\}$ is a relatively weakly compact subset of X.*

PROOF. Suppose (i), (ii) and (iii) are in force. For $g \in L_1(\mu, X)$, define $Tg \in bvca(\Sigma, X)$ by $(Tg)(E) = \int_E g\, d\mu$. Then $T\colon L_1(\mu, X) \to bvca(\Sigma, X)$ is an isometric isomorphism of $L_1(\mu, X)$ onto a subspace of $bvca(\Sigma, X)$. Since X has the Radon-Nikodým property and every member of K is μ-continuous, the range of T includes K. Since K satisfies (i), (ii) and (iii), we see $T^{-1}(K)$ satisfies (i), (ii) and (iii) of Theorem 1. Hence $T^{-1}(K)$ is relatively weakly compact and so is $T(T^{-1}(K)) = K$.

For the converse, let $K \subseteq bvca(\Sigma, X)$ be relatively weakly compact. The truth of (i) is clear. Since $G \to G(E)$ for a fixed $E \in \Sigma$ is a bounded linear operator on $bvca(\Sigma, X)$ to X, (iii) holds. If (ii) does not hold, then there is a sequence (G_n) in K such that $(|G_n|)$ is not a uniformly countably additive sequence. Set $\mu = \sum_{n=1}^{\infty} 2^{-n} |G_n|$ and define $T\colon L_1(\mu, X) \to bvca(\Sigma, X)$ by $(Tf)(E) = \int_E f\, d\mu$ for $f \in L_1(\mu, X)$ and $E \in \Sigma$. As above, T is an isometric isomorphism whose range includes the sequence (G_n). Since $(|G_n|)$ is not uniformly countably additive, $(T^{-1}(G_n))$ is not uniformly integrable and thus $(T^{-1}(G_n))$ is not contained in a weakly compact set, a contradiction which completes the proof.

Through the courtesy of Theorem I.2.4, (ii) above can be replaced by the equivalent condition

(ii′) $\{|G|\colon G \in K\}$ *is uniformly countably additive.*

We end this section by noting that Theorem 5 remains true for finitely additive vector measures on fields of sets. Its proof is an isometric copy of the proof of Theorem 5.

COROLLARY 6 (BROOKS-DINCULEANU). *Let $\mathscr{F}$ be a field of subsets of Ω. Let X and X^* have the Radon-Nikodým property. A subset K of $bva(\mathscr{F}, X)$ is relatively weakly compact if and only if* (i) *K is bounded,* (ii) *there exists a finitely additive finite nonnegative measure $\mu\colon \mathscr{F} \to R$ such that $\lim_{\mu(E)\to 0} |G|(E) = 0$ uniformly in $G \in K$ (or equivalently,* (ii′) *$\{|G|\colon G \in K\}$ is uniformly strongly additive), and* (iii) *for each $E \in \mathscr{F}$, $\{G(E)\colon G \in K\}$ is relatively weakly compact.*

PROOF. Let $\mathscr{F}_1$ be the Stone representation algebra of $\mathscr{F}$. Let $J\colon bva(\mathscr{F}, X) \to bvca(\sigma(\mathscr{F}_1, X))$ be the isometric isomorphism guaranteed by Theorem I.5.7. Since J preserves weakly compact sets, the necessity of (i), (ii) and (iii) is obvious for relatively weakly compact sets $K \subseteq bva(\mathscr{F}, X)$.

On the other hand, if K satisfies (i), (ii) and (iii), $J(K)$ satisfies (i) of Theorem 5. Theorem I.5.7 guarantees $J(K)$ satisfies (ii) of Theorem 5; also, since K satisfies (ii), we see that for each $E \in \sigma(\mathscr{F}_1)$ there exists a sequence $(E_n) \subseteq \mathscr{F}_1$ such that $\lim_n J(G)(E_n) = J(G)(E)$ uniformly in $G \in K$ (the functions $J(G)$ are uniformly continuous with respect to the Fréchet-Nikodým (symmetric difference) metric). It is now immediate that $J(K)$ satisfies (iii) of Theorem 5. Hence $J(K)$ is relatively weakly compact and $K = J^{-1}(J(K))$ is relatively weakly compact.

3. Gel'fand spaces. A classical theorem of Vitali characterizes absolutely continuous real-valued functions on [0, 1] as those functions that are indefinite integrals of their derivatives. This is not true for absolutely continuous functions on [0, 1] with values in arbitrary Banach spaces and it does not take much imagination to observe that the Radon-Nikodým property is the central issue here.

DEFINITION 1. Let X be a Banach space. A function $f\colon [0,1] \to X$ is called *absolutely continuous* if for each $\varepsilon > 0$ there exists a $\delta > 0$ such that if (a_n, b_n) is a sequence of disjoint subintervals of [0, 1] with $\sum_n (b_n - a_n) < \delta$ then

$\sum_n \|f(b_n) - f(a_n)\| < \varepsilon$. A Banach space X will be called a *Gel'fand space* if each absolutely continuous $f: [0,1] \to X$ is differentiable almost everywhere.

THEOREM 2. *A Banach space X is a Gel'fand space if and only if X has the Radon-Nikodým property with respect to Lebesgue measure on the Borel sets in* [0, 1].

PROOF. The proof is a routine exercise in manipulating theorems and definitions. Let Σ be the σ-field of Borel sets in [0,1] and μ be Lebesgue measure on Σ.

If X is a Gel'fand space and $F: \Sigma \to X$ is a μ-continuous measure of bounded variation, define $f: [0,1] \to X$ by $f(t) = F([0,t])$. Then f is absolutely continuous since $|F| \ll \mu$. Let ϕ be the derivative of f and note that if $x^* \in X^*$ and $0 \leqq a \leqq b \leqq 1$, then

$$x^*F([a, b)) = x^*f(b) - x^*f(a) = \int_a^b x^*\phi \, d\mu.$$

In particular, for $x^* \in X$, one has $x^*F(E) = \int_E x^*\phi \, d\mu$ for all intervals $E \subseteq [0,1]$ and all $x^* \in X^*$. By standard facts, this means $x^*F(E) = \int_E x^*\phi \, d\mu$ for all $E \in \Sigma$ and $x^* \in X^*$. It follows that $F(E) =$ Pettis-$\int_E \phi \, d\mu$ for all $E \in \Sigma$. Since F is of bounded variation and ϕ is measurable, it follows that ϕ is Bochner integrable and that X has the Radon-Nikodým property with respect to Lebesgue measure on the Borel sets in [0, 1].

For the converse, suppose X has the Radon-Nikodým property and suppose $f: [0, 1] \to X$ is absolutely continuous. For a subinterval E of [0, 1] with left endpoint a and right endpoint b, define $F(E) = f(b) - f(a)$. Standard procedures show that F has a weakly countably additive extension, still denoted by F, to the field generated by the subintervals of [0, 1]. The extension F is of bounded variation and is μ-continuous. Now appeal to I.5.2 to find a μ-continuous extension $\bar{F}$ of F to Σ. The extension $\bar{F}$ is also of bounded variation on Σ. Since X has the Radon-Nikodým property with respect to μ, there exists a Bochner integrable $\phi: [0,1] \to X$ such that $F(E) = \int_E \phi \, d\mu$ for all $E \in \Sigma$. Accordingly, one has

$$f(t) = \int_0^t \phi \, d\mu + f(0)$$

for all $t \in [0,1]$. Thus f is an indefinite integral. To see that $f' = \phi$ μ-almost everywhere, glance at II.2.9. This completes the proof.

Using similar methods, one can show with the help of the Lebesgue decomposition Theorem I.5.9 that a Banach space X has the Radon-Nikodým property with respect to Lebesgue measure on [0,1] if and only if every function $f: [0,1] \to X$ of bounded variation is differentiable almost everywhere.

It should be mentioned that a Banach space has the Radon-Nikodým property if and only if it has the Radon-Nikodým property with respect to Lebesgue measure on [0,1]. This is proved explicitly in V.3.8, but is also an easy exercise based on the material of Chapter III.

An easy exercise based on the above proof is the fact that an absolutely continuous vector-valued function on [0,1] is norm differentiable off a set of measure zero if and only if it is weakly differentiable off a set of measure zero.

4. Integral operators on $L_p(\mu)$. If (Ω, Σ, μ) is a finite measure space, then a Banach space X has the Radon-Nikodým property with respect to μ if and only if every $T: L_1(\mu) \to X$ has the action

$$T(f) = \int_{\Omega} fg \, d\mu$$

for some fixed $g \in L_\infty(\mu, X)$ and all $f \in L_1(\mu)$. For operators on $L_p(\mu)$, for $p > 1$, there is no analogous statement. In fact, if μ is Lebesgue measure on $[0, 1]$, $1 < p \leqq \infty$ and $T: L_p(\mu) \to L_p(\mu)$ is the identity operator, then there is no measurable function $g: [0, 1] \to L_p(\mu)$ with the property that $T(f) = \text{Pettis-}\int_\Omega fg \, d\mu$ for all $f \in L_p(\mu)$. Indeed, $G(E) = T(\chi_E)$ defines a vector measure whose variation is infinite on every set of positive μ-measure (Example I.1.16). If such a g were to exist, then it would be the Pettis integrable Radon-Nikodým derivative of G, which in view of the local Bochner integrability of measurable Pettis integrable functions is impossible

In this section, we shall study operators $T: L_p(\mu) \to X$ that have the action $T(f) = \text{Pettis-}\int_\Omega fg \, d\mu$ for some fixed measurable $g: \Omega \to X$ and all $f \in L_p(\mu)$. In particular, we shall study compactness properties of these operators and indicate how these operators are related to the classical integral operators. In the course, we shall flirt with order summing operators and some p-summing operators.

Throughout this section (Ω, Σ, μ) is a finite measure space and X is a Banach space.

DEFINITION 1. A bounded linear operator $T: L_p(\mu) \to X$ is called a *vector integral operator* with kernel g if there is a measurable $g: \Omega \to X$ such that

$$x^*T(f) = \int_{\Omega} fx^*g \, d\mu$$

for all $f \in L_p(\mu)$ and all $x^* \in X^*$.

Alternatively, this means

$$T(f) = \text{Pettis-}\int_{\Omega} fg \, d\mu$$

for all $f \in L_p(\mu)$.

It is easily checked that the kernel of a vector integral operator is almost everywhere uniquely defined and that the class of vector integral operators on $L_1(\mu)$ is exactly the class of Riesz representable operators. For $1 \leqq p < \infty$, it is an entertaining closed-graph exercise to show that a measurable $g: \Omega \to X$ is the kernel of a vector integral operator $T: L_p(\mu) \to X$ if and only if $x^*g \in L_q(\mu)$ $(p^{-1} + q^{-1} = 1)$ for all $x^* \in X^*$.

The straightforward proof of the next result is omitted; see III.2.21.

PROPOSITION 2. *Let* $1 \leqq p \leqq \infty$. *A bounded linear operator* $T: L_p(\mu) \to X$ *is a vector integral operator if and only if for each* $E_1 \in \Sigma$ *with* $\mu(E_1) > 0$ *there is an* $E_2 \subseteq E_1$, $E_2 \in \Sigma$ *with* $\mu(E_2) > 0$ *such that the operator* T_{E_2} *defined by* $T_{E_2}(f) = T(f\chi_{E_2})$ *for* $f \in L_p(\mu)$ *has an extension to a compact member of* $\mathscr{L}(L_1(\mu); X)$.

Proposition 2 suggests that vector integral operators have some compactness properties. The next few results reinforce this feeling.

PROPOSITION 3. *The restriction of a Riesz representable operator* $T: L_1(\mu) \to X$ *to* $L_p(\mu)$ $(1 < p \leqq \infty)$ *is a compact vector integral operator in* $\mathscr{L}(L_p(\mu); X)$.

PROOF. The closed unit ball of $L_p(\mu)$ is uniformly integrable and a Riesz representable operator on $L_1(\mu)$ maps uniformly integrable sets into compact sets by Lemma III.2.11.

Vector integral operators can be tested for compactness on the basis of a simple "absolute continuity" condition. Recall $T_E(f) = T(f\chi_E)$.

THEOREM 4. *Suppose* $1 \leqq p \leqq \infty$. *Then a sufficient condition that a vector integral operator* $T\colon L_p(\mu) \to X$ *be compact is that* $\lim_n \|T_{E_n}\| = 0$ *for all sequences* (E_n) *of members of* Σ *that descend to* ϕ.

If $1 < p < \infty$, *this condition is also necessary for an operator* $T\colon L_p(\mu) \to X$ *to be compact.*

PROOF. To prove the first assertion, use Proposition 2 and exhaustion to find an increasing sequence (A_n) of members of Σ with $\bigcup_{n=1}^{\infty} A_n = \Omega$ such that T_{A_n} has an extension to a compact member of $\mathscr{L}(L_1(\mu); X)$. Since a compact member of $\mathscr{L}(L_1(\mu); X)$ is a vector integral operator, Proposition 3 shows that $T_{A_n}\colon L_p(\mu) \to X$ is compact. But

$$\|T - T_{A_n}\| = \|T_{\Omega\setminus A_n}\| \to 0$$

since $(\Omega\setminus A_n)$ descends to ϕ. Hence T is compact.

To prove the second statement, assume $1 < p < \infty$, and assume that $T\colon L_p(\mu) \to X$ is a compact linear operator. Proceeding by contradiction, suppose there is a sequence (E_n) of members of Σ decreasing to ϕ such that $\|T_{E_n}\| > \varepsilon$ for all n and some $\varepsilon > 0$. Then choose (x_n^*) in X^* such that $\|x_n^*\| = 1$ and $\|x_n^* T_{E_n}\| \geqq \varepsilon$ for all n. Now $x_n^* T = T^* x_n^* = g_n \in L_q(\mu)$ $(p^{-1} + q^{-1} = 1)$, and T^* is compact by Schauder's theorem. Hence it can and will be assumed without loss of generality that $\lim_n g_n = g$ exists in $L_q(\mu)$-norm. Therefore, if n_0 is selected such that $\|g_n - g\|_q < \varepsilon/2$ for $n \geqq n_0$, then one obtains

$$\big|\, \|g_n \chi_{E_n}\|_q - \|g\chi_{E_n}\|_q \big| \leqq \|(g_n - g)\chi_{E_n}\|_q \leqq \|g_n - g\|_q < \varepsilon/2$$

for all $n \geqq n_0$. Hence for n as above one has

$$\varepsilon \leqq \|x_n^* T_{E_n}\| = \|g_n \chi_{E_n}\|_q < \varepsilon/2 + \|g\chi_{E_n}\|_q.$$

Therefore $\|g\chi_{E_n}\|_q \geqq \varepsilon/2$ for sufficiently large n, which is impossible since $\lim_n \int_{E_n} |g|^q \, d\mu = 0$. This completes the proof.

The following corollary is immediate.

COROLLARY 5. *A sufficient condition for a vector integral operator* $T\colon L_p(\mu) \to X$ $(1 \leqq p \leqq \infty)$ *to be compact is* $\lim_{\mu(E)\to 0} \|T_E\| = 0$.

If $1 < p < \infty$, *then this condition is also necessary.*

COROLLARY 6. *Suppose* $1 < p \leqq \infty$ *and* $g \in L_q(\mu, X)$ $(p^{-1} + q^{-1} = 1)$. *If* $T\colon L_p(\mu) \to X$ *is defined by* $T(f) = \int_\Omega fg \, d\mu$, *then* T *is a compact vector integral operator.*

PROOF. $\lim_{\mu(E)\to 0} \|g\chi_E\|_q = 0$.

It should be noted that this corollary has a straightforward proof which is based on the fact that simple functions are dense in $L_q(\mu, X)$ for $1 \leqq q < \infty$. A moment's reflection and a glance at the earlier results show that the current proof is also based on this fact.

DEFINITION 7 (DINCULEANU). Let $T\colon L_p(\mu) \to X$ be a bounded linear operator. Set

$$|||T|||_p = \sup\left\{\sum_{i=1}^n \|\alpha_i T(\chi_{E_i})\|_X\right\}$$

where the supremum is taken over all functions $f = \sum_{i=1}^n \alpha_i \chi_{E_i}$ with $E_i \cap E_j = \emptyset$ for $i \neq j$, $E_1, \cdots, E_n \in \Sigma$ and $\|f\|_p \leqq 1$.

As a consequence of Corollary 6 and the next result, we shall see that an operator $T\colon L_p(\mu) \to X$ with $|||T|||_p < \infty$ is compact if $1 < p < \infty$ and X has the Radon-Nikodým property.

THEOREM 8. *Suppose $1 < p < \infty$ and that X has the Radon-Nikodým property. Any one of the following statements about a bounded linear operator $T\colon L_p(\mu) \to X$ implies all the others.*

(a) $|||T|||_p < \infty$.

(b) *T maps positive convergent series in $L_p(\mu)$ into absolutely convergent series (i.e., T is order summing).*

(c) *There exists a function $\phi \in L_q(\mu)$ $(p^{-1} + q^{-1} = 1)$ such that $\|T(f)\| \leqq \int_\Omega |f|\phi \, d\mu$ for all $f \in L_p(\mu)$.*

(d) *There exists $g \in L_q(\mu, X)$ $(p^{-1} + q^{-1} = 1)$ such that $T(f) = \int_\Omega fg \, d\mu$ for all $f \in L_p(\mu)$.*

In case (d) *holds,* $|||T|||_p = \|g\|_q$.

PROOF. The equivalence of (a) and (b) is almost clear. Note that $T\colon L_p(\mu) \to X$ satisfies (b) if and only if

$$\|T\|_L = \sup\left\{\sum_{i=1}^n \|T(f_i)\| : f_i \geqq 0, \left\|\sum_{i=1}^n f_i\right\|_p \leqq 1\right\} < \infty.$$

Certainly $|||T|||_p \leqq \|T\|_L$; so (b) implies (a). To prove the reverse inequality, carefully approximate with simple functions.

Also, it is obvious that (d) implies (c), since $\phi = \|g\|_X$ works. Proving that (c) implies (a) is also simple: If $\|T(f)\| \leqq \int_\Omega |f|\phi \, d\mu$, then

$$\sum_{i=1}^n \|\alpha_i T(\chi_{E_i})\| \leqq \sum_{i=1}^n |\alpha_i| \int_{E_i} \phi \, d\mu$$
$$= \int_\Omega \left(\sum_{i=1}^n |\alpha_i| \chi_{E_i}\right)\phi \, d\mu \leqq \|\phi\|_q$$

whenever $\{E_1, E_2, \cdots, E_n\} \subseteq \Sigma$, $E_i \cap E_j = \emptyset$ for $i \neq j$ and $\|\sum_{i=1}^n \alpha_i \chi_{E_i}\|_p \leqq 1$.

To see that (a) implies (d), define $G\colon \Sigma \to X$ by $G(E) = T(\chi_E)$ for $E \in \Sigma$. The function G is obviously finitely additive. Moreover, one has $\lim_{\mu(E) \to 0} \|G(E)\| \leqq \lim_{\mu(E) \to 0} \|T\| \, \|\chi_E\|_p = 0$. Hence the measure G is μ-continuous and countably additive. Next, choose $\alpha > 0$ so that $\|\alpha \chi_\Omega\|_p = 1$. For any partition π, note that $\alpha\chi_\Omega = \sum_{E \in \pi} \alpha\chi_E$. Accordingly, we have

$$\alpha \sum_{E \in \pi} \|G(E)\| = \sum_{E \in \pi} \|\alpha T(\chi_E)\| \leqq |||T|||_p < \infty.$$

Hence G is of bounded variation. Since X has the Radon-Nikodým property, there exists a Bochner integrable $g: \Omega \to X$ such that $G(E) = \int_E g\, d\mu$ for all $E \in \Sigma$. We will show that $T(f) = \int_\Omega fg\, d\mu$ for all $f \in L_p(\mu)$ and that $g \in L_q(\mu, X)$.

For this let $\{E_1, \ldots, E_n\}$ be a partition of Ω. Clearly, whenever $\| \sum_{i=1}^n \alpha_i \chi_{E_i} \|_p \leqq 1$, one has

$$\sum_{i=1}^n |\alpha_i|\, |G|(E_i) \leqq |\!|\!| T |\!|\!|_p.$$

Hence one has

$$\sum_{1=i}^n \| \alpha_i T(\chi_{E_i}) \| \leqq \sum_{i=1}^n |\alpha_i| \int_{E_i} \|g\|\, d\mu \leqq |\!|\!| T |\!|\!|_p,$$

provided $\| \sum_{i=1}^n \alpha_i \chi_{E_i} \|_p \leqq 1$. Taking appropriate suprema yields the equality

$$|\!|\!| T |\!|\!|_p = \sup\left\{ \int_\Omega |f|\, \|g\|\, d\mu : f \text{ simple}, \|f\|_p \leqq 1 \right\}.$$

From this it follows that $\|g\|_X \in L_q(\mu)$ and that $|\!|\!| T |\!|\!|_p = \|g\|_{L_q(\mu, X)}$. Now since $g \in L_q(\mu, X)$, the Bochner integral $\int_\Omega fg\, d\mu$ exists for all $f \in L_p(\mu)$ by the Holder inequality. Define an operator $T_1: L_p(\mu) \to X$ by

$$T_1(f) = \int_\Omega fg\, d\mu, \qquad f \in L_p(\mu).$$

Then T_1 is continuous. Moreover, T agrees with T_1 on the simple functions. Hence T agrees with T_1 everywhere. This completes the proof.

As a consequence of the next corollary, we shall obtain a relationship between operators $T: L_p(\mu) \to X$ with $|\!|\!| T |\!|\!|_p < \infty$, and absolutely p-summing operators.

COROLLARY 9. *Let $1 < p < \infty$ and let X^* have the Radon-Nikodým property. Any one of the following statements about a bounded linear operator $T: X \to L_p(\mu)$ implies the others.*

(a) $|\!|\!| T^* |\!|\!|_q < \infty$ $(p^{-1} + q^{-1} = 1)$.

(b) *There is a $g \in L_p(\mu, X^*)$ such that $Tx = g(\cdot)(x) \in L_p(\mu)$ for all $x \in X$.*

(c) *There is $\phi \in L_p(\mu)$ such that $|Tx| \leqq \phi \|x\|$ a.e. for each $x \in X$.*

PROOF. The equivalence of (a) and (b) is essentially obvious and is therefore omitted. Setting $\phi = \|g\|_{X^*}$ proves that (b) implies (c). To prove that (c) implies (b), suppose $|T(x)| \leqq \phi \|x\|$ a.e. for some $\phi \in L_p(\mu)$ and all $x \in X$. Let $T^*: L_q(\mu) \to X^*$ be the adjoint of T $(p^{-1} + q^{-1} = 1)$. If $f \in L_q(\mu)$ and $x \in X$, then

$$\begin{aligned} |(T^*f)(x)| &= \left| \int_\Omega T(x) f\, d\mu \right| \leqq \int_\Omega |Tx|\, |f|\, d\mu \\ &\leqq \int_\Omega \phi \|x\|\, |f|\, d\mu. \end{aligned}$$

Hence $\|T^* f\| \leqq \int_\Omega \phi |f|\, d\mu$ for all $f \in L_q(\mu)$. Appealing to Theorem 8, we find a $g \in L_p(\mu, X^*)$ such that $T^*(f) = \int_\Omega fg\, d\mu$ for all $f \in L_q(\mu)$. Accordingly,

$$\int_\Omega fg(x)\, d\mu = (T^* f)(x) = \int_\Omega T(x) f\, d\mu$$

for all $f \in L_q(\mu)$ and $x \in X$. It follows immediately that $Tx = g(\cdot)(x)$. This completes the proof.

Incidentally, an operator $T: X \to L_p(\mu)$ of the form $Tx = g(\cdot)x$ as above is an absolutely p-summing operator. To see this, recall that if $S: X \to Y$ is a bounded linear operator, then S is absolutely p-summing if there exists a real number C_p such that

$$\sum_{j=1}^{n} \|Sx_j\|^p \leqq C_p \sup_{\|x^*\|\leq 1} \sum_{j=1}^{n} |x^*(x_j)|^p$$

for every finite set $\{x_1, \cdots, x_n\}$ in X. Now suppose $T: X \to L_p(\mu)$ has the form $T(x) = g(\cdot)x$ for some $g \in L_p(\mu, X^*)$ and all $x \in X$. If $\{x_1, \cdots, x_n\}$ is a finite set in X, then

$$\begin{aligned}
\sum_{j=1}^{n} \|Tx_j\|^p &= \sum_{j=1}^{n} \int_\Omega |g(\cdot)x_j|^p \, d\mu \\
&= \int_\Omega \sum_{j=1}^{n} \frac{|g(\cdot)(x_j)|^p}{\|g\|_X^p} \|g\|_X^p \, d\mu \qquad \text{(here } 0/0 = 0\text{)} \\
&\leqq \int_\Omega \sup_{\|x^*\|\leq 1} \sum_{j=1}^{n} |x^*(x_j)|^p \, \|g\|_X^p \, d\mu \\
&= \left(\int_\Omega \|g\|_X^p \, d\mu\right) \sup_{\|x^*\|\leq 1} \sum_{j=1}^{n} |x^*(x_j)|^p \\
&= |||T^*|||_p^p \sup_{\|x^*\|\leq 1} \sum_{j=1}^{n} |x^*(x_j)|^p.
\end{aligned}$$

Corollary 9 takes on a pleasingly familiar form when $X = L_r(\sigma)$ for some $1 < r < \infty$ and finite measure space $(\Lambda, \mathscr{F}, \sigma)$. If $T: L_r(\sigma) \to L_p(\mu)$ has $|||T^*|||_q < \infty$, then there is a function $g \in L_p(\mu, L_s(\sigma))$ $(r^{-1} + s^{-1} = 1)$ such that

$$T(f)(\omega) = \int_\Omega g(\omega) f(\lambda) \, d\sigma(\lambda)$$

for all $f \in L_r(\sigma)$ and almost all $\omega \in \Omega$. Now for each $\omega \in \Omega$, the value $g(\omega)$ is an $L_s(\sigma)$-valued function ϕ_ω. Define $K: \Lambda \times \Omega \to$ scalars by $K(\lambda, \omega) = \phi_\omega(\lambda)$, $\lambda \in \Lambda$, $\omega \in \Omega$. It is not difficult to see that K is measurable on the product measure space and that

$$(*) \qquad (Tf)(\omega) = \int_\Lambda K(\lambda, \omega) f(\lambda) \, d\sigma(\lambda)$$

for almost all $\omega \in \Omega$ and that

$$(**) \qquad \begin{aligned}
\left[\int_\Omega \left(\int_\Lambda |K(\lambda, \omega)|^s \, d\sigma(\lambda)\right)^{p/s} d\mu(\omega)\right]^{1/p} &= \left[\int_\Omega \|g(\omega)\|_{L_s(\sigma)}^p \, d\mu(t)\right]^{1/p} \\
&= |||T^*|||_q < \infty.
\end{aligned}$$

Similarly, if $(**)$ is satisfied for some product measurable K and T is defined by $(*)$, then it is obvious that $|Tf(\omega)| \leqq \phi(\omega) \|f\|$ a.e., for some $\phi \in L_p(\mu)$ and consequently $|||T^*|||_q < \infty$.

Thus this little discussion shows that, when X is specialized to the L_p situation, the vector integral operators under current study reduce to the classical integral operators on $L_p(\mu)$. In particular, when $X = L_2(\mu)$, then the class of operators $T: L_2(\mu) \to L_2(\mu)$ with $|||T^*|||_2 < \infty$ is precisely the Hilbert-Schmidt class.

5. The Lewis-Stegall theorem with a dash of Pełczyński. The only complemented infinite dimensional subspaces of $L_1[0,1]$ that come to mind are isomorphic to either l_1, $L_1[0,1]$ or a product of these spaces. The purpose of this section is to show that a complemented infinite dimensional subspace of $L_1[0,1]$ that has the Radon-Nikodým property is indeed isomorphic to l_1.

The starting point is a basic property of the space l_1.

THEOREM 1 (PEŁCZYŃSKI). *Every infinite dimensional subspace of l_1 contains a complemented subspace of l_1 that is isomorphic to l_1.*

PROOF. Let Z be an infinite dimensional closed linear subspace of l_1.

Choose any $z_1 \in Z$ of norm one. Let m_1 be chosen so that the contribution of the coordinates of z_1 beyond m_1 to the norm of z_1 totals not more than 1/4.

Since Z is infinite dimensional, there is a $z_2 \in Z$ for which $\|z_2\| = 1$ and the first m_1 coordinates of z_2 are zero. Let m_2 be chosen so that the contribution of the coordinates of z_2 beyond m_2 to the norm of z_2 totals not more than 1/8.

Again, since Z is infinite dimensional, there is a $z_3 \in Z$ for which $\|z_3\| = 1$ and the first m_2 coordinates of z_3 are zero. Let m_3 be chosen so that the contribution of the coordinates of z_3 beyond m_3 to the norm of z_3 totals not more than 1/16.

The inductive step is clear:

Fix $m_0=0$.

Let $b_n = \sum_{j=m_{n-1}+1}^{m_n} z_{n,j}e_j$ where $z_{n,j}$ denotes the jth coordinate of z_n and e_j denotes the jth unit vector. Note that the closed linear span of $\{b_n\}$, $[b_n]$, is isometric to l_1 and is the range of a norm one projection P. Moreover, we have $\|z_n - b_n\| \leqq 2^{-n-1}$ for each n. Let $(b_n^*) \subseteq [b_n]^*$ be biorthogonal to (b_n); then we have

$$\|b_n^*\| = \frac{1}{\|b_n\|} \leqq \frac{1}{\|z_n\| - \|z_n - b_n\|} = \frac{1}{1-2^{-n-1}}.$$

Consider the operator $T: l_1 \to l_1$ defined by

$$Tx = x - Px + \sum_{n=1}^{\infty} b_n^*(Px)\, z_n.$$

Since $Px \in [b_n]$ and (b_n^*) is biorthogonal to (b_n) and $[b_n]$ is isometric to l_1, we see $(b_n^* Px) \in l_1$; therefore T is well defined continuous and linear. Moreover, if $x \in l_1$, and $\|x\|_1 \leqq 1$ then

$$\begin{aligned}\|x - Tx\| &= \left\|Px - \sum_{n=1}^{\infty} b_n^*(Px) z_n\right\| \\ &\leqq \|P\| \sum_{n=1}^{\infty} \|b_n^*\|\, \|b_n - z_n\| \\ &\leqq \sum_{n=1}^{\infty} \frac{2^{-n-1}}{1-2^{-n-1}} \\ &= \sum_{n=1}^{\infty} (2^{n+1} - 1)^{-1} < 1.\end{aligned}$$

Therefore T is an isomorphism of l_1 onto itself. Clearly T takes $[b_n]$ onto $[z_n]$; hence $[z_n]$ is isomorphic to l_1. Finally, $TPT^{-1} = Q$ is a continuous linear projection of l_1 onto $[z_n] \subseteq Z$.

Before proceeding, some notational conventions will be established. Suppose

(X_n) is a sequence of Banach spaces. Then $(\sum \oplus X_n)_{l_1}$ denotes the Banach space of all sequences (x_n) where $x_n \in X_n$, for each n, $\|(x_n)\| = \sum_n \|x_n\| < \infty$. It is plain that if each X_n is isomorphic to l_1 with a common bound for the norms of the isomorphisms then $(\sum \oplus X_n)_{l_1}$ is isomorphic to l_1. Sometimes $(\sum \oplus X_n)_{l_1}$ is denoted by $(X_1 \oplus X_2 \oplus \ldots)_{l_1}$. Also, if X and Y are Banach spaces, then $X \times Y$ is isomorphic to $(X \oplus Y)_{l_1}$.

COROLLARY 2 (PEŁCZYŃSKI). *Infinite dimensional complemented subspaces of l_1 are isomorphic to l_1.*

PROOF. Let X be an infinite dimensional complemented subspace of l_1 and suppose Y is a complement of X; i.e., $l_1 = (X \oplus Y)$. By Theorem 1, there exist closed linear subspaces Z_1 and Z of X that are complemented in l_1 such that $X = (Z_1 \oplus Z)$ and Z_1 is isomorphic to l_1. Let us agree to use the symbol "$\approx$" to signal the existence of an isomorphism between the left- and right-hand extremities; so $l_1 \approx (X \oplus Y)_{l_1}$ and $X \approx (Z_1 \oplus Z)_{l_1}$ and $Z_1 \approx l_1$.

We now have set the stage for what has come to be known as the Pełczyński decomposition method; all the following are easy to see:

$$\begin{aligned} l_1 \approx (X \oplus Y)_{l_1} &\approx (Z_1 \oplus Z \oplus Y)_{l_1} \approx ((l_1 \oplus Z)_{l_1} \oplus Y)_{l_1} \\ &\approx (l_1 \oplus Z \oplus Y)_{l_1} \approx (l_1 \oplus l_1 \oplus \cdots \oplus Z \oplus Y)_{l_1} \\ &\approx ((X \oplus Y)_{l_1} \oplus (X \oplus Y)_{l_1} \oplus \cdots \oplus Z \oplus Y)_{l_1} \\ &\approx ((X \oplus X \oplus \cdots)_{l_1} \oplus (Y \oplus Y \oplus \cdots)_{l_1} \oplus Z \oplus Y)_{l_1} \\ &\approx ((X \oplus X \oplus \cdots)_{l_1} \oplus (Y \oplus Y \oplus Y \oplus \cdots)_{l_1} \oplus Z)_{l_1} \\ &\approx (((X \oplus Y)_{l_1} \oplus (X \oplus Y)_{l_1} \oplus \cdots)_{l_1} \oplus Z)_{l_1} \\ &\approx ((l_1 \oplus l_1 \oplus \cdots)_{l_1} \oplus Z)_{l_1} \\ &\approx (Z_1 \oplus Z)_{l_1} \approx X. \end{aligned}$$

The main purpose of this section is a fundamental structure theorem for subspaces of $L_1[0,1]$. It is a dramatic strengthening of III.2.16.

THEOREM 3 (D. LEWIS AND STEGALL). *Let (Ω, Σ, μ) be a finite measure space and X be a complemented infinite dimensional subspace of $L_1(\mu)$. If X has the Radon-Nikodým property, then X is isomorphic to l_1.*

PROOF. Let $P: L_1(\mu) \to L_1(\mu)$ be a continuous linear projection onto X. Since X has the Radon-Nikodým property, Theorem III.1.8 guarantees the existence of bounded operators $U: L_1(\mu) \to l_1$ and $V: l_1 \to X$ such that $P = VU$. Since $VU|_X$ is identity on X, the operator U acts as an isomorphism from X to the subspace $U(X)$ of l_1. It follows that UV is a continuous linear projection from l_1 onto $U(X)$ $\approx X$. Hence X is isomorphic to a complemented subspace of l_1 and an application of Corollary 2 finishes the proof.

A close inspection of the above proof reveals that whenever X is the range of a representable continuous linear projection P then X is either finite dimensional or isomorphic to l_1. This is at least formally a stronger statement than Theorem 3. It further accentuates the special character of projections in a Banach space. Indeed, if X is any separable Banach space, then X is the continuous linear image of $L_1[0,1]$ by means of a representable operator T; this is actually a triviality if one notes that X is the continuous linear image of l_1 which in turn is a complemented subspace of $L_1[0,1]$.

6. Notes and remarks.

The space $L_p(\mu, X)$. The identification of $L_p(\mu)^*$ for $1 \leqq p < \infty$ is classical functional analysis in its very best tradition. In the same issue of Comptes Rendues, Fréchet **[1907a]** and F. Riesz **[1907]** announced that $L_2[0,1]$ is self-dual. The first complete proof is in Fréchet **[1907b]**. F. Riesz **[1910]** then found the dual of $L_p[0,1]$ $(1 < p < \infty)$ and Nikodým **[1931]** extended Riesz's representation to $L_p(\mu)$ for finite measures μ as did Dunford **[1938]**. H. Steinhaus **[1919]** proved that $L_\infty[0,1]$ is the dual of $L_1[0,1]$ and Dunford **[1938]** extended Steinhaus's result to finite measures.

Theorem 1.1 for $L_p(\mu, X)^*$ is due to Bochner and Taylor **[1938]** in case μ is Lebesgue measure on [0,1]. They showed that $L_p([0,1], X)^*$ is identifiable with $L_q([0, 1], X^*)$ $(1 \leqq p < \infty, p^{-1} + q^{-1} = 1)$ if and only if X^* satisfies a certain condition they call (D). Spaces satisfying condition (D) are precisely the spaces we call "Gel'fand spaces" in §3 of this chapter. Thus by §3 and Corollary V.3.8 property D is the same as the Radon-Nikodým property. Theorem 1.1 has been extended by Gretsky and Uhl **[1972]** to certain Banach function spaces of vector-valued functions on a σ-finite measure space.

Bochner and Taylor **[1938]** also give a description of $L_p(\mu, X)^*$ when X^* fails to have property D. This description has been generalized very extensively. Most of the generalizations fall into two classes of descriptions neither of which has found concrete application to the structure of $L_p(\mu, X)$ as yet. The first description is in terms of vector measures.

Let (Ω, Σ, μ) be a finite measure space, $1 \leqq p < \infty$ and $p^{-1} + q^{-1} = 1$. For $q < \infty$ the q-variation of a vector measure $F: \Sigma \to X$ is defined by

$$\| F \|_q = \sup_\pi \left\{ \sum_{E \in \pi} \frac{\| F(E) \|^q}{\mu(E)^q} \mu(E) \right\}^{1/q}$$

where π ranges over all finite partitions of Ω into sets from Σ. The convention $0/0 = 0$ is observed here. If $q = \infty$, then $\| F \|_\infty$ is defined to be $\inf\{k > 0: \|F(E)\| \leqq k\mu(E)$ for all $E \in \Sigma\}$. Let $V_q(\mu, X^*)$ be the space of all vector measures $F: \Sigma \to X^*$ with $\| F \|_q < \infty$. It is not difficult to see (the proof is essentially imbedded in the proof of Theorem 1.1) that $L_p(\mu, X)^*$ is isometrically isomorphic to $V_q(\mu, X^*)$ under the correspondence $l \leftrightarrow G$, where $l \in L_p(\mu, X)^*$ and $G \in V_q(\mu, X^*)$, defined by

$$l(f) = \int_\Omega f \, dG, \qquad f \in L_p(\mu, X).$$

There is no trouble finding more about this description of $L_p(\mu, X)^*$: Bogdanowicz **[1965]**, **[1966]**, Dinculeanu **[1965]**, **[1966a]**, **[1973]**, Dinculeanu and Foiaş **[1961]**, Singer **[1958]**, **[1959a]**, **[1960a]**, **[1960b]**. These papers also contain information about operators on $L_p(\mu, X)$. The $V_q(\mu, X)$ spaces (for finitely additive μ) are studied in Bochner **[1939]**, **[1940]**, Bochner and Phillips **[1941]**, Leader **[1953]** and Uhl **[1967]**.

The second description is based on the fact that each member of $V_q(\mu, X^*)$ $(1 < q \leqq \infty)$ admits a weak* density with respect to μ. Here the lifting enters. For more on this, consult Dieudonné **[1941]**, **[1944]**, **[1948]**, **[1951a]**, **[1951b]**, Dinculeanu **[1966c]**, **[1967]**, **[1973]**, Dinculeanu and Foiaş **[1961]**, Ionescu Tulcea **[1962]**, **[1969]**.

Both descriptions are considerably more general than the description given in the text. Any applications of either of these descriptions to an understanding of the structure of $L_p(\mu, X)$ would be very welcome.

The structure of $L_p(\mu, X)$. A basic structure theory for $L_p(\mu, X)$ is still in its infancy. The two basic questions here are what properties of $L_p(\mu, X)$ are inherited from $L_p(\mu)$ and X and what properties of X and $L_p(\mu)$ are consequences of properties of $L_p(\mu, X)$?

In V.4, we shall see that $L_p(\mu, X)$ has the Radon-Nikodým property if and only if both $L_p(\mu)$ and X have the Radon-Nikodým property. Weak compactness in $L_p(\mu, X)$ for certain Banach spaces X can easily be studied by the techniques of §2 with the help of Theorem 1.1.

For another illustration of the first question, let us examine whether a copy of the space c_0 can slip into $L_p(\mu, X)$ $(1 \leqq p < \infty)$ if X contains no copy of c_0. As Hoffman-Jørgensen **[1974]** has pointed out, this is an entertaining exercise if X has the Radon-Nikodým property. Here is how it goes: Take a series $\sum_n f_n$ in $L_p(\mu, X)$ with $\sum_n |l(f_n)| < \infty$ for all $l \in L_p(\mu, X)^*$. Note that $\sum_n |x^* \int_E f_n \, d\mu| < \infty$ for every measurable set E and $x^* \in X^*$. Appeal to I.4.5 to see that $\sum_n \int_E f_n \, d\mu$ is unconditionally convergent in X for each measurable set E. Then make a few computations to see that $\sum_n \int_{(\cdot)} f_n \, d\mu$ is an X-valued μ-continuous vector measure of finite variation whose Radon-Nikodým derivative f belongs to $L_p(\mu, X)$. Verify that $\sum_n f_n = f$ for the $L_q(\mu, X^*)$- $(p^{-1} + q^{-1} = 1)$ topology of $L_p(\mu, X)$. Similarly note that every subseries of $\sum_n f_n$ converges in the $L_q(\mu, X^*)$-topology of $L_p(\mu, X)$. Finally, make the observation that all of this is happening in a separable subspace of $L_p(\mu, X)$ and apply I.4.7 to see that $\sum_n f_n$ is unconditionally convergent. Apply I.4.5 to complete the fun.

Maybe less entertaining but a good deal more exciting than the above is the theorem of Kwapien **[1974]** that says that if X contains no copy of c_0 neither does $L_p(\mu, X)$ $(1 \leqq p < \infty)$.

An important unsolved problem related to the first question is deciding whether $L_p([0,1], X)$ has an unconditional basis if $1 < p < \infty$ and X has an unconditional basis. There is some hope that vector-valued martingales may be brought to bear on this problem.

Studying the structure of X in terms of properties of $L_p(\mu, X)$ is an increasingly fruitful area which is coming into its own. Currently most of this work centers on the derivation of geometric properties of X from analytic properties of martingales in $L_p(\mu, X)$. We shall say more about this in Chapter V.

Extreme points of the ball of $L_p(\mu, X)$. For complete measure spaces (Ω, Σ, μ), J. A. Johnson **[1974]** has characterized the extreme points in the unit ball of $L_p(\mu, X)$ $(1 < p < \infty)$ as those $f \in L_p(\mu, X)$ with $\|f\|_p = 1$ such that $f(\omega)/\|f(\omega)\|_X$ is an extreme point of the unit ball of X for μ-almost all $\omega \in \{\omega : \|f(\omega)\|_X > 0\}$. His theorem is based on the von Neumann **[1949]** Selection Theorem and extends earlier theorems of Sundaresan **[1970]** and Karlin **[1953]**.

Weak compactness in $L_1(\mu, X)$. Dunford **[1939]** characterized the relatively weakly compact subsets of $L_1(\mu)$ (for finite μ) as the bounded uniformly integrable sets. Formalities aside, the proof of Theorem 2.1 given in the text follows Dunford's original proof. In fact, Ionescu Tulcea **[1963]** noted that for reflexive X, Dunford's original proof can be trivially modified to characterize the relatively weakly compact subsets of $L_1(\mu, X)$. A number of authors have uncovered versions of Theorem

2.1, usually in the course of studying other problems. See Batt [**1974**], Batt and Berg [**1969**], Brooks [**1972**], Brooks and Dinculeanu [**1974**], Chatterji [**1963**] and Swartz [**1973e**]. Batt [**1974**] makes the most definitive attack. His paper is a splendid source of counterexamples and should be read by anyone who is interested in working with weak compactness in $L_1(\mu, X)$. His examples illustrate the pitfalls encountered in this direction as does Theorem 2.3. Theorem 2.4 is the natural extension of a similar result of Kadec and Pełczyński [**1962**].

The problem of characterizing the relatively weakly compact subsets of $L_1(\mu, X)$ for general X remains one of the most elusive problems in the theory of vector measures. Diestel [**1976**] has made some progress (see VIII.4.11) by showing that *if* $K \subseteq L_1(\mu, X)$ *is a bounded uniformly integrable family of functions such that for each* $\varepsilon > 0$ *there is a weakly compact set* $K_\varepsilon \subseteq X$ *and a measurable set* Ω_ε *with* $\mu(\Omega \backslash \Omega_\varepsilon) \leqq \varepsilon$ *and such that* $f(\Omega_\varepsilon) \subseteq K_\varepsilon$ *for all* $f \in K$, *then* K *is relatively weakly compact.* Unfortunately, this condition is far from necessary. To see this, let X be any Banach space whose dual has the Radon-Nikodým property. Let (x_n) be *any* bounded sequence in X and let (r_n) be the sequence of Rademacher functions on [0, 1]. If $f_n = x_n r_n \in L_1([0, 1], X)$, an easy computation shows that $\lim_n f_n = 0$ weakly.

Also unknown at present are criteria for conditional weak compactness in $L_1(\mu, X)$. In fact the only results in this connection appear to be those of the text. The work of Odell and Rosenthal [**1975**] and Rosenthal [**1974**], [**1976**] might be helpful in this direction.

The derivation of Theorem 2.5 from Theorem 2.1 follows the path cleared by Bartle, Dunford and Schwartz [**1955**].

In certain concrete situations, Theorem 2.5 is subject to considerable sharpening. Let Ω be a compact Hausdorff space, Σ be the σ-field of Borel subsets of Σ and $bvrca(\Sigma, X)$ be the subspace of $bvca(\Sigma, X)$ consisting of the regular measures. If X and X^* have the Radon-Nikodým property, a bounded set K in $bvrca(\Sigma, X)$ is relatively weakly compact if and only if for every disjoint sequence (O_n) of pairwise disjoint open sets $\lim_n F(O_n) = 0$ uniformly in $F \in K$. This theorem of Grothendieck [**1953**] is an easy consequence of Theorem 2.5 and the regularity lemma of Chapter VI (VI.2.13).

Corollary 2.6 is the main object of Brooks and Dinculeanu [**1974**] and was established for reflexive Banach spaces by Brooks [**1972**].

Finally we remark that the closely related problem of characterizing for which Banach spaces X is $L_1(\mu, X)$ weakly sequentially complete is open. It is not known (even if X has the Radon-Nikodým property) whether $L_1(\mu, X)$ is weakly sequentially complete whenever X is. An easy consequence of the results of §2 is the fact that if X is reflexive then $L_1(\mu, X)$ is weakly sequentially complete.

One should note that all the above problems are of interest primarily in case μ is not purely atomic since for purely atomic measures the answers to the above problems are known and though not always trivial, follow in straightforward fashion from their scalar counterparts.

For Banach lattices X, Cartwright [**1974**] has shown that if X has weakly compact order intervals so does $L_1(\mu, X)$ and consequently $L_1(\mu, X)$ is order complete. He also obtained a partial converse.

The questions of weak sequential completeness, conditional and relative weak compactness criteria also are open for the spaces $L_p(\mu, X)$ when $1 < p < \infty$.

Lipschitz mappings in Banach spaces. Most of our discussion regarding applications of vector measure theory to the structure of Banach spaces centers about the linear topological classification of these spaces. It is natural to ask how finely one can classify a Banach space by means of its topological, uniform or Lipschitz structure.

It is a famous conjecture that two infinite dimensional Banach spaces having the same density character are homeomorphic. This conjecture has been substantiated for separable spaces by Kadec **[1967]** and for reflexive spaces by Bassaga **[1972]**; more on this is to be found in the book of Bessaga and Pełczyński **[1975]**. In any case, the topology of a Banach space is not a rich enough structure to distinguish linear topological properties of the space.

What about the uniform structure of a Banach space? Here there is hope that, at least in the class of Banach spaces with the Radon-Nikodým property, the uniform structure of a Banach space determines its linear topological character. Lindenstrauss **[1964b]** showed that infinite dimensional $C(K)$ spaces are not uniformly homeomorphic to L_p spaces and that if p and q are distinct real numbers each at least as large as two then L_p and L_q are not uniformly homeomorphic. Enflo **[1970c]** showed that for any distinct real numbers $p, q \geqq 1$, L_p and L_q are not uniformly homeomorphic. Moreover, Enflo **[1970a]**, **[1970b]** showed that a Banach space that is uniformly homeomorphic to a subset of a Hilbert space is linearly homeomorphic to a Hilbert space. Results of the same flavor as those of Enflo but with a distinct Radon-Nikodým ingredient have been established by Mankiewicz **[1972]**, **[1973]**, **[1974]**.

There is mounting evidence that the Lipschitz structure of a Banach space with the Radon-Nikodým property determines the space's linear topological nature. We say a Banach space X is *Lipschitz homeomorphic* to a subset of the Banach space Y whenever there is a function $f\colon X \to Y$ and constants $k, K > 0$ such that $k\,\|x - x'\| \leqq \|f(x) - f(x')\| \leqq K\,\|x - x'\|$ for all $x, x' \in X$. Generalizing the classical differentiation theorem of Rademacher **[1919]** to functions whose domain is the Hilbert cube and whose range is a Gel'fand space, Mankiewicz **[1972]**, **[1973]** has proved the

THEOREM (MANKIEWICZ). *If a Banach space X is Lipschitz homeomorphic to a subset of a Banach space Y with the Radon-Nikodým property then X is linearly homeomorphic to a subspace of Y (and therefore has the Radon-Nikodým property).*

It should be noted that some condition such as the Radon-Nikodým condition on Y in the above theorem is necessary. Indeed, Aharoni **[1974]** has shown that every separable Banach space is Lipschitz homeomorphic to a subset of c_0.

It remains unsettled whether Lipschitz homeomorphic Banach spaces (one of which has the Radon-Nikodým property) are linearly homeomorphic.

Integral operators. The study of integral operators from one L_p space to another is a massive subject. We are endeavoring here to give just a hint of how vector measures may be of use in this study. One of the first papers to treat integral operators from one L_p space to another with a distinct vector measure-theoretic flavor was Hille and Tamarkin **[1934]**. Vector measure-theoretic ideas then appeared explicitly in Dunford **[1936a]** and vividly in Dunford and Pettis **[1940]**. Most of the ideas found in this section are present in one form or another in Dunford and Pettis **[1940]**.

Theorem 4.4 is due to Uhl **[1970]** who also treats vector integral operators from a Banach function space to an arbitrary Banach space. This theorem is an extension of a related theorem of Luxemburg and Zaanen **[1963]**. Theorem 4.4 has also been studied by Grobler **[1970]**.

Definition 4.7 is from Dinculeanu **[1966a]**, **[1973]** who showed that $T\colon L_p(\mu) \to X$ has $|||T|||_p < \infty$ if and only if there exists an X-valued μ-continuous measure G of bounded q-variation ($p^{-1} + q^{-1} = 1$) such that $T(f) = \int f\,dG$ for all $\mathscr{F} \in L_p(\mu)$. This fact is embedded in the proof of Theorem 4.8.

Order summing operators were introduced by Schaefer **[1972]** in the context of tensor products of Banach lattices.

Theorems 4.8 and 4.9 have close ancestors in Dunford and Pettis **[1940]**. In the form given here, Theorem 4.8 comes from three sources. Wong **[1971]** essentially established the equivalence of (c) and (d) under the assumption that X is reflexive or X is a separable dual space. About the same time Uhl **[1971b]** proved that (a) and (d) imply each other and shortly thereafter Chaney **[1972]** showed that (b) and (d) are equivalent. It should be noted here that (a), (b) and (c) of Theorem 4.8 are equivalent for arbitrary Banach spaces X; but if for a fixed $p > 1$ one of them implies (d) then X has the Radon-Nikodým property. This remark borders on the trivial since any operator $T\colon L_1(\mu) \to X$ has $|||T|||_p < \infty$ for every $p > 1$.

The equivalence of (b) and (c) of Corollary 4.9 was proved by Wong **[1971]** in the case that X is reflexive or X^* is separable. The equivalence of (a) and (b) in the case $X = L_p(\mu)$ is from Uhl **[1971b]** who showed that an operator $T\colon L_p(\mu) \to L_p(\mu)$ ($1 < p < \infty$) has finite double norm (Zaanen **[1953]**) if and only if its adjoint $T^*\colon L_q(\mu) \to L_q(\mu)$ ($p^{-1} + q^{-1} = 1$) has $|||T^*|||_q < \infty$. In this case the double norm of T coincides with $|||T^*|||_q$. For another look at this, see Grobler **[1973]** and Van Eldik and Grobler **[1973]**.

Wong **[1971]** noted that the operators under scrutiny in Corollary 4.9 are absolutely p-summing. For more on p-summing operators consult the notes and remarks section of Chapter VIII. The discussion about the classical integral operators at the end of §4 is lifted from Uhl **[1971b]** and Wong **[1971]**.

The types of operators described by Theorems 4.8 and 4.9 have direct relations with some modern abstract theory of operators on Banach spaces. Wong **[1971]** has observed that if X^* has the Radon-Nikodým property and $T\colon X \to L_p(\mu)$ ($1 < p < \infty$) is such that $|||T^*|||_q < \infty$ ($p^{-1} + q^{-1} = 1$) then T is a p-nuclear operator and its adjoint is p-nuclear (Persson and Pietsch **[1969]** and Persson **[1971]**). In this connection Persson **[1971]** has proven that if X^* has the Radon-Nikodým property and $1 \leqq p \leqq \infty$ for any Banach space Y, then the p-integral and p-nuclear operators from X to Y are identical classes. Persson uses *ad hoc* vector measure methods that can be replaced by theorems from this survey.

Those who like the class of operators described by Corollary 4.9(b) will like the classes of p-decomposed operators and p-decomposing operators. An operator $T\colon X \to L_p(\mu)$ ($1 \leqq p \leqq \infty$) is called *p-decomposed* if there is a function $f \in L_p(\mu, X^*)$ such that, for each $x \in X$, $(Tx)(t) = f(t)(x)$ for μ-almost all t. If X and Y are Banach spaces, a bounded linear operator $S\colon X \to Y$ is called *p-decomposing* if for any $L_p(\mu)$ and any bounded linear operator $R\colon Y \to L_p(\mu)$ the composition $T = RS\colon X \to L_p(\mu)$ is p-decomposed. These terms are from Schwartz **[1970]**. Generalizing results of Schwartz **[1969]**, Kwapien **[1970]** and Saphar **[1972]**, Gordon and Saphar

[1976] have shown that if $1 < p < \infty$, a bounded linear operator is p-decomposing if and only if its adjoint is p-absolutely summing. For $p = 1$, this is the case if the dual of the domain of the operators has the Radon-Nikodým property. This last result can easily be viewed in the context of VI.4.

The Lewis-Stegall Theorem. Theorem 5.1 and Corollary 5.2 are due to Pełczyński **[1960]**. Each holds for c_0 or l_p $(1 < p < \infty)$ as well as l_1 as a glance at the proof in the text reveals.

Theorem 5.3 is due to D. R. Lewis and Stegall **[1973]**. It suggests the following unsolved problem: If P is a bounded linear projection on $L_1[0, 1]$ and the representing measure of P is differentiable on no set of positive measure then is the range of P isomorphic to $L_1[0, 1]$?

Theorem 5.3 is a variation of the following remarkable theorem of Lewis and Stegall **[1973]**.

THEOREM. *Let X be a Banach space. Then X^* is a copy of $l_1(\Gamma)$ for some set Γ if and only if for each Banach space Y every absolutely summing operator from X to Y is nuclear.*

Consequently, if μ is any measure, then an infinite dimensional complemented subspace of $L_1(\mu)$ with the Radon-Nikodým property is a copy of $l_1(\Gamma)$ for some set Γ.

Previously D. R. Lewis **[1972c]** had established the isometric version of the above theorem. Stegall **[1973]** improved the above theorem by showing that X^* is a copy of $l_1(\Gamma)$ for some set Γ if every compact absolutely summing operator with domain X is nuclear. For more on this refer to Stegall and Retherford **[1972]** as well as the above references.

V. MARTINGALES

Continuously studied since its introduction more than thirty years ago, martingale theory is one of the central components of probability theory. Today martingale theory has become recognized as an important tool in a diversity of topics in mathematical analysis. At this writing, martingale theory is having an increasingly important impact in Banach space theory. In this chapter the basic theory of martingales of Bochner integrable functions will be discussed. Then we shall try to demonstrate that martingales of Bochner integrable functions provide effective ways of studying the internal structure of arbitrary Banach spaces. The chapter opens with a discussion of conditional expectation operators on $L^p(\mu, X)$ and with some examples of martingales of Bochner integrable functions.

In the second section, the basic mean convergence and pointwise convergence theorems for martingales of Bochner integrable functions are established. Here we notice a certain dual personality of martingales of Bochner integrable functions. When they behave, they behave very well—exactly as their scalar-valued counterparts behave. When they do not behave well, they can bounce around wildly. The third section is mainly an exploitation of divergent martingales. Here the notions of σ-dentability and dentability are introduced from the point of view of divergent martingales. Then a remarkable transubstantiation takes place. With the help of martingales, we see the Radon-Nikodým property transform itself into an internal geometric property of Banach spaces. Throughout the chapter, X is a Banach space and (Ω, Σ, μ) is a finite measure space.

1. Conditional expectations and martingales. Basic to the theory of martingales is a process of averaging measurable functions over sub-σ-fields. This averaging operation is called a conditional expectation and is the major topic of this section. Once a certain familiarity with conditional expectations is obtained, we shall move on to define martingales of Bochner integrable functions. In the course, we shall see that trees in Banach spaces are easily viewed as martingales.

A sub-σ-field of Σ is a subset of Σ that contains Ω and that is a σ-field in its own right. If $f \in L_1(\Omega, \Sigma, \mu, X)$ and B is a sub-σ-field of Σ, then f is called B-measurable if $f \in L_1(\Omega, B, \mu \mid B, X)$.

DEFINITION 1. Let B be a sub-σ-field of Σ and $f \in L_1(\mu, X)$. An element g of

$L_1(\mu, X)$ is called the *conditional expectation of f relative to B* if g is B-measurable and

$$\int_E g\, d\mu = \int_E f\, d\mu \quad \text{for all } E \in B.$$

In this case g is denoted by $E(f \mid B)$.

It is clear that $E(f \mid B)$ is uniquely defined whenever it is defined. It is not so clear that $E(f \mid B)$ is defined for all $f \in L_1(\mu, X)$. Before clearing up this matter we shall look at a simple illustration.

EXAMPLE 2. *A conditional expectation operator.* Let (A_n) be a sequence of disjoint sets in Σ with $\bigcup_{n=1}^{\infty} A_n = \Omega$. Let B be the σ-field of all unions of members of (A_n). If $f \in L_1(\mu, X)$ a straightforward check verifies the equality

$$E(f \mid B) = \sum_{n=1}^{\infty} \frac{\int_{A_n} f\, d\mu}{\mu(A_n)} \chi_{A_n} \qquad (0/0 = 0).$$

In view of this example, it is plain that the operators E_π used in III.2.1 are nothing but trivial conditional expectations. The next lemma helps us establish the existence of conditional expectations on $L_1(\mu, X)$.

LEMMA 3. *Let B be a sub-σ-field of Σ. Then $E(f \mid B)$ exists for every $f \in L_1(\mu)$ $(= L_1(\mu, \mathbf{R}))$. In fact if $f \in L_p(\mu)$ $(1 \leqq p < \infty)$, then*

$$\| E(f \mid B) \|_p \leqq \| f \|_p.$$

Consequently $E(\cdot \mid B)$ is a linear contractive projection on each $L_p(\mu)$, $1 \leqq p < \infty$.

PROOF. Let $f \in L_1(\mu)$, and define a scalar measure λ on B by $\lambda(A) = \int_A f\, d\mu$ for $A \in B$. Then λ is obviously a $\mu \mid B$-continuous finite measure. By the (scalar) Radon-Nikodým theorem, there exists a B-measurable $g \in L_1(\mu)$ such that

$$\lambda(A) = \int_A g\, d\mu \quad \text{for all } A \in B.$$

A glance at the definition of λ shows that $g = E(f \mid B)$. Furthermore, note that this construction shows that $E(E(f \mid B) \mid B) = E(f \mid B)$ for $f \in L_1(\mu)$ and that $E(\cdot \mid B)$ is a linear projection whose range plainly coincides with the B-measurable functions in $L_1(\mu)$.

To complete the proof, it remains to show that $\| E(f \mid B) \|_p \leqq \| f \|_p$ for f in $L_p(\mu)$. We shall prove this by proving Jensen's inequality. To this end, note first that $E(\cdot \mid B)$ maps the nonnegative functions into nonnegative functions and preserves the constant functions. Hence if $at + b$ is a support line for the convex function $\Phi(t) = t^p$ (t real) and $f \in L_p(\mu)$,

$$a(E(f \mid B)) + b = E(af + b \mid B) \leqq E(|f|^p \mid B).$$

Taking suprema over appropriate support lines gives $|E(f \mid B)|^p \leqq E(|f|^p \mid B)$. Consequently one obtains

$$\int_\Omega |E(f \mid B)|^p\, d\mu \leqq \int_\Omega E(|f|^p \mid B)\, d\mu = \int_\Omega |f|^p\, d\mu$$

since $\Omega \in B$. It follows that the operator $E(\cdot \mid B)$ is a contraction. This completes the proof.

One comment on the proof of Lemma 3 should be made: If X has the Radon-Nikodým property, roughly the same proof would work with $L_p(\mu, X)$ replacing $L_p(\mu)$. Even if X does not have the Radon-Nikodým property, the conclusions of Lemma 3 can be parlayed from the context of $L_p(\mu)$ to the context of $L_p(\mu, X)$.

THEOREM 4. *Let B be a sub-σ-field of Σ. Then $E(f\,|\,B)$ exists for every $f \in L_1(\mu, X)$. In fact if $f \in L_p(\mu, X)$ $(1 \leqq p < \infty)$, then*

$$\|E(f\,|\,B)\|_p \leqq \|f\|_p.$$

Consequently $E(\cdot\,|\,B)$ is a linear contractive projection on $L_p(\mu, X)$, $1 \leqq p < \infty$.

PROOF. Let $f = \sum_{i=1}^n x_i \chi_{E_i}$ where $x_i \in X$, $E_i \in \Sigma$, and $E_i \cap E_j = \emptyset$ for $i \neq j$, be a simple function in $L_p(\mu, X)$ $(1 \leqq p < \infty)$. Define $E(f\,|\,B)$ by

$$E(f\,|\,B) = \sum_{i=1}^n x_i E(\chi_{E_i}\,|\,B)$$

where $E(\chi_E\,|\,B)$ is the conditional expectation of $\chi_E \in L_p(\mu, R)$ whose existence is ensured by Lemma 3. There should be no confusion arising from the dual use of the symbol $E(\cdot\,|\,B)$. A routine security check reveals that $E(\cdot\,|\,B)$ is unambiguously defined and is linear on the dense subset of simple functions on $L_p(\mu, X)$.

Moreover, if f is as above, then

$$\begin{aligned}
\|E(f\,|\,B)\|_p &= \left(\int_\Omega \|E(f\,|\,B)\|^p \,d\mu\right)^{1/p} \leqq \left(\int_\Omega \left(\sum_{i=1}^n \|x_i\| E(\chi_{E_i}\,|\,B)\right)^p d\mu\right)^{1/p} \\
&= \left(\int_\Omega E\left(\sum_{i=1}^n \|x_i\| \chi_{E_i}\,|\,B\right)^p d\mu\right)^{1/p} \\
&\leqq \left\|\sum_{i=1}^n \|x_i\| \chi_{E_i}\right\|_p \quad \text{by Lemma 3} \\
&= \left(\int_\Omega \|f\|^p \,d\mu\right)^{1/p} = \|f\|_p.
\end{aligned}$$

Hence $E(\cdot\,|\,B)$ has a contractive linear extension to all of $L_p(\mu, X)$, still denoted by $E(\cdot\,|\,B)$. That the extended $E(\cdot\,|\,B)$ has the required properties is clear to anyone who has read this far.

Now that the existence of conditional expectations has been dispensed with, it is possible to define the main objects of study for this chapter.

DEFINITION 5. Let $(B_\tau, \tau \in T)$ be a monotone increasing net of sub-σ-fields of Σ $(B_{\tau_1} \subseteq B_{\tau_2}$ for $\tau_1 \leqq \tau_2$ in $T)$. A net $(f_\tau, \tau \in T)$ in $L_p(\mu, X)$ $(1 \leqq p < \infty)$ over the same directed set T is a *martingale* if

$$E(f_\tau\,|\,B_{\tau_1}) = f_{\tau_1}$$

for all $\tau \geqq \tau_1$.

Usually a martingale of the above form will be denoted by $(f_\tau, B_\tau, \tau \in T)$ to display both the functions and sub-σ-fields involved.

Probably the easiest way to produce a martingale in $L_p(\mu, X)$ is given in

EXAMPLE 6. *Martingales in $L_p(\mu, X)$*. Let $(B_\tau, \tau \in T)$ be an increasing net of sub-σ-fields of Σ, and $f \in L_p(\mu, X)$ $(1 \leqq p < \infty)$. Then $(E(f\,|\,B_\tau), B_\tau, \tau \in T)$ is a martingale. Later it will be seen that all $L_p(\mu, X)$-norm convergent martingales arise in this manner.

So far our work with conditional expectations and martingales has differed from the classical scalar theory only by notation. The next example allows us to cast certain structures in Banach spaces as martingales.

EXAMPLE 7. *An infinite tree in X as a martingale in $L_1([0, 1), X)$ and a nonconvergent bounded martingale in $L_1([0, 1), X)$.* Recall that an infinite tree J in X is a sequence (x_n) in X with the property that $x_n = (x_{2n} + x_{2n+1})/2$ for all $n \geqq 1$.

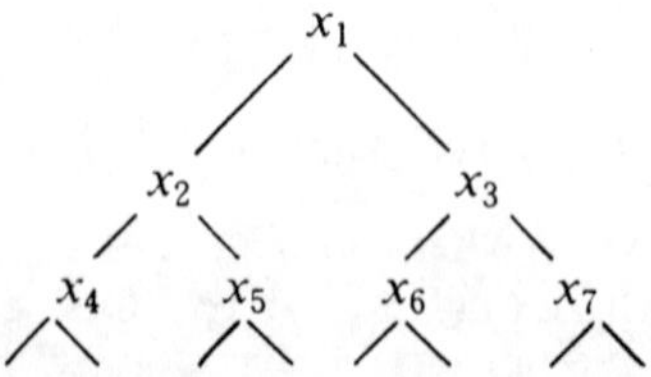

Sometimes it is helpful to think of x_1 as the trunk and $x_j, j = 2^k, 2^k + 1, \cdots, 2^{k+1} - 1$ as the kth year's growth.[2] To realize a tree $J = (x_n)$ as a martingale in $L_1([0, 1), X)$, let μ be Lebesgue measure and write

$$f_1 = x_1\chi_{[0,1)}, \qquad f_2 = x_2\chi_{[0,1/2)} + x_3\chi_{[1/2,1)},$$

and

$$f_k = \sum_{i=2^{k-1}}^{2k-1} x_i\chi_{I_{k,i}}$$

where $I_{k,i} = [(i - 2^{k-1})/2^{k-1}, (i - 2^{k-1} + 1)/2^{k-1})$ for $i = 2^{k-1}, 2^{k-1} + 1, \cdots, 2^k - 1$ and $k \geqq 1$. Immediately note that $\int_{[0,1)} f_1\, d\mu = \int_{[0,1)} f_2\, d\mu$ since $(x_2 + x_3)/2 = x_1$. For the same reason, it follows that $\int_{I_{k,i}} f_{k+1}\, d\mu = \int_{I_{k,i}} f_k\, d\mu$ for each $i = 2^{k-1}, \cdots, 2^k - 1$ and $k \geqq 1$. Thus if B_0 is the trivial σ-field consisting of $\emptyset$ and $[0,1)$ and B_k is the finite σ-field generated by $\{I_{k,i}, i = 2^{k-1}, 2^{k-1} + 1, \cdots, 2^k - 1\}$, $k \geqq 1$, then (f_k, B_k) is a martingale in $L_1([0, 1), X)$.

The prototype example of a nontrivial infinite tree can be found in the familiar space $L_1[0, 1)$: Let $x_1 = \chi_{[0,1)}$, $x_2 = 2\chi_{[0,1/2)}$, $x_3 = 2\chi_{[1/2,1)}$ and $x_i = 2^{k-1}\chi_{I_{k,i}}$ for $i = 2^{k-1}, \cdots, 2^k - 1$ and $k \geqq 1$ where the intervals $I_{k,i}$ are exactly as above. Furthermore, note that

$$\|x_1 - x_2\| = \int_{[0,1)} |\chi_{[0,1)} - 2\chi_{[0,1/2)}|\, d\mu = \int_{[0,1)} 1\, d\mu = 1.$$

For the same reason, one has

$$\|x_n - x_{2n}\| = \|x_n - x_{2n+1}\| = 1 \quad \text{for all } n.$$

Transferring to the martingale in $L_1([0, 1), X)$ with $X = L_1[0, 1)$ constructed above, we find that $\|f_n(t)\|_{L_1} = 1$ for all $t \in [0, 1)$ since $\|x_k\| = 1$ for all k and $\|f_n(t) - f_{n+1}(t)\|_{L_1} = 1$ for all $t \in [0, 1)$ (since if $t \in [0, 1)$ and $I_{n,i}$ is selected such that $t \in I_{n,i}$, then $f_n(t) = x_i$ and $f_{n+1}(t)$ is either x_{2i} or x_{2i+1}). Accordingly (f_n, B_n) is martingale of uniformly bounded functions in $L_1([0, 1), L_1[0, 1))$ with the property $\|f_n - f_{n+1}\|_{L_1[0,1)} \equiv 1$ for all n. It follows immediately that (f_n, B_n) is not convergent in $L_1([0, 1), L_1[0, 1))$-norm. Those familiar with scalar-valued

[2]These trees have great economic potential. Cuttings from it immediately take root and grow instantly into trees of the same size as the parent tree.

martingale theory will immediately realize that such an example is impossible in the scalar-valued context.

A property of the infinite tree in $L_1[0, 1)$ constructed above is isolated in

DEFINITION 8. An infinite tree (x_n) in X is called an *infinite δ-tree* if there is a $\delta > 0$ such that $\| x_n - x_{2n} \| \geqq \delta$ and $\| x_n - x_{2n+1} \| \geqq \delta$ for all n.

Example 7 displays an infinite 1-tree in $L_1[0, 1)$. Naturally if a bounded infinite δ-tree in X is transferred to a martingale in $L_1([0, 1), X)$ as in the first part of Example 7, the resulting martingale (f_n, B_n) is a martingale of uniformly bounded functions with the property that $\| f_n(t) - f_{n+1}(t) \|_X \geqq \delta$ for all $t \in [0, 1)$. Thus whenever there is a bounded infinite δ-tree in a Banach space X, there is a non-convergent $L_\infty(\mu, X)$-bounded martingale in $L_1([0, 1), X)$.

Obviously no finite dimensional Banach space contains a bounded infinite δ-tree and this property is shared by many infinite dimensional Banach spaces. This fact is a consequence of the convergence theorems for martingales of Bochner integrable functions which we shall presently study.

2. Convergence theorems. This section is devoted to the important problems of deciding when a martingale $(f_\tau, B_\tau, \tau \in T)$ in $L_p(\mu, X)$ $(1 \leqq p < \infty)$ is $L_p(\mu, X)$-norm convergent and when a martingale (f_n, B_n) in $L_1(\mu, X)$ is almost everywhere convergent in X-norm. We shall learn that Radon-Nikodým theorems for vector measures play the central role in solving these problems. Further, we shall see that the proofs of the martingale convergence theorem are formally the same as some of the classical convergence proofs for martingales of scalar-valued functions. Again X is a Banach space and (Ω, Σ, μ) is a finite measure space.

A simple but crucial property of a martingale $(f_\tau, B_\tau, \tau \in T)$ in $LB_p(\mu, X)$ $(1 \leqq p < \infty)$ is that if $E \in \bigcup_\tau B_\tau$, then

$$\lim_\tau \int_E f_\tau \, d\mu = F(E)$$

exists trivially. To see this let $E \in \bigcup_\tau B_\tau$. Since $(B_\tau, \tau \in T)$ is a monotone increasing net of sub-σ-fields of Σ, there is $\tau_1 \in T$ such that $E \in B_\tau$ for all $\tau \geqq \tau_1$. Consequently for $\tau \geqq \tau_1$ one has

$$\int_E f_\tau \, d\mu = \int_E E(f_\tau | B_{\tau_1}) \, d\mu = \int_E f_{\tau_1} \, d\mu$$

by the martingale property. Hence the net $(\int_E f_\tau \, d\mu, \tau \in T)$ is eventually constant and therefore convergent. This simple fact tells us the full story of norm convergent martingales and relates norm convergent martingales to Radon-Nikodým derivatives.

THEOREM 1. *Let $1 \leqq p < \infty$. A martingale $(f_\tau, B_\tau, \tau \in T)$ in $L_p(\mu, X)$ converges in $L_p(\mu, X)$-norm if and only if there exists $f \in L_p(\mu, X)$ such that for each $E \in \bigcup_\tau B_\tau$ one has*

$$\lim_\tau \int_E f_\tau \, d\mu = F(E) = \int_E f \, d\mu.$$

PROOF. Suppose $\lim_\tau f_\tau = f$ in $L_p(\mu, X)$-norm. Since the operation of integration of an $L_p(\mu, X)$ function over a set in Σ defines a bounded linear operator from

$L_p(\mu, X)$ to X, one has $F(E) = \lim_\tau \int_E f_\tau \, d\mu = \int_E f \, d\mu$ for all $E \in \Sigma$ and hence for all $E \in \bigcup_\tau B_\tau$.

For the converse, suppose there is $f \in L_p(\mu, X)$ with $\lim_\tau \int_E f_\tau \, d\mu = F(E) = \int_E f \, d\mu$ for all $E \in \bigcup_\tau B_\tau$. Let B_∞ be the σ-field generated by $\bigcup_\tau B_\tau$ and set $f_\infty = E(f \mid B_\infty)$. Then $F(E) = \int_E f_\infty \, d\mu$ for all $E \in \bigcup_\tau B_\tau$. In addition, one has $E(f_\infty \mid B_\tau) = f_\tau$ for all $\tau \in T$.

Now it will be shown that $\lim_\tau \| f_\tau - f_\infty \|_p = 0$. At this point note that there is no loss of generality in assuming $B_\infty = \Sigma$. Further note that $\bigcup_\tau B_\tau$ is a field since $(B_\tau, \tau \in T)$ is monotone increasing. By virtue of the fact that $\sigma(\bigcup_\tau B_\tau) = \Sigma$, simple functions of the form $\sum_{i=1}^n x_i \chi_{E_i}$ where $x_i \in X$ and $E_i \in \bigcup_\tau B_\tau$ are dense in $L^p(\mu, X)$ (this is precisely the stage that the hypothesis $1 \leqq p < \infty$ is used). Consequently for each $\varepsilon > 0$ there is a simple function $f_\varepsilon = \sum_{i=1}^n x_i \chi_{E_i}$ with $x_i \in X$ and $E_i \in \bigcup_\tau B_\tau$ such that $\| f_\varepsilon - f_\infty \|_p < \varepsilon/2$. Again since $(B_\tau, \tau \in T)$ is a monotone increasing net, there is an index τ_0 such that $E_i \in B_{\tau_0}$ for all $i = 1, \cdots, n$. Thus f_ε is B_τ-measurable for all $\tau \geqq \tau_0$ and

$$E(f_\varepsilon \mid B_\tau) = f_\varepsilon \quad \text{for } \tau \geqq \tau_0.$$

Now if $\tau \geqq \tau_0$, then

$$\begin{aligned} \| f_\tau - f_\infty \|_p &\leqq \| f_\tau - f_\varepsilon \|_p + \| f_\varepsilon - f_\infty \|_p \\ &= \| E(f_\infty - f_\varepsilon \mid B_\tau) \|_p + \| f_\varepsilon - f_\infty \|_p \\ &\leqq 2 \| f_\infty - f_\varepsilon \|_p < \varepsilon. \end{aligned}$$

This completes the proof.

The next corollary is simply a translation of Theorem 1 into a form familiar to many.

COROLLARY 2. *A martingale $(f_\tau, B_\tau, \tau \in T)$ in $L_p(\mu, X)$ $(1 \leqq p < \infty)$ is convergent in $L_p(\mu, X)$-norm if and only if there exists $f \in L_p(\mu, X)$ such that $E(f \mid B_\tau) = f_\tau$ for all $\tau \in T$.*

Before making the statement of the next corollary, which happens to be the main result of the section, let us examine $F(E) = \lim_\tau \int_E f_\tau \, d\mu$ for a martingale $(f_\tau, B_\tau, \tau \in T)$. In order that $\lim_\tau f_\tau = f$ in $L_p(\mu, X)$, $F(E)$ must be equal to $\int_E f \, d\mu$ for $E \in \bigcup_\tau B_\tau$. Hence two necessary conditions must be satisfied.

(i) $\sup \| f_\tau \|_p$ must be finite (since $\| f_\tau \|_p = \| E(f \mid B_\tau) \|_p \leqq \| f \|_p$) and

(ii) F must be μ-continuous on $\bigcup_\tau B_\tau$ (since $\lim_{\mu(E) \to 0} F(E) = \lim_{\mu(E) \to 0} \int_E f \, d\mu = 0$).

When $p > 1$, (i) guarantees (ii) by the Hölder inequality. Specifically designed to meet the needs of $L_1(\mu, X)$ bounded martingales is

DEFINITION 3. A martingale $(f_\tau, B_\tau, \tau \in T)$ in $L_1(\mu, X)$ is *uniformly integrable* if

$$\lim_{\mu(E) \to 0,\, E \in B_\tau} \int_E \| f_\tau \| \, d\mu = 0$$

uniformly in $\tau \in T$.

COROLLARY 4 (MARTINGALE MEAN CONVERGENCE THEOREM). *Let X have the Radon-Nikodým property, let $1 \leqq p < \infty$, and let $(f_\tau, B_\tau, \tau \in T)$ be a martingale in $L_p(\mu, X)$. Then $\lim_\tau f_\tau$ exists in $L_p(\mu, X)$-norm if and only if*

(i) $p = 1$, $\sup_\tau \|f_\tau\|_1 < \infty$ *and* $(f_\tau, B_\tau, \tau \in T)$ *is uniformly integrable, or*
(ii) $1 < p < \infty$ *and* $\sup_\tau \|f_\tau\|_p < \infty$.

PROOF. To prove the sufficiency of (i), for $E \in \bigcup_\tau B_\tau$, set $F(E) = \lim_\tau \int_E f_\tau \, d\mu$. Since $(f_\tau, B_\tau, \tau \in T)$ is uniformly integrable, it is plain that $\lim_{\mu(E)\to 0} F(E) = 0$ on $\bigcup_\tau B_\tau$. Furthermore if $\pi \subseteq \bigcup_\tau B_\tau$ is a partition of Ω, then there is an index τ_0 such that $\pi \subseteq B_{\tau_0}$. Consequently one has

$$\sum_{E\in\pi} \|F(E)\| = \sum_{E\in\pi} \left\| \int_E f_{\tau_0} \, d\mu \right\| \leqq \int_\Omega \|f_{\tau_0}\| \, d\mu \leqq \sup_\tau \|f_\tau\|_1 < \infty.$$

Hence F is of bounded variation on $\bigcup_\tau B_\tau$. An appeal to I.5.2 produces a μ-continuous vector measure G of bounded variation on Σ_0, the σ-field generated by $\bigcup_\tau B_\tau$, such that $G(E) = F(E)$ for all $E \in \bigcup_\tau B_\tau$. Since X has the Radon-Nikodým property, there is $f \in L_1(\mu | \Sigma_0, X)$ such that $G(E) = \int_E f \, d\mu$ for all $E \in \Sigma_0$. But if $E \in \bigcup_\tau B_\tau$, then

$$\lim_\tau \int_E f_\tau \, d\mu = F(E) = G(E) = \int_E f \, d\mu.$$

An invocation of Theorem 1 completes the proof of statement (i).

To prove the sufficiency of statement (ii), let $1 < p < \infty$ and suppose $\sup_\tau \|f_\tau\|_p < \infty$. An application of the Hölder inequality shows that $(f_\tau, B_\tau, \tau \in T)$ is also a uniformly integrable bounded martingale in $L_1(\mu, X)$. By the sufficiency of (i), there is $f \in L_1(\mu, X)$ such that $\lim_\tau \|f_\tau - f\|_1 = 0$. Consequently we obtain

$$\int_E f \, d\mu = \lim_\tau \int_E f_\tau \, d\mu = F(E)$$

for all $E \in \bigcup_\tau B_\tau$. Now, if it can be shown that $f \in L_p(\mu, X)$, then an appeal to Theorem 1 will complete the proof. To this end, select a sequence (τ_n) in T such that $\lim_n \|f_{\tau_n} - f\|_1 = 0$ and such that $\lim_n f_{\tau_n} = f$ μ-almost everywhere as well. By Fatou's lemma, we have

$$\int_\Omega \|f\|^p \, d\mu \leqq \underline{\lim}_n \int_\Omega \|f_{\tau_n}\|^p \, d\mu \leqq \sup_\tau \|f_\tau\|_p^p < \infty.$$

This completes the proof of the sufficiency of (ii).

The necessity of (i) and (ii) should be clear on the basis of the discussion preceding the statement of this corollary.

Now let us see what the martingale mean convergence theorem says in terms of infinite δ-trees.

COROLLARY 5. *No Banach space with the Radon-Nikodým property contains a bounded infinite δ-tree.*

PROOF. Consult Example 1.7 and the discussion following it.

The above corollary is a harbinger of the geometric concept of dentability which will be discussed in §3.

The martingale mean convergence theorem has a converse that ties mean martingale convergence to the Radon-Nikodým property.

THEOREM 6. *Suppose X is a Banach space such that for every finite measure space*

(Ω, Σ, μ) every bounded uniformly integrable martingale in $L_1(\mu, X)$ converges in $L_1(\mu, X)$-norm. Then X has the Radon-Nikodým property.

PROOF. The proof should be clear to anyone familiar with Chapter III. Most of the Radon-Nikodým theorems in Chapter III are, in fact, martingale convergence theorems. If the proof is still not clear, read on. Let (Ω, Σ, μ) be a fixed finite measure space. Let P be the class of all partitions of Ω into Σ sets and direct P by refinement. If $F: \Sigma \to X$ is a μ-continuous vector measure of bounded variation, define

$$f_\pi = \sum_{E \in \pi} \frac{F(E)}{\mu(E)} \chi_E$$

observing the convention that $0/0 = 0$. Let B_π be the (trivial) σ-field generated by π. Evidently $(f_\pi, B_\pi, \pi \in P)$ is a martingale in $L_1(\mu, X)$ with $F(E) = \lim_\pi \int_E f_\pi \, d\mu$ for every $E \in \Sigma$. A quick computation shows that

$$\int_\Omega \| f_\pi \| \, d\mu \leqq |F|(\Omega) < \infty$$

for every $\pi \in P$. Also, since $F \ll \mu$, we have $|F| \ll \mu$. Hence for each $\varepsilon > 0$, there is $\delta > 0$ such that $|F|(E) < \varepsilon$ whenever $\mu(E) < \delta$. Now if $E \in B_\pi$ and $\mu(E) < \delta$, then

$$\int_E \| f_\pi \| \, d\mu \leqq |F|(E) < \varepsilon.$$

Thus $(f_\pi, B_\pi, \pi \in P)$ is uniformly integrable. By hypothesis $\lim_\pi \| f_\pi - f \|_1 = 0$ for some $f \in L_1(\mu, X)$. Thus we have $F(E) = \lim_\pi \int_E f_\pi \, d\mu = \int_E f \, d\mu$ for all $E \in \Sigma$. This completes the proof.

Virtually all of the Radon-Nikodým theorems of Chapter III involve testing a martingale $(f_\pi, B_\pi, \pi \in P)$ of the form used in the proof of Theorem 6 for convergence. The alert reader will note that by replacing the special martingales $(f_\pi, B_\pi, \pi \in P)$ by more general martingales, each of the Radon-Nikodým theorems of Chapter III can be recast as a mean martingale convergence theorem. On the other hand, Corollary 4 allows us to prove that a Banach space X lacks the Radon-Nikodým property if we can construct an $L_1([0, 1], X)$ bounded uniformly integrable martingale that does not converge. Use will be made of this fact in the next section.

Next, the subject of almost everywhere convergence of martingales of the form $(f_n, B_n, n \in N)$ indexed by the positive integers will be studied. The full story is a consequence of

LEMMA 7 (MAXIMAL LEMMA). *Let (f_n, B_n) be a martingale in $L_1(\mu, X)$ and let $\delta > 0$. If $S_\delta = \{\omega: \sup_n \| f_n(\omega) \| > \delta\}$, then*

$$\limsup_n \int_{S_\delta} (\| f_n \| - \delta) \, d\mu \geqq 0.$$

Consequently

$$\mu\left(\left\{\omega: \sup_n \| f_n(\omega) \| > \delta\right\}\right) \leqq \frac{1}{\delta} \sup_n \| f_n \|_1.$$

PROOF. To prove the first assertion, write for each positive integer m

$$S_\delta^m = \{\omega: \|f_m(\omega)\| > \delta, \|f_j(\omega)\| \leqq \delta \quad \text{for } j < m\}.$$

Then $S_\delta = \bigcup_{m=1}^\infty S_\delta^m$, $S_\delta^n \cap S_\delta^m = \emptyset$ for $m \neq n$ and $S_\delta^m \in B_m$ for each m.
Accordingly we have

$$\limsup_n \int_{S_\delta} (\|f_n\| - \delta)\, d\mu \geqq \limsup_n \lim_k \sum_{m=1}^k \int_{S_\delta^m} (\|f_n\| - \delta)\, d\mu.$$

But now, if k is fixed and $n \geqq k$, then the facts that $E(f_n | B_m) = f_m$ for $m = 1, \cdots, k$ and that $E(\cdot | B_m)$ is a contraction on $L_1(\mu | A, X)$ for any $A \in B_m$ imply

$$\int_{S_\delta^m} (\|f_n\| - \delta)\, d\mu \geqq \int_{S_\delta^m} (\|f_m\| - \delta)\, d\mu.$$

It follows that

$$\begin{aligned}\limsup_n \int_{S_\delta} (\|f_n\| - \delta)\, d\mu &\geqq \limsup_n \lim_k \sum_{m=1}^k \int_{S_\delta^m} (\|f_m\| - \delta)\, d\mu \\ &= \sum_{m=1}^\infty \int_{S_\delta^m} (\|f_m\| - \delta)\, d\mu \geqq 0.\end{aligned}$$

This proves the first assertion.
For the second assertion, write

$$\begin{aligned}\sup_n \frac{\|f_n\|_1}{\delta} &\geqq \limsup_n \frac{1}{\delta} \int_{S_\delta} \|f_n\|\, d\mu \geqq \frac{\delta}{\delta} \mu(S_\delta) \\ &= \mu\left(\left\{\omega: \sup_n \|f_n(\omega)\| > \delta\right\}\right).\end{aligned}$$

This completes the proof.

THEOREM 8. *An $L_1(\mu, X)$ convergent martingale $(f_n, B_n, n \in N)$ converges to its $L_1(\mu, X)$-limit almost everywhere.*

PROOF. Let $\lim_n f_n = f$ in $L_1(\mu, X)$-norm. If $\varepsilon, \delta > 0$, then there is n_0 such that if $n, m \geqq n_0$, then $\|f_n - f_m\|_1 < \varepsilon\delta$. Now fix $m \geqq n_0$ and note that $(f_n - f_m, B_n, n \geqq m)$ is an $L_1(\mu, X)$-martingale. According to the maximal lemma, we have

$$\mu\left\{\omega: \sup_{n \geqq m} \|f_n(\omega) - f_m(\omega)\| > \varepsilon\right\} \leqq \frac{1}{\varepsilon} \sup_{n \geqq m} \|f_n - f_m\|_1 \leqq \frac{1}{\varepsilon} \varepsilon\delta = \delta.$$

It follows immediately that (f_n) is almost uniformly Cauchy. Since $\lim_n f_n = f$ in $L_1(\mu, X)$-norm, it is clear that $\lim_n f_n = f$ almost everywhere as well.

The stage is now set for an examination of the almost everywhere convergence of $L_1(\mu, X)$-martingales that may fail to be $L_1(\mu, X)$-norm convergent. Let $(f_n, B_n, n \in N)$ be an $L_1(\mu, X)$-bounded martingale. As we have done in the past, define $F: \bigcup_n B_n \to X$ by

$$F(E) = \lim_n \int_E f_n\, d\mu, \qquad E \in \bigcup_n B_n.$$

As in the discussion preceding Definition 3, the fact that (f_n, B_n) is $L_1(\mu, X)$-bounded ensures F is of bounded variation. If $F \ll \mu$, then we can hunt for a Radon-Nikodým derivative as before. The case of interest now is the case in which F is *not* μ-continuous. This case arises when (f_n, B_n) is $L_1(\mu, X)$-bounded but not uniformly integrable. Now in this case, the Lebesgue decomposition Theorem I.5.9 produces unique finitely additive measures G and H on $\bigcup_n B_n$ such that $F = G + H$, G and H are both of bounded variation, and such that $|G|$ and $|H|$ are mutually singular with $|G| \ll \mu$ and H μ-singular.

THEOREM 9 (MARTINGALE POINTWISE CONVERGENCE THEOREM). *Let (f_n, B_n) be an $L_1(\mu, X)$-bounded martingale. Let*

$$F(E) = \lim_n \int_E f_n \, d\mu, \qquad E \in \bigcup_n B_n,$$

and $F = G + H$, $|G| \ll \mu$, $|H| \perp \mu$, be the Lebesgue decomposition of F with respect to μ. Then $\lim_n f_n$ *exists almost everywhere if and only if G has an $L_1(\mu, X)$-Radon-Nikodým derivative g.*

In this case, we have

$$\lim_n f_n = E\left(g \,\middle|\, \sigma\left(\bigcup_n B_n\right)\right)$$

where $\sigma(\bigcup_n B_n)$ is the σ-field generated by $\bigcup_n B_n$.

PROOF. Throughout it will be assumed that $\Sigma = \sigma(\bigcup_{n=1}^{\infty} B_n)$. For the sufficiency, let $g \in L_1(\mu, X)$ be the Radon-Nikodým derivative of G with respect to μ and set $g_n = E(g \mid B_n)$. Then (g_n, B_n) is an $L_1(\mu, X)$-bounded martingale which converges to g both in $L_1(\mu, X)$-norm and μ-almost everywhere. Next write $h_n = f_n - g_n$. Then (h_n, B_n) is an $L_1(\mu, X)$-bounded martingale. If we can show that $\lim_n h_n = 0$ μ-almost everywhere, then we will be done.

To this end, let $E \in B_n$ and note that

$$\begin{aligned} G(E) + H(E) = F(E) &= \int_E f_n \, d\mu = \int_E g_n \, d\mu + \int_E h_n \, d\mu \\ &= G(E) + \int_E h_n \, d\mu. \end{aligned}$$

Thus $H(E) = \int_E h_n \, d\mu$ for all $E \in B_n$. Keep this in mind for a moment and recall that μ and $|H|$ are mutually singular measures on the field $\bigcup_n B_n$. Accordingly if $\varepsilon, \delta > 0$ and $\varepsilon < 1$, then there is a set $A \in \bigcup_n B_n$ such that

$$\mu(\Omega \backslash A) + |H|(A) < \varepsilon\delta/2.$$

Choose n_0 such that $A \in B_{n_0}$ and use the maximal lemma to show that

$$\begin{aligned} &\mu\left(\left\{\omega : \sup_{n \geqq n_0} \|h_n(\omega)\| > \varepsilon\right\}\right) \\ &= \mu\left(\left\{\omega : \sup_{n \geqq n_0} \|h_n(\omega)\| > \varepsilon\right\} \backslash A\right) + \mu\left(\left\{\omega : \sup_{n \geqq n_0} \|h_n(\omega)\| > \varepsilon\right\} \cap A\right) \\ &\leqq \varepsilon\delta/2 + (1/\varepsilon) \sup_{n \geqq n_0} \int_A \|h_n\| \, d\mu \\ &\leqq \varepsilon\delta/2 + (1/\varepsilon)|H|(A) < \varepsilon\delta/2 + \delta/2 < \delta. \end{aligned}$$

It follows that $\lim_n h_n = 0$ μ-almost uniformly and that $\lim_n f_n = \lim_n g_n + \lim_n h_n = g$ μ-almost everywhere. This proves the sufficiency.

For the converse, suppose $\lim_n f_n = \phi \in L_1(\mu, X)$ μ-almost everywhere. Write

$g_n = E(\phi | B_n)$ and note that $(f_n - g_n, B_n)$ is an $L_1(\mu, X)$-bounded martingale. Define H_1 on $\bigcup_n B_n$ by

$$H_1(E) = \lim_n \int_E (f_n - g_n)\, d\mu \quad \text{for } E \in \bigcup_n B_n.$$

Also let $H_{11} + H_{12} = H_1$ be the Lebesgue decomposition of H_1 with respect to μ where $H_{11} \ll \mu$ and H_{12} is μ-singular. If H_{11} is not identically zero, there is $x_0^* \in X^*$ such that $x_0^* H_{11}$ is not identically zero. In addition, we have

$$\lim_n \int_E x_0^*(f_n - g_n)\, d\mu = x_0^* H_{11}(E) + x_0^* H_{12}(E)$$

for all $E \in \bigcup_n B_n$. By the (scalar) Radon-Nikodým theorem, $x_0^* H_{11}$ has a nonzero Radon-Nikodým derivative $h \in L_1(\mu)$. By the sufficiency part of the proof above, $\lim_n x_0^*(f_n - g_n) = h$ almost everywhere. But $\lim_n f_n - g_n = \lim_n f_n - \phi + \phi - g_n = 0$ in X-norm almost everywhere. This contradiction proves that H_1 is μ-singular. Now set $G_1(E) = \int_E \phi\, d\mu$ for $E \in \bigcup_n B_n$. Then for $E \in \bigcup_n B_n$, we obtain

$$\begin{aligned} F(E) &= \lim_n \int_E f_n\, d\mu = \lim_n \int_E g_n\, d\mu + H_1(E) \\ &= G_1(E) + H_1(E). \end{aligned}$$

Since $G_1 \ll \mu$ and H_1 is μ-singular, it follows that $G_1 = G$ and $H_1 = H$. Thus G has a Radon-Nikodým derivative ϕ and $\lim_n f_n = \phi$ μ-almost everywhere, and the proof is over.

3. Dentable sets and the Radon-Nikodým property. Suppose we are given a Banach space X and are told to prove quickly that X lacks the Radon-Nikodým property. One decisive course of action would be to retort with a martingale (f_n, B_n) of countably valued functions in $L_1([0, 1), X)$ with the properties that

(i) $\sup_n \|f_n\|_\infty < \infty$ and

(ii) $\|f_n(t) - f_{n+1}(t)\|_X \geqq \varepsilon$

for some $\varepsilon > 0$ and all $t \in [0, 1)$. Such a martingale is plainly uniformly integrable and nonconvergent. Let us now see what would be involved in the construction of this sort of martingale.

First, since each f_n is countably valued, each f_n has the form

$$f_n = \sum_{E \in \Delta_n} x_E \chi_E$$

where $x_E \in X$ $(f_n(E) = x_E)$ and each Δ_n is a sequence of disjoint sets in Σ of positive μ-measure with $\Omega = \bigcup_{E \in \Delta_n} E$. In addition we may and do assume that B_n is the σ-field generated by Δ_n. In this case the inclusion $B_n \subseteq B_{n+1}$ means that each $E \in \Delta_n$ can be written as

$$E = \bigcup_{A \in \Delta_{n+1};\, A \subseteq E} A.$$

With these notational formalities settled, note that (i) means that the set $D = \{x_A : A \in \Delta_n, n = 1, 2, \cdots\}$ is bounded while (ii) means that $\|x_A - x_E\| \geqq \varepsilon$ whenever $E \in \Delta_n$, $A \in \Delta_{n+1}$ and $A \subseteq E$. In addition the martingale property means that if $E \in \Delta_n$, then

$$x_E\mu(E) = \int_E f_n \, d\mu = \int_E f_{n+1} \, d\mu = \sum_{A \subseteq E; A \in \Delta_{n+1}} \int_A f_{n+1} \, d\mu = \sum_{A \subseteq E; A \in \Delta_{n+1}} x_A \mu(A).$$

Thus we have

$$x_E = \sum_{A \subseteq E; \, A \in \Delta_{n+1}} \frac{\mu(A)}{\mu(E)} x_A$$

for each $E \in \Delta_n$. Note that the sum on the right-hand side is an (infinite) convex sum. The properties of the set $\{x_E : E \in \Delta_n, n = 1, 2, \cdots\}$ are isolated in

DEFINITION 1. A subset D of a Banach space is *not σ-dentable* if there exists an $\varepsilon > 0$ such that each $x \in D$ has the form $x = \sum_{i=1}^{\infty} \alpha_i x_i$ where $\sum_{i=1}^{\infty} \alpha_i = 1$, $\alpha_i > 0$, $x_i \in D$ and $\| x - x_i \| \geqq \varepsilon$ for all i.

Any bounded infinite δ-tree furnishes a quick example of a non-σ-dentable set. Also any Banach space containing a bounded infinite δ-tree is a Banach space without the Radon-Nikodým property. The following theorem is true for roughly the same reasons.

THEOREM 2 (MAYNARD). *Suppose X is a Banach space containing a bounded non-σ-dentable set D. Then there exists a $\delta > 0$ and a martingale (f_n, B_n) in $L_1([0, 1), X)$ such that $f_n([0, 1)) \subseteq D$ and $\| f_n(t) - f_{n+1}(t) \| \geqq \delta$ for all $n \in N$ and $t \in [0, 1)$.*

Consequently a Banach space containing a bounded non-σ-dentable set is a Banach space without the Radon-Nikodým property.

PROOF. Let X be a Banach space and D be a bounded non-σ-dentable subset of X. The proof is a realization of the discussion before Definition 1. We shall build a martingale (f_n, B_n) in $L^1([0, 1), X)$ with $f_n([0, 1)) \subseteq D$ and $\| f_n(t) - f_{n+1}(t) \|_X \geqq \varepsilon$ for some fixed $\varepsilon > 0$ and all $t \in [0, 1)$. By Corollary 2.4, this will be enough.

Toward the construction of (f_n, B_n), pick $\varepsilon > 0$ such that for each $x \in D$ there is a sequence of positive real numbers $(\alpha_n(x))$ with $\sum_{n=1}^{\infty} \alpha_n(x) = 1$ and a sequence $(x_n(x))$ in D with $\| x - x_n(x) \| \geqq \varepsilon$ for all n such that

$$(*) \qquad x = \sum_{n=1}^{\infty} \alpha_n(x) x_n(x).$$

Pick $\bar{x} \in D$ arbitrarily and set $f_1 = \bar{x}\chi_{[0,1)}$ and $B_1 = \{\phi, [0, 1)\}$. Then we have

$$f_1 = \sum_{n=1}^{\infty} \alpha_n(\bar{x}) x_n(\bar{x}) \chi_{[0,1)}.$$

Let $\beta_m = \sum_{n=1}^{m} \alpha_n(\bar{x})$ ($\beta_0 = 0$) and set $I_m = [\beta_{m-1}, \beta_m)$ for $m \geqq 1$. Define

$$f_2 = \sum_{n=1}^{\infty} x_n(\bar{x}) \chi_{I_n}.$$

Since $\mu(I_n) = \alpha_n(\bar{x})$ (here μ is Lebesgue measure), it follows that

$$\int_{[0,1)} f_2 \, d\mu = \int_{[0,1)} f_1 \, d\mu = \bar{x}.$$

Hence the conditional expectation $E(f_2 | B_1) = f_1$. Let B_2 be the σ-field generated by (I_n) and note that $\| f_2(t) - f_1(t) \|_X \geqq \varepsilon$ for all $t \in [0, 1)$.

Instead of giving a formal inductive proof, we shall be satisfied by showing how to construct f_3 and B_3 from f_2 and B_2. Write $\Delta_2 = \{I_n\}_{n=1}^{\infty}$. Then f_2 can be written as $f_2 = \sum_{E \in \Delta_2} x_E \chi_E$ where $x_E = x_n(\bar{x})$ when $E = I_n$. The construction of f_3 from f_2 is similar to the construction of f_2 from f_1 except that we must work inside each of the E's $\in \Delta_2$. Fix $E \in \Delta_2$ and note that by (*)

$$x_E = \sum_{n=1}^{\infty} \alpha_n(x_E) x_n(x_E)$$

with $\alpha_n(x_E)$ and $x_n(x_E)$ as in (*). Then one has

$$f_2 \chi_E = \left(\sum_{n=1}^{\infty} \alpha_n(x_E) x_n(x_E) \right) \chi_E.$$

For the moment, let $E = [a, b)$, and $\beta_n = (b - a) \sum_{m=1}^{n} \alpha_n(x_E)$ with $\beta_0 = 0$. Also for the moment, let $I_n = [a + \beta_{n-1}, a + \beta_n)$ and define f_3 on E by

$$f_3 \chi_E = \sum_{n=1}^{\infty} x_n(x_E) \chi_{I_n}.$$

Since $\mu(I_n) = \alpha_n(x_E)(b - a)$, we infer

$$\int_E f_3 \, d\mu = (b - a) x_E = \int_E f_2 \, d\mu.$$

Define f_3 on each $E \in \Delta_2$ as above; it follows immediately from the last line that $E(f_3 | B_2) = f_2$. Further from (*) we obtain $f_3([0, 1)) \subseteq D$ and $\| f_3(t) - f_2(t) \|_X \geqq \varepsilon$ for all $t \in [0, 1)$. Finally, let B_3 be the σ-field generated by all the intervals I_n constructed above as E ranges over Δ_2. The construction of f_3 and B_3 is now complete and so is the proof.

Next we shall alter Definition 1 a bit and then make corresponding adjustments to the proof of Theorem 2.

DEFINITION 3. A subset D of a Banach space is *not dentable* if there exists an $\varepsilon > 0$ such that, for each $x \in D$, $x \in \overline{\text{co}}(D \backslash B_\varepsilon(x))$ where $B_\varepsilon(x) = \{y : \| y - x \| < \varepsilon\}$ and $\overline{\text{co}}(D \backslash B_\varepsilon(x))$ is the closed convex hull of $(D \backslash B_\varepsilon(x))$.

Naturally a non-σ-dentable set is nondentable, i.e., dentability implies σ-dentability. Consequently the following theorem is an apparent generalization of Theorem 2.

THEOREM 4 (HUFF-DAVIS-PHELPS). *A Banach space containing a bounded nondentable set is a Banach space without the Radon-Nikodým property.*

PROOF. The proof is a variant of the proof of Theorem 2 and ultimately involves a nonconvergent martingale. Let X be a Banach space containing a bounded nondentable set D. Choose $\varepsilon > 0$ to satisfy the criterion of Definition 3. We shall demonstrate the existence of a sequence of (countable) partitions π_n of $[0, 1)$ into half-open intervals and a sequence (f_n) of countably valued functions on $[0, 1)$ such that

(i) Each f_n has the form $f_n = \sum_{E \in \pi_n} x_E \chi_E$ where $x_E \in D$ for all $E \in \pi_n$.

(ii) π_{n+1} refines π_n in the sense that each interval in π_n is a union of intervals in π_{n+1}.

(iii) The σ-field of Borel sets in $[0, 1)$ is the smallest σ-field containing $\bigcup_{n=1}^{\infty} \pi_n$.

(iv) $\| f_n(t) - f_{n+1}(t) \| \geqq \varepsilon$ for all n and $t \in [0, 1)$.

(v) If μ is Lebesgue measure, then $\| \int_E (f_m - f_n)\, d\mu \| < \mu(E)/2^n$ for all $E \in \pi_n$ and all $m \geqq n$.

Note that (v) is a relaxation of the corresponding condition $\int_E f_m\, d\mu = \int_E f_n\, d\mu$, for $m \geqq n$ and $E \in \pi_n$, found in the proof of Theorem 2. Now given that we can arrange to verify statements (i)—(v), the proof proceeds by associating a martingale with (f_n) as follows: By (v), we see that $F(E) = \lim_n \int_E f_n\, d\mu$ exists for all $E \in \bigcup_n \pi_n$. Write

$$g_n = \sum_{E \in \pi_n} \frac{F(E)}{\mu(E)} \chi_E \qquad (0/0 = 0).$$

If B_n is the σ-field generated by π_n, then (g_n, B_n) is a martingale in $L_1([0, 1), X)$. Moreover since D is bounded, for each $E \in \bigcup_n \pi_n$, one has

$$\| F(E)/\mu(E) \| = \left\| \lim_n \int_E f_n\, d\mu \right\| \Big/ \mu(E) \leqq K$$

where K is a bound for D. Thus $\sup_{t \in [0,1);\, n \in N} \| g_n(t) \|_X \leqq K$ and (g_n, B_n) is a uniformly integrable $L_1([0, 1), X)$-bounded martingale. Now note that

$$\begin{aligned} \int_{[0,1]} \| f_n - g_n \|\, d\mu &= \sum_{E \in \pi_n} \| x_E \mu(E) - F(E) \| \\ &= \lim_m \sum_{E \in \pi_n} \left\| \int_E (f_n - f_m)\, d\mu \right\| \leqq \sum_{E \in \pi_n} \mu(E)/2^n \leqq 1/2^n. \end{aligned}$$

Hence $(f_n - g_n)$ is a Cauchy sequence in $L_1([0, 1), X)$. Glancing at (iv), we see that (f_n) is not Cauchy. Thus (g_n, B_n) is a nonconvergent uniformly integrable $L_1([0, 1), X)$-bounded martingale. This fact combined with Theorem 2.4 implies that X lacks the Radon-Nikodým property.

Therefore to complete the proof only statements (i)—(v) above need be verified.

To this end, note that on the basis of our selection of ε, for each $\delta > 0$, there is for each $x \in D$ a sequence of positive reals $(\alpha_n(x, \delta))$ with $\sum_{n=1}^{\infty} \alpha_n(x, \delta) = 1$ and a sequence$(x_n(x, \delta))$ in $D \backslash B_\varepsilon(x)$ such that

$$(*) \qquad \left\| x - \sum_{m=1}^{\infty} \alpha_n(x, \delta) x_n(x, \delta) \right\| < \delta.$$

Note that by repeating some of the $x_n(x, \delta)$'s and decreasing the corresponding $\alpha_n(x, \delta)$'s, we can arrange to have $\alpha_n(x, \delta) < \delta$. This will be used to establish (iii).

Now to construct (f_n) and (π_n), choose $\bar{x}$ arbitrarily in D and set $f_1 = \bar{x}\chi_{[0,1)}$ and $\pi_1 = \{[0, 1)\}$. Suppose π_n and $f_n = \sum_{E \in \pi_n} x_E \chi_E$ have been defined with $x_E \in D$ for all $E \in \pi_n$ and with each $E \in \pi_n$ a half-open interval. Apply $(*)$ to obtain for each E

$$\left\| x_E - \sum_{m=1}^{\infty} \alpha_m(x_E, 1/2^{n+1}) x_m(x_E, 1/2^{n+1}) \right\| < 1/2^{n+1}$$

with $\alpha_m(x_E, 1/2^{n+1})$, $x_m(x_E, 1/2^{n+1})$ as in $(*)$ with $\delta = 1/2^{n+1}$. For the moment write $E = [a, b)$ and

$$\beta_n = (b - a) \sum_{m=1}^{n} \alpha_m(x_E, 1/2^{n+1})$$

with $\beta_0 = 0$. Also let $I_m = [a + \beta_{m-1}, a + \beta_m)$. Define f_{n+1} on E by

$$f_{n+1}\chi_E = \sum_{m=1}^{\infty} x_m(x_E, 1/2^{n+1})\chi_{I_m}.$$

Do this for every $E \in \pi_n$ and note that $\|f_n(t) - f_{n+1}(t)\| \geqq \varepsilon$ for all $t \in [0, 1)$. This establishes statement (iv).

Furthermore note that

$$\left\| \int_E (f_n - f_{n+1})\, d\mu \right\| = \left\| x_E - \sum_{m=1}^{\infty} x_m(x_E, 1/2^{n+1})\alpha_m(x_E, 1/2^{n+1}) \right\| \mu(E)$$
$$\leqq \mu(E)/2^{n+1}.$$

This establishes statement (v) above. To establish statements (i) and (ii), let π_{n+1} be the countable partition of $[0, 1)$ consisting of half-open intervals obtained from all the intervals I_n constructed above as E ranges over π_n.

Finally statement (iii), which will be used later in this section, follows from the fact that $\alpha_j(x_E, 1/2^{n+1}) < 1/2^{n+1}$ for all E and n. This completes the proof.

The following example seems to indicate that Theorem 4 is an honest strengthening of Theorem 2.

EXAMPLE 5. *The closed unit ball of $L_\infty[0, 1]$ is σ-dentable but not dentable.* Let D be the closed unit ball of $L_\infty[0, 1]$. To see that D is σ-dentable, note that if $\chi_{[0,1]} = \sum_{n=1}^{\infty} \alpha_n f_n$ with $\|f_n\|_\infty \leqq 1$, $0 < \alpha_n \leqq 1$, and $\sum_{n=1}^{\infty} \alpha_n = 1$, then $f_n = \chi_{[0,1]}$ a.e. for all n. Thus D is σ-dentable. To see that D is not dentable, let $f \in D$ and $\varepsilon > 0$. If $\|f\|_\infty > \varepsilon$, then for a positive integer m there are disjoint measurable sets $E_1, E_2, \cdots, E_m$ such that $\|f\chi_{E_n}\|_\infty > \varepsilon$, for each $n = 1, 2, \cdots, m$. Setting $f_n = f - f\chi_{E_n}$, one sees that

$$\|f - f_n\|_\infty = \|f\chi_{E_n}\| > \varepsilon, \qquad n = 1, 2, \cdots, m.$$

Moreover one has

$$\left\| f - \sum_{n=1}^{m} \frac{1}{m} f_n \right\|_\infty \leqq \frac{1}{m}\|f\|_\infty.$$

Since $1/m$ can be made as small as we please, we see that, for $0 < \varepsilon < 1$, $f \in \overline{\mathrm{co}}(D \backslash B_\varepsilon(f))$ provided $\|f\|_\infty > \varepsilon$. On the other hand, if $\|f\|_\infty \leqq \varepsilon < 1/3$, then $\|f + 2\varepsilon\chi_{[0,1]} - f\| = 2\varepsilon$ and $\|f - 2\varepsilon\chi_{[0,1]} - f\| = 2\varepsilon$. Setting $f_1 = f + 2\varepsilon\chi_{[0,1]}$ and $f_2 = f - 2\varepsilon\chi_{[0,1]}$ implies $\|f_i\|_\infty < 3\varepsilon < 1$ and $\|f - f_i\| = 2\varepsilon$, $i = 1, 2$. Therefore $f_1, f_2 \in D\backslash B_\varepsilon(f)$ for $i = 1, 2$ but $f = \frac{1}{2}f_1 + \frac{1}{2}f_2$. Thus $f \in D$ implies $f \in \overline{\mathrm{co}}(D\backslash B_\varepsilon(f))$ for every $f \in D$. Therefore D is not dentable.

The reader should keep in mind that, in a sense, Example 5 is a bit misleading. We shall see why after proving a crude Radon-Nikodým theorem.

LEMMA 6. *Let (Ω, Σ, μ) be a finite measure space and $F: \Sigma \to X$ be a μ-continuous vector measure of bounded variation. There exists $f \in L_1(\mu, X)$ such that $F(E) = \int_E f\, d\mu$ for all $E \in \Sigma$ provided that for each $\varepsilon > 0$ and $A \in \Sigma$* with $\mu(A) > 0$, *there is a set $B \subseteq A$, $B \in \Sigma$ and $\mu(B) > 0$ such that the set $\{F(E)/\mu(E): E \in \Sigma, E \subseteq B, \mu(E) > 0\}$ has diameter $< \varepsilon$.*

PROOF. Fix $\varepsilon > 0$. By the Exhaustion Lemma III.2.4, there is a disjoint sequence $(E_n(\varepsilon)) \subseteq \Sigma$ with $\mu(E_n(\varepsilon)) > 0$ such that

$$\mu\left(\Omega \setminus \bigcup_{n=1}^{\infty} E_n(\varepsilon)\right) = 0$$

and each set $\{F(E)/\mu(E): E \subseteq E_n(\varepsilon),\ \mu(E) > 0\}$ has diameter $\leqq \varepsilon$. Define $f_\varepsilon: \Omega \to X$ by writing

$$f_\varepsilon = \sum_{n=1}^{\infty} \frac{F(E_n(\varepsilon))}{\mu(E_n(\varepsilon))} \chi_{E_n(\varepsilon)}.$$

If $F_\varepsilon(\cdot) = \int_{(\cdot)} f_\varepsilon \, d\mu$, then for a partition π of Ω,

$$\begin{aligned}
\sum_{E\in\pi} \|F(E) - F_\varepsilon(E)\| &= \sum_{E\in\pi} \left\| F(E) - \int_E f_\varepsilon \, d\mu \right\| \\
&\leqq \sum_{E\in\pi} \sum_{n=1}^{\infty} \left\| F(E \cap E_n(\varepsilon)) - \int_{E\cap E_n(\varepsilon)} f_\varepsilon \, d\mu \right\| \\
&= \sum_{E\in\pi} \sum_{n=1}^{\infty} \left\| \frac{F(E \cap E_n(\varepsilon))}{\mu(E \cap E_n(\varepsilon))} - \frac{F(E_n(\varepsilon))}{\mu(E_n(\varepsilon))} \right\| \mu(E \cap E_n(\varepsilon)) \\
&\qquad \text{(here } 0/0 = 0) \\
&\leqq \sum_{E\in\pi} \sum_{n=1}^{\infty} \varepsilon\mu(E \cap E_n(\varepsilon)) \leqq \varepsilon\mu(\Omega).
\end{aligned}$$

Hence $\lim_{\varepsilon\to 0} |F - F_\varepsilon|(\Omega) = 0$. It follows that $\lim_{\delta\to 0;\, \varepsilon\to 0} |F_\varepsilon - F_\delta|(\Omega) = 0$. Hence

$$\lim_{\delta\to 0;\, \varepsilon\to 0} \int_\Omega \|f_\varepsilon - f_\delta\| \, d\mu = 0$$

by II.2.4. Consequently if $f_n = f_\varepsilon$ with $\varepsilon = 1/n$, then (f_n) is a Cauchy sequence in $L_1(\mu, X)$. Let $\lim_n f_n = f$ in $L_1(\mu, X)$; then obviously one has

$$F(E) = \int_E f \, d\mu \quad \text{for all } E \in \Sigma.$$

This completes the proof.

The next result is the central result of this section. It provides us with our first concrete evidence that the Radon-Nikodým property is a geometric property of Banach spaces.

THEOREM 7 (RIEFFEL-MAYNARD-HUFF-DAVIS-PHELPS). *Any one of the following statements about a Banach space X implies all the others.*

(a) *Every bounded subset of X is dentable.*

(b) *Every bounded subset of X is σ-dentable.*

(c) *The space X has the Radon-Nikodým property.*

PROOF. The fact that (c) implies (b) is Theorem 2; while the fact that (c) implies (a) is Theorem 4. Since dentability implies σ-dentability the theorem will be proved if it can be shown that (b) implies (c).

To this end, let (Ω, Σ, μ) be a finite measure space and $F: \Sigma \to X$ be a μ-continuous vector measure of bounded variation. Since $|F|(\Omega)$ is finite, there is a disjoint sequence (A_n) in Σ with $\bigcup_{n=1}^{\infty} A_n = \Omega$ and such that $|F|(A)/\mu(A)$ is bounded for $A \subseteq A_n$ and n fixed. (To see this let $\phi \in L_1(\mu)$ be the Radon-Nikodým derivative of $|F|$ with respect to μ and set $A_n = [n-1 \leqq \phi < n]$, $n = 1, 2, \cdots$.) To prove that F has a Radon-Nikodým derivative in $L_1(\mu, X)$, we are going to apply Lemma 6. Let $A \in \Sigma$ have positive μ-measure. Then for some n, one has $\mu(A \cap A_n) > 0$. Hence there exists a set $A' \subseteq A$, $A' \in \Sigma$ with $\mu(A') > 0$ such that $\Phi = \{F(E)/\mu(E): E \subseteq A', \mu(E) > 0\}$ is bounded. By (b), Φ is σ-dentable. Thus if $\varepsilon > 0$, there is a set $C \subseteq A'$, $\mu(C) > 0$ such that if

$$\frac{F(C)}{\mu(C)} = \sum_{n=1}^{\infty} \alpha_n \frac{F(E_n)}{\mu(E_n)}$$

with $\alpha_n > 0$, $\sum_{n=1}^{\infty} \alpha_n = 1$ and $E_n \subseteq C$, then

$$\left\| \frac{F(E_{n_0})}{\mu(E_{n_0})} - \frac{F(C)}{\mu(C)} \right\| \leqq \varepsilon$$

for at least one choice of n_0. Now if

$$\sup\left\{ \left\| \frac{F(E)}{\mu(E)} - \frac{F(C)}{\mu(C)} \right\| : E \subseteq C \right\} \leqq \varepsilon,$$

then stop and let $B = C$ in Lemma 6. Otherwise let j_1 be the smallest integer $\geqq 2$ for which there is $C_1 \subseteq C$ with $\mu(C_1) \geqq 1/j_1$ and $\| F(C_1)/\mu(C_1) - F(C)/\mu(C) \| > \varepsilon$. Also note that

$$\frac{F(C)}{\mu(C)} = \frac{F(C_1)}{\mu(C_1)} \frac{\mu(C_1)}{\mu(C)} + \frac{F(C \backslash C_1)}{\mu(C \backslash C_1)} \frac{\mu(C \backslash C_1)}{\mu(C)}.$$

Now if $\sup\{\|F(E)/\mu(E) - F(C)/\mu(C)\| : E \subseteq C\backslash C_1\} \leqq \varepsilon$, then stop and apply Lemma 6 with $B = C\backslash C_1$. Otherwise let j_2 be the smallest positive integer $\geqq 2$ for which there is a set $C_2 \subseteq C\backslash C_1$ with $\mu(C_2) > 1/j_2$ and

$$\|F(C_2)/\mu(C_2) - F(C)/\mu(C)\| > \varepsilon.$$

Note that

$$\frac{F(C)}{\mu(C)} = \frac{F(C_1)}{\mu(C_1)} \frac{\mu(C_1)}{\mu(C)} + \frac{F(C_2)}{\mu(C_2)} \frac{\mu(C_2)}{\mu(C)} + \frac{F(C \backslash C_1 \backslash C_2)}{\mu(C \backslash C_1 \backslash C_2)} \frac{\mu(C \backslash C_1 \backslash C_2)}{\mu(C)}.$$

Continue in this way. If the process comes to a halt in a finite number of steps, say at n iterations, the assertion is established by appealing to Lemma 6 with $B = C\backslash C_1\backslash C_2\backslash \cdots \backslash C_{n-1}$.

If the process does not stop, continue in this way to produce a disjoint sequence (C_n) of subsets of C, all of positive μ-measure, a nondecreasing sequence (j_n) of positive integers such that

(i) $\| F(C_n)/\mu(C_n) - F(C)/\mu(C) \| > \varepsilon$ for all n,

(ii) if $E \subseteq C\backslash \bigcup_{n=1}^{m} C_n$, $E \in \Sigma$ and $\| F(E)/\mu(E) - F(C)/\mu(C) \| > \varepsilon$ then $\mu(E) < 1/(j_m - 1)$, and

(iii) $$\frac{F(C)}{\mu(C)} = \sum_{n=1}^{m} \frac{F(C_n)}{\mu(C_n)} \frac{\mu(C_n)}{\mu(C)} + \frac{F(C\backslash \bigcup_{n=1}^{m} C_n)}{\mu(C\backslash \bigcup_{n=1}^{m} C_n)} \frac{\mu(C\backslash \bigcup_{n=1}^{m} C_n)}{\mu(C)}$$

for all m.

At this point glance at (iii) and recall $F(C\backslash \bigcup_{n=1}^{m} C_n)/\mu(C\backslash \bigcup_{n=1}^{m} C_n)$ is bounded as $m \to \infty$. Hence $\lim_m \mu(C \setminus \bigcup_{n=1}^m C_n) = \mu(C \setminus \bigcup_{n=1}^\infty C_n) \neq 0$; for otherwise one has

$$\frac{F(C)}{\mu(C)} = \sum_{n=1}^{\infty} \frac{F(C_n)}{\mu(C_n)} \frac{\mu(C_n)}{\mu(C)}$$

with $C_n \subseteq C$, $\| F(C_n)/\mu(C_n) - F(C)/\mu(C) \| > \varepsilon$ and $\sum_{n=1}^\infty \mu(C_n)/\mu(C) = 1$, which contradicts the choice of $F(C)/\mu(C)$. Therefore $B = C \setminus \bigcup_{n=1}^\infty C_n$ has positive μ-measure. Now if there exists $E \subseteq B$ with $\mu(E) > 0$ and $\| F(E)/\mu(E) - F(C)/\mu(C) \| > \varepsilon$ then $E \subseteq C \setminus \bigcup_{n=1}^m C_n$ for every m and accordingly $\mu(E) < 1/(j_m - 1)$ for all m by (ii). But $\mu(C_m) \geqq 1/j_m$ for all m and (C_m) is a disjoint sequence. Thus $\sum_{m=1}^\infty (1/j_m) \leqq \sum_{m=1}^\infty \mu(C_m) < \infty$ and hence

$$\mu(E) \leqq \lim_m \frac{1}{j_m - 1} = 0.$$

This shows $\| F(E)/\mu(E) - F(C)/\mu(C) \| \leqq \varepsilon$ for every $E \subseteq B \mu(E) > 0$; this together with Lemma 6 completes the proof.

We can now easily see why Example 5 is a bit misleading. By Theorem 7, such an example is possible only in a Banach space without the Radon-Nikodým property.

According to III.3.2, the Radon-Nikodým property for a Banach space X is determined by the separable subspaces of X. This is also a consequence of the next corollary.

COROLLARY 8. *A Banach space has the Radon-Nikodým property if and only if it has the Radon-Nikodým property with respect to Lebesgue measure on* [0, 1).

PROOF. The construction in the proof of Theorem 4 is (by (iii) of that proof) executed in $L_1([0, 1), X)$.

We hasten to remark that Corollary 8 does not require the material of this section for a proof. Indeed, Corollary 8 can be proved as a routine exercise based on the material of Chapter III.

A careful look at the proof of Theorem 7 results in a new Radon-Nikodým theorem for a single measure with values in an arbitrary Banach space.

COROLLARY 9 (RIEFFEL). *Let (Ω, Σ, μ) be a finite measure space and $F: \Sigma \to X$ be a μ-continuous vector measure of bounded variation. If for each $A \in \Sigma$ with $\mu(A) > 0$ there is $B \in \Sigma$ with $B \subseteq A$ and $\mu(B) > 0$ such that $\{F(E)/\mu(E): E \subseteq B, \mu(E) > 0\}$ is σ-dentable, then there exists $f \in L_1(\mu, X)$ such that $F(E) = \int_E f\, d\mu$ for all $E \in \Sigma$.*

The converse of Corollary 9 is true. Its proof is an easy consequence of the next result.

THEOREM 10. *Let D be a bounded subset of X.*

(i) *If $\overline{\text{co}}(D)$ is dentable, then D is dentable.*

(ii) *If D is relatively weakly compact, then D is dentable.*

(iii) *If D has an exposed point x_0 (i.e., there is $x_0^* \in X^*$, $x_0^*(x_0) > x_0^*(x)$ for all $x \in D \setminus \{x_0\}$), then D is σ-dentable.*

(iv) *If D has a strongly exposed point x_0 (i.e., if there is $x_0^* \in X^*$ such that $x_0^*(x_0) > x_0^*(x)$ for all $x \in D \setminus \{x_0\}$ and such that $\lim_n x_0^*(x_n) = x_0^*(x_0)$ for $(x_n) \subseteq D$ implies $\lim_n x_n = x_0$), then D is dentable.*

PROOF. (i) Suppose $\overline{\text{co}}(D)$ is dentable and suppose $\varepsilon > 0$. Then there is $x_\varepsilon \in \overline{\text{co}}(D)$ such that $x_\varepsilon \notin \overline{\text{co}}(\overline{\text{co}}(D)\backslash B_{\varepsilon/2}(x_\varepsilon)) = Q$. Then $x_\varepsilon \in \overline{\text{co}}(D)$ but $x_\varepsilon \notin Q$. Next note that $D\backslash Q$ is not empty; for if $D \subseteq Q$, then $\overline{\text{co}}(D) \subseteq Q$ since Q is convex. But $x_\varepsilon \in \overline{\text{co}}(D)$ and $x_\varepsilon \notin Q$, a quick contradiction.

Now select $d \in D\backslash Q$. We shall establish that

$$d \notin \overline{\text{co}}(D\backslash B_\varepsilon(d))$$

and thus prove that D is dentable. To this end, note that $d \in B_{\varepsilon/2}(x_\varepsilon)$. For otherwise $d \in D\backslash B_{\varepsilon/2}(x_\varepsilon) \subseteq \overline{\text{co}}(D\backslash B_{\varepsilon/2}(x_\varepsilon)) \subseteq Q$; which is impossible since $d \notin Q$. Since $d \in D\backslash Q$ is unspecified otherwise, we have $D\backslash Q \subseteq B_{\varepsilon/2}(x_\varepsilon)$. From this inclusion, the inclusion $D\backslash B_\varepsilon(d) \subseteq Q$ obtains since if $d_0 \in D$ and $\|d_0 - d\| \geqq \varepsilon$ and $d_0 \notin Q$, then $d_0, d \in D\backslash Q$ implies

$$\|d_0 - d\| \leqq \|d_0 - x_\varepsilon\| + \|x_\varepsilon - d\| < 2\varepsilon/2 = \varepsilon.$$

Recalling that Q is closed and convex, we see that $\overline{\text{co}}(D\backslash B_\varepsilon(d)) \subseteq Q$. Since $d \in D\backslash Q$, it follows that $d \notin \overline{\text{co}}(D\backslash B_\varepsilon(d))$. (Note. Anyone who considers this proof unmotivated is urged to draw the appropriate pictures and follow his instincts.)

(ii) Weakly compact sets are dentable because the construction used to prove Theorem 4 cannot be executed inside a weakly compact set. Let us see why. Suppose D is a subset of X that is contained in a weakly compact convex subset W of X. If D is not dentable, then the martingale (g_n) constructed in the proof of Theorem 4 is not convergent. On the other hand, it is easily seen that $g_n([0, 1))$ is a subset of W for all n. It is equally easy to see that if $\mathscr{F}$ is the field generated by $\bigcup_n \pi_n$ (the π_n's are as in the proof of Theorem 4) then

$$\lim_n \int_E g_n \, d\mu/\mu(E) \in W$$

for all $E \in \mathscr{F}$ with $\mu(E) > 0$. Define $G: \mathscr{F} \to X$ by $G(E) = \lim_n \int_E g_n \, d\mu$ for $E \in \mathscr{F}$. Since the g_n's are uniformly bounded, this limit exists for all E in the σ-field Σ generated by $\mathscr{F}$ and defines a countably additive extension $\bar{G}: \Sigma \to X$ of G. Evidently $\bar{G}(E)/\mu(E) \in W$ for all $\bar{E} \in \Sigma$. By the Dunford-Pettis-Phillips Theorem III.2.18, there is a Bochner integrable g such that $\bar{G}(E) = \int_E g \, d\mu$ for all $E \in \Sigma$. In particular $\lim_n \int_E g_n \, d\mu = \int_E g \, d\mu$ for all $E \in \Sigma$. According to Theorem 2.1, this means (g_n) is $L_1(\mu, X)$ convergent, a contradiction which proves that weakly compact sets are dentable.

(iii) If $x_0 \in D$ is an exposed point and $(x_n) \subseteq D$ is a sequence such that there is a sequence of reals (α_n) with $0 < \alpha_n \leqq 1$ and $\sum_{n=1}^\infty \alpha_n = 1$ such that $x_0 = \sum_{n=1}^\infty \alpha_n x_n$, then one has

$$\sum_{n=1}^\infty \alpha_n x_0^*(x_0) = x_0^*(x_0) = \sum_{n=1}^\infty \alpha_n x_0^*(x_n),$$

i.e., $\sum_{n=1}^\infty \alpha_n(x_0^*(x_0) - x_0^*(x_n)) = 0$. Since this last series has nonnegative entries, each entry must be zero. Since $\alpha_n > 0$ for all $n \in N$, we see that $x_0^*(x_0) = x_0^*(x_n)$ for all $n \in N$. Hence $x_0 = x_n$ for all $n \in N$. It follows immediately that D is σ-dentable.

(iv) Suppose x_0 is strongly exposed and suppose $x_0 \in \overline{\text{co}}(D\backslash B_\varepsilon(x_0))$. There must be convex sums $\sum_{n=1}^m \alpha_n x_n$ with $0 < \alpha_n \leqq 1$, $\sum_{n=1}^m \alpha_n = 1$ and $x_n \in D\backslash B_\varepsilon(x_0)$ that are as close to x_0 as we please. Since $x_0^*(x_n) < x_0^*(x_0)$ for each n, a slight refinement of the argument used in (c) shows that there must be a sequence (y_n) in $D \backslash B_\varepsilon(x_0)$ such that $x_0^*(y_n) \to x_0^*(x_0)$. Hence $\lim_n y_n = x_0$, and this is a contradiction.

Providing the converse to Corollary 9 is

COROLLARY 11. *Let (Ω, Σ, μ) be a finite measure space. If $f \in L_1([0, 1), X)$ and $F(E) = \int_E f\, d\mu$ for $E \in \Sigma$, then for each $\varepsilon > 0$ there is $E_\varepsilon \in \Sigma$ with $\mu(\Omega \backslash E_\varepsilon) < \varepsilon$ such that $\{F(E)/\mu(E): E \subseteq E_\varepsilon, \mu(E) > 0\}$ is dentable.*

PROOF. By III.2.7, there exists $E_\varepsilon \in \Sigma$ with $\mu(\Omega \backslash E_\varepsilon) < \varepsilon$ such that $\{F(E)/\mu(E) : E \subseteq E_\varepsilon, \mu(E) > 0\}$ is relatively compact. By Theorem 10, compact sets are dentable.

4. The Radon-Nikodým property for $L_p(\mu, X)$. According to Corollary IV.1.3, $L_p(\mu, X)$ $(1 < p < \infty)$ has the Radon-Nikodým property if there exists a Banach space Y such that $Y^* = X$ and every separable subspace of Y has a separable dual space. This was derived as a direct consequence of III.3.6. In this section martingale methods will be used to prove

THEOREM 1. *Let (Ω, Σ, μ) be a nonatomic finite measure space and X be a Banach space. Then $L_p(\mu, X)$ has the Radon-Nikodým property if and only if $1 < p < \infty$ and X has the Radon-Nikodým property.*

PROOF. Since $L_p(\mu, X)$ contains isometric copies of both $L_p(\mu)$ and X, it is clear that if $L_p(\mu, X)$ has the Radon-Nikodým property then X also has the Radon-Nikodým property and $1 < p < \infty$.

For the converse, let $(S, \mathscr{F}, \lambda)$ be a finite measure space and $F: \mathscr{F} \to L_p(\mu, X)$ be a λ-continuous vector measure of bounded variation. There is no loss of generality in assuming that $\|F(E)\|_{L_p(\mu, X)} \leqq \lambda(E)$ for all $E \in \mathscr{F}$. Let π be a partition of S into a finite number of members of $\mathscr{F}$ and Δ be a partition of Ω into a finite number of members of Σ. Write

$$f_{\pi,\Delta}(s, \omega) = \sum_{E \in \pi} \sum_{I \in \Delta} \frac{\int_I F(E)\, d\mu}{\lambda(E)\mu(I)} \chi_I(\omega)\chi_E(s)$$

for $(s, \omega) \in S \times \Omega$. (Here $0/0 = 0$.) Since the X-valued set function $\int_I F(E)\, d\mu$ is finitely additive in both $E \in \mathscr{F}$ and $I \in \Sigma$, it is clear that $(f_{\pi,\Delta}, \mathscr{B}_{\pi,\Delta})$ (where $\mathscr{B}_{\pi,\Delta}$ is the trivial σ-field generated by sets of the form $E \times I$ with $E \in \pi$ and $I \in \Delta$) is a martingale in $L_p(\lambda \times \mu, X)$.

Now since $1 < p < \infty$ and X has the Radon-Nikodým property, Corollary 2.4 guarantees that this martingale is $L_p(\lambda \times \mu, X)$-convergent if it is $L_p(\lambda \times \mu, X)$-bounded. To see that it is $L_p(\lambda \times \mu, X)$-bounded, first note that

$$\begin{aligned} \left\| \int_I F(E)\, d\mu \right\|_X^p &= \left\| \int_I F(E)\chi_I\, d\mu \right\|_X^p \\ &\leqq \|F(E)\chi_I\|_{L_p(\mu, X)}^p \mu(I)^{p/q} \end{aligned}$$

$(p^{-1} + q^{-1} = 1)$ by the Hölder inequality. Thus

$$\begin{aligned} \|f_{\pi,\Delta}\|_{L_p(\lambda \times \mu, X)} &= \sum_{E \in \pi} \sum_{I \in \Delta} \frac{\| \int_I F(E)\, d\mu \|_X^p}{\lambda(E)^p \mu(I)^p} \mu(I)\lambda(E) \\ &\leqq \sum_{E \in \pi} \sum_{I \in \Delta} \frac{\|F(E)\chi_I\|_{L_p(\mu, X)}^p}{\lambda(E)^p} \mu(I)^{1 + p/q - p} \lambda(E) = \sum_{E \in \pi} \sum_{I \in \Delta} \frac{\|F(E)\chi_I\|_{L_p(\mu, X)}^p}{\lambda(E)^p} \lambda(E) \end{aligned}$$

since $1 + p/q - p = 0$. Now note that $\|F(E)\chi_I\|_{L_p(\mu,X)}^p$ is an additive function of $I \in \Delta$. Hence

$$\begin{aligned}\|f_{\pi,\Delta}\|_{L_p(\lambda\times\mu,X)} &\leqq \sum_{E\in\pi} \frac{\|F(E)\|_{L_p(\mu,X)}^p}{\lambda(E)^p}\lambda(E)\\ &\leqq \sum_{E\in\pi}\lambda(E) = \lambda(S),\end{aligned}$$

since $\|F(E)\|_{L_p(\mu,X)} \leqq \lambda(E)$ for all $E \in \mathscr{F}$. Therefore $\lim_{\pi,\Delta} f_{\pi,\Delta}$ exists in $L_p(\lambda \times \mu, x)$-norm. Let its limit be f.

Now note that

$$\int \|f(s, \omega)\|_X^p \, d\mu(\omega)\, d\lambda(s) < \infty.$$

Hence $f(s, \cdot) \in L_p(\mu, X)$ for λ-almost all $s \in S$. Redefine f to be zero on the exceptional set and set $g(s) = f(s, \cdot)$ for $s \in S$. It is not difficult to use a simple functions argument to prove that g is measurable. Thus g is an $L_p(\mu, X)$-valued λ-Bochner integrable function.

Finally if $A \in \mathscr{F}$, we have

$$\begin{aligned}\int_A g\, d\lambda &= \lim_{\pi,\Delta}\int_A \sum_{E\in\pi}\sum_{I\in\Delta}\frac{\int_I F(E)\,d\mu}{\mu(I)\lambda(E)}\chi_I\chi_E\,d\lambda\\ &= \lim_{\pi}\int_A \sum_{E\in\pi}\lim_{\Delta}\left(\sum_{I\in\Delta}\frac{\int_I F(E)\,d\mu}{\mu(I)}\chi_I\right)\frac{\chi_E}{\lambda(E)}\,d\lambda\\ &= \lim_{\pi}\int_A \sum_{E\in\pi}\frac{F(E)}{\lambda(E)}\chi_E\,d\lambda,\end{aligned}$$

since $\sum_{I\in\Delta}\int_I F(E)\,d\mu/\mu(I)\chi_I$ is a martingale in $L_p(\mu, X)$ converging to $F(E)$ in $L_p(\mu, X)$ by Corollary 2.2. But since

$$\lim_{\pi}\int_A \sum_{E\in\pi}\frac{F(E)}{\lambda(E)}\chi_E\,d\lambda = F(A),$$

we have $F(A) = \int_A g\,d\lambda$ for all $A \in \mathscr{F}$. This completes the proof.

5. Notes and remarks. Detailed studies of martingales were initiated by Doob **[1950]**. Their impact on mathematical analysis has not been softened by the passage of time. No doubt martingales are fundamental to many parts of mathematical analysis outside Banach space theory and it is perhaps a bit surprising that Banach space theorists have not given martingales more attention in the past. Today, however, a new trend is developing as workers in Banach space theory are beginning to exploit the profitable interplay between Banach space theory and martingale theory, an interplay which has only begun to realize its potential.

The martingale mean convergence theorem. Vector-valued martingales first appeared (implicitly) in the early work of Dunford and Pettis **[1940]** and Phillips **[1940]**, **[1943]**. Vector-valued martingales seem to have been studied first for their own merits in the independent papers of Chatterji **[1960]**, **[1964]** and Scalora **[1961]**. A few years later a spate of independent papers cemented the relationship between the Radon-Nikodým theorem and the martingale mean convergence theorem. The best known of these papers is the definitive paper of Chatterji **[1968]** but he was not alone. The relationship between the Radon-Nikodým theorem and the martingale mean convergence theorem is also apparent in Metivier **[1967]**, Ronnow **[1967]** and Uhl **[1969b]**, **[1969c]**.

The proof of the martingale mean convergence theorem used here is essentially that of Helms **[1958]**. Our proof for the vector-valued case differs only notationally from Helms's original proof for the scalar-valued case. For arbitrary Banach spaces X, mean convergent martingales in $L_p(\mu, X)$ were studied by Metivier **[1967]** and characterized by Uhl **[1969b]**. Mean convergent martingales of measurable Pettis integrable functions were also characterized by Uhl **[1972b]**.

The martingale pointwise convergence theorem. Unlike the proof of the martingale mean convergence theorem, the proof of the martingale pointwise convergence theorem for vector-valued martingales is not just a simple notational modification of its scalar-valued ancestor. The most evident reason for this is that Doob's classical upcrossing has no clearcut reinterpretation in the vector-valued case. Nevertheless some early pointwise convergence theorems were proved by Scalora **[1961]** and Chatterji **[1960]**, **[1964]** who skillfully used Banach's theorem on the convergence of measurable functions found in Dunford and Schwartz **[1958**, IV.11.2]. Shortly thereafter using arguments from ergodic theory, Ionescu Tulcea **[1963]** proved Theorem 2.9 under the assumption that X is either reflexive or a separable dual space. About the same time, Neveu **[1964]** proved the maximal lemma (Lemma 2.7) and deduced Theorem 2.8 from it. This set the stage for Theorem 2.9 which is due to Chatterji **[1968]**. For the pointwise convergence theorem for martingales with values in arbitrary Banach space, see Uhl **[1969b]** who supplies conditions that ensure that the measure G involved in Theorem 2.9 has a Radon-Nikodým derivative. See also Metivier **[1967]**. For a treatment of martingales of functions with values in Banach lattices see Szulga and Woyczynski **[1975]**.

Dentability and σ-dentability. On the basis of Chapter III the news that the Radon-Nikodým property can be thought of as a purely geometric property may be a bit of a shock. If so then the shock can be mollified by the fact that the concept of uniform convexity was introduced by Clarkson **[1936]** for the purposes of proving a Radon-Nikodým theorem for the Bochner integral. Geometry and Radon-Nikodým theorems then went their separate ways until Rieffel's fundamental paper **[1967]** some thirty years later. In that paper Rieffel introduced the notion of a dentable set and proved Corollaries 3.9 and 3.11. At the time of Rieffel's paper, it was known (Lindenstrauss **[1963]**) that separable weakly compact sets are dentable but it was not known whether weakly compact sets were dentable. Thus Corollaries 3.9 and 3.11 seemed to mesh with Theorem III.2.18 only in the separable case. The final link between Theorem III.2.18 and Corollaries 3.9 and 3.11 was provided by Troyanski **[1971]** who proved that all weakly compact sets are dentable. Thus at this stage of life, Theorem III.2.18 could be viewed as a consequence of Corollary 3.11.

On the other hand, the problem of characterizing Banach spaces with the Radon-Nikodým property was still wide open.

Then it was Maynard **[1973b]** who in his pregnant paper introduced the notion of σ-dentability and characterized Banach spaces with the Radon-Nikodým property as spaces whose bounded sets are σ-dentable. In this paper Maynard also showed that dentability of a set is a property determined by countable subsets. Maynard's work must be regarded as a signal achievement for it provided the foundation for much important work on the geometry of Banach spaces and the Radon-Nikodým property (see Chapter VII). Acting on Maynard's lead, Davis and Phelps **[1974]** and Huff **[1974]** independently completed the cycle of components of Theorem 3.7. Huff's work appears in the text as Theorem 3.4. Davis and Phelps showed that if a Banach space has a non-σ-dentable bounded subset then it has a nondentable bounded subset. This fact can be viewed as an immediate consequence of the proof of Theorem 3.4 and the proof of Theorem 3.7. In fact a close inspection of the proofs of the above theorems shows that if every subset of a bounded set A is σ-dentable then A is dentable without regard to the ambient space of A.

The story of the Radon-Nikodým property as a geometric notion does not end here. The works of Rieffel, Maynard, Huff, Davis and Phelps ignited interest in the search for a concrete correlation between extreme point structure and the Radon-Nikodým property. The search has been quite successful and is chronicled in some detail in the notes and remarks to Chapter VII.

The Radon-Nikodým property for $L_p(\mu, X)$. Theorem 4.1 was first proved by Sundaresan **[1976]** who used the fact that $L_p[0, 1]$ $(1 < p < \infty)$ has an unconditional basis. Our proof is from Turett and Uhl **[1976]** who note that this proof can be modified to show that if $L_\Phi(\mu)$ is an Orlicz space and X is a Banach space then $L_\Phi(\mu, X)$ has the Radon-Nikodým property if and only if there exists a $K > 0$ such that $\Phi(2x) \leqq K\Phi(x)$ for $x \geqq K$, $\lim_{t\to\infty} \Phi(t)/t = \infty$, and X has the Radon-Nikodým property.

Martingales and Schauder bases in $L_p(\mu)$. Here we shall take a brief look at some folklore interpretations of martingales as Schauder decompositions. Let $(f_n, \mathscr{B}_n)$ be a martingale in $L_p(\mu)$ $(1 < p < \infty)$. Let $d_1 = f_1$ and $d_k = f_k - f_{k-1}$ for $k \geqq 2$. Note that if (α_k) is a sequence of scalars, and $g_n = \sum_{k=1}^n \alpha_k d_k$, then $(g_n, \mathscr{B}_n)$ is also a martingale which is convergent if it is bounded. Moreover $\|g_n\|_p \leqq \|g_{n+1}\|_p$ since $E(g_{n+1}|\mathscr{B}_n) = g_n$. These simple observations are the key to the fact that if none of the d_k's are zero, then (d_k) is a boundedly complete monotone basis of its span. What is not so clear is that (d_k) is also an unconditional basis of its closed linear span in $L_p(\mu)$. This deep fact is due to Burkholder **[1966]**. Extensions of this to the vector-valued case seem to be nonexistent. It is possible that vector-valued extensions of Burkholder's theorem could be used to prove that $L_p([0, 1], X)$ has an unconditional basis if $1 < p < \infty$ and X has an unconditional basis.

Superreflexive and super-Radon-Nikodým spaces. We already know that no Banach space with the Radon-Nikodým property contains an infinite δ-tree. How about finite δ-trees? This question (and its answer) is intimately related to the notion of superreflexivity introduced by James **[1972a]**, **[1972b]**.

Let X, Y be Banach spaces. We say Y is *finitely representable* in X if each finite dimensional subspace of Y fits almost isometrically in X. If every Banach space Y which is finitely representable in X is reflexive, then we call X *superreflexive*. According to a theorem of James **[1972b]**, a Banach space X is not superreflexive precisely when for each $0 < \delta < 1$ and each positive integer k there is a (δ, k)-tree in the unit ball of X (a set $\{x_1, \cdots, x_{2^k-1}\}$ is called a (δ, k)-tree if, for each n, $x_n =$

$(x_{2n} + x_{2n+1})/2$, $\|x_n - x_{2n}\| \geqq \delta$, $\|x_n - x_{2n+1}\| \geqq \delta$). The crux of an observation due to Pisier [**1975**] is that if X is not superreflexive then a suitable grafting produces a Banach space Y which is finitely represented in X and admits an infinite δ-tree in its unit ball. Therefore a Banach space is superreflexive if and only if it has the super-Radon-Nikodým property.

Martingale inequalities and martingales with values in uniformly convex spaces. Martingale inequalities for scalar-valued martingales have proved to be of compelling importance through the work of Burkholder, Davis and Gundy [**1972**] (see also Burkholder [**1973**]). For a long while martingale inequalities for vector-valued martingales remained elusive. Most likely this was because distribution function methods are of crucial importance in the scalar case and there was no natural analogue of distribution functions in the vector case. In fact virtually no progress was made in this direction until the stunning achievements of Pisier [**1975**] who was the first to demonstrate that martingale inequalities for X-valued martingales intrinsically depend on the geometry of X. As in the scalar case, martingale inequalities have resulted in powerful new techniques yielding correspondingly powerful theorems. At this writing this area has just been opened by Pisier and this area of inquiry seems to have tremendous potential.

Let us now attempt to give a bit of the flavor of Pisier's work. By martingale inequality methods Pisier [**1975**] has proved the super theorem of Enflo [**1972**] that says that a Banach space is superreflexive if and only if it is uniformly convexifiable (which in light of the fact that X is superreflexive if and only if X^* is shows that a Banach space is uniformly convexifiable if and only if it is uniformly smoothable; see Day [**1973**]). Further Pisier goes on to prove that a uniformly convex Banach space can be renormed to have a modulus of convexity of power type (see Day [**1973**] for the definition and basic facts about the modulus of convexity of a Banach space). The basic inequality here deals with martingales of the type found in Example 1.7 which Pisier calls Walsh-Paley martingales. Pisier shows that the existence of an equivalent uniformly convex norm of power type is intimately tied to proving an inequality of the following form for all Walsh-Paley martingales $(f_n, \mathscr{B}_n)$ in $L_p([0, 1], X)$:

$$\|f_1\|_p^p + \sum_{n \geqq 2} \|f_n - f_{n-1}\|_p^p \leqq C \sup_n \|f_n\|^p$$

and

$$\sup_n \|f_n\|_q^q \leqq K^q\left(\|f_1\|_q^q + \sum_{n \geqq 2} \|f_n - f_{n-1}\|_q^q\right)$$

where C and K are universal constants depending on p, q and X. The reader is referred to Pisier [**1975**] for the proof that these inequalities obtain if and only if X is superreflexive and for related martingale inequalities.

For a comprehensive survey of this topic as well as other topics in the interchange between martingales and geometry, see Woyczynski [**1975**].

Choquet-type theorems. The classical Choquet-Bishop-de Leeuw theorem (see Phelps [**1966**]) states that if K is a compact convex subset of a locally convex space E and $x \in K$, then there is a probability measure μ defined on the Baire subsets of K such that for each $x^* \in E^*$

$$x^*(x) = \int_K x^*(w)\, d\mu(w)$$

and such that μ vanishes on every Baire set in K which is disjoint from the extreme points of K. Until 1974, every proof of this theorem, even in the case of weakly compact subsets of separable Banach spaces, depended crucially on the compactness properties of K. In 1974 Edgar **[1975]** proved the following theorem which shows that compactness seems to play *no* role whatsoever.

THEOREM (EDGAR). *Let C be a separable closed bounded convex subset of a Banach space X. If every subset of C is dentable (or σ-dentable) then each $x \in C$ is the barycenter of a probability measure μ defined on the universally measurable subsets of C so that*

$$x = \textit{Bochner-}\int_C c\, d\mu(c)$$

and μ (extreme points of C) $= 1$.

The proof is a beautiful mixture of martingale methods, transfinite induction, a selection theorem of Kuratowski and Ryll-Nardzewski with a twist of continuous functions on ordered rays. Here is the basic idea: Let $x_1 \in C$. If x_1 is a convex combination of extreme points then the obvious convex combination of point masses works as a representing measure. Otherwise, one can sprout a tree from x_1 : x_1 is the midpoint of two distinct points x_2, x_3 of C each of which is the midpoint of two distinct points of C, $\cdots$. Mimicking Example 1.7, one builds a C-valued martingale which (and here is the point of the dentability assumption) converges pointwise and in mean. The hope is that the values of the limit of the martingale are all extreme points. Though this need not happen, if one takes a few more steps it will come about. The steps involve splitting the first limit function in a measurable fashion so that any value of the limit function which is not properly split is already extreme. This is where the Kuratowski-Ryll-Nardzewski Selection Theorem enters. As in the case of the x_n's we build a bigger martingale with values in C such that stopping times have extreme values. This takes us through $\omega + n$; the martingale converges to give an $\omega + \omega$ value. Continual applications of Kuratowski-Ryll-Nardzewski, the martingale convergence theorem and transfinite induction give a "long" martingale $(f_\alpha, \mathscr{B}_\alpha)$ indexed by the countable ordinals with values at stopping times extreme. The martingale constructed is pointwise a continuous C-valued function on the ordinal ray $[0, \Omega)$, where Ω is the first uncountable ordinal. Thus pointwise it stops after some countable ordinal, i.e., eventually the values of the martingale are all extreme points of C. If p is the product probability measure on $\{0, 1\}^\Omega$ where each coordinate is given the uniform distribution measure, then the measure μ defined by

$$\mu(A) = \int \chi_A \cdot \lim f_\alpha(\omega)\, dp(\omega)$$

is a probability measure supported by the extreme points of C which represents x_1.

Since Edgar's theorem there has been considerable effort expended on extending the result to nonseparable situations. Thus far these efforts have not been rewarded with a general result; it is not out of the question that a Choquet-type theorem holds in arbitrary Banach spaces with the Radon-Nikodým property. The form of such a result will not be quite as pleasing as the above result of Edgar; this is apparent from the results obtained by Edgar **[1976]**. The question of uniqueness of representing measures has been beautifully and successfully treated by Bourgin and Edgar **[1976]** and Saint Raymond **[1976]**.

VI. OPERATORS ON SPACES OF CONTINUOUS FUNCTIONS

The study of the behavior of bounded linear operators on spaces of continuous functions is one of the central applications of the theory of vector measures. This should come as no surprise. For if Ω is a compact Hausdorff space and $C(\Omega)$ is the space of all scalar-valued continuous functions on Ω, vector measures should be of use in studying operators on $C(\Omega)$ in the same way that scalar-valued measures can be used to study linear functionals on $C(\Omega)$. In this chapter we shall attempt to make the reader believe this.

Basic to this chapter is the following question: What is the Riesz Representation Theorem for operators on $C(\Omega)$? Specifically, given a bounded linear operator T from $C(\Omega)$ to a Banach space X when is there a regular vector measure G on the Borel sets of Ω with values in X such that

$$T(f) = \int_{\Omega} f \, dG$$

for all $f \in C(\Omega)$? This question is so basic that there are those who argue that this question motivates the whole theory of vector measures. The purpose of this chapter is to look at this question and natural questions spawned by this question.

Before we proceed into the chapter, it is a good idea to ask what dividends should be paid by such a representation for operators on $C(\Omega)$. A pure representation theory may be a beautiful theory in its own right, but an important representation theory should provide concrete structural information about the objects being represented. The Riesz Representation Theorem for operators on $C(\Omega)$ as presented in this chapter yields a great deal of information concerning the action of the operator being represented. We shall find that the Riesz Representation Theorem as above holds for an operator T on $C(\Omega)$ if and only if T is weakly compact; T is compact if and only if G has a relatively compact range; T is absolutely summing if and only if G is of bounded variation, and T is nuclear if and only if G has a Bochner integrable Radon-Nikodým derivative with respect to its variation $|G|$. These facts, in turn, allow us to examine the action of the various classes of operators on $C(\Omega)$ and, as we shall see presently, allow one to examine the action of an operator T on $C(\Omega)$ by examining its representing measure G and the range space of the operator T.

The first section is a bit of a digression. Here operators on spaces of bounded measurable functions are examined. Many of the results of this section should be considered as prototypes for the theorems in §2 which deals with weakly compact operators on $C(\Omega)$. §§3 and 4 deal with absolutely summing, integral, and nuclear operators on $C(\Omega)$ and the role of the Radon-Nikodým theorem in the theory of integral and nuclear operators.

1. Operators on $B(\Sigma)$ and $L_\infty(\mu)$. This section is mainly an exercise in translating the basic measure-theoretic theorems of Chapter 1 into the language operators on spaces of bounded measurable functions. Although this translation procedure is not difficult, it will give us a good feeling for the behavior of operators on spaces of measurable functions. More importantly, this section allows one to gain intuition for the seemingly more difficult problems that arise in the context of operators on $C(\Omega)$. Throughout this section X is a Banach space, $\mathscr{F}$ is a field of subsets of a point set S, Σ is a σ-field of subsets of S, and (S, Σ, μ) is a finite measure space; $L_\infty(\mu)$ is as usual. The space $B(\Sigma)$ (resp. $B(\mathscr{F})$) is the Banach space of all scalar-valued functions f on S that can be uniformly approximated by a function of the form $\sum_{n=1}^m \alpha\chi_{E_n}$ where the α_n's are scalars and $E_n \in \Sigma$ (resp. $\mathscr{F}$) for all n.

Now let T: $B(\mathscr{F}) \to X$ be a bounded linear operator and define $G: \mathscr{F} \to X$ by $G(E) = T(\chi_E)$ for $E \in \mathscr{F}$. It is clear that $G: \mathscr{F} \to X$ is a finitely additive vector measure. G will be termed the *representing* measure of T. A glance at Theorem I.1.13 guarantees that G is bounded, that $\|T\| = \|G\|(S)$, and that $T(f) = \int_S f\,dG$ for all $f \in B(\mathscr{F})$. Of course, the same line of statements holds for operators on $L_\infty(\mu)$.

Most of this section is devoted to showing that the correspondence $T \to G$ is not just an idle representation, but rather that this representation can be used to reveal some important properties of operators on the spaces $B(\mathscr{F})$, $B(\Sigma)$, and $L_\infty(\mu)$. The first theorem relates weakly compact operators to strongly additive vector measures.

THEOREM 1. *A bounded linear operator $T: B(\mathscr{F}) \to X$ ($L_\infty(\mu) \to X$) is weakly compact if and only if its representing measure G is strongly additive.*

PROOF. If T is weakly compact, then $G(\mathscr{F}) = \{T(\chi_E): E \in \mathscr{F}\}$ is contained in a weakly compact set since $\|\chi_E\| \leqq 1$ for all $E \in \mathscr{F}$. Hence the range of G is contained in a weakly compact set. A glance at I.5.3 shows G is strongly additive.

To prove the converse, suppose G is strongly additive. Another appeal to I.5.3 shows that $G(\mathscr{F})$ is relatively weakly compact. Define $T: B(\mathscr{F}) \to X$ by $T(f) = \int_S f\,dG$. To show T is a weakly compact operator, it is plainly sufficient to show that the collection of all sums of the form $\sum_{n=1}^m \alpha_n G(E_n)$, $0 \leqq \alpha_1 \leqq \alpha_2 \leqq \cdots \leqq \alpha_m \leqq 1$, $E_1, \cdots, E_m \in \mathscr{F}$, $E_i \cap E_j = \emptyset$ for $i \neq j$ lie in the convex hull of $G(\mathscr{F})$. To this end, consider a sum $\sum_{n=1}^m \alpha_n G(E_n)$ of the above form and sum it by parts by writing $A_1 = \bigcup_{n=1}^m E_n$, $A_2 = \bigcup_{n=2}^m E_n$, $\cdots$, $A_{m-1} = E_{m-1} \cup E_m$, and $A_m = E_m$. Since $0 \leqq \alpha_1 \leqq \alpha_2 \leqq \cdots \leqq \alpha_m \leqq 1$, we have

$$\sum_{n=1}^m \alpha_n G(E_n) = \alpha_1 G(A_1) + \sum_{n=2}^m (\alpha_n - \alpha_{n-1}) G(A_n).$$

Also since $0 \leqq \alpha_1 + \sum_{n=2}^{m} (\alpha_n - \alpha_{n-1}) = \alpha_m \leqq 1$, the right-hand sum belongs to the convex hull of $G(\mathscr{F})$. (Note $0 \in G(\mathscr{F})$ since $G(\emptyset) = 0$.) This completes the proof.

In spite of its simplicity, Theorem 1.1 has a wealth of corollaries, most of which follow from Rosenthal's lemma.

COROLLARY 2. *If $T: B(\mathscr{F}) \to X$ is a bounded linear operator that is not weakly compact, there is a subspace of $B(\mathscr{F})$ that is an isometric copy of c_0 on which T acts as an isomorphism.*

In particular if X contains no copy of c_0 every bounded linear operator from $B(\mathscr{F})$ to X is weakly compact.

PROOF. If T is not weakly compact, then G is not strongly additive. According to I.4.2 there exists an isomorphism $U: c_0 \to X$ and a sequence of disjoint sets $\{E_n\} \subset \mathscr{F}$ such that $U(e_n) = G(E_n)$ where (e_n) is the unit vector basis of c_0. Set $V = \{f \in B(\mathscr{F}): f = \sum_{n=1}^{\infty} \alpha_n \chi_{E_n}, \lim_n \alpha_n = 0\}$. V is a closed subspace of $B(\mathscr{F})$ and is isometric to c_0. Moreover

$$U((\alpha_n)) = \sum_{n=1}^{\infty} \alpha_n G(E_n) = \int_S \sum_{n=1}^{\infty} \alpha_n \chi_{E_n} \, dG = T\left(\sum_{n=1}^{\infty} \alpha_n \chi_{E_n} \right).$$

Hence T acts as an isomorphism on V.

If $\mathscr{F}$ is a σ-field, more can be said.

COROLLARY 3. *If Σ is a σ-field of subsets of S, and $T: B(\Sigma) \to X$ is a bounded linear operator that is not weakly compact, then there is a linear subspace of $B(\Sigma)$ that is an isometric copy of l_∞ on which T acts as an isomorphism.*

In particular if X contains no copy of l_∞ every bounded linear operator $T: B(\Sigma) \to X$ is weakly compact.

The same statements remain true if "$B(\Sigma)$" is changed to read "$L_\infty(\mu)$".

PROOF. If T is not weakly compact, then its representing measure G is not strongly additive. According to I.4.2. there is a disjoint sequence (E_n) in Σ and an isomorphism $U: l_\infty \to X$ such that if (α_n) is a sequence of zeroes and ones, then $U((\alpha_n)) = G(\bigcup\{E_n: \alpha_n = 1\})$. Set $V = \{f: f \in B(\Sigma), f = \sum_{n=1}^{\infty} \alpha_n \chi_{E_n}\}$. Then V is isometric to l_∞, and

$$U((\alpha_n)) = \int_S \sum_{n=1}^{\infty} \alpha_n \chi_{E_n} \, dG = T\left(\sum_{n=1}^{\infty} \alpha_n \chi_{E_n} \right)$$

for all finitely valued sequences (α_n), i.e., $U((\alpha_n)) = T(\sum_{n=1}^{\infty} \alpha_n \chi_{E_n})$ whenever $\sum_{n=1}^{\infty} \alpha_n \chi_{E_n}$ is a simple function. Since simple functions are dense in V, $U((\alpha_n)) = T(\sum_{n=1}^{\infty} \alpha_n \chi_{E_n})$ for every $\sum_{n=1}^{\infty} \alpha_n \chi_{E_n} \in V$. Thus T acts as an isomorphism on V. This completes the proof.

It is well known that $L_1[0, 1]$ contains reflexive subspaces. It is not so well known that $L_1[0, 1]$ contains no infinite dimensional nonreflexive second dual subspaces. For, if X^{**} is isomorphic to a subspace of $L_1[0, 1]$, then X^{***} is a quotient of $L_\infty[0, 1]$, i.e., there is a bounded linear operator T from $L_\infty[0, 1]$ onto X^{***}. Since X^* is complemented in X^{***}, there is a bounded linear operator mapping $L_\infty[0, 1]$ onto the separable space X^*. By Corollary 4, this operator is weakly compact and X^* is reflexive by the interior mapping principle.

Before moving to the next corollary, recall that a series $\sum_n x_n$ in a Banach space is called weakly unconditionally Cauchy if

$$\sup_{\Delta}\left\{\left\| \sum_{n\in\Delta} x_n \right\| : \Delta \subset N, \Delta \text{ finite}\right\}$$

is finite.

COROLLARY 4. *Any one of the following statements about a bounded linear operator* $T: B(\mathcal{F}) \to X$ *(or* $T: L_\infty(\mu) \to X$*) implies all the others.*

(a) *T is weakly compact.*

(b) *T maps weakly unconditionally Cauchy series into unconditionally convergent series.*

(c) *If $\sum_{n=1}^{\infty} f_n$ is a weakly unconditionally Cauchy series in $B(\mathcal{F})$, then* $\lim_n T(f_n) = 0$.

PROOF. To prove that (c) implies (a), note that if $(E_n) \subset \mathcal{F}$ is disjoint, then $\sum_{n=1}^{\infty} \chi_{E_n}$ is weakly unconditionally Cauchy. Hence $\lim_n T(\chi_{E_n}) = 0$. Therefore, if G is the representing measure of T, then $\lim_n G(E_n) = \lim_n T(\chi_{E_n}) = 0$ for all disjoint sequences $\{E_n\} \subset \mathcal{F}$. An appeal to I.1.17 shows G is strongly additive. Hence T is weakly compact.

The implication (a) $\Rightarrow$ (b) is true in general and is an easy consequence of the Orlicz-Pettis Theorem I.4.4.

Finally the implication (b) $\Rightarrow$ (c) is trivial.

The Vitali-Hahn-Saks theorem teamed with Theorem 1 results in

COROLLARY 5. *Let Σ be a σ-field of subsets of S. If (T_n) is a sequence of weakly compact operators from $B(\Sigma)$ to X that converges in the strong operator topology to an operator T, then T is also weakly compact and $\{T_n^*\}$ is an equiweakly compact sequence in the sense that $\bigcup_{n=1}^{\infty} T_n^*(U^*)$ is contained in a weakly compact set. (Here U^* is the closed unit ball of X^*.) The same statement is true if "$B(\Sigma)$" is changed to read "$L_\infty(\mu)$".*

PROOF. Let G_n be the representing measure of T_n and G be the representing measure of T. Then for $E \in \Sigma$ one has

$$\lim_n G_n(E) = \lim_n T_n(\chi_E) = T(\chi_E) = G(E).$$

Since Σ is a σ-field, I.4.8 guarantees that G is strongly additive and that the family $\{G_n: n \in N\}$ is uniformly additive. Hence $\{x^*G_n: n \in N, \|x^*\| \leq 1\}$ is uniformly additive and is thus weakly compact in $L_\infty(\mu)^*$ by IV.2.6. But $\{x^*G_n: n \in N, \|x^*\| \leq 1\} = \bigcup_{n=1}^{\infty} T_n^*(U^*)$. Hence the set $\{T_n^*\}$ is set of equiweakly compact operators.

In preparation for the next result, let us agree that $T: X^* \to Y$ is weak*-weakly continuous if T is continuous for X^* equipped with its weak* topology and Y equipped with its weak topology.

LEMMA 6. *Let Σ be a σ-field of subsets of S and μ be a finite nonnegative countably additive measure on Σ. Any one of the following statements about a bounded linear operator $T: L_\infty(\mu) \to X$ implies all the others.*

(a) *T is weak*-weakly continuous.*

(b) *The representing measure of T is countably additive.*

(c) *The representing measure of T is μ-continuous.*

PROOF. Suppose $T: L_\infty(\mu) \to X$ is weak*-weakly continuous and has representing measure G. If $(E_n) \subset \Sigma$ is a sequence that satisfies $E_{n+1} \subset E_n$ for all n and $\bigcap_{n=1}^\infty E_n = \varnothing$, then $\lim_n \int_S \chi_{E_n} g\, d\mu = 0$ for all $g \in L_1(\mu)$. Hence $\lim_n \chi_{E_n} = 0$ in the weak*-topology of $L_\infty(\mu)$. Hence $\lim_n G(E_n) = \lim_n T(\chi_{E_n}) = 0$ weakly in X. Thus G is weakly countably additive and is therefore countably additive by the Orlicz-Pettis Theorem I.4.4. This proves that (a) implies (b).

To verify the implication (b) $\Rightarrow$ (c) note that if $G: \Sigma \to X$ is the representing measure of T, G must vanish on μ-null sets. Since G is countably additive, an appeal to I.2.1 shows G is μ-continuous as well.

To prove that (c) implies (a), suppose G is the representing measure of T and that G is countably additive. Since G vanishes on μ-null sets, there is for each $x^* \in X^*$ a $g_{x^*} \in L_1(\mu)$ such that $x^*G(E) = \int_E g_{x^*}\, d\mu$ for all $E \in \Sigma$. Now if $(f_\alpha, \alpha \in A)$ is a net in $L_\infty(\mu)$ that converges to zero in the weak*-topology, and $x^* \in X^*$, then we have

$$\lim_\alpha x^*T(f_\alpha) = \lim_\alpha x^* \int_S f_\alpha\, dG = \lim_\alpha \int_S f_\alpha\, dx^*G = \lim_\alpha \int_S f_\alpha g_{x^*}\, d\mu = 0.$$

Hence T is weak*-weakly continuous.

The following theorem is a direct consequence of Theorem 1 and the decomposition Theorem I.5.9.

THEOREM 7. *If $T: L_\infty(\mu) \to X$ is a weakly compact operator, then there exist operators T_c and $T_s: L_\infty(\mu) \to X$ such that*

(i) *T_c is weak*-weakly continuous;*

(ii) *if x^*T_s is a weak*-continuous linear functional on $L_\infty(\mu)$, then $x^*T_s = 0$, and*

(iii) *$T = T_c + T_s$.*

PROOF. Let G be the representing measure of T. Since T is weakly compact, G is strongly additive. By virtue of Theorem I.5.9, there are (strongly additive) vector measures G_c and G_s on Σ such that $G = G_c + G_s$, $G_c \ll \mu$, and $x^* G_s \perp \mu$ for all $x^* \in X^*$. Define T_c and T_s on $L_\infty(\mu)$ by $T_c(f) = \int_S f\, dG_c$ and $T_s(f) = \int_S f\, dG_s$ for all $f \in L_\infty(\mu)$. Then for $f \in L_\infty(\mu)$, one has

$$\begin{aligned} T(f) &= \int_S f\, dG = \int_S f\, d(G_c + G_s) \\ &= \int_S f\, dG_c + \int_S f\, dG_s = T_c(f) + T_s(f). \end{aligned}$$

Thus $T = T_c + T_s$. Since $G_c \ll \mu$, Lemma 6 ensures that T_c is weak*-weakly continuous. Moreover if x^*T_s is a weak*-weakly continuous linear functional for some $x^* \in X^*$, then Lemma 6 asserts that $x^*G_s \ll \mu$. Since $x^*G_s \perp \mu$, it follows immediately that $x^*T_s = 0$.

2. Weakly compact operators on $C(\Omega)$ and the Riesz Representation Theorem. This section is devoted to a study of weakly compact operators on $C(\Omega)$. The first part deals with the Riesz Representation Theorem for operators. Here we shall see that if X is a Banach space, a bounded linear operator $T: C(\Omega) \to X$ is weakly compact if and only if there exists a countably additive X-valued vector measure G on the Borel sets in Ω such that $T(f) = \int_\Omega f\, dG$ for all $f \in C(\Omega)$. After this is accomplished, the relatively simple theory of weakly compact operators on $B(\mathscr{F})$ spaces will be applied to operators on $C(\Omega)$. This will link the first section with this section. Finally some properties of families of regular scalar measures will be established and used to examine some of the more elusive properties of weakly compact operators on $C(\Omega)$.

As this section progresses, we should try to keep in mind the following principle: Although operators on the $C(\Omega)$ spaces are generally more difficult to analyze than operators on $B(\mathscr{F})$ spaces, many theorems dealing with operators on $B(\mathscr{F})$ spaces remain true for operators on $C(\Omega)$ spaces. Operating from this intuitive point of view, we shall prove some analogues of the results of §1 for operators on $C(\Omega)$.

One of the reasons operators on $B(\mathscr{F})$ are easy to study is that it is trivial to write down representing measures for them. Since indicator functions are sometimes scarce in $C(\Omega)$ spaces, writing down representing measures for operators on $C(\Omega)$ can be difficult. In fact we shall find that this difficulty is a constant harassment, but it pays some handsome dividends.

THEOREM 1. *Let Ω be a compact Hausdorff space and T: $C(\Omega) \to X$ be a bounded linear operator. There exists a weak*-countably additive measure G defined on the Borel sets in Ω with values in X^{**} such that*

(i) *$G(\cdot)x^*$ is a regular countably additive Borel measure for each $x^* \in X^*$;*

(ii) *the mapping $x^* \to G(\cdot)x^*$ of X^* into $C(\Omega)^*$ is weak*- to weak*-continuous;*

(iii) *$x^*T(f) = \int_K f\, d(x^*G)$, for each $f \in C(\Omega)$ and each $x^* \in X^*$; and*

(iv) *$\|T\| = \|G\|(\Omega)$.*

Conversely, if G is an X^{**}-valued vector measure defined on the Borel sets of Ω for which (i) *and* (ii) *hold, then* (iii) *defines a bounded linear operator from $C(\Omega)$ to X which satisfies* (iv).

PROOF. Suppose E is a Borel set and let φ_E be the element of $C(\Omega)^{**}$, the second adjoint of $C(\Omega)$, defined by $\varphi_E(\mu) = \mu(E)$ for $\mu \in C(\Omega)^*$ (= all regular Borel measures on Ω). Define a set function G on the Borel sets by $G(E) = T^{**}(\varphi_E)$ for each Borel set E. By the Riesz Representation Theorem for linear functionals on $C(\Omega)$, $T^*(x^*)$ is a regular Borel measure μ_{x^*} defined on the Borel sets in Ω. Moreover if $x^* \in X^*$ and E is a Borel set, then

$$\begin{aligned}\mu_{x^*}(E) &= \varphi_E(\mu_{x^*}) = \varphi_E(T^*(x^*)) \\ &= T^{**}(\varphi_E)[x^*] = G(E)(x^*).\end{aligned}$$

Clearly (i) and (iii) follow immediately. Statement (ii) is true since this equation shows that $T^*(x^*) = x^*G$ for all $x^* \in X^*$. To prove (iv), note that

$$\begin{aligned}\|T\| &= \sup_{\|x^*\| \leq 1} \|T^*(x^*)\| = \sup_{\|x^*\| \leq 1} |x^*G|(\Omega) \\ &\leqq \sup_{\|x^{***}\| \leq 1} |x^{***}G|(\Omega) = \|G\|(\Omega).\end{aligned}$$

But G may be viewed as the vector measure defined on the σ-field Σ of Borel sets of Ω which represents the operator $\tilde{T}^{**}$: $B(\Sigma) \to X^{**}$ defined by the restriction T^{**} to $B(\Sigma) \subseteq C(\Omega)^{**}$.

By I.1.13, one has

$$\|G\|(\Omega) = \|\tilde{T}^{**}\| \leqq \|T^{**}\| = \|T\|.$$

This completes the proof of the necessity of the conditions (i)—(iv).

Conversely, if (i) and (ii) are satisfied for the mapping which sends x^* into x^*G, then it follows that for each fixed $f \in C(\Omega)$, the mapping $x^* \to \int_\Omega f\, dx^*G$ is weak*-

continuous on X^* and is therefore generated by some $x_f \in X$. Thus the mapping $T: C(\Omega) \to X$ defined by $T(f) = x_f$ is a linear operator mapping $C(\Omega)$ into X. It is straightforward to verify that T is continuous and has the stated properties.

DEFINITION 2. If $T: C(\Omega) \to X$ is a bounded linear operator, then the measure G satisfying (i)—(iv) of the statement of Theorem 1 will be termed the *representing measure* of T.

Showing that Theorem 1 is the best possible is

EXAMPLE 3. *A bounded linear operator on* $C[0, 1]$ *whose representing measure is neither (norm) countably additive nor X-valued.* Define $T: C[0, 1] \to c_0$ by

$$T(f) = (f(1/n) - f(0))$$

for $f \in C[0, 1]$. It is easily checked that the representing measure G of T is given by

$$G(E) = (\chi_E(1/n) - \chi_E(0)) \in l^\infty = c_0^{**}$$

for each Borel set $E \subset [0, 1]$. Setting $E = \{1/2n: n \in N\}$ reveals that G takes at least one value outside c_0. Taking $E_n = \{1/n\}$ shows that $\sum_{n=1}^{\infty} G(E_n)$ is far from Cauchy and therefore G is not countably additive.

Theorem 1 provides a convenient link between the work of §1 on operators on $B(\mathscr{F})$ spaces and our current study of operators on $C(\Omega)$ spaces. Let $T: C(\Omega) \to X$ be a bounded linear operator with representing measure G. Let Σ be the Borel σ-field of subsets of Ω and define $\hat{T}: B(\Sigma) \to X^{**}$ by $\hat{T}f = \int_\Omega f\, dG$ for $f \in B(\Sigma)$. (The integral is the elementary Bartle integral of I.1.12.) If $x^* \in X^*$ and $f \in C(\Omega)$, then $f \in B(\Sigma)$ and

$$\hat{T}f(x^*) = x^* \int_\Omega f\, dG = \int_\Omega f\, d(x^*G)$$

by I.1.13. Since this holds for all $x^* \in X^*$, a glance at Theorem 1(iii) reveals that $\hat{T}f = Tf$ for all $f \in C(\Omega)$. Thus the operator $\hat{T}$ so defined is an extension of T to $B(\Sigma)$, and will be called the natural extension of T to $B(\Sigma)$. (Another way to construct $\hat{T}$ is to inject isometrically $B(\Sigma)$ into $C(\Omega)^{**}$ in the natural way and let $\hat{T}$ be T^{**} restricted to $B(\Sigma)$; this, in effect, is what has been done above.) Since $\hat{T}$ extends T, the operator T is weakly compact whenever $\hat{T}$ is weakly compact; this leads to a cheap proof of

PROPOSITION 4. *If X^{**} contains no copy of l_∞, then every bounded linear operator $T: C(\Omega) \to X$ is weakly compact.*

PROOF. Since X^{**} contains no copy of l^∞, Corollary 1.3 guarantees that every bounded linear operator $S: B(\Sigma) \to X^{**}$ is weakly compact. Therefore $\hat{T}$ is weakly compact; thus T is weakly compact. More precisely, T viewed as an operator into X^{**} is weakly compact. Therefore T viewed as an operator into X is weakly compact.

The next theorem is the key to the understanding of the properties of weakly compact operators on $C(\Omega)$.

THEOREM 5 (BARTLE-DUNFORD-SCHWARTZ). *Let $T: C(\Omega) \to X$ be a bounded linear operator with representing measure G. Any one of the following statements implies all the others.*

(a) *The operator T is weakly compact.*
(b) *The measure G takes all its values in X.*
(c) *The measure G is countably additive.*
(d) *The measure G is strongly additive.*

PROOF. To prove (a) implies (b), note that if T is weakly compact, T^{**} has all its values in the closed subspace X of X^{**}. A glance at the construction of G in the proof of Theorem 1 reveals that G takes all its values in X.

To prove (b) implies (c), note that if the range of G is contained in X, then Theorem 1(ii) guarantees that G is a weakly countably additive measure defined on the Borel σ-field. By I.4.4, G is countably additive. This proves that (b) implies (c).

The fact that (c) implies (d) is obvious from the fact that G is defined on a σ-field.

To prove that (d) implies (a), suppose G is strongly additive. Since G is also the representing measure of $\hat{T}: B(\Sigma) \to X^{**}$, an appeal to Theorem 1 shows that $\hat{T}$ is weakly compact. Thus T, viewed as an operator into X^{**}, is weakly compact since T is a restriction of $\hat{T}$. It follows that $T: C(\Omega) \to X$ is weakly compact.

Theorem 5 has some simple consequences.

COROLLARY 6 (GROTHENDIECK AND BARTLE-DUNFORD-SCHWARTZ). *A weakly compact linear operator from $C(\Omega)$ to X sends weakly Cauchy sequences into norm convergent sequences. Consequently a weakly compact linear operator from $C(\Omega)$ to X maps weakly compact sets into norm compact sets.*

PROOF. Let $T: C(\Omega) \to X$ be a weakly compact linear operator with representing measure G. If $(f_n) \subset C(\Omega)$ is a weakly Cauchy sequence, then f_n is pointwise convergent to some $f_0 \in B(\Sigma)$ and $\sup_n \|f_n\| < \infty$. Since T is weakly compact, G is countably additive and by the bounded convergence Theorem II.4.1,

$$\lim_n T(f_n) = \lim \int_\Omega f_n \, dG = \int_\Omega f_0 \, dG = \hat{T}(f_0).$$

This proves the first statement.

The second assertion is now a direct consequence of Eberlein's theorem.

According to Corollary III.2.17, the conclusion of the last corollary holds for weakly compact operators on $L^1(\mu)$ spaces as well as for weakly compact operators on $C(\Omega)$ spaces. Since the truth of this statement for operators on $L^1(\mu)$ spaces follows from the Dunford-Pettis theorem, it is natural to call spaces with this property "spaces with the Dunford-Pettis property". Another look at spaces with the Dunford-Pettis property can be found in the notes and remarks section.

The next application of Theorem 5 will be to operators on $C(\Omega)$ when Ω is a Stonean space. In fact we shall see that, from the point of view of weak compactness, operators on $C(\Omega)$ for Stonean Ω behave the same way as operators on $B(\Sigma)$ behave. First some preparatory definitions and facts are needed.

DEFINITION 7. *A topological space is called Stonean* (*or extremally disconnected*) *if the closure of every open set is open.*

LEMMA 8. *As a vector lattice, $C(\Omega)$ is order complete* (*i.e., each decreasing net in $C(\Omega)$ that is bounded from below by a member of $C(\Omega)$ has a greatest lower bound*) *if and only if Ω is Stonean.* (*We are considering only the real Banach space $C(\Omega)$.*)

PROOF. Suppose that Ω is Stonean. To show that $C(\Omega)$ is order complete, it is sufficient, by translation, to show that every collection $\{f_\alpha : \alpha \in A\}$ of nonnegative functions in $C(\Omega)$ has an infimum. For each $\alpha \in A$ and each positive number r, set

$$G_{\alpha r} = \{\omega \in \Omega : f_\alpha(\omega) < r\}.$$

Then each $G_{\alpha r}$ is open and the set $G_r = \bigcup_{\alpha \in A} G_{\alpha r}$ is also open for each $r > 0$. Since Ω is Stonean, the closure $\bar{G}_r$ of G_r is open for each $r > 0$. In addition $\Omega = \bigcup_{r>0} \bar{G}_r$ and $\bar{G}_r \subseteq \bar{G}_s$ whenever $r \leqq s$. Therefore if $\omega \in \Omega$, then either $\omega \in \bigcap_{r>0} \bar{G}_r$ or there is an $r > 0$ ($r = r(\omega)$) such that $\omega \in \bar{G}_{r+\varepsilon}$ and $\omega \notin G_{r-\varepsilon}$ for all $\varepsilon > 0$. Define g on Ω by writing

$$g(\omega) = \begin{cases} r & \text{if } \omega \in \bar{G}_{r+\varepsilon} \backslash \bar{G}_{r-\varepsilon} \text{ for all } \varepsilon > 0, \\ 0 & \text{if } w \in \bigcap_{r>0} \bar{G}_r. \end{cases}$$

Next note that the set

$$E_r = \{\omega \in \Omega : g(\omega) < r\} = \bigcup_{0<\varepsilon<r} \bar{G}_{r-\varepsilon}$$

is open since Ω is Stonean and the set

$$F_r = \{\omega \in \Omega : g(\omega) \leqq r\} = \bigcap_{\varepsilon>0} \bar{G}_{r+\varepsilon}$$

is closed. Accordingly, if $0 \leqq r_1 < r_2$ then $\{\omega : r_1 < g(\omega) < r_2\} = E_{r_2} \backslash F_{r_1}$ is open in Ω. Thus g is continuous on Ω.

To prove g is a lower bound of $\{f_\alpha : \alpha \in A\}$, note that, for any fixed α and $r > 0$, one has $\{\omega \in \Omega : f_\alpha(\omega) < r - \varepsilon\} \subseteq E_r$ for each $\varepsilon > 0$. Therefore $g \leqq f_\alpha$ for each $\alpha \in A$.

To prove that g is the greatest lower bound of $\{f_\alpha : \alpha \in A\}$, let $h \in C(\Omega)$ and $h \leqq f_\alpha$ for all $\alpha \in A$. Note that $\{\omega \in \Omega : h(\omega) < r - \varepsilon\} \supseteq G_{r-\varepsilon}$ for all $r > 0$ and $\varepsilon > 0$. Since h is continuous, we have $\{\omega \in \Omega : h(\omega) \leqq r - \varepsilon\} \supseteq \bar{G}_{r-\varepsilon}$ for all $r > 0$ and $0 < \varepsilon$. It follows that

$$\{\omega \in \Omega : h(\omega) < r\} \supseteq \bigcup_{\varepsilon>0} \bar{G}_{r-\varepsilon} = \{\omega \in \Omega : g(\omega) < r\}$$

for each $r > 0$. Hence $h \leqq g$ and $g = \inf\{f_\alpha : \alpha \in A\}$. This proves $C(\Omega)$ is order complete.

For the converse, suppose $C(\Omega)$ is order complete. If G is a nonempty open set in Ω, then for each $\tau \in G$ there is $f_\tau \in C(\Omega)$ with $f_\tau(\Omega) \subset [0, 1]$ such that $f_\tau(\tau) = 0$ and $f_\tau(\omega) = 1$ for $\omega \notin G$. Set $g = \inf\{f_\tau : \tau \in G\}$ and note that g vanishes on G. Since g is continuous, the function g also vanishes on $\bar{G}$. Now, for each $\omega \notin \bar{G}$, there is $h_\omega \in C(\Omega)$ with $h_\omega(\Omega) \subseteq [0, 1]$ such that h_ω vanishes on $\bar{G}$ and $h_\omega(\omega) = 1$. Clearly $h_\omega \leqq f_\tau$ for all τ and ω. Therefore $h_\omega \leqq g$. But then $1 = h_\omega(\omega) \leqq g(\omega) \leqq 1$ and $g(\omega) = 1$ for each $\omega \notin \bar{G}$. It follows that $\bar{G}$ is open. This proves that Ω is Stonean.

Now consider, for the moment, real Banach spaces Y and Z and suppose Y is a closed subspace of Z. Let Ω be Stonean and $C_R(\Omega)$ be the Banach space of real-valued continuous functions on Ω. Let $T: Y \to C_R(\Omega)$ be a bounded linear operator. The number $\|T\|$ is the value of the smallest constant function k such that $|T(y)| \leqq k$ for all $y \in Y$ with $\|y\| \leqq 1$. Armed with the fact that $C_R(\Omega)$ is order complete, mimic the standard proof of the Hahn-Banach theorem to obtain a bounded linear extension $S: Z \to C_R(\Omega)$ of T such that $|S(z)| \leqq k$ for all $z \in Z$ with $\|z\| \leqq 1$. Hence S is a bounded linear extension of T to all of Z with $\|S\| = \|T\|$.

In particular, suppose $C_R(\Omega)$ is a subspace of Z. Taking T equal to the identity operator above reveals that S is a norm one projection from Z onto $C_R(\Omega)$. This proves

LEMMA 9. *If Ω is Stonean and $C_R(\Omega)$ is a subspace of another real Banach space Z, then $C_R(\Omega)$ is complemented in Z by a norm one linear projection.*

Lemma 9 also holds for complex Banach spaces by more complicated techniques that will not be given here. If the reader is willing to believe this, he will believe the following theorem which translates Corollary 1.3 from operators on $B(\Sigma)$ to operators on $C(\Omega)$ for Stonean Ω.

THEOREM 10 (ROSENTHAL). *If X contains no copy of l_∞ and Ω is Stonean, then every bounded linear operator $T: C(\Omega) \to X$ is weakly compact.*

PROOF. Let Σ be the σ-field of Borel sets in Ω and note that the natural injection of $C(\Omega)$ into $B(\Sigma)$ is an isometry. By Lemma 9, there is a norm one projection P of $B(\Sigma)$ onto $C(\Omega)$. Now define $S: B(\Sigma) \to X$ by $S(f) = TP(f)$ for $f \in B(\Sigma)$. Since X contains no copy of l_∞, an appeal to Corollary 1.3 shows that S is weakly compact. Therefore T is also weakly compact.

A powerful corollary follows.

COROLLARY 11 (ROSENTHAL). *If Ω is Stonean, then an infinite dimensional complemented subspace of $C(\Omega)$ must contain a copy of l_∞.*

PROOF. Suppose X is a subspace of $C(\Omega)$ and suppose P is a projection of $C(\Omega)$ onto X. If X contains no copy of l_∞, then Theorem 10 guarantees that P is weakly compact. If U denotes the closed unit ball of $C(\Omega)$, then $P(U) = P^2(U) = P(P(U))$. An appeal to Corollary 2.6 reveals that $P(P(U))$ is a compact set since P is a weakly compact set. Hence P is a compact projection and $P(C(\Omega)) = X$ is finite dimensional.

An immediate specialization of Corollary 11 is

COROLLARY 12 (GROTHENDIECK). *If Ω is Stonean and X is a separable Banach space, then every bounded linear operator $T: C(\Omega) \to X$ is weakly compact.*

PROOF. The space l_∞ is not separable.

The spaces $C(\Omega)$ for Stonean Ω are by no means rare creatures in Banach space theory. Indeed the order complete spaces $l_\infty(\Gamma)$ (Γ infinite), $L_\infty(\mu)$ (when μ is σ-finite or more generally "localizable") are $C(\Omega)$ spaces for Stonean Ω. Moreover many Banach spaces of a somewhat more general nature than $C(\Omega)$ for Stonean Ω share the operator-theoretic property enunciated in Corollary 12. This broader class of spaces, the so-called "Grothendieck spaces", will be discussed in the notes and remarks section.

At this point, it is a good idea to examine where we stand. We have obtained some general theorems about weakly compact operators on $C(\Omega)$ and we have some feeling for special weak compactness properties of operators on $C(\Omega)$ for Stonean Ω. On the other hand, we do not have much information about weakly compact operators for spaces as familiar as $C[0, 1]$. Examining operators on $C[0, 1]$ for weak compactness is a much more delicate matter than examining operators on $C(\Omega)$ for weak compactness if Ω is Stonean. Before we can advance much farther, we need to know some regularity properties of representing measures for operators on $C(\Omega)$. Recall that a vector measure G defined on the Borel σ-field of subsets of a compact Hausdorff space is *regular* if for each Borel set E and $\varepsilon > 0$ there exists a compact set K and an open set O such that $K \subset E \subset O$ and $\|G\|(O \backslash K) < \varepsilon$.

The following lemma is the key to the understanding of properties of weakly compact operators on $C(\Omega)$.

LEMMA 13. *Let Σ be the σ-field of Borel sets of Ω. Let M be a family of regular (countably additive) scalar measures defined on Σ. Each of the following statements about M implies all the others.*

(a) *$\mathrm{Lim}_n \mu(O_n) = 0$ uniformly in $\mu \in M$ for each disjoint sequence (O_n) of open subsets of Ω.*

(b) *$\mathrm{Lim}_n |\mu|(O_n) = 0$ uniformly in $\mu \in M$ for each disjoint sequence (O_n) of open subsets of Ω.*

(c) *M is uniformly inner regular on the open sets, i.e., if O is an open set and $\varepsilon > 0$, then there exists a compact $K \subset O$ such that $\sup_{\mu \in M} |\mu|(O \backslash K) \leqq \varepsilon$.*

(d) *M is uniformly inner regular, i.e., if $E \in \Sigma$ and $\varepsilon > 0$, there exists a compact set $F \subset E$ such that $\sup_{\mu \in M} |\mu|(E \backslash F) \leqq \varepsilon$.*

(e) *M is uniformly countably additive.*

(f) *M is uniformly regular, i.e., if $E \in \Sigma$ and $\varepsilon > 0$ then there exists an open set O and a compact set K such that $K \subseteq E \subseteq O$ and $\sup_{\mu \in M} |\mu|(O \backslash K) \leqq \varepsilon$.*

PROOF. The implication (a) $\Rightarrow$ (b) is a direct consequence of the regularity of each $\mu \in M$. If there is a disjoint sequence (O_n) of open sets and measures $(\mu_n) \subset M$ such that $\inf_n |\mu_n|(O_n) > 0$, then using regularity of each μ_n one can find a sequence of open sets (O'_n) with $O'_n \subseteq O_n$ and $\inf_n |\mu_n(O'_n)| > 0$.

To prove that (b) implies (c), let O be an open set in Ω and $\varepsilon > 0$. If there exists no compact subset K of O such that $\sup_{\mu \in M} |\mu|(O \backslash K) \leqq \varepsilon$, then there is $\mu_1 \in M$ such that $|\mu_1|(O) > \varepsilon$, for otherwise $K = \emptyset$ will provide a contradiction. Since $|\mu_1|$ is regular, there exists a compact subset K_1 of O such that $|\mu_1|(K_1) > \varepsilon$. Since the compact Hausdorff space Ω is normal, there is an open set O_1 with

$$O \supseteq \bar{O}_1 \supseteq O_1 \supseteq K_1.$$

Moreover $|\mu_1|(O_1) \geqq |\mu_1|(K) > \varepsilon$. Now $\bar{O}_1$ is a compact subset of O. In view of our assumption, there is $\mu_2 \in M$ such that $|\mu_2|(O \backslash \bar{O}_1) > \varepsilon$. Since $|\mu_2|$ is regular there exists a compact set K_2 with $O \supseteq K_2 \supseteq \bar{O}_1$ and $|\mu_2|(K_2 \backslash \bar{O}_1) > \varepsilon$. Again there is an open set O_2 such that

$$O \supseteq \bar{O}_2 \supseteq O_2 \supseteq K_2 \supseteq \bar{O}_1 \supseteq O_1.$$

Accordingly $|\mu_2|(O_2 \backslash \bar{O}_1) \geqq |\mu_2|(K_2 \backslash \bar{O}_1) > \varepsilon$. Next by our assumption there exists $\mu_3 \in M$ such that $|\mu_3|(O \backslash \bar{O}_2) > \varepsilon$. By regularity, of $|\mu_3|$, there is a compact set K_3 with $O \supseteq K_3 \supseteq \bar{O}_2$ and $|\mu_3|(K_3 \backslash \bar{O}_2) > \varepsilon$. Again there exists an open set O_3 with

$$O \supseteq \bar{O}_3 \supseteq O_3 \supseteq K_3 \supseteq \bar{O}_2 \supseteq O_2$$

and hence, $|\mu_3|(O_3 \backslash \bar{O}_2) \geqq |\mu_3|(K_3 \backslash O_2) > \varepsilon$.

Continue this process to produce an increasing sequence of $\{O_n\}$ of open subsets of Ω and a sequence of measures $(\mu_n) \subset M$ such that

$$O \supseteq \bar{O}_{n+1} \supseteq O_{n+1} \supseteq \bar{O}_n \supseteq O_n$$

and $|\mu_{n+1}|(O_{n+1}\backslash\bar{O}_n) > \varepsilon$ for all $n \geqq 1$. Write $G_{n+1} = O_{n+1}\backslash\bar{O}_n$ for $n \geqq 1$ and note that (G_{n+1}) is a sequence of disjoint open sets with $|\mu_{n+1}|(G_{n+1}) > \varepsilon$ for all $n \geqq 1$. This contradicts (ii) and proves that (b) implies (c).

To prove that (c) implies (d), call a Borel set E M-measurable if for each $\varepsilon > 0$ there is a compact set K such that $E \cap K$ is compact and $\sup_{\mu\in M}|\mu|(\Omega\backslash K) \leqq \varepsilon$. Every compact set is M-measurable by default. Also if O is open and $\varepsilon > 0$, then an appeal to (c) produces a compact subset K_1 of O such that $\sup_{\mu\in M}|\mu|(O\backslash K_1) \leqq \varepsilon$. Let $K = K_1 \cup (\Omega\backslash O)$ and note that $\sup_{\mu\in M}|\mu|(\Omega\backslash K) = \sup_{\mu\in M}|\mu(O\backslash K_1)| \leqq \varepsilon$ and $K \cap O = K_1$ is compact. Thus each open set is M-measurable.

Next it will be shown that the collection of M-measurable sets is a σ-field. Let (E_n) be a sequence of M-measurable sets and let $\varepsilon > 0$. Then there exists a sequence (K_n) of compact sets such that $E_n \cap K_n$ is compact and $\sup_{\mu\in M}|\mu|(\Omega\backslash K_n) \leqq \varepsilon 2^{-n}$ for each $n \geqq 1$. Also the set

$$\left(\bigcap_{n=1}^{\infty} E_n\right) \cap \left(\bigcap_{n=1}^{\infty} K_n\right) = \bigcap_{n=1}^{\infty} (E_n \cap K_n)$$

is compact and

$$\sup_{\mu\in M} |\mu| \left(\Omega\backslash\left(\bigcap_{n=1}^{\infty} K_n\right)\right) \leqq \sum_{n=1}^{\infty} \sup_{\mu\in M} |\mu| (\Omega\backslash K_n) \leqq \sum_{n=1}^{\infty} \varepsilon 2^{-n} = \varepsilon.$$

Thus the collection of M-measurable sets is closed under countable intersections.

Next it will be shown that the collection of M-measurable sets is closed under complementation. If E is M-measurable and $\varepsilon > 0$, there is a compact set K_1 such that $E \cap K_1$ is compact and $\sup_{\mu\in M}|\mu|(\Omega\backslash K_1) \leqq \varepsilon/2$. Now $\Omega\backslash(E \cap K_1)$ is open and is therefore M-measurable. Thus there is a compact set K_2 such that $K_2 \cap (\Omega\backslash(E \cap K_1))$ is compact and $\sup_{\mu\in M}|\mu|(\Omega\backslash K_2) \leqq \varepsilon/2$.

Therefore $K_1 \cap K_2 \cap (\Omega\backslash(E \cap K_1)) = (K_1 \cap K_2) \cap (\Omega\backslash E)$ is compact and we have

$$\sup_{\mu\in M} |\mu| (\Omega\backslash(K_1 \cap K_2)) \leqq \sup_{\mu\in M} |\mu|(\Omega\backslash K_1) + \sup_{\mu\in M} |\mu|(\Omega\backslash K_2) \leqq \varepsilon/2 + \varepsilon/2 = \varepsilon.$$

Thus the collection of M-measurable sets is closed under complementation.

It follows that the collection of M-measurable sets is a σ-field of subsets of Ω containing the compact sets. Therefore Σ coincides with the class of M-measurable sets. A glance at the definition of an M-measurable set shows μ is uniformly inner regular on the σ-field of Borel sets. This proves that (c) implies (d).

The proof that (d) implies (e) is immediate: Let (E_n) be a sequence of Borel sets such that $E_n \supseteq E_{n+1}$ for all $n \in N$ and $\bigcap_{n=1}^{\infty} E_n = \emptyset$. Let $\varepsilon > 0$. Then there exists a sequence of compact sets (K_n) such that $K_n \subseteq E_n$ for all $n \in N$ and $\sup_{\mu\in M}|\mu| \cdot(E_n\backslash K_n) < \varepsilon/2^n$. Now since $\bigcap_{n=1}^{\infty} K_n = \emptyset$ and since each K_n is compact, there exists n_0 such that $\bigcap_{n=1}^{m} K_n = \emptyset$ for $m \geqq n_0$. Hence for $m \geqq n_0$ one has

$$\begin{aligned}\sup_{\mu\in M}|\mu|(E_m) &= \sup_{\mu\in M}|\mu|\Big(E_m\Big\backslash\bigcap_{n=1}^m K_n\Big) = \sup_{\mu\in M}|\mu|\Big(\bigcup_{n=1}^m (E_m\backslash K_n)\Big)\\ &\leqq \sum_{n=1}^m \sup_{\mu\in M}|\mu|(E_m\backslash K_n) \leqq \sum_{n=1}^m \sup_{\mu\in M}|\mu|(E_n\backslash K_n)\\ &\leqq \sum_{n=1}^m \varepsilon/2^n \leqq \varepsilon.\end{aligned}$$

Hence M is uniformly countably additive.

To prove that (e) implies (f), suppose that M is uniformly countably additive; let $E\in\Sigma$ and $\varepsilon>0$. Let (μ_n) be a sequence of members of M, and pick sequences (O_n) and (K_n) of open sets and compact sets, respectively, such that $K_n\subseteq E\subseteq O_n$ and such that $|\mu_n|(O_n\backslash K_n)<\varepsilon$ for all n. If $G=\bigcap_{n=1}^\infty O_n$ and $F=\bigcup_{n=1}^\infty K_n$, then one has

$$F\subseteq E\subseteq G \quad\text{and}\quad |\mu_n|(G\backslash F)<\varepsilon$$

for all n. Since M is uniformly countably additive, one obtains

$$\lim_m |\mu_k|\Big(\bigcap_{n=1}^m G_n\Big\backslash\Big(\bigcup_{n=1}^m K_n\Big)\Big) = |\mu_k|(G\backslash F)<\varepsilon$$

uniformly in k. Hence there exists $m_0\in N$ such that

$$|\mu_k|\Big(\bigcap_{n=1}^{m_0} G_n\Big\backslash\bigcup_{n=1}^{m_0} K_n\Big)<\varepsilon$$

for all k. Since $\bigcap_{n=1}^{m_0} G_n$ is an open set containing E and $\bigcup_{n=1}^{m_0} K_n$ is a compact set contained in E, the fact that (e) implies (f) is established.

Finally, to prove that (a) follows from (f), let (O_n) be a sequence of disjoint open sets. Then $\bigcup_{n=1}^\infty O_n$ is an open set, and for $\varepsilon>0$, the fact that M is uniformly regular produces a compact set K such that $K\subseteq\bigcup_{n=1}^\infty O_n$ and $|\mu|((\bigcup_{n=1}^\infty O_n\backslash K)\leqq\varepsilon$ for all $\mu\in M$. Since (O_n) is an open cover for K, there is n_0 such that $K\subseteq\bigcup_{n=1}^{n_0} O_n$. Consequently for $m\geqq n_0+1$, one has

$$|\mu|(O_m)\leqq|\mu|\Big(O_m\cup\Big(\bigcup_{n=1}^{n_0} O_n\backslash K\Big)\Big)\leqq|\mu|\Big(\bigcup_{n=1}^{\infty} O_n\backslash K\Big)\leqq\varepsilon$$

for all $\mu\in M$. This completes the proof.

COROLLARY 14. *If T: $C(\Omega)\to X$ is a bounded linear operator with representing measure G, then T is weakly compact if and only if G is regular.*

PROOF. Apply Lemma 13 to the family $\{x^*G\colon \|x^*\|\leqq 1,\ x^*\in X^*\}$ and invoke Theorem 5.

The next theorem shows that intuition about weakly compact operators on $B(\mathscr{F})$ (for a field $\mathscr{F}$) is roughly the same as intuition about weakly compact operators on $C(\Omega)$ when no special assumptions about Ω are made. The next result should be compared to Corollary 2 and Corollary 12.

THEOREM 15. *Let T: $C(\Omega)\to X$ be a nonweakly compact bounded linear operator. Then $C(\Omega)$ contains an isometric copy of c_0 on which T acts as an isomorphism. Con-*

sequently, if X contains no copy of c_0, then every bounded linear operator $T\colon C(\Omega) \to X$ is weakly compact.

PROOF. Suppose $T\colon C(\Omega) \to X$ is not weakly compact. By Theorem 5 its representing measure G is not countably additive.

Consequently the family $\{|x^*G|\colon x^* \in X^*,\ \|x^*\| \leqq 1\}$ of regular nonnegative measures is not uniformly countably additive. An appeal to Lemma 13 produces a sequence (O_n) of disjoint open sets, an $\varepsilon > 0$ and a sequence $(x_n^*) \subset X^*$ with $\|x_n^*\| \leqq 1$ such that $|x_n^*G|(O_n) \geqq \varepsilon$ for all n. At this point, the proof becomes similar to the proof of I.4.2. By Rosenthal's lemma the sequences (x_n^*) and (O_n) may have been chosen such that $|x_n^*G|(O_n) \geqq \varepsilon$ while $|x^*G|(\bigcup_{m\neq n} O_m) < \varepsilon/2$ for all n. Now there is a sequence (f_n) in $C(\Omega)$ such that

(a) $\|f_n\| = 1$,

(b) f_n vanishes outside O_n, and

(c) $x_n^*T(f_n) = \int_\Omega f_n\, dx_n^*G > \varepsilon$

for all n.

Let $Y = \{\sum_{n=1}^\infty \alpha_n f_n\colon (\alpha_n) \in c_0\}$. Since $\|f_n\| = 1$ for all n, it is plain that Y is an isometric copy of c_0. Moreover if $f = \sum_{n=1}^\infty \alpha_n f_n$ for some sequence $(\alpha_n) \in c_0$, then for any n

$$\begin{aligned} |x_n^*T(f)| &= \left|\int_\Omega f\, dx_n^*G\right| = \left|\alpha_n \int_{O_n} f_n\, dx_n^*G + \int_{\bigcup_{m\neq n} O_m} f\, dx_n^*G\right| \\ &\geqq |\alpha_n|\varepsilon - \int_{\bigcup_{m\neq n} O_m} |f|\, d|x_n^*G| \\ &\geqq |\alpha_n|\varepsilon - |x_n^*G|\Big(\bigcup_{m\neq n} O_n\Big)\|f\| \\ &\geqq |\alpha_n|\varepsilon - \|f\|\varepsilon/2. \end{aligned}$$

But $\|f\| = \sup_n |\alpha_n|$; hence

$$\|T(f)\| \geqq \sup_n |x_n^*T(f)| \geqq \|f\|\varepsilon - \|f\|\varepsilon/2 = (\varepsilon/2)\|f\|$$

and T is an isomorphism on Y.

Paralleling Corollary 11 is

COROLLARY 16. *A complemented infinite dimensional subspace of $C(\Omega)$ contains a copy of c_0.*

PROOF. Take the proof of Corollary 11 and substitute c_0 for l_∞.

Paralleling Corollary 1.4 is

COROLLARY 17. *Any one of the following statements about a bounded linear operator $T\colon C(\Omega) \to X$ implies all the others.*

(a) *T is unconditionally converging, i.e., T maps weakly unconditionally Cauchy series into unconditionally convergent series.*

(b) *T is weakly compact.*

(c) *T maps sequences that tend to zero weakly into norm convergent sequences.*

(d) *T maps weakly Cauchy sequences into norm convergent sequences.*

(e) *If (f_n) is a bounded sequence in $C(\Omega)$ with $f_n \cdot f_m = 0$ for $m \neq n$, then $\lim_n T(f_n) = 0$.*

PROOF. To prove that (a) implies (b), note that if $T\colon C(\Omega) \to X$ is not weakly compact, then T operates as an isomorphism on an isometric copy of c_0. Since c_0 contains plenty of nonconvergent weakly unconditionally Cauchy series, T is not unconditionally converging.

The converse (b) implies (a) is true in general without reference to the domain of T; see the proof of Corollary 4.

The fact that (b) implies (e) is just Corollary 6.

It is trivial that (d) implies (c) while the fact that (c) implies (e) is clear since if $\sup_n \|f_n\| < \infty$ and $f_n \cdot f_m = 0$ for $m \neq n$, then $\lim_n f_n = 0$ weakly.

Finally the proof that (e) implies (a) is embedded in the proof of Theorem 15.

It is interesting to note that while (c) above characterizes weakly compact operators on $C(\Omega)$, statement (c) does not characterize weakly compact operators on any infinite dimensional $L^1(\mu)$ space. Indeed, by the method of Example III.2.22 one can see that any infinite dimensional $L_1(\mu)$ space supports a representable operator that is not weakly compact. On the other hand, all representable operators on $L_1(\mu)$ map weakly convergent sequences into norm convergent sequences.

This section will be concluded with a brief look at compact operators on $C(\Omega)$ spaces and their representing measures.

THEOREM 18. *A bounded linear operator T: $C(\Omega) \to X$ is compact if and only if its representing measure has a relatively norm compact range.*

PROOF. If T: $C(\Omega) \to X$ is compact with representing measure G, then T is also weakly compact and G has its values in X. Also since $\hat{T}$: $B(\Sigma) \to X$ (Σ is the σ-field of Borel sets) is a restriction of T^{**}, $\hat{T}$ is compact. Accordingly, one has

$$\{G(E): E \in \Sigma\} = \{\hat{T}(\chi_E): E \in \Sigma\} \subseteq \{\hat{T}(f): f \in B(\Sigma), \|f\| \leqq 1\},$$

and therefore G has a relatively compact range.

Conversely if the representing measure G: $\Sigma \to X^{**}$ of T has a relatively compact range, then G is strongly additive by Theorem 5, and hence G has its values in X. Moreover according to the proof of Theorem 1.1, the set $\{\sum_{i=1}^n \alpha_i G(E_i): 0 \leqq \alpha_i \leqq 1, E_i \in \Sigma, E_i \cap E_j = \emptyset \text{ for } i \neq j\}$ is in the convex hull of $G(\Sigma)$. Then

$$\{T(f): f \in C(\Omega), \|f\| \leqq 1\} \subseteq \overline{\text{co}}\,(G(\Sigma) - G(\Sigma)),$$

a set which is compact by Mazur's theorem.

3. Absolutely summing operators on $C(\Omega)$. The weakly compact operators on $C(\Omega)$ are precisely those operators that map weakly unconditionally Cauchy series into unconditionally convergent series. A more imposing requirement is to demand that an operator map weakly unconditionally Cauchy series into absolutely convergent series. Operators that satisfy this requirement are called *absolutely summing operators* and are the objects of study of this section. We shall see that absolutely summing operators on $C(\Omega)$ are precisely those whose representing measures are of bounded variation. This fact will be used to relate the class of absolutely summing operators on $C(\Omega)$ to the classes of integral operators in the sense of Pietsch. Throughout this section Ω is a compact Hausdorff space, Σ is the σ-field of Borel sets in Ω and X and Y are Banach spaces.

DEFINITION 1. *A bounded linear operator $T : X \to Y$ is called absolutely summing if T maps weakly unconditionally Cauchy series in X into absolutely convergent series in Y.*

The following proposition gives some equivalent descriptions of absolutely summing operators.

PROPOSITION 2. *Any one of the following statements about a bounded linear operator $T: X \to Y$ implies all the others.*

(a) *T is absolutely summing.*

(b) *T maps unconditionally convergent series in X into absolutely convergent series in Y.*

(c) *There exists a constant $K \geqq 0$ such that for any finite set $x_1, x_2, \cdots, x_n \in X$ the following inequality obtains:*

$$\sum_{m=1}^{n} \| T x_m \| \leqq K \sup \left\{ \sum_{i=1}^{n} |x^* x_i| : x^* \in X^*, \|x^*\| \leqq 1 \right\}.$$

(d) *T maps strongly additive X-valued vector measures into Y-valued vector measures of bounded variation.*

PROOF. The proof is straightforward if the closed graph theorem is used at the proper time.

The least constant K such that the inequality in (c) holds is called the *absolutely summing norm* of T and will be denoted by $\|T\|_{\text{as}}$. It is easy to prove that the class of absolutely summing operators from X to Y is a Banach space under this norm. This Banach space will be denoted by AS(X, Y). Another trivial comment will be useful: If W, X, Y and Z are Banach spaces and $T: W \to X$, $S: X \to Y$ and $R: Y \to Z$ are bounded linear operators and S is absolutely summing, then $RST: W \to Z$ is absolutely summing and

$$\|RST\|_{\text{as}} \leqq \|R\| \, \|S\|_{\text{as}} \, \|T\|.$$

The fundamental result about absolutely summing operators on $C(\Omega)$ is

THEOREM 3. *A bounded linear operator $T: C(\Omega) \to X$ is absolutely summing if and only if its representing measure G is of bounded variation. In this case $\|T\|_{\text{as}} = |G|(\Omega)$.*

PROOF. First suppose G is of bounded variation. If $f_1, f_2, \cdots, f_n \in C(\Omega)$, then

$$\begin{aligned} \sum_{m=1}^{n} \|T(f_m)\| &= \sum_{m=1}^{n} \left\| \int_\Omega f_m \, dG \right\| \\ &\leqq \sum_{m=1}^{n} \int_\Omega |f_m| \, d|G| = \int_\Omega \sum_{m=1}^{n} |f_m| \, d|G| \\ &\leqq \left\| \sum_{m=1}^{n} |f_m| \right\| |G|(\Omega). \end{aligned}$$

But a moment's reflection shows that

$$\begin{aligned} \left\| \sum_{m=1}^{n} |f_m| \right\| &= \sup_{|\varepsilon_m|=1} \left\| \sum_{m=1}^{n} \varepsilon_m f_m \right\| \\ &= \sup_{|\varepsilon_m|=1} \sup \left\{ \int_\Omega \sum_{m=1}^{n} \varepsilon_m f_m \, d\mu : \mu \text{ is} \right. \\ &\qquad\qquad \left. \text{a regular Borel measure and } |\mu|(\Omega) \leqq 1 \right\} \\ &= \sup \left\{ \sum_{m=1}^{n} \left| \int_\Omega f_m \, d\mu \right| : \mu \text{ is} \right. \\ &\qquad\qquad \left. \text{a regular Borel measure and } |\mu|(\Omega) \leqq 1 \right\}. \end{aligned}$$

Hence

$$\sum_{m=1}^{n} \|T(f_m)\| \leqq |G|(\Omega) \sup\left\{ \sum_{m=1}^{n} \left| \int_{\Omega} f_n \, d\mu \right| : \mu \text{ is a regular Borel measure and } |\mu|(\Omega) \leqq 1 \right\}.$$

The (scalar) Riesz Representation Theorem teamed with the definition of $\|T\|_{\text{as}}$ shows T is absolutely summing and shows that

$$\|T\|_{\text{as}} \leqq |G|(\Omega).$$

For the converse, suppose T is absolutely summing. If $\{O_m\}_{m=1}^{n}$ is a finite disjoint family of open sets in Ω and $\{f_m\}_{m=1}^{n}$ is a corresponding finite family of members $C(\Omega)$ such that $\|f_m\| \leqq 1$ and each f_m vanishes outside O_m, then we have

$$\sum_{m=1}^{n} \left| \int_{\Omega} f_m \, d\mu \right| \leqq |\mu|(\Omega)$$

for all regular Borel measures μ. Accordingly, one has

$$\sum_{m=1}^{n} \|T(f_m)\| \leqq \|T\|_{\text{as}}.$$

Moreover if $\{x_m^*\}_{m=1}^{n}$ is in the closed unit ball of X^*,

$$\sum_{m=1}^{n} \left| \int_{O_m} f_m \, d(x_m^* G) \right| = \sum_{m=1}^{n} |x_m^* T(f_m)| \leqq \sum_{m=1}^{n} \|T(f_m)\| \leqq \|T\|_{\text{as}}.$$

Next note that the bound on the right is independent of $f_1, \cdots, f_n$ as long as $\|f_m\| \leqq 1$ and f_m vanishes outside O_m for $m = 1, 2, \cdots, n$. From the regularity of each measure $x_m^* G$, it follows that

$$\sum_{m=1}^{n} |x_m^* G|(O_m) \leqq \|T\|_{\text{as}}$$

provided $\|x_m^*\| \leqq 1$ for $m = 1, \cdots, n$.

Now let $\{E_m\}_{m=1}^{n}$ be a finite family of disjoint Borel sets and let $\varepsilon > 0$. Choose $x_1^*, x_2^*, \cdots, x_n^*$ in the closed unit ball of X^* such that

$$\sum_{m=1}^{n} |x_m^* G(E_m)| \geqq \sum_{m=1}^{n} \|G(E_m)\| - \varepsilon/2.$$

The regularity of each of the measures $x_m^* G$ produces disjoint compact sets $F_1, \cdots, F_n$ such that $F_m \subset E_m$ for $m = 1, \cdots, n$ and

$$\sum_{m=1}^{n} |x_m^* G(F_m)| \geqq \sum_{m=1}^{n} |x_m^* G(E_m)| - \varepsilon/2 \geqq \sum_{m=1}^{n} \|G(E_m)\| - \varepsilon.$$

Since $\{F_m\}_{m=1}^{n}$ is a disjoint family of compact subsets of the compact Hausdorff space Ω, there are disjoint open sets $O_1, \cdots, O_n$ with $F_m \subset O_m$ for $m = 1, \cdots, n$.

Now

$$\sum_{m=1}^{n} \| G(E_m) \| - \varepsilon \leqq \sum_{m=1}^{n} | x_m^* G(F_m) | \leqq \sum_{m=1}^{n} | x_m^* G | (F_m)$$

$$\leqq \sum_{m=1}^{n} | x_m^* G | (O_m) \leqq \| T \|_{\mathrm{as}}.$$

Since $\varepsilon > 0$ is arbitrary, it follows that $\sum_{m=1}^{n} \| G(E_m) \| \leqq \| T \|_{\mathrm{as}}$ and consequently that $|G|(\Omega) \leqq \| T \|_{\mathrm{as}}$. This completes the proof.

A few comments on the proof of Theorem 3 are in order. Note that if the theorem had considered operators from $B(\Sigma)$ to X^{**}, the measure-theoretic details involving regularity would have evaporated and the resulting proof would have been shorter and cleaner than the proof presented above. Such an approach is possible if it is known that an operator T is absolutely summing if and only if T^{**} is absolutely summing. Such a result does hold and is an easy consequence of the so-called "principle of local reflexivity". In any case we have

COROLLARY 4. *Theorem* 3 *remains true if the symbol* "$C(\Omega)$" *is replaced by* "$L_\infty(\mu)$" *or* "$B(\mathscr{F})$".

An incidental corollary and an important example follow.

COROLLARY 5. *An absolutely summing operator on* $C(\Omega)$ *is weakly compact.*

PROOF. Note that, by I.1.15, a vector measure of bounded variation is strongly additive and apply Theorem 2.5.

EXAMPLE 6. *Let* μ *be a nonnegative regular Borel measure on* Ω *and* J: $C(\Omega) \to L_1(\mu)$ *be the natural inclusion. Then* J *is absolutely summing and* $\|J\|_{\mathrm{as}} = \mu(\Omega)$.

PROOF. Verify that the representing measure G of J satisfies $G(E) = \chi_E$, and note that if π is a partition, then

$$\sum_{E \in \pi} \|G(E)\| = \sum_{E \in \pi} \|\chi_E\| = \sum_{E \in \pi} \mu(E) = \mu(\Omega).$$

COROLLARY 7. *A bounded linear operator* $T : C(\Omega) \to X$ *is absolutely summing if and only if there exists a nonnegative Borel measure* μ *on the Borel sets of* Ω *and a bounded linear operator* S: $L_1(\mu) \to X$ *such that* T *admits the following factorization*:

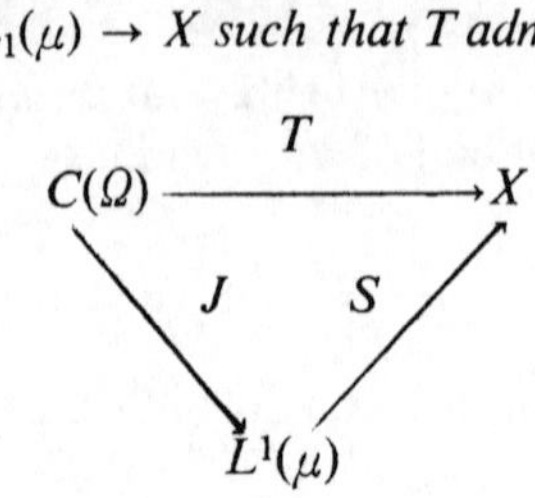

where J *is the natural inclusion of* $C(\Omega)$ *into* $L_1(\mu)$.

In this case μ *and* S *can be chosen such that* $\mu(\Omega) = \|T\|_{\mathrm{as}}$ *and* $\|S\| \leqq 1$.

PROOF. Let T: $C(\Omega) \to X$ be absolutely summing with representing measure G. Define S: $L_1(|G|) \to X$ by $S(\sum_{i=1}^{n} \alpha_i \chi_{E_i}) = \sum_{i=1}^{n} \alpha_i G(E_i)$ for a simple function $\sum_{i=1}^{n} \alpha_i \chi_{E_i}$ (where $E_i \cap E_j = \emptyset$, $i \neq j$ and $E_i \in \Sigma$) and note

$$\left\|\sum_{i=1}^{n} \alpha_i G(E_i)\right\| \leqq \sum_{i=1}^{n} |\alpha_i| \, \|G(E_i)\| \leqq \sum_{i=1}^{n} |\alpha_i| \, |G|(E_i)$$

$$= \left\|\sum_{i=1}^{n} \alpha_i \chi_{E_i}\right\|_{L_1(|G|)}$$

Extend S to all of $L_1(|G|)$ and note that $\|S\| \leqq 1$ and that $|G|(\Omega) = \|T\|_{\text{as}}$.

For the converse suppose the indicated factorization exists. To show T is absolutely summing, it is enough to show the natural inclusion $J\colon C(\Omega) \to L_1(\mu)$ is absolutely summing, and this is just Example 6.

The alert reader will note that the factorization of an absolutely summing operator on $C(\Omega)$ through an $L_1(\mu)$ space allows for the application of the results on representable operators on $L_1(\mu)$ found in Chapter 3 to absolutely summing operators on $C(\Omega)$. This idea will be investigated in the next section.

DEFINITION 8. A bounded linear operator $T\colon X \to Y$ is called *Pietsch integral* if there exists a Y-valued countably additive vector measure G of bounded variation defined on the Borel (for the weak*-topology) sets of the closed unit ball U_{X^*} of X^* such that for each $x \in X$

$$T(x) = \int_{U_{X^*}} x^*(x) \, dG(x^*).$$

It is straightforward to verify that the class of Pietsch integral operators from X to Y becomes a Banach space under the norm

$$\|T\|_{\text{pint}} = \inf\{|G|(U_{X^*})\}$$

where the infimum is taken over all measures G that satisfy the above definition. The Banach space of Pietsch integral operators from X to Y will be denoted by $\text{PI}(X, Y)$. It is evident that $\|T\| \leqq \|T\|_{\text{pint}}$.

Exhibiting the "ideal" structure of the class of Pietsch integral operators is

PROPOSITION 9. *Let W, X, Y and Z be Banach spaces and $T\colon W \to X$, $S\colon X \to Y$ and $R\colon Y \to Z$ be bounded linear operators. If S is Pietsch integral, so is RST and* $\|RST\|_{\text{pint}} \leqq \|R\| \, \|S\|_{\text{pint}} \, \|T\|$.

PROOF. Suppose $\varepsilon > 0$ and let F be a Y-valued countably additive vector measure of bounded variation on the Borel sets in U_{X^*} such that

$$S(x) = \int_{U_{X^*}} x^*(x) \, dF(x^*)$$

and

$$\|S\|_{\text{pint}} \leqq |F|(U_{X^*}) \leqq \|S\|_{\text{pint}} + \varepsilon.$$

Define an auxiliary operator $U\colon C(U_{W^*}) \to Z$ by

$$U(f) = \|T\| \int_{U_{X^*}} f\left(\frac{T^*(x^*)}{\|T\|}\right) dR \circ F(x^*)$$

for $f \in C(U_{W^*})$, where U_{W^*} is the closed unit ball of W^* in its weak*-topology. Then we have

$$\|U(f)\| \leqq \|T\| \int_{U_{x^*}} |f|\, d|R \circ F|$$
$$\leqq \|T\|\, \|R\| \int_{U_{x^*}} |f|\, d|F|$$

for all $f \in C(U_{W^*})$. Since F is of bounded variation, U is continuous. Moreover if $f_1, \cdots, f_n \in C(U_{W^*})$, then

$$\sum_{m=1}^{n} \|U(f_m)\| \leqq \|T\|\, \|R\| \sum_{m=1}^{n} \int_{U_{x^*}} \left| f_m\left(\frac{T^*(x^*)}{\|T\|}\right)\right| d|F|(x^*)$$
$$\leqq \|T\|\, \|R\| \left\| \sum_{m=1}^{n} |f_m| \right\| |F|(U_{X^*}),$$

an inequality which shows U is absolutely summing on $C(U_{W^*})$ by a computation used in the first part of the proof of Theorem 3. Furthermore this inequality shows

$$\|U\|_{\mathrm{as}} \leqq \|T\|\, \|R\|\, |F|(U_{X^*})$$
$$\leqq \|T\|\, \|R\|\, (\|S\|_{\mathrm{pint}} + \varepsilon).$$

Now by Theorem 3 there is a Z-valued countably additive vector measure G on the weak*-Borel sets of U_{W^*} with $|G|(U_{W^*}) = \|U\|_{\mathrm{as}}$ such that $U(f) = \int_{U_{w^*}} f\,dG$ for $f \in C(U_{W^*})$. In addition if $w \in W$, then $w \in C(U_{W^*})$ and we have

$$U(w) = \int_{U_{x^*}} T^*(x^*)(w)\, d(R \circ F)(x^*)$$
$$= R \int_{U_{x^*}} x^*T(w)\, dF(x^*) = RST(w).$$

Therefore

$$RST(w) = \int_{U_{w^*}} w^*(w)\, dG(w^*)$$

for all $w \in W$. Consequently RST is Pietsch integral and

$$\|RST\|_{\mathrm{pint}} \leqq |G|(U_{W^*}) = \|U\|_{\mathrm{as}}$$
$$\leqq \|T\|\, \|R\| (\|S\|_{\mathrm{pint}} + \varepsilon).$$

Hence $\|RST\|_{\mathrm{pint}} \leqq \|T\|\, \|R\|\, \|S\|_{\mathrm{pint}}$ and the proposition is proved.

With the help of Proposition 9, a useful example comes to life.

EXAMPLE 10. *If μ is a regular Borel measure on Ω, then the natural injection J of $C(\Omega)$ into $L_1(\mu)$ is Pietsch integral. Moreover* $\|J\|_{\mathrm{pint}} = \|J\|_{\mathrm{as}} = \mu(\Omega)$. To verify this, consider the natural (evaluation) homeomorphism of Ω into $U_{C(\Omega)^*}$ (weak*-topology) that takes $\omega \in \Omega$ into the point evaluation $\delta_\omega \in U_{C(\Omega)^*}$. Define D: $C(U_{C(\Omega)^*}) \to C(\Omega)$ by $(Df)(\omega) = f(\delta_\omega)$ for $f \in C(U_{C(\Omega)^*})$. Evidently D is a bounded linear operator. Since J is absolutely summing by Example 6, the operator JD: $C(U_{C(\Omega)^*}) \to C(\Omega) \to L_1(\mu)$ is absolutely summing. An appeal to Theorem 3 produces an $L_1(\mu)$-valued countably additive vector measure G of bounded variation on the (weak*) Borel sets of $U_{C(\Omega)^*}$ such that

$$JD(g) = \int_{U_{C(\Omega)^*}} g(\lambda)\, dG(\lambda)$$

for all $g \in U_{C(\Omega)^{**}}$. In particular for each $f \in C(\Omega)$ (viewed as a member $\hat{f}$ of $C(U_{C(\Omega)^*}))$, one has

$$JD(\hat{f}) = \int_{U_{C(\Omega)^*}} \lambda(f)\, dG(\lambda).$$

But now note that $JD(\hat{f})$ coincides with $J(f)$ for $f \in C(\Omega)$. Hence we have $J(f) = \int_{U_{C(\Omega)^*}} \lambda(f)\, dG(\lambda)$. Thus J is Pietsch integral. To check that $\|J\|_{\text{pint}} = \|J\|_{\text{as}}$, note that

$$\begin{aligned} \|J\|_{\text{pint}} &\leqq |G|(U_{C(\Omega)^*}) = \|JD\|_{\text{as}} \\ &\leqq \|D\|\, \|J\|_{\text{as}} \leqq \|J\|_{\text{as}} = \mu(\Omega) \\ &= \|J\| \leqq \|J\|_{\text{pint}}. \end{aligned}$$

Example 10 is no idle example for it allows us to view Pietsch integral operators in the context of absolutely summing operators on $C(\Omega)$. As a consequence of Proposition 9 and Example 10, one sees that if a bounded linear operator T: $X \to Y$ admits a factorization

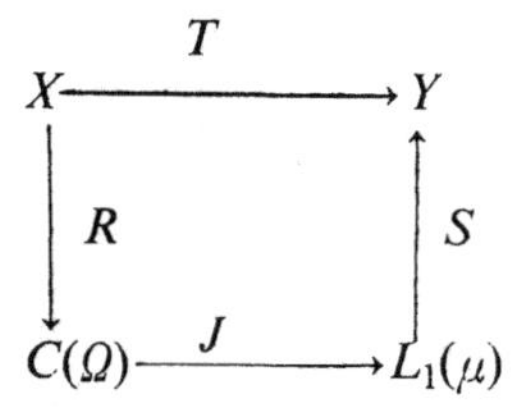

for some compact Hausdorff space Ω, some nonnegative regular Borel measure μ on Ω, bounded linear operators R: $X \to C(\Omega)$, T: $L_1(\mu) \to Y$ and natural inclusion J: $C(\Omega) \to L_1(\mu)$, then T must be Pietsch integral.

On the other hand if T is Pietsch integral, it is easy to see that T has such a factorization. Indeed, suppose T: $X \to Y$ is Pietsch integral and suppose $\varepsilon > 0$. Then there exists a Y-valued countably additive vector measure G on the (weak*) Borel sets of U_{X^*} such that $\|T\|_{\text{pint}} \leqq |G|(U_{X^*}) \leqq \|T\|_{\text{pint}} + \varepsilon$. Define V: $C(U_{X^*}) \to Y$ by $V(f) = \int_{U_{X^*}} f\, dG$. Then by Theorem 3, the operator V is absolutely summing and Corollary 7 guarantees that V admits the factorization

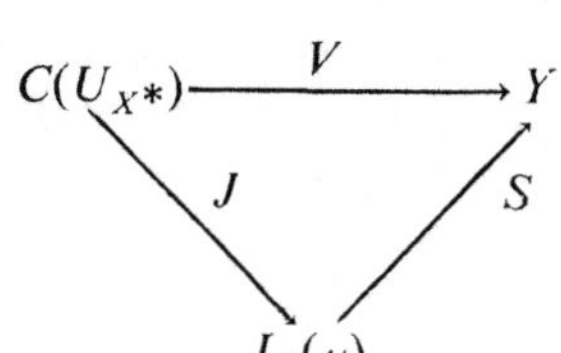

with $\mu(\Omega) = \|V\|_{\text{as}} = |G|(U_{X^*})$, $\|S\| \leqq 1$ and natural inclusion J. Now let R be the natural injection of X into $C(U_{X^*})$. Then $VR(x) = \int_{U_{X^*}} x^*(x)\, dG(x^*) = T(x)$ for all $x \in X$. Accordingly T admits the factorization

$$\begin{array}{ccc} X & \xrightarrow{\quad T \quad} & Y \\ {\scriptstyle R}\downarrow & & \uparrow{\scriptstyle S} \\ C(U_{X^*}) & \xrightarrow{\quad J \quad} & L_1(\mu) \end{array}$$

with $\|R\| \leqq 1$, $\|S\| \leqq 1$ and $\|T\|_{\text{pint}} \leqq \mu(U_{X^*}) \leqq \|T\|_{\text{pint}} + \varepsilon$. Summarizing this short discussion is

THEOREM 11. *A bounded linear operator $T: X \to Y$ is Pietsch integral if and only if T admits a factorization*

$$\begin{array}{ccc} X & \xrightarrow{\quad T \quad} & Y \\ {\scriptstyle R}\downarrow & & \uparrow{\scriptstyle S} \\ C(\Omega) & \xrightarrow{\quad J \quad} & L_1(\mu) \end{array}$$

where Ω is some compact Hausdorff space, μ is some regular Borel measure on Ω, $R: X \to C(\Omega)$ and $S: L_1(\mu) \to X$ are bounded linear operators and J is the natural inclusion of $C(\Omega)$ into $L_1(\mu)$.

In this case, for each fixed $\varepsilon > 0$, the measure μ and the operators R and S may be chosen such that

$$\|T\|_{\text{pint}} \leqq \mu(\Omega) \leqq \|T\|_{\text{pint}} + \varepsilon$$

and such that $\|R\|$ and $\|S\| \leqq 1$.

In particular T is Pietsch integral if and only if T admits a factorization

$$\begin{array}{ccc} X & \xrightarrow{\quad T \quad} & Y \\ & {\scriptstyle R}\searrow \quad \nearrow{\scriptstyle S} & \\ & C(\Omega) & \end{array}$$

where $R: X \to C(\Omega)$ is bounded and $S: C(\Omega) \to Y$ is absolutely summing. In this case, for each fixed $\varepsilon > 0$, the space Ω and the operators R and S may be chosen such that $\|R\| \leqq 1$ and $\|T\|_{\text{pint}} \leqq \|S\|_{\text{as}} \leqq \|T\|_{\text{pint}} + \varepsilon$.

Consequently a Pietsch integral operator is absolutely summing.

Glancing at Theorem 11 one is lead to ask what happens when $X = C(\Omega)$ for some compact Hausdorff space Ω? Another glance at Theorem 11 shows that a bounded linear $T: C(\Omega) \to X$ is absolutely summing if and only if T is Pietsch integral. Not so obvious is that in this case $\|T\|_{\text{as}} = \|T\|_{\text{pint}}$, a fact which combines the theory of absolutely summing operators on $C(\Omega)$ and the theory of Pietsch integral operators on $C(\Omega)$ into one theory.

THEOREM 12. *A bounded linear operator $T: C(\Omega) \to X$ is absolutely summing if and only if it is Pietsch integral. In this case $\|T\|_{as} = \|T\|_{pint}$.*

PROOF. If $T: C(\Omega) \to X$ is absolutely summing, then T admits the factorization

$$\begin{array}{ccc} C(\Omega) & \xrightarrow{\quad T \quad} & X \\ & {\scriptstyle J}\searrow \quad \nearrow{\scriptstyle S} & \\ & L_1(\mu) & \end{array}$$

for some nonnegative regular Borel measure with $\mu(\Omega) = \|T\|_{as}$, a bounded linear operator $S: L_1(\mu) \to X$ with $\|S\| \leqq 1$ and natural inclusion $J: C(\Omega) \to L_1(\mu)$. Appeals to Proposition 9 and Example 10 yield the inequalities

$$\|T\|_{pint} \leqq \|S\| \, \|J\|_{pint} \leqq \mu(\Omega) = \|T\|_{as}.$$

For the converse and reverse inequality, suppose $T: C(\Omega) \to X$ is Pietsch integral. If $\varepsilon > 0$, there is an X-valued countably additive vector measure G of bounded variation on the (weak*) Borel sets in $U_{C(\Omega)^*}$ such that for $f \in C(\Omega)$ we have

$$T(f) = \int_{U_{C(\Omega)^*}} \lambda(f) \, dG(\lambda)$$

and such that

$$\|T\|_{pint} \leqq |G|(U_{C(\Omega)^*}) \leqq \|T\|_{pint} + \varepsilon.$$

Now let R be the natural injection of $C(\Omega)$ into $C(U_{C(\Omega)^*})$ and note that $\|R\| = 1$. Also let

$$V(\varphi) = \int_{U_{C(\Omega)^*}} \varphi \, dG$$

for $\varphi \in C(U_{C(\Omega)^*})$. The operator $V: C(U_{C(\Omega)^*}) \to X$ is absolutely summing and $\|V\|_{as} = |G| \, (U_{C(\Omega)^*})$ by Theorem 3. In addition, we have $T(f) = VR(f)$ for all $f \in C(\Omega)$. Hence T is absolutely summing and

$$\begin{aligned} \|T\|_{as} &\leqq \|V\|_{as} \|R\| = \|V\|_{as} \\ &= |G|(U_{C(\Omega)^*}) \leqq \|T\|_{pint} + \varepsilon. \end{aligned}$$

Since $\varepsilon > 0$ was not specified, we see that $\|T\|_{as} \leqq \|T\|_{pint}$. Combining this with the inequality $\|T\|_{pint} \leqq \|T\|_{as}$ above, one obtains $\|T\|_{pint} = \|T\|_{as}$. This completes the proof.

4. Nuclear operators on $C(\Omega)$. The Radon-Nikodým theorems of Chapters III and V form the base for the theory of nuclear operators on $C(\Omega)$ spaces. The connection between Radon-Nikodým derivatives and nuclear operators on $C(\Omega)$ is so simple that it is scarcely more than manipulation of appropriate definitions. Nevertheless Radon-Nikodým theorems for vector measures give a complete description of nuclear operators on $C(\Omega)$ and of Pietsch integral operators that are nuclear. The basic result of this section says that an operator on $C(\Omega)$ is nuclear if and only if its representing measure arises as an indefinite Bochner integral. Again X and Y are Banach spaces.

DEFINITION 1. A bounded linear operator $T : X \to Y$ is called *nuclear* if there exist sequences (x_n^*) in X^* and (y_n) in Y such that $\sum_{n=1}^{\infty} \|x_n^*\| \, \|y_n\| < \infty$ and such that

$$T(x) = \sum_{n=1}^{\infty} x_n^*(x) y_n$$

for all $x \in X$. If T is a nuclear operator the nuclear norm of T is defined by

$$\|T\|_{\text{nuc}} = \inf\left\{\sum_{n=1}^{\infty} \|x_n^*\| \, \|y_n\|\right\}$$

where the infimum is taken over all sequences (x_n^*) and (y_n) such that $x_n^* \in X^*$, $y_n \in Y$ and $T(x) = \sum_{n=1}^{\infty} x_n^*(x) y_n$ for all $x \in X$.

The class of nuclear operators is intimately related to the topological theory of tensor products, especially the theory of the projective tensor product of two Banach spaces. Some salient features of this theory are discussed in Chapter VIII where the relationship between nuclear operators and tensor products is developed and exploited.

The next result collects some elementary facts about nuclear operators.

PROPOSITION 2. (i) *A nuclear operator is a compact Pietsch integral operator. In addition* $\|T\| \leqq \|T\|_{\text{pint}} \leqq \|T\|_{\text{nuc}}$.

(ii) *If W, X, Y, and Z are Banach spaces and T: $W \to X$, S: $X \to Y$ and R: $Y \to Z$ are bounded linear operators with S nuclear then RST is nuclear and*

$$\|RST\|_{\text{nuc}} \leqq \|R\| \, \|S\|_{\text{nuc}} \, \|T\|.$$

(iii) *The nuclear operators from X to Y form a normed linear space under* $\|\cdot\|_{\text{nuc}}$.

(iv) *A bounded linear operator $T : X \to Y$ is nuclear if and only if T admits a factorization*

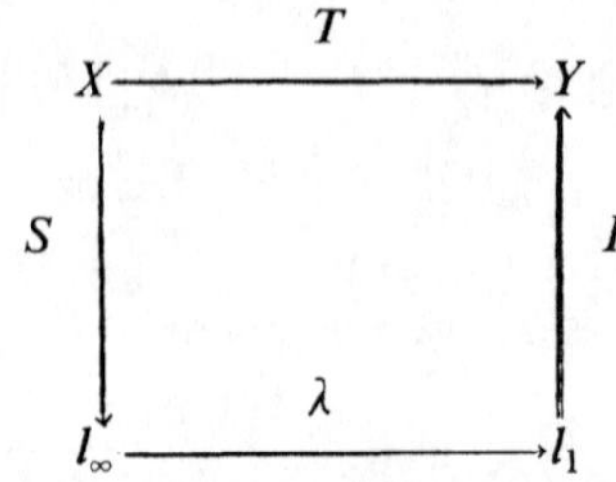

where S: $X \to l_\infty$ and R: $l_1 \to Y$ are bounded linear operators, and λ: $l_\infty \to l_1$ is a nuclear operator. In this case for any $\varepsilon > 0$, S and R may be chosen with $\|S\| \leqq 1$, $\|R\| \leqq 1$ and $\lambda : l_\infty \to l_1$ may be chosen to have action $\lambda((\alpha_n)) = (\lambda_n \alpha_n)$ for some sequence (λ_n) of nonnegative reals with $\|\lambda\|_{\text{nuc}} \leqq \sum_{n=1}^{\infty} \lambda_n < \|T\|_{\text{nuc}} + \varepsilon$.

PROOF. The triangle inequality guarantees that nuclear operators are compact. To see that every nuclear operator is Pietsch integral, suppose $T: X \to Y$ has action

$$T(x) = \sum_{n=1}^{\infty} x_n^*(x) y_n$$

with $\sum_{n=1}^{\infty} \|x_n^*\| \, \|y_n\| \leqq 1$. Let U_{X^*} be a closed unit ball of X^* in its weak*-topology. Define vector measure G on the Borel sets of U_{X^*}.

$$G(E) = \sum_{n=1}^{\infty} \|x_n^*\| \, \delta_{x_n^*/\|x_n^*\|}(E) y_n$$

where δ_{x^*} denotes the point mass concentrated at x^* and E is a (weak*) Borel set. Evidently G is a countably additive measure with

$$|G|(U_{X^*}) \leqq \sum_{n=1}^{\infty} \|x_n^*\| \, \|y_n\|.$$

Moreover for each $x \in X$ we have

$$T(x) = \int_{U_{X^*}} x^*(x) \, dG(x^*).$$

Thus T is Pietsch integral. That $\|T\| \leqq \|T\|_{\text{nuc}}$ is obvious. To prove that $\|T\|_{\text{pint}} \leqq \|T\|_{\text{nuc}}$, let $\varepsilon > 0$ and choose (x_n^*) and (y_n) as above with the added requirement that $\sum_{n=1}^{\infty} \|x_n^*\| \, \|y_n\| < \|T\|_{\text{nuc}} + \varepsilon$. Then

$$\|T\|_{\text{pint}} \leqq |T|(U_{X^*}) \leqq \sum_{n=1}^{\infty} \|x_n^*\| \, \|y_n\| \leqq \|T\|_{\text{nuc}} + \varepsilon.$$

This proves (i).

The statement (ii) is true as a straightforward computation shows, and (iii) is obvious.

To check (iv) note that if T has such a factorization, then (ii) guarantees that T is nuclear. To check the converse implication in (iv), suppose

$$T(x) = \sum_{n=1}^{\infty} x_n^*(x) y_n$$

where $(x_n^*) \subset X^*$, $(y_n) \subset Y$ and $\sum_{n=1}^{\infty} \|x_n^*\| \, \|y_n\| \leqq \|T\|_{\text{nuc}} + \varepsilon$. Define $S : X \to l_\infty$ by $S(x) = (x_n^*(x)/\|x_n^*\|)$, $\lambda: l_\infty \to l_1$ by $((\alpha_n)) = (\alpha_n \|x_n^*\| \, \|y_n\|)$ and $R : l_1 \to Y$ by $R((\beta_n)) = \sum_{n=1}^{\infty} \beta_n y_n/\|y_n\|$. Then λ is nuclear with

$$\|\lambda\|_{\text{nuc}} \leqq \sum_{n=1}^{\infty} \|x_n^*\| \|y_n\|,$$

and

$$R\lambda S(x) = R\lambda\left(\frac{x_n^*(x)}{\|x_n\|}\right) = R(x_n^*(x)\|y_n\|)$$
$$= \sum_{n=1}^{\infty} x_n^*(x) y_n = Tx$$

for all $x \in X$, as required.

The main theorem about nuclear operators on $C(\Omega)$ follows from the following elementary fact. Its connection with nuclear operators on $C(\Omega)$ is almost obvious.

LEMMA 3. *Let* (S, Σ, μ) *be a finite measure space and* $f: S \to X$ *be Bochner integrable. For each* $\varepsilon > 0$ *there is a sequence* (x_n) *in* X *and a (not necessarily disjoint) sequence* (E_n) *in* Σ *such that*

(i) *the series* $\sum_{n=1}^{\infty} x_n \chi_{E_n}$ *converges to* f *absolutely* μ-*a.e. and*

(ii) $\int_S \|f\| \, d\mu \leqq \sum_{n=1}^{\infty} \|x_n\| \mu(E_n) \leqq \int_S \|f\| \, d\mu + \varepsilon$.

PROOF. The proof is based on the definition of a measurable function and is essentially the same as the proof of III.1.8. By Corollary II.1.3 to Pettis's Measurability Theorem there is a sequence (g_n) of countably valued functions such that $\lim_n g_n = f$ uniformly off a set of μ-measure zero. Accordingly we shall assume that $\lim_n g_n = f$ uniformly on S. Moreover by discarding (if necessary) some of the members of the sequence (g_n), we can and do assume that

$$\|f(s) - g_1(s)\| \leqq \varepsilon(2\mu(S))^{-1}$$

and

$$\|g_n(s) - g_{n-1}(s)\| \leqq \varepsilon(2^n \mu(S))^{-1}$$

for all $s \in S$ and all $n \geqq 2$. Next write $g_0 = 0$ and

$$g_n - g_{n-1} = \sum_{m=1}^{\infty} y_{n,m} \chi_{A_{n,m}}$$

where $(y_{n,m})$ is a sequence in X and $(A_{n,m})$ is a *disjoint* sequence in Σ for each fixed $n \geqq 1$. Telescoping yields

$$f(s) = \sum_{n=1}^{\infty} g_n(s) - g_{n-1}(s) = \sum_{n=1}^{\infty} \sum_{m=1}^{\infty} y_{n,m} \chi_{A_{n,m}}(s)$$

for each $s \in S$. The choice of (g_n) ensures

$$\begin{aligned} \sum_{n=1}^{\infty} \sum_{m=1}^{\infty} \|y_{n,m}\| \chi_{A_{n,m}}(s) &\leqq \|g_1(s)\| + \sum_{n=2}^{\infty} \varepsilon/(2^n \mu(S)) \\ &= \|g_1(s)\| + \varepsilon/(2\mu(S)) \end{aligned}$$

for all $s \in S$. Integrating through the last inequality gives

$$\begin{aligned} \sum_{n=1}^{\infty} \sum_{m=1}^{\infty} \|y_{n,m}\| \mu(A_{n,m}) &\leqq \int_S \|g_1\| \, d\mu + \varepsilon/2 \\ &\leqq \int_S (\|f\| + \varepsilon/(2\mu(S))) \, d\mu + \varepsilon/2 \\ &= \int_S \|f\| \, d\mu + \varepsilon. \end{aligned}$$

On the other hand,

$$\begin{aligned} \int_S \|f\| \, d\mu &\leqq \int_S \sum_{n=1}^{\infty} \sum_{m=1}^{\infty} \|y_{n,m}\| \chi_{A_{n,m}} \, d\mu \\ &= \sum_{n=1}^{\infty} \sum_{m=1}^{\infty} \|y_{n,m}\| \mu(A_{n,m}). \end{aligned}$$

Hence if the double series $\sum_{n=1}^{\infty} \sum_{m=1}^{\infty} y_{n,m} \chi_{A_{n,m}}$ is written as a series $\sum_{n=1}^{\infty} x_n \chi_{E_n}$, the advertised sequences (x_n) and (E_n) are produced.

THEOREM 4. *A bounded linear operator $T: C(\Omega) \to X$ is nuclear if and only if its representing measure G is of bounded variation and has a Bochner integrable derivative g with respect to its variation $|G|$. In this case* $\|T\|_{\text{nuc}} = |G|(\Omega) = \|g\|_{L_1(|G|, X)}$.

PROOF. Suppose there exists a $|G|$-Bochner integrable function g such that

$$G(E) = \int_E g \, d|G|$$

for every Borel set E. Let $\varepsilon > 0$. According to Lemma 3 there is a sequence (x_n) in X and a sequence of Borel sets (E_n) such that

$$g = \sum_{n=1}^{\infty} x_n \chi_{E_n} \quad |G|\text{-a.e.}$$

and

$$\int_\Omega \|g\| \, d|G| \leqq \sum_{n=1}^{\infty} \|x_n\| \, |g|(E_n) \leqq \int_\Omega \|g\| \, d|G| + \varepsilon.$$

Moreover if $\varphi \in C(\Omega)$, we have

$$\begin{aligned} T(\varphi) &= \int_\Omega \varphi \, dF = \int_\Omega \varphi g \, d|F| \\ &= \int_\Omega \varphi \sum_{n=1}^{\infty} x_n \chi_{E_n} \, d|G| = \sum_{n=1}^{\infty} \int_{E_n} \varphi \, d|G| \, x_n. \end{aligned}$$

But $\varphi \to \int_{E_n} \varphi \, d|G|$ is a bounded linear functional on $C(\Omega)$ whose norm is $|G|(E_n)$. Thus if we write $l_n(\varphi) = \int_{E_n} \varphi \, d|G|$ then $\|l_n\| = |G|(E_n)$. Also $T(\varphi) = \sum_{n=1}^{\infty} l_n(\varphi) x_n$ and

$$\sum_{n=1}^{\infty} \|l_n\| \, \|x_n\| \leqq \int_\Omega \|g\| \, d\mu + \varepsilon = |G|(\Omega) + \varepsilon.$$

This proves that T is nuclear and that $\|T\|_{\text{nuc}} \leqq |G|(\Omega)$.

To prove the converse and the reverse inequality, suppose that $T: C(\Omega) \to X$ is nuclear. If $\varepsilon > 0$, there is a sequence of regular Borel measures (μ_n) and a sequence (x_n) in X such that for $\varphi \in C(\Omega)$ we have

$$T(\varphi) = \sum_{n=1}^{\infty} \int_\Omega \varphi \, d\mu_n \, x_n \quad \text{and} \quad \sum_{n=1}^{\infty} |\mu_n|(\Omega) \|x_n\| < \|T\|_{\text{nuc}} + \varepsilon.$$

Write $G(E) = \sum_{n=1}^{\infty} \mu_n(E) x_n$. It is plain that G is a countably additive regular X-valued measure on the Borel sets. Equally plain is the fact that

$$|G|(\Omega) \leqq \sum_{n=1}^{\infty} |\mu_n|(\Omega) \|x_n\| \leqq \|T\|_{\text{nuc}} + \varepsilon.$$

In addition, if $\varphi \in C(\Omega)$, we have

$$T(\varphi) = \int_\Omega \sum_{n=1}^{\infty} \varphi \, d\mu_n \, x_n = \int_\Omega \varphi \, dG$$

by the Dominated Convergence Theorem. It follows immediately that G is the representing measure F of T and that $|G|(\Omega) \leqq \|T\|_{\text{nuc}}$.

Now all that is left is to show that G has a $|G|$-Bochner integrable derivative. For this, note that

$$G(E) = \sum_{n=1}^{\infty} \mu_n(E)x_n$$

for all Borel sets E. Let $\bar{\mu}_n$ be the $|G|$-continuous part of μ_n and $\bar{\bar{\mu}}_n$ be the $|G|$-singular part of μ_n for each n. Then

$$G(E) = \sum_{n=1}^{\infty} \bar{\mu}_n(E)x_n + \sum_{n=1}^{\infty} \bar{\bar{\mu}}_n(E)x_n$$

for each Borel set E. Since each $\bar{\bar{\mu}}_n$ is $|G|$-singular and

$$\sum_{n=1}^{\infty} |\bar{\bar{\mu}}_n|(E)\|x_n\| \leqq \sum_{n=1}^{\infty} |\mu_n|(E)\|x_n\|,$$

it follows that the scalar measure $E \to \sum_{n=1}^{\infty} |\bar{\bar{\mu}}_n|(E)\|x_n\|$ is $|G|$-singular. Consequently there are disjoint Borel sets S_1 and S_2 with $S_1 \cup S_2 = \Omega$ such that $|G|(S_1) = |G|(\Omega)$, $|G|(S_2) = 0$, $\sum_{n=1}^{\infty} |\bar{\bar{\mu}}_n|(S_1)\|x_n\| = 0$, and

$$\sum_{n=1}^{\infty} |\bar{\bar{\mu}}_n|(S_2)\|x_n\| = \sum_{n=1}^{\infty} |\bar{\bar{\mu}}_n|(\Omega)\|x_n\|.$$

From this it follows that for each Borel set E

$$\sum_{n=1}^{\infty} \bar{\bar{\mu}}_n(E \cap S_1)x_n = 0.$$

Since $G \ll |G|$, we have

$$\begin{aligned} G(E) &= G(E \cap S_1) \\ &= \sum_{n=1}^{\infty} \bar{\mu}_n(E \cap S_1)\, x_n + \sum_{n=1}^{\infty} \bar{\bar{\mu}}_n(E \cap S_1)\, x_n = \sum_{n=1}^{\infty} \bar{\mu}_n(E)x_n \end{aligned}$$

for every Borel set E. Now let f_n be the Radon-Nikodým derivative of $\bar{\mu}_n$ with respect to $|G|$ and consider the formal series $\sum_{n=1}^{\infty} f_n x_n$, and note that

$$\begin{aligned} \int_\Omega \left\| \sum_{n=k}^{m} f_n x_n \right\| d|G| &\leqq \sum_{n=k}^{m} \int_\Omega |f_n|\, \|x_n\|\, d|G| \\ &\leqq \sum_{n=k}^{m} |\bar{\mu}_n|(\Omega)\|x_n\| \leqq \sum_{n=k}^{m} |\mu_n|(\Omega)\|x_n\| \end{aligned}$$

for all $k \leqq m$. Therefore $\sum_{n=1}^{\infty} f_n x_n$ converges in $L_1(|G|, X)$-norm to a function f which satisfies $G(E) = \int_E f\, d|G|$ for every Borel set E. This completes the proof.

COROLLARY 5. *If X has the Radon-Nikodým property, then every absolutely summing operator T from $C(\Omega)$ to X is nuclear with $\|T\|_{\text{nuc}} = \|T\|_{\text{as}}$. Consequently* $\text{PI}(C(\Omega), X)$, $\text{AS}(C(\Omega), X)$ *and* $\text{N}(C(\Omega), X)$ *are identical classes with identical norms.*

PROOF. Let $T: C(\Omega) \to X$ be absolutely summing. If G is the representing measure of T then $|G|(\Omega) < \infty$ and G has its values in X since T is weakly compact. Since X has the Radon-Nikodým property, there exists a $|G|$-Bochner integrable function $f: \Omega \to X$ such that

$$G(E) = \int_E f\, d|G|$$

for every Borel set E. According to Theorem 4, T is nuclear. Moreover, by Theorems 3 and 4, $\|T\|_{as} = |G|(\Omega) = \|T\|_{nuc}$. This fact together with Theorem 3.12 shows that $\mathrm{PI}(C(\Omega), X)$, $\mathrm{AS}(C(\Omega), X)$ and $\mathrm{N}(C(\Omega), X)$ are identical classes with identical norms.

A full converse to Corollary 5 is not possible, since if the compact Hausdorff space Ω has no perfect subsets (i.e., Ω is "scattered") then all regular Borel measures on Ω are purely atomic. But, as was seen in III.1, every vector measure of bounded variation that is continuous with respect to a purely atomic finite scalar measure has a Bochner integrable derivative with respect to the scalar measure. This fact combines with Theorem 4 to show that *if Ω is scattered, every absolutely summing operator on $C(\Omega)$ is nuclear without regard to the range of the operator*, a fact which is nothing but a mere formality.

A partial converse to Corollary 5 is

COROLLARY 6. *If X does not have the Radon-Nikodým property, then there is an absolutely summing operator from $C[0, 1]$ to X that is not nuclear.*

PROOF. If X lacks the Radon-Nikodým property there is a bounded set $K \subset X$ that is not dentable. Appealing to V.3.8 one finds an X-valued countably additive vector measure G on the Borel sets in $[0, 1]$ such that G is of bounded variation, G is continuous with respect to Lebesgue measure and G admits no Bochner integrable derivative with respect to Lebesgue measure. Since the variation $|G|$ is continuous with respect to Lebesgue measure, the measure G admits no Bochner integrable derivative with respect to $|G|$. Accordingly the operator $\varphi \to \int_{[0,1]} \varphi \, dG$ is an absolutely summing operator on $C[0,1]$ which is not nuclear by Theorem 4.

It is worth noting that $[0, 1]$ could be replaced above by any compact Hausdorff space Ω that carries a nonatomic regular Borel measure (i.e., any compact Hausdorff space containing a perfect subset). In any case, summarizing the corollaries is

COROLLARY 7. *The space X has the Radon-Nikodým property if and only if for every compact Hausdorff space Ω every absolutely summing operator from $C(\Omega)$ to X is nuclear.*

Now let us look at Pietsch integral operators for a moment. Since a Pietsch integral operator $T: X \to Y$ always admits a factorization

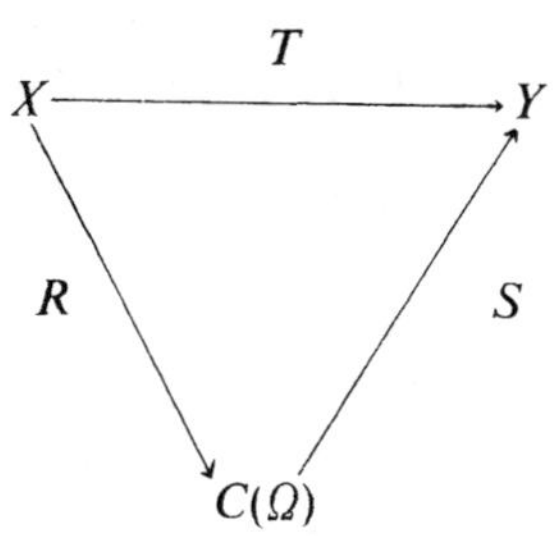

where S is absolutely summing, we are compelled to apply the last few results to obtain information about the nuclearity of Pietsch integral operators.

THEOREM 8. *A Banach space Y has the Radon-Nikodým property if and only if* *for every Banach space X, every Pietsch integral operator from X to Y is nuclear. In this case* $\mathrm{PI}(X, Y)$ *and* $\mathrm{N}(X, Y)$ *are identical classes with identical norms.*

PROOF. If every Pietsch integral $T: C[0, 1] \to Y$ is nuclear, then every absolutely summing operator $T: C[0, 1] \to Y$ is nuclear. A glance at Corollary 6 reveals that Y has the Radon-Nikodým property.

Conversely suppose Y has the Radon-Nikodým property. If $T: X \to Y$ is Pietsch integral and $\varepsilon > 0$, then T admits a factorization

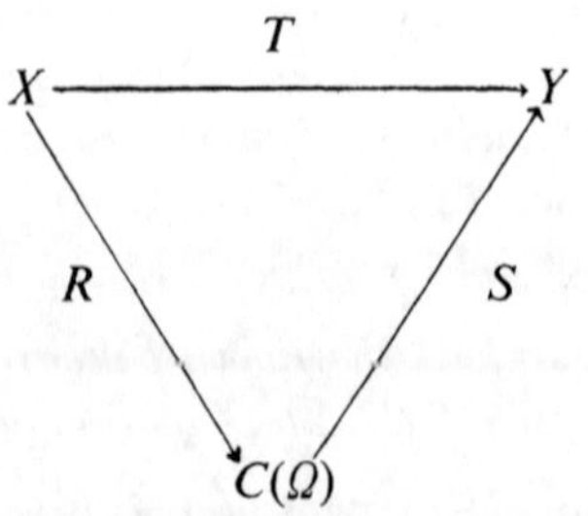

where Ω is some compact Hausdorff space $\|R\| \leqq 1$, S is absolutely summing and $\|T\|_{\text{pint}} \leqq \|S\|_{\text{as}} \leqq \|T\|_{\text{pint}} + \varepsilon$. Now since Y has the Radon-Nikodým property, S is nuclear and $\|S\|_{\text{as}} = \|S\|_{\text{nuc}}$. Consequently T is nuclear and $\|T\|_{\text{nuc}} \leqq \|R\| \|S\|_{\text{nuc}} \leqq \|S\|_{\text{nuc}} = \|S\|_{\text{as}} \leqq \|T\|_{\text{pint}} + \varepsilon$. Thus $\|T\|_{\text{nuc}} \leqq \|T\|_{\text{pint}}$. On the other hand, $\|T\|_{\text{pint}} \leqq \|T\|_{\text{nuc}}$ holds for any nuclear operator. This completes the proof.

5. Notes and remarks. The first substantial studies of operators on spaces of continuous functions were made independently in the fundamental papers of Grothendieck [**1953**] and Bartle, Dunford and Schwartz [**1955**]. Vector measures make an implicit appearance in the former and a crucially explicit appearance in the latter. Although most of our European friends seem to prefer the Grothendieck approach and our American friends seem to have opposite feelings, there is no reason to regard these independent approaches as competing theories. They complement each other beautifully. Since our approach is vector measure-theoretic, our starting point is the paper of Bartle, Dunford and Schwartz whose influence has already been felt in Chapter I. The heavy hand of Grothendieck has also influenced the exposition of this chapter. In fact it may not be an overstatement to say that the central theme of this chapter is motivated by Grothendieck [**1953**], [**1955a**].

§1 consists mainly of folklore results and is motivational in character. Perhaps Theorem 1.1 was first observed by Diestel [**1973a**].

§2 is an effort to amalgamate the Bartle, Dunford, Schwartz and Grothendieck studies. It begins along the lines of Bartle, Dunford and Schwartz (the influence of Dunford and Schwartz [**1958**] is also seen here) and finishes along the lines of Grothendieck [**1953**]. Theorems 2.1 and 2.5 are due to Bartle, Dunford and Schwartz; Example 2.3 can be found in Gel'fand [**1938**] and Pełczyński [**1960**].

The Dunford-Pettis property. A Banach space X has the *Dunford-Pettis property* if every weakly compact operator defined on X takes weakly compact sets into norm compact sets. Dunford and Pettis (III.2.14) proved that the $L_1(\mu)$ spaces have this property. Grothendieck **[1953]** and Bartle, Dunford and Schwartz **[1955]** independently established that $C(\Omega)$ spaces also have this property and thus proved Corollary 2.6. Grothendieck seems to have the priority in this matter and the term "Dunford-Pettis property" was coined by him. The Bartle, Dunford and Schwartz proof appears in the text. Brace **[1953]** and Grothendieck gave some valuable characterizations of Banach spaces with the Dunford-Pettis property.

THEOREM (BRACE-GROTHENDIECK). *Any of the following statements about a Banach space X implies all the others:*

(i) *The space X has theDunford-Pettis property.*

(ii) *For all Banach spaces Y, every weakly compact operator from X to Y sends weakly convergent sequences onto norm convergent sequences.*

(iii) *For all Banach spaces Y, every weakly compact operator from X to Y sends weakly Cauchy sequences onto norm convergent sequences.*

(iv) *If* (x_n) *and* (x_n^*) *are sequences in X and* X^* *respectively and* $\lim_n x_n = 0$ *weakly and* $\lim_n x_n^* = 0$ *weakly then* $\lim_n x_n^* x_n = 0$.

PROOF. The equivalence of (i) and (ii) is a trivial consequence of the Eberlein-Smulian theorem and (ii) is an equally trivial consequence of (iii). The proof that (ii) implies (iii) is an easy consequence of the observation that a sequence (z_n) in a Banach space is weakly Cauchy (respectively norm Cauchy) if and only if for each pair of strictly increasing sequences (k_n) and (j_n) of positive integers $\lim_n z_{k_n} - z_{j_n} = 0$ weakly (respectively in norm).

To see that (i) implies (iv), suppose X has the Dunford-Pettis property and that (x_n) and (x_n^*) are as in the hypothesis of (iv). Define operators $T\colon l_1 \to X$ and $S\colon X \to c_0$ by

$$T(\lambda_n) = \sum_n \lambda_n x_n, \qquad Sx = (x_n^* x).$$

Note that since T takes the closed unit ball of l_1 into the absolutely closed convex hull of $\{x_n\}$, T is weakly compact. Since $S^*\colon l_1 \to X^*$ takes the closed unit ball of l_1 into the absolutely closed convex hull of $\{x_n^*\}$ the operator S^*, and therefore S, is weakly compact. Now since X has the Dunford-Pettis property, $TS\colon l_1 \to c_0$ is a compact operator. This fact and a simple calculation show that $\lim_n x_n^* x_n = 0$. This proves that (i) implies (iv).

To see that (iv) implies (ii), suppose that there exist a sequence (x_n) in X and a weakly compact operator $T\colon X \to Y$ such that $\lim_n x_n = 0$ weakly and such that $\inf_n \|Tx_n\| = \delta > 0$. For each n, select $y_n^* \in X^*$ such that $\|y_n^*\| = 1$ and $\|y_n^* Tx_n\| = \|Tx_n\|$. Let $x_n^* = y_n^* T = T^* y_n^*$. Since T is weakly compact, T^* is weakly compact. By the Eberlein-Smulian theorem (and passing to a suitable subsequence if necessary) we may assume $\lim_n x_n^* = x^*$ exists weakly. Now since $(x_n^* - x^*)$ converges to zero weakly and (x_n) converges to zero weakly, $\lim_n (x_n^* - x^*)(x_n) = 0$ by (iv) and $\lim_n x^* x_n = 0$ by hypothesis. On the other hand, we have

$$\delta \leqq \underline{\lim}_n\, x_n^* x_n = \underline{\lim}_n\, [(x_n^* - x^*)(x_n) + x^*(x_n)] = 0,$$

a contradiction which proves that (iv) implies (ii).

COROLLARY (GROTHENDIECK **[1953]**). *If X is a Banach space whose dual has the Dunford-Pettis property then X has the Dunford-Pettis property.*

For some time the converse of the above corollary was an unsolved problem. Finally Stegall **[1972]** gave an example of a Banach space X with the Schur property (weakly convergent sequences are norm convergent) but such that X^* lacks the Dunford-Pettis property. The list of spaces with the Dunford-Pettis property (other than $C(\Omega)$ spaces and $L_1(\mu)$ spaces) is presently quite short. It has been shown recently by Kisliakov **[1975]** that the disk algebra and H^∞ have the Dunford-Pettis property. An open question is whether $L_1(\mu, X)$ has the Dunford-Pettis property whenever X does. It is also unknown if the spaces $C^k(I^n)$ of k-times continuously differentiable functions defined on the n cube ($n \geqq 2$) have the Dunford-Pettis property; it ought to be remarked here that the representation theory of operators on $C^k(I^n)$ is crying be developed.

Here is a striking application of the Dunford-Pettis property.

THEOREM (GROTHENDIECK **[1954]**). *If X is a linear subspace of $L_\infty(\mu)$ for some finite μ and X is closed in some $L_p(\mu)$ for $1 \leqq p < \infty$, then X is finite dimensional.*

PROOF. First note the identity map of X is a $L_\infty(\mu)$-to-$L_p(\mu)$ isomorphism by the open mapping theorem. Evidently X is reflexive in both the $L_\infty(\mu)$- and $L_p(\mu)$-topologies. But as a $C(\Omega)$ space $L_\infty(\mu)$ has the Dunford-Pettis property. Thus the inclusion map of $L_\infty(\mu)$ into $L_p(\mu)$ maps the unit ball of X onto a compact set with nonempty interior, since the $L_\infty(\mu)$- and $L_p(\mu)$-topologies on X are the same. This proves that X is finite dimensional.

Lemma 2.8 is due to Nakano **[1941]** and Stone **[1948]**. Both studied other order-theoretic closure properties of $C(\Omega)$ spaces and their topological characterizations. Goodner **[1950]** and Nachbin **[1950]** were the first to prove that $C(\Omega)$ spaces are injective for Stonean spaces Ω. Their results extended Phillips's **[1940]** observation that $l_\infty(\Gamma)$ is injective for any set Γ. Goodner and Nachbin also studied the converse to Lemma 2.9 and, with a strong helping hand from Kelley **[1952]**, showed that a real Banach space is norm one complemented in every superspace if and only if X is isometric to a $C(\Omega)$ space for some Stonean space Ω. The nontrivial complex version of the same theorem is due to Hasumi **[1958]**.

Naturally, the study of injectivity is closely related to problems of extension of operators and so has attracted a great deal of attention. Nachbin based his work on a careful study of intersection properties of balls. Lindenstrauss **[1964a]** used this approach very handily to find conditions on a Banach space that permit compact extensions of compact operators; many of the Lindenstrauss results were obtained with approximation assumptions earlier by Grothendieck **[1956a]** who used considerably different methods. The Lindenstrauss theorems, in turn, have been improved in Lindenstrauss and Rosenthal **[1969]** and Stegall and Retherford **[1972]**. The Goodner-Nachbin-Kelley-Hasumi characterization pertains only to the *isometric* theory of Banach spaces. The problem of characterizing those Banach spaces (called P_λ spaces) that are complemented by a continuous linear projection in any superspace remains open. Rosenthal **[1970b]** has given the deepest analysis of this class and among other things he has shown that every such space contains an isomorphic copy of l_∞. It is not known whether every P_λ space is isomorphic to a $C(\Omega)$ space or, if so, what properties Ω must have. To our knowledge the works of Amir **[1962a]**, **[1962b]**, **[1964]** and Wolfe **[1971]**, **[1974]** best address these issues. Sobczyk **[1940]** proved that c_0 is complemented in every separable space in which it resides although the norm of the projection need not be one. No other separable space with this property is currently known;* Amir [**1962b**] has shown that if X is isomorphic to a $C(\Omega)$ space and X is separably injective then X is isomorphic to c_0. Pełczyński [**1960**] has given a particularly elegant proof of Sobczyk's result and Veech [**1971**] has given a beautiful proof of the fact that c_0 is complemented in any weakly compactly generated space in which it resides. Of course, no such weakly compactly generated space can be a dual space since c_0 lacks the Radon-Nikodým property. Related work is found in Baker [**1973**], Baker and Wolfe [**1976**], Cohen, Labbé and Wolfe [**1972**], and Isbell and Semadeni [**1963**].

Theorem 2.10 is due to Rosenthal [**1970a**] as is Corollary 2.11. Those interested in $C(\Omega)$ when Ω is σ-Stonean or when Ω is an F-space should consult the Rosenthal paper above [**1970a**] as well as the work of Seever [**1968**]. The starting point for theorems resembling 2.10 is the work of Grothendieck [**1953**] who proved Corollary 2.12.

Grothendieck spaces. A Banach space X with the property that separably valued bounded linear operators on X are weakly compact is called a *Grothendieck space.* According to Corollary 2.12 and the remarks following it, $B(\Sigma)$ is a Grothendieck space if Σ is a σ-field. Grothendieck spaces are close to the heart of every vector measure theorist. A close relationship exists between $B(\mathscr{F})$ spaces that are Grothendieck spaces, the validity of the Vitali-Hahn-Saks theorem for measures on the field $\mathscr{F}$ and the validity of the Nikodým Boundedness Theorem for measures defined on $\mathscr{F}$ (see Faires [**1976**] and Seever [**1968**]). This fact further accentuates the importance of the study of the interrelationships between Grothendieck spaces and vector measure theory. Here is a list of reformulations of the definition of Grothendieck spaces. For proofs, see Diestel [**1973c**], Faires [**1974b**], Grothendieck [**1953**].

THEOREM. *Any one of the following statements about a Banach space X implies all the others.*

(i) *The space X is a Grothendieck space.*

(ii) *For all Banach spaces Y such that Y^* has a weak* sequentially compact unit ball, every bounded linear operator $T: X \to Y$ is weakly compact.*

(iii) *For all weakly compactly generated Banach spaces Y, every bounded linear operator $T: X \to Y$ is weakly compact.*

(iv) *For any Banach space Y, the weakly compact linear operators from X to Y are closed under the process of taking pointwise weak sequential limits.*

(v) *For any Banach space Y, the weakly compact linear operators from X to Y are closed under the process of taking pointwise norm sequential limits.*

(vi) *Weak* and weak sequential convergence in X^* coincide.*

(vii) *Every bounded linear operator $T: X \to c_0$ is weakly compact.*

Some of the known Grothendieck spaces are the $C(\Omega)$ spaces if Ω is an F-space (Seever [**1968**]) and the bounded Baire functions on [0, 1] of order at least one (Dashiell [**1976**]). Prior to these results, Ando [**1961**] had shown that, if Ω is σ-Stonean, then $C(\Omega)$ is a Grothendieck space, a fact also noted by Semadeni [**1964**].

*ADDED IN PROOF. During the early summer 1976, M. Zippin proved the delightful fact that c_0 *is* the only separable space with the Sobczyck property.

A basic unresolved question that is crying out for a solution is the following: If Y contains no copy of l_∞ and X is a Grothendieck space need every bounded linear operator from X to Y be weakly compact? Another basic question is what are necessary and sufficient conditions on a Boolean algebra $\mathscr{F}$ that $B(\mathscr{F})$ is a Grothendieck space? The basic intuition here is that if $\mathscr{F}$ is rich enough to behave something like a σ-algebra then $B(\mathscr{F})$ is a Grothendieck space. Again, we remark that this question is intimately related to the validity of the Vitali-Hahn-Saks theorem for measures defined on $\mathscr{F}$. Finally, there is some evidence (Akemann **[1967]**, **[1968]**) that the space $\mathscr{L}(H; H)$ of bounded linear operators on a Hilbert space H is a Grothendieck space and that more generally the space $\mathscr{L}(X; X)$ is a Grothendieck space for any reflexive Banach space X.

Lemma 2.13 follows the general train of thought of Grothendieck **[1953]** and Pełczyński **[1962]**. Our proof is not the shortest possible, but it is very simple. Theorem 2.15 is due to Pełczyński as is Corollary 2.16. Our method of proof is from Rosenthal **[1970a]**. Early versions of Theorem 2.15 took the form "every operator from a $C(\Omega)$ space to a weakly sequentially complete space is weakly compact." These can be found in Gel'fand **[1938]**, Pettis **[1939b]**, Grothendieck **[1953]** and Bartle, Dunford and Schwartz **[1955]**. Pełczyński **[1960]** modified the Bartle, Dunford and Schwartz proof to obtain the second assertion of Theorem 2.15; both these proofs involve a direct verification: that the representing measure has values in X. We recommend their reading to our readers.

The equivalences of (b) through (e) of Theorem 2.17 are due essentially to Grothendieck **[1953]** and the inclusion of (a) is Pełczyński's **[1962]**.

Theorem 2.18 is from Bartle, Dunford and Schwartz **[1955]** who give a different proof from ours. A forerunner is due to Gel'fand **[1938]** in the case that $\Omega = [0, 1]$.

The results of §2 are naturally the objects of considerable generalization. D. Lewis **[1970]** has carried out the general representation theory of weakly compact operators from $C(\Omega)$ to locally convex spaces and has included the case in which Ω is only locally compact (in which case $C_0(\Omega)$, the space of continuous functions vanishing at ∞, is considered). In the course of his work, Lewis has shown that when $C_0(\Omega)$ is equipped with any of the frequently encountered locally convex topologies, $C_0(\Omega)$ has the Dunford-Pettis property. Most noteworthy is the fact that $C_0(\Omega)$, equipped with the strict topology, has the Dunford-Pettis property. An alternate approach to the study of weakly compact operators on $C(\Omega)$ has been studied at length by Thomas **[1970]**. By some results of Tumarkin **[1970]**, the work of Thomas proves Theorem 2.15 for range spaces that are sequentially complete locally convex spaces. Extensions of Theorem 2.10 (including nonlocally convex range spaces) have been obtained by Drewnowski **[1976a]**, **[1976b]**, Kalton **[1974b]**, **[1975]** and Labuda **[1976a]**, **[1976b]**; Labuda **[1976a]** has extended Theorem 2.10 to range spaces that are sequentially complete locally convex spaces.

Recognizing $C(\Omega)$ as a C^* algebra one may wonder if some of the results of §2 extend to operators on C^* algebras. We are happy to say that the answer is yes. Akemann, Dodds and Gamelin **[1972]** have shown that if X is any C^* algebra and Y contains no copy of c_0 then every bounded linear operator $T: X \to Y$ is weakly compact. In addition, Akemann **[1968]** has proved that if X is a W^* algebra and (x_n^*) is a sequence of *positive* elements of X^* that converge to zero in the weak* topology then (x_n^*) converges to zero in the weak topology. The question of whether the positivity assumption can be omitted is important and unsolved.

More generalizations of the results of §2 have been studied by Dodds **[1975a]**, **[1975b]**. Dodds has been able to characterize the linear operators from a Riesz space L to a Banach space X that map order intervals into relatively weakly compact subsets of X. His arguments are steeped in the tradition of vector measures but do not involve, in the strictest sense, any vector measure theory. In fact, Dodds is able to deduce quickly and smoothly some of the basic results of Chapter I and of §2. Playing the role of the devils' advocate, we should mention that some of Dodds's results can be obtained alternatively from the vector measure theory found in this book. Here is the basic idea: If L is a Riesz space (with certain properties) and X is a Banach space and $T: L \to X$ is a linear operator mapping an order interval $[-l, l]$ into a weakly compact set then if L_∞ is the span of $[-l, l]$ and L_∞ is normed by $\|x\| = \inf\{t > 0: |x| \leqq tl\}$, L_∞ is an abstract M space with unit l and T is a weakly compact operator on L_∞. It is now possible, in principle, to apply the theorems of §2 to the study of the behavior of T. We should mention also that, as appealing as this approach may seem, Dodds's approach is the more efficient and illuminating of the two in the study of operators on Riesz spaces.

A predecessor of Dodds's paper is the important paper of Kluvanek **[1965]** in which he executes the theory of the Daniell integral for operators. Much of the basic representation theory of weakly compact operators on $C(\Omega)$ can be derived from Kluvanek's vectorial Daniell integral.

Nonlinear operators. Recently efforts at extending the Riesz Representation Theorem to nonlinear operators have met with some success. Chaçon and Friedman **[1965]** and Friedman and Katz **[1966]**, **[1969]** have proved representation theorems for functionals that are additive over disjointly supported functions and satisfy certain continuity conditions. Friedman and Tong **[1971]** have extended many of these results to operators with similar properties. More recently Batt **[1972]**, **[1973b]** has established some representation theorems for a class of nonlinear operators that are closely related to Hammerstein operators. Batt's work seems to include much of the work of Friedman and his co-authors. Most recently, Batt's work was extended by Rothenberger **[1973]** to include operators on spaces of vector-valued bounded measurable functions. An extensive survey of the work on nonlinear operators on $C(\Omega)$ is contained in Batt **[1973a]**. Also see Mizel and Sundaresan **[1971]**.

Operators on $C(\Omega, X)$. Let Ω be a compact Hausdorff space with Borel σ-field Σ and X and Y be Banach spaces and $\mathscr{L}(X; Y)$ be the Banach space of all bounded linear operators from X to Y. Let $G: \Sigma \to \mathscr{L}(X; Y^{**})$ be a finitely additive vector measure and, for each $y^* \in Y^*$ and $E \in \Sigma$, let $G_{y^*}(E)(x) = y^*G(E)(x)$. Then $G_{y^*}: \Sigma \to X^*$ is a finitely additive vector measure. For each $E \in \Sigma$, let $\|G\|(E) = \sup\{|G_{y^*}|(E): \|y^*\| \leqq 1\}$. The quantity $\|G\|(E)$ is called the semivariation of G on $E \in \Sigma$. The vector measure G is called *weakly regular* if G_{y^*} is regular for each $y^* \in Y^*$. Note that if G_{y^*} is regular for $y^* \in Y^*$ then G_{y^*} is countably additive (Corollary 2.14).

Proceeding in a fashion similar to the proof of Theorem 2.1 yields the following representation theorem which seems to have been first observed by Singer **[1957]**,

[1958], **[1959]** in the case Y is the scalars and in general by Dinculeanu **[1965a]**, **[1965b]**. For a long look at this theorem Dinculeanu's monograph **[1967]** is recommended.

THEOREM (DINCULEANU-SINGER). *Every bounded linear operator $T: C(\Omega, X) \to Y$ determines a unique vector measure $G: \Sigma \to \mathscr{L}(X; Y^{**})$ such that*

(i) *G is finitely additive and $\|G\|(\Omega) < \infty$;*

(ii) *G is weakly regular;*

(iii) *the mapping $y^* \to G_{y^*}$ is weak*-weak* continuous from Y^* to rcabv $(\Sigma, X^*) = C(\Omega, X)^*$;*

(iv) *$T(f) = \int_\Omega f\, dG$ for all $f \in C(\Omega, X)$;*

(v) *$\|G\|(\Omega) = \|T\|$; and*

(vi) *$T^*y^* = G_{y^*}$ for all $y^* \in Y^*$.*

Conversely, any vector measure $G: \Sigma \to \mathscr{L}(X; Y^{**})$ that satisfies (i), (ii) *and* (iii) *defines a bounded linear operator $T: C(\Omega, X) \to Y$ by means of* (iv) *and both* (v) *and* (vi) *follow.*

As in the case of the representation of operators on $C(\Omega)$, the test of any representation theorem lies in its applicability to the structure and behavior of the objects being represented. The representation theorem for operators on $C(\Omega)$ has paid handsome dividends both in enriching the structure theory for operators on $C(\Omega)$ and in the enriching of the $C(\Omega)$ structure theory itself. The above theorem has not yet achieved its maturity but some early results based on it are encouraging.

Dobrakov **[1971]** has noted that if $T: C(\Omega, X) \to Y$ has representing measure G then G has all its values in $\mathscr{L}(X; Y)$ if and only if for each $x \in X$ the operator $T_x: C(\Omega) \to Y$ defined by $T_x(f) = T(xf)$, $f \in C(\Omega)$, is weakly compact. In this case, G is regular in the weak operator topology and countably additive in the strong operator topology of $\mathscr{L}(X; Y)$. Batt and Berg **[1969]** have shown that a representing measure $G: \Sigma \to \mathscr{L}(X; Y^{**})$ is norm regular only if all the values of G lie in $\mathscr{L}(X; Y)$ and in this case $G: \Sigma \to \mathscr{L}(X; Y)$ is norm regular. They also show that if the measures $\{G_{y^*}: \|y^*\| \leq 1\}$ are uniformly countably additive then G is norm regular and hence has its values in $\mathscr{L}(X; Y)$ and go on to show that if T is weakly compact then $\{G_{y^*}: \|y^*\| \leq 1\}$ is uniformly countably additive.

Dobrakov **[1971]** and Swartz **[1972a]** have looked at unconditionally converging operators $T: C(\Omega, X) \to Y$. Dobrakov showed that if T is unconditionaly lconverging, then G has its values in $\mathscr{L}(X; Y)$, each value of G is an unconditionally converging operator and that if (E_n) is a decreasing sequence of Borel sets with empty intersection then $\lim_n \|G\|(E_n) = 0$ (and thus G is regular). Swartz **[1976]** established the converse.

In the same paper Swartz studied completely continuous operators from $C(\Omega, X)$ to Y. (A completely continuous operator is one that maps weakly convergent sequences into norm convergent sequences.) In this paper it is claimed that $C(\Omega, X)$ has the Dunford-Pettis property if X has the Dunford-Pettis property. Unfortunately there seems (to us) to be a serious gap in the proof as stated in this paper. Thus this question remains open.

Weakly compact and compact operators $T: C(\Omega, X) \to Y$ have been studied in terms of their representing measures by Batt **[1969]**, Batt and Berg **[1969]** and

Brooks and P. Lewis **[1974]**. Aside from the case in which X is reflexive (or nearly so) most of these results are preliminary in that they do not yet reveal structural properties of weakly compact or compact operators on $C(\Omega, X)$. This is an area ripe for more study. No doubt a mature theory of weakly compact operators on $C(\Omega, X)$ will use some of the theorems of the above papers and will probably involve at least partial solution to the problem of characterizing the weakly compact subsets of $L_1(\mu, X)$.

Pełczyński **[1962]** studied Banach spaces X such that for every Banach space Y a bounded linear operator $T: X \to Y$ is weakly compact if and only if T is unconditionally converging. Such a Banach space is said to have property V. The $C(\Omega)$ spaces have property V; the $L_1(\mu)$ spaces of infinite dimension lack property V. Pełczyński showed that if X is reflexive then $C(\Omega, X)$ has property V. Accordingly, if X is reflexive and Y contains no copy of c_0, every bounded linear operator from $C(\Omega, X)$ to Y is weakly compact. This theorem was also proved by Batt and Berg **[1969]**. Currently there is no known example of a Banach space X with property V such that $C(\Omega, X)$ lacks property V. Related to the results of Batt-Berg and Pełczyński is an interesting theorem proved by Gamlen **[1974]**: If X^* has the Radon-Nikodým property and Y is weakly sequentially complete then every operator from $C(\Omega, X)$ to Y is weakly compact.

The class of absolutely summing operators was introduced by Grothendieck **[1955a]**, **[1956a]**; in **[1955a]**, they are called "semi-integrale à droit". Dinculeanu **[1967a]**, **[1967b]** studied a class of operators on $C(\Omega)$ called dominated operators. The dominated operators on $C(\Omega)$ are precisely the absolutely summing operators on $C(\Omega)$; this fact did not seem to be noticed for some time. Most of the theorems presented in §3 on absolutely summing operators are implicit in the work of Grothendieck although it was Pietsch **[1965]**, **[1969]** who gave the first explicit formulation of Theorem 3.3. The operators that we call Pietsch integral appear in Persson and Pietsch **[1969]** and the basic properties of these operators are discussed there. Singer **[1959b]**, **[1959c]** seems to have been the first to identify Pietsch integral operators on $C(\Omega)$ as the absolutely summing operators. Essentially Singer's proof depends on showing that integral operators (see Chapter VIII) coincide with the dominated operators of Dinculeanu.

A bounded linear operator $T: C(\Omega, X) \to Y$ is called p-dominated $(1 \leqq p < \infty)$ if there is a regular Borel measure μ on Ω such that $\|Tf\| \leqq \|f\|_{L_p(\mu, X)}$ for all $f \in C(\Omega, X)$. By the regularity of μ, $C(\Omega, X)$ is dense in $L_p(\mu, X)$ and therefore, T has an extension $\hat{T}$ to all of $L_p(\mu, X)$ and T admits a factorization

$$\begin{array}{ccc} C(\Omega, X) & \xrightarrow{\quad T \quad} & Y \\ {\scriptstyle I}\searrow & & \nearrow{\scriptstyle \hat{T}} \\ & L_p(\mu, X) & \end{array}$$

where I is the natural inclusion of $C(\Omega, X)$ into $L_p(\mu, X)$. From this it follows that if $1 \leqq p < \infty$ and X is reflexive the p-dominated operators are weakly compact since $L_p(\mu, X)$ $(1 < p < \infty)$ is reflexive; while if $p = 1$, the inclusion $I: C(\Omega, X) \to L_1(\mu, X)$ factors through $L_2(\mu, X)$. This fact was proved first by Batt and Berg **[1969]** who used a somewhat different proof. If X is the scalars, p-dominated operators on $C(\Omega, X)$ coincide with the p-absolutely summing operators (Pietsch **[1969]**). If X is not finite dimensional this is not generally true. Swartz **[1973d]** has observed that absolutely summing operators on $C(\Omega, X)$ are 1-dominated for arbitrary X and proved the

THEOREM (SWARTZ). *A bounded linear operator $T: C(\Omega, X) \to Y$ with representing measure G is absolutely summing if and only if each of the values of G is an absolutely summing operator from X to Y and G is of bounded variation as a measure into the space of absolutely summing operators from X to Y under the absolutely summing norm.*

Thus, if every operator from X to Y is absolutely summing, then the 1-dominated operators from $C(\Omega, X)$ to Y are absolutely summing and conversely. This holds in particular, whenever X is an $L_1(\mu)$ space and Y is a Hilbert space; see Lindenstrauss and Pełczyński **[1968]**.

The notion of a nuclear operator is also due to Grothendieck **[1955a]**. He realized the importance of this class and we speculate that he understood how vector measures are fundamental to the understanding of this class. We owe to him all the ideas and most of the results (at least implicitly) of §4. However, the understanding of the methods of Grothendieck was not immediate. For example, Theorems 3 and 4 of Gil de la Madrid **[1965]**, **[1966]** that deal with the property of Phillips and Proposition 9 of Grothendieck **[1955]** are variants of Theorem 4.4. Uhl **[1969a]** proved Theorem 4.4 but did not state it. Theorem 4.4 also appears in various forms in Diestel **[1972]**, Persson **[1969]**, Swartz **[1973]**, and Tong **[1971]**. We feel obliged to emphasize that the proof of Theorem 4.4 is nothing but an application of the definition of measurability.

Thus far the study of integral and nuclear operators on $C(\Omega, X)$ has not moved out of its infancy. However, Alexander **[1976]** has initiated the study of nuclear operators on $C(\Omega, X)$ and has given some indications of the inherent difficulties.

Further discussion of the classes of integral, absolutely summing and nuclear operators will be delayed until the notes and remarks section of Chapter VIII.

Vector measures and local unconditional structure. Gordon and Lewis **[1974]** have shown that if a Banach space Y has local unconditional structure (Dubinsky, Pełczyński and Rosenthal **[1972]**, Grothendieck **[1956a]**) then for every Banach space X every absolutely summing operator $T: Y \to X^*$ admits a factorization

$$\begin{array}{ccc} Y & \xrightarrow{\quad T \quad} & X^* \\ & {\scriptstyle R}\searrow \quad \nearrow{\scriptstyle S} & \\ & L_1(\mu) & \end{array}$$

for some finite measure μ and operators $R: Y \to L_1(\mu)$ and $S: L_1(\mu) \to X^*$. For this reason Radon-Nikodým theorems are of obvious potential use in studying global properties of spaces with local unconditional structure. An example:

THEOREM (PEŁCZYŃSKI). *The disk algebra A of functions analytic on the unit disk and continuous on the boundary lacks local unconditional structure.*

PROOF. Let H^1 be the usual Hardy space and note that the natural inclusion mapping $T: A \to H^1$ is absolutely summing. (This follows from VI.3.) Note that each of the functions e^{int} appears in the image of the unit ball of A under T. It follows that T is not a compact operator.

If A has local unconditional structure then the Gordon-Lewis theorem would allow us to factor $T: A \to H^1$ as follows:

$$\begin{array}{ccc} A & \xrightarrow{T} & H^1 \\ & {}_R\searrow \quad \nearrow_S & \\ & L_1(\mu) & \end{array}$$

where μ is a finite measure and $R: A \to L_1(\mu)$, $S: L_1(\mu) \to H^1$ are bounded linear operators. Since the separable space H^1 is a dual space (Hoffman [**1962**, p. 137]), Theorem III.3.1 implies that the operator $S: L_1(\mu) \to H^1$ is representable. Moreover, $R: A \to L_1(\mu)$ is weakly compact. In fact, $R^*: L_\infty(\mu) \to A^*$ is weakly compact. In fact, the F. and M. Riesz theorem (Hoffman [**1962**]) has as a consequence that $A^* = (L_1[0, 2\pi]/A^\perp] \oplus M$ where M is the subspace of $C[0, 2\pi]^*$ consisting of the singular measures. But neither $L_1[0, 2\pi]/A^\perp$ nor M can contain an isomorphic copy of l_∞; the first space because it is separable, the second because it is an abstract L space. Therefore $(L_1[0, 2\pi]/A^\perp) \oplus M$ contains no copy of l_∞. Hence an appeal to Theorem IV.2.10 shows that $R^*: L_\infty(\mu) \to A^*$ is weakly compact.

Now $R: A \to L_1(\mu)$ is weakly compact and $S: L_1(\mu) \to H^1$ is representable so $T = SR$ is compact by III.2.14. This contradiction implies that A must lack local unconditional structure.

It is not difficult to abstract the above proof to see that if X^* contains no copy of l_∞, Y^* has the Radon-Nikodým property and there is an absolutely summing noncompact operator $T: X \to Y^*$ then X fails to have local unconditional structure.

VII. GEOMETRIC ASPECTS OF THE RADON-NIKODÝM PROPERTY

The Radon-Nikodým property began to emerge as an internal geometric property of Banach spaces in Chapter V. The present chapter builds on this theme, and by the end of this chapter the Radon-Nikodým property, as a geometric property, will take on an air of towering importance. §1 opens with a discussion of extreme points of convex sets in Banach spaces and shows that a Banach space X having the Radon-Nikodým property has the Kreĭn-Mil′man property in the sense that the conclusion of the Kreĭn-Mil′man theorem describes each closed bounded convex subset of X.

In §2, dual spaces with the Radon-Nikodým property are characterized in two ways, one purely analytic, the other purely geometric. Here it will be seen that if X is a Banach space, X^* has the Radon-Nikodým property if and only if each separable subspace of X has a separable dual. This solves a problem left open in Chapter III. §2 also contains an exposition of the fact that a dual Banach space has the Radon-Nikodým property if and only if it has the Kreĭn-Mil′man property. As the section progresses, we shall observe the beautiful fact that although these two theorems seem unrelated, they are consequences of the same underlying construction in dual spaces.

More powerful results are obtained in §3 where it is shown that a Banach space X has the Radon-Nikodým property if and only if each closed bounded convex subset of X is the norm closed convex hull of its strongly exposed points.

Further spectacular results are obtained in §4. For example, here it is seen that a Banach space X lacks the Radon-Nikodým property if and only if there is a closed bounded set A in X and a bounded open convex set K in X such that A is contained in K but both A and K have the same closed convex hull. In the same vein, we shall learn that a Banach space lacks the Radon-Nikodým property if and only if it has a closed bounded subset A such that no bounded linear functional achieves a maximum value on A.

1. The Kreĭn-Mil′man theorem and the Radon-Nikodým property. A connection between extreme point structure and the Radon-Nikodým property is intuitively suggested by the characterization of Banach spaces with the Radon-Nikodým property in terms of their dentable subsets. In this section we shall take a step in the direction of making this connection precise by proving that a closed bounded convex set in a Banach space with the Radon-Nikodým property can be written as the closed convex hull of its extreme points. In the course, the Bishop-Phelps theorem will be obtained.

The first block of results leads to the Bishop-Phelps theorem. For notational convenience define for each $x^* \in X^*$ and $M > 0$ a closed convex cone $K(x^*, M)$ by

$$K(x^*, M) = \{x \in X: \|x\| \leqq Mx^*(x)\}.$$

LEMMA 1. *Let C be a closed convex subset of X. If $x^* \in X^*$ and x^* is bounded on C and $M > 0$, then for each $z \in C$ there exists $x_0 \in C$ with $x_0 - z \in K(x^*, M)$ and such that $x_0 + K(x^*, M)$ supports C at x_0 in the sense that*

$$C \cap (x_0 + K(x^*, M)) = \{x_0\}.$$

PROOF. Partially order C by writing $x \geqq y$ whenever $x, y \in C$ and $x - y \in K(x^*, M)$. Preparing to use Zorn's lemma, let $\mathscr{Z}$ be the collection of those $x \in C$ with $x \geqq z$ and let W be a chain in $\mathscr{Z}$. Since $(x^*(w): w \in W)$ is a bounded monotone real-valued net, it converges to its supremum α. Since then $(x^*(w): w \in W)$ is Cauchy, the defining property of $K(x^*, M)$ and the definition of the partial ordering combine to imply that $(w: w \in W)$ converges in norm to an element x_1. The continuity of x^* guarantees that $x_1 \in C$ and that $x_1 \in \mathscr{Z}$. Thus every chain in $\mathscr{Z}$ has an upper bound in $\mathscr{Z}$ and by Zorn's lemma $\mathscr{Z}$ has a maximal element x_0. Moreover, the fact that $x_0 - z \in K(x^*, M)$ is clear and it is equally clear that

$$x_0 \in C \cap (x_0 + K(x^*, M)).$$

In addition if $y \in C \cap (x_0 + K(x^*, M))$, then $y - x_0 \in K(x^*, M)$, and $y \geqq x_0$. But $x_0 \geqq z$ so $y \geqq z$ and our choice of x_0 requires that $y = x_0$. This completes the proof.

The next lemma formulates a fact which is fairly obvious from geometric considerations.

LEMMA 2. *Let x^* and $y^* \in X^*$ with $\|x^*\| = 1 = \|y^*\|$. If $\varepsilon > 0$ and $|y^*(x)| \leqq \varepsilon/2$ whenever $\|x\| \leqq 1$ and $x^*(x) = 0$, then either $\|x^* - y^*\| \leqq \varepsilon$ or $\|x^* + y^*\| \leqq \varepsilon$.*

PROOF. Restrict y^* to the null space of x^* and then let z^* be any Hahn-Banach (norm-preserving) extension of this functional back to a member of X^*. Evidently, $\|z^*\| \leqq \varepsilon/2$. Moreover, $y^* - z^*$ vanishes whenever x^* does and therefore $y^* - z^* = \alpha x^*$ for some α. Now

$$|1 - |\alpha|| = |\|y^*\| - \|y^* - z^*\|| \leqq \|z^*\| \leqq \varepsilon/2.$$

Thus if $\alpha \geqq 0$, we have

$$\|x^* - y^*\| = \|(1 - \alpha)x^* - z^*\| \leqq |1 - \alpha| + \|z^*\| \leqq \varepsilon;$$

while, if $\alpha < 0$, we have

$$\|x^* + y^*\| = \|(1 + \alpha)x^* + z^*\| \leqq |1 + \alpha| + \|z^*\| \leqq \varepsilon.$$

In either case, the lemma is proved.

LEMMA 3. *Let x^* and $y^* \in X^*$ with $\|x^*\| = 1 = \|y^*\|$. If $0 < \varepsilon < 1$ and $M > 1 + 2\varepsilon^{-1}$ then $\|x^* - y^*\| \leqq \varepsilon$ whenever y^* is nonnegative on $K(x^*, M)$.*

PROOF. Choose $x \in X$ so that $\|x\| = 1$ and $1 + 2\varepsilon^{-1} < Mx^*(x)$. If $y \in X$, $\|y\| < 2\varepsilon^{-1}$ and $x^*(y) = 0$, then we have

$$\|x \pm y\| \leqq 1 + 2\varepsilon^{-1} < Mx^*(x) = Mx^*(x \pm y).$$

Accordingly, $x \pm y \in K(x^*, M)$. By hypothesis, $y^*(x \pm y) \geqq 0$; so $|y^*(y)| \leqq y^*(x) \leqq \|x\| = 1$. A glance at Lemma 2 now ensures that either $\|x^* + y^*\| \leqq \varepsilon$ or $\|x^* - y^*\| \leqq \varepsilon$. To rule out the former case, note that, since ε and $M^{-1} < 1$, there is $z \in X$ such that $\|z\| = 1$ and $\max(\varepsilon, M^{-1}) < x^*(z)$. But then $\|z\| \leqq Mx^*(z)$ and $z \in K(x^*, M)$. Again $y^*(z) \geqq 0$ and hence

$$\varepsilon < (x^* + y^*)(z) \leqq \|x^* + y^*\|$$

and the proof is complete.

THEOREM 4 (BISHOP-PHELPS). *Let C be a closed bounded convex subset of a Banach space X. The collection of functionals that achieve their maximum values on C is norm dense in X^*.*

PROOF. It is plainly sufficient to approximate $x^* \in X^*$ with $\|x^*\| = 1$ by functionals that achieve their maximum values on C. Further it can and will be assumed that $0 \in C$.

Let $0 < \varepsilon < 1$ and choose $M > 1 + 2\varepsilon^{-1}$. Since $M > 1$, we see that $K(x^*, M)$ is a closed convex cone with nonempty interior (if $x_0 \in X$ is chosen so that $x^*(x_0\|x_0\|^{-1}) > M^{-1}$ then $K(x^*, M)$ contains an open ball centered at $x_0\|x_0\|^{-1}$). Now apply Lemma 1 to C with $z = 0$ to obtain $x_0 \in C \cap (x_0 + K(x^*, M))$ such that $x_0 + K(x^*, M)$ supports C at x_0 in the sense of Lemma 1. Next, separate $x_0 + K(x^*, M)$ from C by $y^* \in Y^*$ chosen such that

$$\begin{aligned}\sup_{x \in C} y^*(x) = y^*(x_0) &= \inf_{x \in K(x^*,M)} y^*(x + x_0) \\ &= \inf_{x \in K(x^*,M)} y^*(x) + y^*(x_0).\end{aligned}$$

With this y^* we find that

$$y^*(x) \geqq 0 \quad \text{for } x \in K(x^*, M).$$

Now appeal directly to Lemma 3 to deduce that $\|x^* - y^*\| \leqq \varepsilon$. Since y^* achieves its maximum value at $x_0 \in C$, the proof is complete.

There is an intuitive connection between the statements that "a set S has an extreme point" and "a set S is dentable". Although these ideas offer little intuitive resistance in finite dimensional spaces, it is impossible to make this connection precise since in certain Banach spaces there are bounded dentable sets with no extreme points. On the other hand, if a bounded subset D of a Banach space is subset dentable, then for each $\varepsilon > 0$ there is $x \in D$ such that $x \notin \overline{\text{co}}(D \backslash B_\varepsilon(x))$. The separation theorem and the Bishop-Phelps theorem combine to produce an $x^* \in X^*$ that assumes a maximum value on D at $x_0 \in D$ such that $x^*(x_0) > x^*(y)$ for all $y \in D \backslash B_\varepsilon(x)$. Next, intersect D with the hyperplane $x^*(x) = x^*(x_0)$. If the result is a singleton, that singleton is an extreme point for D. If the result is not a singleton, replace D by this intersection and repeat the argument. The idea is that if we slice long enough an extreme point should survive. This is the intuition underlying

THEOREM 5 (LINDENSTRAUSS). *Each nonempty closed bounded convex subset of a Banach space with the Radon-Nikodým property has an extreme point.*

PROOF. Let X have the Radon-Nikodým property and D be a nonempty closed bounded convex subset of X. Then every nonempty subset of D is dentable. In particular, there is $x_1 \in D$ such that

$$x_1 \notin \overline{\text{co}}(D \backslash B_{2^{-1}}(x_1)) = C_1$$

where $B_\delta(x_1)$ is the open ball of radius $\delta > 0$ centered at x_1. By the geometric form of the Hahn-Banach theorem, there is an $x^* \in X^*$ such that $\sup_{x \in C_1} x^*(x) < x^*(x_1)$. By jiggling x^* a bit with the Bishop-Phelps theorem, we can find an $x_1^* \in X^*$ such that

$$\sup x_1^*(C_1) < \sup x_1^*(D) = x_1^*(z_0)$$

for some $z_0 \in D$.

Set $D_1 = \{x \in D: x_1^*(x) = x_1^*(z_0)\}$. The set D_1 is dentable and has diameter $\leqq 1$. Now there is $x_2 \in D_1$ such that

$$x_2 \notin \overline{\text{co}}(D_1 \setminus B_{2^{-2}}(x_2)) = C_2.$$

As above select $x_2^* \in X^*$ such that

$$\sup x_2^* C_2 < \sup x_2^*(D_1) = x_2^*(z_1)$$

for some $z_1 \in D_1$.

Next, let $D_2 = \{x \in D_1: x_2^*(x) = x^*(z_1)\}$. Then D_2 is a nonempty closed bounded convex dentable subset of D_1 with diameter $\leqq 2^{-1}$.

The inductive procedure is now clear by fiat. By induction, produce a sequence

$$D \supseteq D_1 \supseteq D_2 \supseteq \ldots \supseteq D_n \supseteq \ldots$$

such that the diameter of D_n is $\leqq n^{-1}$ and D_{n+1} is a "face" of D_n in the sense that there is $x_{n+1}^* \in X^*$ such that

$$D_{n+1} = \{x \in D_n: x_{n+1}^*(x) = \sup x_{n+1}^*(D_n)\} \neq \varnothing.$$

Use the completeness of X and the diameters of the D_n's to write $\bigcap_{n=1}^\infty D_n = \{x\}$. If $x = \alpha y_1 + (1 - \alpha) y_2$ for some y_1 and $y_2 \in D$ and $0 < \alpha < 1$, then

$$\sup x_{n+1}^*(D_n) = x_{n+1}^*(x) = \alpha x_{n+1}^*(y_1) + (1 - \alpha) x_{n+1}^*(y_2)$$

for all $n \geqq 1$. Hence both y_1 and y_2 belong to $\bigcap_{n=1}^\infty D_n$ and $x = y_1 = y_2$. Thus x is an extreme point of D.

Now let us formalize the conclusion of Theorem 5.

DEFINITION 6. A Banach space X has the *Kreĭn-Mil'man property* if each closed bounded convex subset of X is the norm closed convex hull of its extreme points.

As in the classical situation, the assumption that each closed bounded convex set contains an extreme point leads to the existence of many extreme points.

THEOREM 7 (LINDENSTRAUSS). *If each nonempty closed bounded convex subset of a Banach space X contains an extreme point then X has the Kreĭn-Mil'man property.*

In particular, a Banach space with the Radon-Nikodým property has the Kreĭn-Mil'man property.

PROOF. Let B be a nonempty closed bounded convex subset of X and let E be the closed convex hull of the set of extreme points of B. If $E \neq B$ then, by the separation theorem and the Bishop-Phelps theorem, there exists $x^* \in X^*$ such that

$$\sup x^*(E) < \sup x^*(B) = x^*(b_0)$$

for some $b_0 \in B$. By the choice of x^* and b_0, the set $C = \{b \in B: x^*(b) = \sup x^*(B)\}$ is a nonempty closed bounded convex subset of X. By hypothesis, C has an extreme point and by the choice of x^* one sees quickly that an extreme point of C is also an extreme point of B. This contradicts the assumption that $C \cap E = \emptyset$ and proves the theorem.

2. Separable dual spaces, the Kreĭn-Mil'man property and the Radon-Nikodým property. Two results of fundamental importance are the goals of this section. We already know that if X is a Banach space then X^* has the Radon-Nikodým property provided every separable subspace of X has a separable dual space. In this section, the converse is established. The second theorem of prime importance proved here is that a dual Banach space has the Radon-Nikodým property if and only if it has the Kreĭn-Mil'man property. Further, we shall see that, even though one of these facts seems purely analytic while the other seems purely geometric, both depend on a single construction.

LEMMA 1. *Let Z be a nonseparable Banach space and Ω be the first uncountable ordinal number. Then for each $\varepsilon > 0$, there exist indexed sets $\{z_\alpha: \alpha<\Omega\}$ in Z and $\{z_\alpha^*: \alpha<\Omega\}$ in Z^* such that for all ordinals α, $\beta < \Omega$*

$$\|z_\alpha\| = 1, \qquad \|z_\alpha^*\| < 1 + \varepsilon,$$

and

$$z_\beta^*(z_\alpha) = \begin{cases} 0 & \text{if } \alpha < \beta, \\ 1 & \text{if } \alpha = \beta. \end{cases}$$

PROOF. The proof is a simple transfinite induction argument. Select $z_1 \in Z$ and $z_1^* \in Z^*$ such that

$$1 = \|z_1\| = \|z_1^*\| = z_1^*(z_1).$$

Let $\beta < \Omega$ and assume that, for all $\alpha < \beta$, z_α and z_α^* have been chosen such that $\|z_\alpha\| = 1$, $\|z_\alpha^*\| < 1 + \varepsilon$, and $z_\gamma^*(z_\alpha) = \delta_{\gamma\alpha}$ where $\delta_{\gamma\alpha}$ is Kronecker's δ and $\gamma \leqq \alpha$. Since $\{z_\alpha: \alpha < \beta\}$ spans a separable subspace of the nonseparable space Z, there is, by the Hahn-Banach theorem, $z_\beta^* \in Z^*$ such that $z_\beta^*(z_\alpha) = 0$ for all $\alpha < \beta$ and $\|z_\beta^*\| = 1 + \varepsilon 2^{-1}$. All that is left is to choose $z_\beta \in Z$ such that $z_\beta^* z_\beta = 1 = \|z_\beta\|$. This completes the construction.

Moving now to an interlude of point-set topology, recall that if $A \subseteq X^*$ then $x^* \in X^*$ is a weak*-condensation point of A if each weak*-neighborhood of x^* contains uncountably many points of A. A simple consequence of the fact bounded subsets of the dual of a separable Banach space are weak*-Lindelöf is

LEMMA 2. *If A is a bounded uncountable subset of the dual of a separable Banach space, then all but countably many points of A are weak*-condensation points of A.*

DEFINITION 3. *A pre-Haar system of sets* is a sequence (A_n) of nonempty sets such that

$$A_{2n} \cup A_{2n+1} \subseteq A_n$$

for all n and such that, for each $k \geqq 0$, $A_{2^k}, \cdots, A_{2^{k+1}-1}$ are disjoint. If the sequence (A_n) also satisfies $A_{2n} \cup A_{2n+1} = A_n$ for all n, then (A_n) is called a *Haar system of sets*.

The basic construction of this section is contained in the next lemma.

LEMMA 4 (STEGALL). *Let X be a separable Banach space whose dual is nonseparable. Then for each $\varepsilon > 0$, there is a pre-Haar system (A_n) in $\{x^* \in X^*: \|x^*\| = 1\}$ and an associated sequence (x_n) in X such that*

$$\|x_n\| < 1 + \varepsilon \quad \textit{for all } n$$

and

$$|x^*(x_n) - \chi_{A_n}(x^*)| < \varepsilon 2^{-k}$$

for all $k = 0, 1, 2, \ldots$, all n with $2^k \leqq n < 2^{k+1}$ and all $x^* \in \bigcup_{j=2^k}^{2^{k+1}-1} A_j$.

PROOF. Let $\varepsilon > 0$ and let Ω be the first uncountable ordinal. With the help of Lemma 1 find sets $A = \{x_\alpha^*: \alpha < \Omega\}$ in X^* and $\{x_\alpha^{**}: \alpha < \Omega\}$ in X^{**} such that $\|x_\alpha^*\| = 1$, $\|x_\alpha^{**}\| < 1 + \varepsilon$ and

$$x_\beta^{**} x_\alpha^* = \begin{cases} 0 & \text{if } \alpha < \beta, \\ 1 & \text{if } \alpha = \beta. \end{cases}$$

For $x \in X$, $x^* \in X^*$ and $\delta > 0$, let $W(x^*; x, \delta)$ be the weak*-neighborhood of x^* given by

$$W(x^*; x, \delta) = \{y^* \in X^*: |y^*(x) - x^*(x)| < \delta\}.$$

Step 1. With the help of Lemma 2, let $x_{\alpha_1}^*$ be a weak*-condensation point of A and use Goldstine's theorem to produce $x_1 \in X$ such that $\|x_1\| < 1 + \varepsilon$ and $x_{\alpha_1}^*(x_1) = 1$. Clearly $A_1 = W(x_{\alpha_1}^*; x_1, \varepsilon) \cap A$ is uncountable. Also, if $x^* \in A_1$, then

$$\begin{aligned} |x^*(x_1) - \chi_{A_1}(x^*)| &= |x^*(x_1) - 1| \\ &= |x^*(x_1) - x_{\alpha_1}^*(x_1)| < \varepsilon. \end{aligned}$$

Step 2. Since A_1 is uncountable, there exist weak*-condensation points $x_{\beta_1}^*$ and $x_{\alpha_2}^*$ of A_1 where $\alpha_1 < \beta_1 < \alpha_2$. Switching these indices to $\{x_\alpha^*: \alpha < \Omega\}$ and $\{x_\alpha^{**}: \alpha < \Omega\}$ produced above, we find that

$$x_{\alpha_2}^{**}(x_{\beta_1}^*) = 0 \quad \text{and} \quad x_{\alpha_2}^{**}(x_{\alpha_2}^*) = 1 \quad \text{and} \quad \|x_{\alpha_2}^{**}\| < 1 + \varepsilon.$$

An appeal to Helly's theorem produces $x_2 \in X$ with $\|x_2\| < 1 + \varepsilon$ and such that

$$x_{\beta_1}^*(x_2) = x_{\alpha_2}^{**}(x_{\beta_1}^*) = 0 \quad \text{and} \quad x_{\alpha_2}^*(x_2) = x_{\alpha_2}^{**}(x_{\alpha_2}^*) = 1.$$

Now $x_{\beta_1}^*$ is a weak*-condensation point of A_1 and $x_{\beta_1}^*(x_2) = 0$. By Lemma 2, there is a weak*-condensation point $x_{\alpha_3}^*$ in A_1 where $\alpha_2 < \alpha_3$ and such that $|x_{\alpha_3}^*(x_2)| < \varepsilon 2^{-2}$. Again switching indices to $\{x_\alpha^*: \alpha < \Omega\}$ and $\{x_\alpha^{**}: \alpha < \Omega\}$ we find that

$$x_{\alpha_3}^{**}(x_{\alpha_2}^*) = 0, \qquad x_{\alpha_3}^{**}(x_{\alpha_3}^*) = 1$$

and $\|x_{\alpha_3}^{**}\| < 1 + \varepsilon$. Again Helly's theorem ensures the existence of $x_3 \in X$ with $\|x_3\| < 1 + \varepsilon$ and $x_{\alpha_2}^*(x_3) = 0$, $x_{\alpha_3}^*(x_3) = 1$.

Now we take a larger step: Consider the weak*-neighborhoods

$$U = W(x^*_{\alpha_2}; x_2, \varepsilon/2) \cap W(0; x_3, \varepsilon/2)$$

and

$$V = W(x^*_{\alpha_3}; x_3, \varepsilon/2) \cap W(0; x_2, \varepsilon/2).$$

The neighborhoods U and V are disjoint and the sets $A_2 = A_1 \cap U$ and $A_3 = A_1 \cap V$ each have uncountably many weak*-condensation points. Another property of A_2 and A_3 will be checked presently; we shall see that if $x^* \in A_2 \cup A_3$ then $|x^*(x_2) - \chi_{A_3}(x^*)|$, $|x^*(x_3) - \chi_{A_3}(x^*)| < \varepsilon 2^{-1}$. For, if $x^* \in A_2$, then $\chi_{A_2}(x^*) = 1$ and

$$|x^*(x_2) - \chi_{A_2}(x^*)| = |x^*(x_2) - 1| < \varepsilon 2^{-1},$$

since $x^* \in W(x^*_{\alpha_2}; x_2, \varepsilon/2)$; while $\chi_{A_3}(x^*) = 0$ and hence $|x^*(x_3) - \chi_{A_3}(x^*)| = |x^*(x_3)| < \varepsilon 2^{-1}$ since $x^* \in W(0; x_3, \varepsilon/2)$. The case $x^* \in A_3$ is handled similarly. At this point, we shall relax for a moment and note that $x_1, x_2, x_3, A_1, A_2, A_3$ have all been chosen appropriately.

Step 3. Here the process of choosing $x_4, x_5, x_6, x_7, A_4, A_5, A_6$, and A_7 will be detailed. By fiat, the general procedure will be clear from this.

To this end, select weak*-condensation points $x^*_{\beta_i}$ of A_i, $i = 2, 3$, with $\beta_2 < \beta_3$. Then select a weak*-condensation point $x^*_{\alpha_4}$ of A_2 with $\beta_3 < \alpha_4$. There is no problem in finding $x^*_{\beta_2}$, $x^*_{\beta_3}$ and $x^*_{\alpha_4}$ if Lemma 2 is used. Preparing for an invocation of Helly's theorem, note that $x^{**}_{\alpha_4}(x^*_{\beta_2}) = x^{**}_{\alpha_4}(x^*_{\beta_3}) = 0$, $x^{**}_{\alpha_4}(x^*_{\alpha_4}) = 1$, and $\| x^{**}_{\alpha_4} \| < 1 + \varepsilon$. With the help of Helly's theorem, select $x_4 \in X$ such that

$$x^*_{\beta_2}(x_4) = x^*_{\beta_3}(x_4) = 0, \qquad x^*_{\alpha_4}(x_4) = 1,$$

and $\| x_4 \| < 1 + \varepsilon$.

Also, since $x^*_{\beta_2}$ is a weak*-condensation point of A_2 and $x^*_{\beta_2}(x_4) = 0$, there is a weak*-condensation point $x^*_{\alpha_5}$ of A_2 such that $\alpha_4 < \alpha_5$ and such that $|x^*_{\alpha_5}(x_4)| < \varepsilon 2^{-2}$. Of course, $x^{**}_{\alpha_5}(x^*_{\beta_3}) = x^{**}_{\alpha_5}(x^*_{\alpha_4}) = 0$, $x^{**}_{\alpha_5}(x^*_{\alpha_5}) = 1$, and $\| x^{**}_{\alpha_5} \| < 1 + \varepsilon$. Again by Helly's theorem, select $x_5 \in X$ such that

$$x^*_{\beta_3}(x_5) = x^*_{\alpha_4}(x_5) = 0, \qquad x^*_{\alpha_5}(x_5) = 1$$

and $\| x_5 \| < 1 + \varepsilon$.

Now use the facts that $x^*_{\beta_3}$ is a weak*-condensation point of A_3, and that $x^*_{\beta_3}(x_4) = x^*_{\beta_3}(x_5) = 0$ to find a weak*-condensation point $x^*_{\alpha_6}$ of A_3 such that $\alpha_5 < \alpha_6$ and $|x^*_{\alpha_6}(x_4)|$, $|x^*_{\alpha_6}(x_5)| < \varepsilon 2^{-2}$. Since $x^{**}_{\alpha_6}(x^*_{\beta_3}) = x^{**}_{\alpha_6}(x^*_{\alpha_4}) = x^{**}_{\alpha_6}(x^*_{\alpha_5}) = 0$, $x^{**}_{\alpha_6}(x^*_{\alpha_6}) = 1$ and $\| x^{**}_{\alpha_6} \| < 1 + \varepsilon$, Helly's theorem guarantees the existence of $x_6 \in X$ with $\| x_6 \| < 1 + \varepsilon$ and with the same action on $x^*_{\beta_3}$, $x^*_{\alpha_4}$, $x^*_{\alpha_5}$ and $x^*_{\alpha_6}$ as $x^{**}_{\alpha_6}$.

Next, glance back to note that $x^*_{\beta_3}(x_i) = 0$ for $i = 4, 5, 6$. Thus, since $x^*_{\beta_3}$ is a weak*-condensation point of A_3, there is a weak*-condensation point $x^*_{\alpha_7}$ of A_3 with $\alpha_6 < \alpha_7$ such that $|x^*_{\alpha_7}(x_i)| < \varepsilon 2^{-2}$ for $i = 4, 5, 6$. Also $x^{**}_{\alpha_7}(x^*_{\alpha_i}) = 0$ for $i = 4, 5, 6$; $x^{**}_{\alpha_7}(x^*_{\alpha_7}) = 1$ and $\| x^{**}_{\alpha_7} \| < 1 + \varepsilon$. Yet another appeal to Helly's theorem gives us $x_7 \in X$ such that $\| x_7 \| < 1 + \varepsilon$ and such that x_7 and $x^{**}_{\alpha_7}$ have identical behavior on $x^*_{\alpha_4}$, $x^*_{\alpha_5}$, $x^*_{\alpha_6}$ and $x^*_{\alpha_7}$.

Now define

$$A_4 = A_2 \cap W(x^*_{\alpha_4}; x_4, \varepsilon/4) \cap \Big(\bigcap_{i=5,6,7} W(0; x_i, \varepsilon/4)\Big),$$

$$A_5 = A_2 \cap W(x^*_{\alpha_5}; x_5, \varepsilon/4) \cap \Big(\bigcap_{i=4,6,7} W(0; x_i, \varepsilon/4)\Big),$$

$$A_6 = A_3 \cap W(x^*_{\alpha_6}; x_6, \varepsilon/4) \cap \Big(\bigcap_{i=4,5,7} W(0; x_i, \varepsilon/4)\Big),$$

$$A_7 = A_3 \cap W(x^*_{\alpha_7}; x_7, \varepsilon/4) \cap \Big(\bigcap_{i=4,5,6} W(0; x_i, \varepsilon/4)\Big).$$

It is now only bookkeeping to check that $\{x_i : i = 1,\cdots,7\}$ and $\{A_i : i = 1,\cdots,7\}$ have been built correctly. This completes the discussion.

The next lemma builds on the work of Lemma 4 by employing some basic limiting arguments to enable one to produce a Haar system of clopen sets with the same properties as the pre-Haar system of Lemma 4.

LEMMA 5 (STEGALL). *Let X be a separable Banach space whose dual is nonseparable. Then for each $\varepsilon > 0$ there is a nonempty weak*-compact subset Δ of the unit ball of X^*, a Haar system of clopen subsets (C_n) of Δ and a sequence (x_n) in X such that*

$$\|x_n\| \leqq 1 + \varepsilon \quad \textit{for all } n$$

and

$$|x^*(x_n) - \chi_{C_n}(x^*)| < \varepsilon 2^{-k}$$

for all $k = 0, 1, 2, \cdots$, all n with $2^k \leqq n < 2^{k+1}$ and all $x^* \in \Delta$.

In addition, the sequence (C_n) may be chosen so that the weak-diameter of C_n tends to zero as n tends to infinity.*

PROOF. Assume $0 < \varepsilon < \frac{1}{2}$. Select a pre-Haar system (A_n) of subsets of $\{x^* \in X^* : \|x^*\| = 1\}$ and associated sequence (x_n) as given by Lemma 4. First, let B_n be the weak*-closure of A_n. Now the fact that (B_n) constitutes a pre-Haar system will be established. For this it suffices to show that $B_m \cap B_n = \varnothing$ whenever $2^k \leqq m < n < 2^{k+1}$. If $x^* \in B_m \cap B_n$, then there is a net $(x^*_\sigma : \sigma \in S)$ in A_m and a net $(y^*_\tau : \tau \in T)$ in A_n such that

$$\lim_\sigma x^*_\sigma = x^* = \lim_\tau y^*_\tau$$

in the weak*-topology. But this is impossible since

$$\begin{aligned} |x^*(x_m) - 1| &= \lim_\sigma |x^*_\sigma(x_m) - 1| \\ &= \lim_\sigma |x^*_\sigma(x_m) - \chi_{A_m}(x^*_\sigma)| \\ &\leqq \varepsilon < \tfrac{1}{2}; \end{aligned}$$

while

$$\begin{aligned} |x^*(x_m)| &= \lim_\tau |y^*_\tau(x_m)| \\ &= \lim_\tau |y^*_\tau(x_m) - \chi_{A_m}(y^*_\tau)| \\ &\leqq \varepsilon < \tfrac{1}{2}. \end{aligned}$$

Thus (B_n) is a pre-Haar system.

Next, note that if $2^k \leqq n < 2^{k+1}$ and $x^* \in \bigcup_{m=2^k}^{2^{k+1}-1} B_m$ then $|x^*(x_n) - \chi_{B_n}(x^*)| < \varepsilon 2^{-k}$. Indeed, if $x^* \in B_n$, there is a net $(x^*_\alpha : \alpha \in S)$ in A_n such that $\lim_\alpha x^*_\alpha = x^*$ in the weak*-topology. But then

$$\begin{aligned}|x^*(x_n) - \chi_{B_n}(x^*)| = |x^*(x_n) - 1| \\ &= \lim_\alpha |x_\alpha^*(x_n) - \chi_{A_n}(x_\alpha^*)| \\ &\leqq \varepsilon 2^{-k}.\end{aligned}$$

On the other hand, if $x^* \in B_m$, $m \neq n$ and $2^k \leqq m < 2^{k+1}$, then there is a net $(y_\tau^*: \tau \in T)$ in A_m converging to x^* in the weak*-topology. Thus in this case, we have

$$\begin{aligned}|x^*(x_n) - \chi_{B_n}(x^*)| = |x^*(x_n)| = \lim_\tau |y_\tau^*(x_n)| \\ = \lim_\tau |y_\tau^*(x_n) - \chi_{A_n}(y_\tau^*)| \leqq \varepsilon 2^{-k}.\end{aligned}$$

Recapitulating, we have just seen that

$$(*) \qquad |x^*(x_n) - \chi_{B_n}(x^*)| \leqq \varepsilon 2^{-k}$$

provided $2^k \leqq n < 2^{k+1}$ and $x^* \in \bigcup_{m=2^k}^{2^{k+1}-1} B_m$. Now set

$$\Delta = \bigcap_{k=1}^{\infty} \left(\bigcap_{m=2^k}^{2^{k+1}-1} B_m \right).$$

The set Δ is evidently a nonempty weak*-compact set in X^*. For each n, set $C_n = \Delta \cap B_n$. Obviously, (C_n) is a pre-Haar system and, since

$$C_n \backslash (C_{2n} \cup C_{2n+1}) = [B_n \backslash (B_{2n} \cup B_{2n+1})] \cap \Delta = \varnothing,$$

the system (C_n) is in fact a Haar system. Moreover, for each $k \geqq 0$, $\{C_n\}_{n=2^k}^{2^{k+1}-1}$ is a partition of Δ into closed (and hence open) subsets of Δ. Finally, by $(*)$, we have $|x^*(x_n) - \chi_{C_n}(x^*)| \leqq \varepsilon 2^{-k}$ if $2^k \leqq n < 2^{k+1}$ and $x^* \in \Delta$.

To arrange for the weak*-diameters of the C_n's to go to zero, modify the construction in an obvious way to force this condition.

We are now in a superb position to prove a fundamental result in the theory of Banach spaces and vector measures.

THEOREM 6 (STEGALL). *If a Banach space X has a separable subspace whose dual is not separable, then there is a bounded infinite δ-tree in X^*.*

Consequently, if a Banach space X has a separable subspace whose dual is not separable, then X^* lacks the Radon-Nikodým property.

PROOF. Appeal to Lemma 5 to produce a nonempty weak*-compact subset Δ of X^*, a Haar system (C_n) of subsets of Δ (with $C_1 = \Delta$) and a sequence (y_n) in Y such that $\|y_n\| \leqq 9/8$ for all n and such that $|y^*(y_n) - \chi_{C_n}(y^*)| < 2^{-k-3}$ for all $k = 0, 1, 2, \cdots$ and all n with $2^k \leqq n < 2^{k+1}$ and all $y^* \in \Delta$. In addition, choose (C_n) so that the limit of the weak*-diameter of C_n tends to zero as n tends to infinity.

Now let Σ be the σ-field generated by (C_n) and let μ be the unique countably additive finite measure on Σ satisfying $\mu(C_n) = 2^{-k}$ for $2^k \leqq n < 2^{k+1}$. Note that since $\lim_n \delta(C_n) = 0$, each $\phi \in C(\Delta)$ is Σ-measurable. Accordingly, if $y \in Y$, then $(Ty)(y^*) = y^*(y)$, $y^* \in \Delta$, defines a linear operator $T: Y \to L_\infty(\mu)$ which is evidently bounded. Since $L_\infty(\mu)$ is injective (Corollary VI.2.9 and the remarks following Corollary VI.2.11), T has a bounded linear extension, still called T, to all of X.

Under this notation, the condition $|y^*(y_n) - \chi_{C_n}(y^*)| < 2^{-k-3}$ for all $k = 0, 1, 2, \cdots$ and all n with $2^k \leqq n < 2^{k+1}$ and all $y^* \in \Delta$ translates into

$$\|T(y_n) - \chi_{C_n}\|_\infty \leqq \mu(C_n)/8,$$

for all n. Considering $L_1(\mu)$ as a subspace of $L_\infty(\mu)^*$, look at the sequence

$(T^*(\chi_{C_n})/\mu(C_n))$. This sequence is evidently bounded. The fact that (C_n) is a Haar system and the definition of the measure μ guarantee that this sequence is a tree in X^*. To complete the proof, we shall show that this tree is a 7/18-tree in X^*.

To this end, note that

$$\begin{aligned}
\left\| \frac{T^*(\chi_{C_j})}{\mu(C_j)} - \frac{T^*(\chi_{C_{2j+1}})}{\mu(C_{2j+1})} \right\| &= \frac{1}{\mu(C_j)} \| T^*(\chi_{C_j}) - 2T^*(\chi_{C_{2j+1}}) \| \\
&= \frac{1}{\mu(C_j)} \| T^*(\chi_{C_{2j}}) - T^*(\chi_{C_{2j+1}}) \| \\
&\geqq \frac{8}{9\mu(C_j)} | T^*(\chi_{C_{2j}} - \chi_{C_{2j+1}})(y_{2j}) | \\
&= \frac{8}{9\mu(C_j)} \int_{C_j} T(y_{2j})(\chi_{C_{2j}} - \chi_{C_{2j+1}})\, d\mu \\
&\geqq \frac{8}{9\mu(C_j)} \left[\int_{C_j} \chi_{C_{2j}}(\chi_{C_{2j}} - \chi_{C_{2j+1}})\, d\mu \right. \\
&\qquad \left. - \int_{\Delta} | T(y_{2j}) - \chi_{C_{2j}} | \, | \chi_{C_{2j}} - \chi_{C_{2j+1}} | \, d\mu \right] \\
&\geqq \frac{8}{9\mu(C_j)} \left[\frac{\mu(C_j)}{2} - \frac{1}{2} \frac{\mu(C_j)}{8} \cdot \mu(C_j) \right] \\
&= \frac{4}{9} - \frac{1}{18}\mu(C_j) \geqq \frac{7}{18}.
\end{aligned}$$

The proof that $\| T^*(\chi_{C_j})/\mu(C_j) - T^*(\chi_{C_{2j}})/\mu(C_{2j}) \| \geqq 7/18$ is similiar and will be omitted. This completes the proof of the first assertion. To prove the second statement, appeal to V.2.5. This completes the proof of the theorem.

The proof of Theorem 6 provides the basis for a theorem apparently more powerful than Theorem 6.

THEOREM 7 (HUFF-MORRIS). *If a Banach space X has a separable subspace Y whose dual Y^* is nonseparable, then X^* lacks the Kreĭn-Mil'man property.*

PROOF. The proof is a variation and extension of the theme of the proof of Theorem 6. In this proof we shall make free use of the sequences (y_n) and (C_n) and the operator $T: X \to L_\infty(\mu)$ as constructed in the proof of Theorem 6. For the purposes of this proof we shall regard $L_1(\mu)$ as a subsapce of $L_\infty(\mu)^*$ and we shall regard $L_\infty(\mu)^*$ as the space of all finitely additive measures on Σ that vanish when μ vanishes, equipped with the variation norm. First, let C be the weak*-closed convex hull of $(\chi_{C_n}/\mu(C_n))$ in $L_\infty(\mu)^*$; let $x_n^* = T^*(\chi_{C_n}/\mu(C_n))$ and let D be the weak*-closed convex hull of $\{x_n^*\}$ in X^*. Finally, let $K = \{x^* \in D: \lim_n x^*(y_n) = 0\}$. Plainly C and D are weak*-compact convex subsets of their ambient spaces.

Not so obvious is the fact that K is a nonempty norm closed bounded convex set in X^* which is without extreme points. Let us understand why this is true. First, it is clear that K is bounded and convex. To see that K is nonempty, note that

$$\begin{aligned}
|x_n^*(y_m)| &= |T^*(\chi_{C_n}/\mu(C_n))(y_m)| \\
&= \frac{1}{\mu(C_n)} \left| \int_{C_n} T(y_m)\, d\mu \right| \\
&\leqq \left| \frac{1}{\mu(C_n)} \int_{C_n} | T(y_m) - \chi_{C_m} | \, d\mu + \frac{1}{\mu(C_n)} \int_{C_n} \chi_{C_m}\, d\mu \right| \\
&\leqq \frac{\mu(C_m)}{8} \frac{\mu(C_n)}{\mu(C_n)} + \frac{\mu(C_m \cap C_n)}{\mu(C_n)}
\end{aligned}$$

by the fact that $\| Ty_m - \chi_{C_m} \|_\infty < \mu(C_m)/8$. Since $\lim_m \mu(C_m) = 0$, it is evident that $\lim_m x_n^*(y_m) = 0$. Hence each $x_n^* \in K$ and K is nonempty.

Checking the fact that K is norm closed is also easy. If x^* is a norm cluster point of K and $\varepsilon > 0$, there is $y^* \in K$ such that $\| x^* - y^* \| < \varepsilon/2 \| y_n \|$ for all n. Since $y^* \in K$ there is a positive integer m such that $| y^*(y_j) | < \varepsilon/2$ for $j \geqq m$. Hence when $j \geqq m$, we have

$$\begin{aligned} |x^*(y_j)| &\leqq |x^*(y_j) - y^*(y_j)| + |y^*(y_j)| \\ &\leqq \varepsilon/2 + \varepsilon/2 = \varepsilon. \end{aligned}$$

Thus $x^* \in K$ and K is norm closed.

Finally, we shall demonstrate that K has no extreme points. To this end, note that

$$\begin{aligned} x_n^* y_m &= \frac{1}{\mu(C_n)} \int_{C_n} T y_m \, d\mu \\ &= \frac{1}{\mu(C_n)} \left[\int_{C_n} (T(y_m) - \chi_{C_m}) \, d\mu + \mu(C_m \cap C_n) \right] \\ &\geqq - \frac{\mu(C_m)}{8} \end{aligned}$$

since $\| Ty_m - \chi_{C_m} \|_\infty < \mu(C_m)/8$. Since $\lim_n \mu(C_n) = 0$, it follows that

$$\liminf_m x^*(y_m) \geqq 0$$

for all $x^* \in D$. From this it follows quickly that any extreme point of K is also an extreme point of the larger set D. The proof will be complete if we can prove that none of the extreme points of D lie in K.

For this, let e^* be an extreme point of D. Since $T^*(C) = D$, we see that $C \cap (T^*)^{-1}(\{e^*\})$ is a nonempty convex weak*-closed subset of C. The Kreĭn-Mil'man theorem produces an extreme point β of $C \cap (T^*)^{-1}(\{e^*\})$. A simple algebraic check reveals that any extreme point of this set is an extreme point of C as well. Next, note that the weak*-closure of $\{\chi_{C_n}/\mu(C_n)\}$ in $L_\infty(\mu)^*$ is a weak*-compact set whose closed convex hull is C. Since β is an extreme point of C, Mil'man's theorem guarantees that the finitely additive measure β is in the weak*-closure of $\{\chi_{C_n}/\mu(C_n)\}$. Hence there is a net $(\chi_{C_\alpha}/\mu(C_\alpha) \colon \alpha \in A)$ in the set $\{\chi_{C_n}/\mu(C_n)\}$ such that

$$\beta(E) = \lim_\alpha \int_E [\chi_{C_\alpha}/\mu(C_\alpha)] \, d\mu$$

for all $E \in \Sigma$. In particular,

$$\beta(C_m) = \lim_\alpha \int_{C_m} [\chi_{C_\alpha}/\mu(C_\alpha)] \, d\mu$$

for all m. But now $\chi_{C_\alpha}/\mu(C_\alpha)$ is not an extreme point of C since

$$\chi_{C_k}/\mu(C_k) = \tfrac{1}{2} (\chi_{C_{2k}}/\mu(C_{2k}) + \chi_{C_{2k+1}}/\mu(C_{2k+1}))$$

disqualifies $\chi_{C_k}/\mu(C_k)$ as an extreme point of C. From this it follows that the net $(\int_{C_m} \chi_{C_\alpha}/\mu(C_\alpha) \, d\mu \colon \alpha \in A)$ is a convergent net of 0's or 1's. Hence $\beta(C_m) = 0$ or 1 for all m. In addition, for any k,

$$\sum_{n=2^k}^{2^{k+1}-1} \beta(C_n) = \beta(\Delta) = \lim_\alpha \int_\Delta \chi_{C_\alpha}/\mu(C_\alpha)\, d\mu = 1.$$

Therefore $\beta(C_m) = 1$ for infinitely many m. If $\beta(C_m) = 1$, then

$$\begin{aligned} |e^*(y_m) - 1| &= |e^*(y_m) - \beta(C_m)| = |T^*\beta(y_m) - \beta(C_m)| \\ &= \left|\int_\Delta Ty_m - \chi_{C_m}\, d\beta\right| \leqq \|\beta\| \, \|Ty_m - \chi_{C_m}\| \leqq 1/8, \end{aligned}$$

since $\|\beta\| \leqq 1$ and $\|T(y_m) - \chi_{C_m}\| \leqq \mu(C_m)/8 \leqq 1/8$. Consequently $e^*(y_m) \geqq 7/8$ for infinitely many m. Thus $\lim_m e^*(x_m) \neq 0$ and $e^* \notin K$, a fact which completes the proof.

Let us now consolidate our position.

COROLLARY 8. *Any one of the following statements about a Banach space X implies all the others.*

(a) *The space X^* has the Radon-Nikodým property.*

(b) *The space X^* has the Kreĭn-Mil'man property.*

(c) *Every separable subspace of X has a separable dual.*

(d) *Every separable subspace of X^* is a subspace of a separable dual space.*

PROOF. The proof is an easy consequence of Theorem 1.5, Theorems 6 and 7 and Theorem III.3.2 arranged in the appropriate order.

Here is an application of Corollary 8. It is a special case of V.4.1 with a radically different proof.

COROLLARY 9. *Let (Ω, Σ, μ) be a finite measure space and $1 < p < \infty$. If X is a Banach space, $L_p(\mu, X^*)$ has the Radon-Nikodým property if and only if X^* has the Radon-Nikodým property.*

PROOF. Since $L_p(\mu, X^*)$ contains copies of X^*, X^* has the Radon-Nikodým property whenever $L_p(\mu, X^*)$ does. For the converse, use Corollary 8 to see that the hypothesis of Corollary IV.1.3 is satisfied; keep in mind the fact established in Chapter IV that if X^* has the Radon-Nikodým property then $L_p(\mu, X^*) = L_q(\mu, X)^*$ where $p^{-1} + (q')^{-1} = 1$.

An apparently deeper application is a stability result for preduals of spaces with the Radon-Nikodým property.

COROLLARY 10 (STEGALL). *Let X be a Banach space and Y be a Banach space which is a continuous linear image of a closed subspace of X. If X^* has the Radon-Nikodým property, then Y^* has the Radon-Nikodým property.*

PROOF. If Z is a separable closed subspace of Y then Z is the continuous linear image of a separable closed subspace W of X. By Theorem 6, W^* is separable. Since Z is a quotient of W, Z^* is a subspace of W^* and hence Z^* is separable. By Corollary 8, Y^* has the Radon-Nikodým property.

COROLLARY 11. *If X is a weakly sequentially complete Banach space and X^* has the Radon-Nikodým property, then X is reflexive.*

PROOF. Let (x_n) be a bounded sequence in X. Let Y be the closed linear span of $\{x_n : n \in N\}$. By Corollary 8, Y^* has the Radon-Nikodým property. Since Y is separable, Y^* is separable by Corollary 8. But (x_n) is a bounded sequence in Y so there is a subsequence (x_{n_k}) of (x_n) which is weakly Cauchy. Since Y is a closed linear subspace of X, Y is weakly sequentially complete, so (x_{n_k}) is weakly convergent in Y, hence is also weakly convergent in X.

3. Strongly exposed points and the Radon-Nikodým property. As an internal structural property of Banach spaces, the Radon-Nikodým property can be viewed as a purely geometric phenomenon. This fact became evident in Chapter V where it was seen that Banach spaces with the Radon-Nikodým property are characterized by their lack of nondentable bounded sets. In §§1 and 2 of this chapter, the geometric nature of this property became even more graphic; a dual space lacks the Radon-Nikodým property if and only if it has a closed bounded convex subset with no extreme points. The purpose of this section is to prove that Banach spaces with the Radon-Nikodým property have a stronger extreme point structure than we have seen so far. Indeed, we shall see that in a Banach space having the Radon-Nikodým property closed bounded convex sets are the norm closed convex hull of their *strongly exposed points.*

It is convenient to agree on some terminology and notation. Throughout this section, X is a Banach space. A *slice* of a closed bounded convex set C in X is a set of the form

$$S(x^*, \alpha, C) = \{x \in C: x^*(x) + \alpha \geqq \sup x^*(C)\}$$

where $x^* \in X^*$, $\|x^*\| = 1$ and $\alpha > 0$. Slices are not new to this chapter in the sense that they have implicitly appeared in §1. With the help of the separation theorem, it is easily seen that *a closed bounded subset of X is dentable if and only if it has slices of arbitrarily small diameter.*

Recall that a point x_0 of a closed bounded set C is *strongly exposed* if there is an $x_0^* \in X$ such that $x_0^* x_0 = \sup x^*(C)$ and such that $\lim_n \|x_n - x_0\| = 0$ whenever (x_n) is a sequence in C with $\lim_n x_0^*(x_n) = x_0^*(x)$. In this case, x_0^* *strongly exposes* x_0 in C.

The first lemma provides the basic mechanism for producing strongly exposed points.

LEMMA 1 (BISHOP). *Let C be a closed bounded convex subset of X. Suppose that for each slice $S(x^*, \alpha, C)$ of C and each $\varepsilon > 0$ there is a slice $S(y^*, \beta, C)$ of C such that*

(i) $S(y^*, \beta, C) \subseteq S(x^*, \alpha, C)$,

(ii) $S(y^*, \beta, C)$ *has diameter less than* ε, *and*

(iii) $\|x^* - y^*\| < \varepsilon$.

Then each slice of C contains a strongly exposed point of C. Consequently, C is the closed convex hull of its strongly exposed points.

PROOF. It is enough to prove the result under the assumption that C lies inside the closed unit ball of X. Make this assumption; let $\varepsilon > 0$, and let $S(x^*, \alpha, C)$ be a slice of C. Set $y_0^* = x^*$ and $\beta_0 = \alpha$. Use (i), (ii) and (iii) to construct a sequence (y_n^*) in X^* and a sequence (β_n) of positive reals such that

(a) $\|y_n^*\| = 1$,

(b) $\|y_{n+1}^* - y_n^*\| < \beta_n 2^{-n}$,

(c) $\beta_{n+1} < \beta_n 2^{-1}$,

(d) $S(y_{n+1}^*, \beta_{n+1}, C)$ has diameter less than $\beta_n 2^{-n}$, and

(e) $S(y_{n+1}^*, \beta_{n+1}, C) \subseteq S(y_n^*, \beta_n, C)$, for all n.

Combining (a) and (b), one finds the sequence (y_n^*) converges to some $x_0^* \in X^*$ with $\|x_0^*\| = 1$. On the other hand, (d) and (e) combine to produce a single point $x_0 \in C$ such that $\{x_0\} = \bigcap_{n=1}^\infty S(y_n^*, \beta_n, C)$. To complete the proof of the first assertion it is enough to show that x_0 is strongly exposed by x_0^*.

To this end, note that, by (b), $\|x_0^* - y_n^*\| < \beta_n 2^{-n+1}$. Hence, for each $x \in C$ and each $n \geqq 3$, we have

$$|x_0^*(x) - y_n^*(x)| < \beta_n/4.$$

Accordingly, if $n \geqq 3$ and $x \in S(x_0^*, \beta_n/4, C)$, then

$$\begin{aligned} \sup y_n^*(C) - 3\beta_n/4 &\leqq \sup x_0^*(C) - \beta_n/4 - \beta_n/4 \\ &\leqq x_0^*(x) - \beta_n/4 \leqq y_n^*(x). \end{aligned}$$

Thus

$$S(x_0^*, \beta_n/4, C) \subseteq S(y_n^*, \beta_n, C)$$

for $n \geqq 3$. This establishes that

$$\bigcap_{n=3}^{\infty} S(x_0^*, \beta_n/4, C) = \{x_0\},$$

and that the diameter of $S(x_0^*, \beta_n/4, C) \to 0$ as $n \to \infty$. The fact that x_0^* strongly exposes x_0 in C is a quick consequence of these facts.

For the second assertion, let C_0 be the closed convex hull of the strongly exposed points of C and suppose $C_0 \subsetneqq C$. Using the separation theorem one finds a slice of C disjoint from C_0. But by the first part, every slice of C contains a strongly exposed point of C. This contradicts the definition of C_0. This proves Lemma 1.

The proof of the next lemma contains the basic construction that allows Lemma 1 to be used to find strongly exposed points in closed bounded convex subsets of Banach spaces with the Radon-Nikodým property.

LEMMA 2 (PHELPS). *Let X have the Radon-Nikodým property, K be a nonempty closed bounded convex subset of X, and let $x^* \in X^*$ have norm 1. If $K \subseteq \{x \in X : x^*(x) \geqq 0\}$ and x^* is not identically zero on K, then for each $\varepsilon > 0$ with $\varepsilon < 1$ there is a slice $S(y^*, \beta, K)$ of diameter less than ε such that*

(i) *$S(y^*, \beta, K) \subseteq \{x \in K: x^*(x) > 0\}$ and*

(ii) *$\|x^* - y^*\| < \varepsilon$.*

PROOF. The proof will proceed in two thrusts. First, a slice $S(y^*, \beta, K)$ of K of diameter less than ε satisfying (i) will be produced. After this it will be shown how to modify the construction to obtain a slice $S(y^*, \beta, K)$ that also satisfies (ii). Throughout the proof N is the null space $\{x \in X: x^*(x) = 0\}$ of x^*.

Toward finding a slice $S(y^*, \beta, K)$ of diameter less than ε and satisfying (i), note that if $K \cap N = \varnothing$ then the dentability of K produces a slice $S(y^*, \beta, K)$ of K with diameter less than ε. Thus $S(y^*, \beta, K)$ automatically has acceptable diameter and a fortiori satisfies (i).

The case $K \cap N \neq \varnothing$ is more interesting and complicated. In this case, select $z \in K$ with $x^*(z) > 0$. For each $x \in K \cap N$ define a bounded linear operator T_x: $X \to X$ by

$$T_x(y) = y - (2x^*(y)[z - x]/x^*(z))$$

for $y \in X$. The following properties of T_x are easy to verify:

(1) $x = \frac{1}{2}(z + T_x z)$ for all $x \in K \cap N$;

(2) T_x^2 is the identity on X; thus $T_x^{-1} = T_x$ for all $x \in K \cap N$;

(3) T_x is the identity on N; and

(4) $\|T_x\| \leqq 1 + (4/x^*(z)) \sup\{\|x\|: x \in K\} = M_0$.

Next define a family $\mathscr{K}$ of sets by

$$\mathscr{K} = \{K\} \cup \{T_x(K): x \in K \cap N\}.$$

Let K_1 be the closed convex hull of the union of all the members of $\mathscr{K}$. Note that (4) ensures that K_1 is bounded. In addition z and $T_x z$ belong to K_1 whenever $x \in K \cap N$. Also (1) guarantees that each $x \in K \cap N$ is the midpoint of the line segment from z to $T_x z$. This line segment has length $\|z - T_x z\| = 2\|z - x\| \geqq 2x^*(z) > 0$. We shall return to these line segments shortly.

But first, note that since X has the Radon-Nikodým property, K_1 is dentable. Accordingly, there is a slice $S(y^*, \alpha, K_1)$ of K_1 whose diameter is strictly less than $d = \min(x^*(z), \varepsilon/M_0)$. Also since

$$\sup y^*(K_1) = \sup y^*(\bigcup \{K': K' \in \mathscr{K}\}),$$

there is at least one $K_0 \in \mathscr{K}$ such that $\sup y^*(K_0) > \sup y^*(K_1) - \alpha$. Let

$$\beta = \sup y^*(K_0) - [\sup y^*(K_1) - \alpha].$$

Plainly $\alpha \geqq \beta > 0$ and $S(y^*, \beta, K_0) \subseteq S(y^*, \alpha, K_1)$. Consequently the diameter of $S(y^*, \beta, K_0)$ is no greater than d.

Next, we shall see that $S(y^*, \alpha, K_1)$ misses $K \cap N$. For, if $x \in K \cap N \cap S(y^*, \alpha, K_1)$, then the line segment from z to $T_x z$ is a line segment from z to $T_x z$ with midpoint x which intersects $S(y^*, \alpha, K_1)$. By the defining inequality for $S(y^*, \alpha, K_1)$, either the line segment from z to x or the line segment from x to $T_x z$ lies wholly inside $S(y^*, \alpha, K_1)$. But we have already seen that each of these line segments has length no less than $x^*(z)$, a conclusion which contradicts the choice of d. Consequently, $S(y^*, \alpha, K_1)$ as well as the smaller slice $S(y^*, \beta, K_0)$ contains no points of $K \cap N$.

Now we shall perturb the slice $S(y^*, \beta, K_0)$ to achieve (i). If $K_0 = K$, there is nothing to do; $S(y, \beta, K_0)$ has an acceptable diameter and satisfies (i). If $K_0 \neq K$, then there is $x \in K \cap N$ with $K_0 = T_x(K)$. Therefore we have

$$\begin{aligned} T_x^{-1}(S(y^*, \beta, K_0)) &= T_x^{-1}(S(y^*, \beta, T_x(K))) \\ &= \{T_x^{-1}(y): y \in T_x(K), y^*(y) \geqq \sup y^* T_x(K) - \beta\} \\ &= \{k \in K: y^* T_x(k) \geqq \sup y^* T_x(K) - \beta\} \\ &= \{k \in K: (T_x^* y^*)(k) \geqq \sup T_x^* y^*(K) - \beta\} \end{aligned}$$

is a slice of K of diameter at most

$$\|T_x^{-1}\|d = \|T_x\|d \leqq M_0 d < \varepsilon.$$

In addition this slice also misses $K \cap N$ since (3) guarantees N is fixed by T_x^{-1}. Thus in this case, we have found a slice with diameter less than ε that satisfies (i).

To complete the proof we shall show how to modify the construction above to obtain a slice with diameter less than ε that satisfies (i) and (ii). First, let F be the closed convex hull of $K \cup \{x \in N: \|x\| \leqq \lambda\}$ where $\lambda = 4M\varepsilon^{-1}$ and $M = \sup\{\|x\|: x \in K\}$. Apply the construction in the previous part of the proof to the set F instead of K to obtain a slice $S(y^*, \beta, F)$ of F of diameter less than ε that misses $F \cap N$. Since $S(y^*, \beta, F)$ misses $F \cap N$,

$$y^*(x) \geqq \sup y^*(K) - \beta = \sup y^*(F) - \beta$$

whenever $x \in S(y^*, \beta, K)$. Thus $S(y^*, \beta, K) \subseteq S(y^*, \beta, F)$ and it remains only to show that $\|x^* - y^*\| < \varepsilon$.

Let $u \in S(y^*, \beta, F)$. Since $S(y^*, \beta, F)$ misses $F \cap N$, we have

$$y^*(u) > \sup\{y^*(x): x \in N, \|x\| \leqq \lambda\} = \lambda\|y^*\|_N.$$

But now by Lemma 1.2 we have either

$$\|x^* + y^*\| \leqq 2y^*(u)/\lambda \quad \text{or} \quad \|x^* - y^*\| \leqq 2y^*(u)/\lambda.$$

Since $x^*(u) > 0$, the assumption that $\|x^* + y^*\| \leqq 2y^*(u)/\lambda$ yields

$$\begin{aligned} y^*(u)/M &\leqq y^*(u)/\|u\| < (x^* + y^*)(u)/\|u\| \\ &\leqq \|x^* + y^*\| \leqq 2y^*(u)/\lambda. \end{aligned}$$

Hence $2M > \lambda$ and

$$\lambda = 4M\varepsilon^{-1} > 4M > 2\lambda,$$

a blatant contradiction. The only possibility now left open is $\|x^* - y^*\| \leqq 2y^*(u)/\lambda < \varepsilon$. This completes the proof.

The following theorem which is the main result of this section is a striking improvement of the Kreĭn-Mil'man theorem in spaces with the Radon-Nikodým property.

THEOREM 3 (PHELPS). *If a Banach space X has the Radon-Nikodým property, then each nonempty closed bounded convex subset of X is the closed convex hull of its strongly exposed points.*

PROOF. Let C be a nonempty closed bounded convex subset of X and $S(y^*, \alpha, C)$ be any slice of C. By translating C we can and do assume that $\sup x^*(C) = \alpha$. Set

$$K = \{x \in C: x^*(x) \geqq 0\} = S(x^*, \alpha, C).$$

Since $\alpha > 0$, we have $K \cap \{x \in X: x^*(x) > 0\} \neq \varnothing$. If $0 < \varepsilon < 1$, Lemma 2 produces a slice $S(y^*, \beta, K)$ of K with diameter no greater than ε such that $S(y^*, \beta, K) \subseteq \{x \in X: x^*(x) > 0\}$ and $\|y^* - x^*\| < \varepsilon$. Now note that $S(y^*, \beta, K) \subseteq S(y^*, \alpha, C)$. Thus, if the inclusion $S(y^*, \beta, C) \subseteq S(y^*, \beta, K)$ can be established, an appeal to Lemma 1 will complete the proof.

To prove that $S(y^*, \beta, C) \subseteq S(y^*, \beta, K)$, suppose there is $x \in C\backslash S(y^*, \beta, K)$. If $x \in K$, then

$$y^*(x) < \sup y^*(K) - \beta \leqq \sup y^*(C) - \beta.$$

Thus $x \notin S(y^*, \beta, C)$. On the other hand, if $x \notin K$, then $x^*(x) < 0$. Select any $z \in S(y^*, \beta, K)$ and note that $x^*(z) > 0$. Therefore there is a point w on the line segment from x to z at which $x^*(w) = 0$. But $w \notin S(y^*, \beta, K)$. Thus $y^*(w) < \sup y^*(K) - \beta \leqq y^*(z)$. This yields the inequalities

$$y^*(x) < \sup y^*(K) - \beta \leqq \sup y^*(C) - \beta,$$

and again $x \notin S(y^*, \beta, C)$.

In any case, $x \in C \backslash S(y^*, \beta, K)$ excludes the possibility $x \in S(y^*, \beta, C)$. Hence $S(y^*, \beta, C) \subseteq S(y^*, \beta, K)$ and the proof is complete.

COROLLARY 4 (PHELPS-RIEFFEL). *A Banach space X has the Radon-Nikodým property if and only if each nonempty closed bounded convex set of X is the norm closed convex hull of its strongly exposed points.*

PROOF. Apply Theorem 3, V.3.10(iv) and V.3.7.

Perhaps it is time for a bit of philosophy. Theorem 3 together with the analytic Radon-Nikodým theorems of Chapter III usually provide the simplest way of verifying the existence of strongly exposed points in closed bounded convex sets. For instance, this line of reasoning trivially yields the fact that a nonempty closed bounded convex set in a separable dual space is the norm closed convex hull of its strongly exposed points. The link between the analytic conditions of Chapter III and the geometric conditions under study in this chapter is the integration theory of Chapter V.

4. The Radon-Nikodým property and the existence of extreme points for nonconvex closed bounded sets. By now it is evident that Banach spaces with the Radon-Nikodým property enjoy a very rich extreme point structure. In this section, we are going to build on this theme by proving theorems that are even more striking than most of the theorems of the first three sections. One example: We shall see that in a Banach space without the Radon-Nikodým property there is always a *closed* bounded set A and a convex bounded *open* set B containing A such that A and B have the same closed convex hull, a fact which graphically exhibits the dramatic lack of extreme point structure in Banach spaces without the Radon-Nikodým property. As consequences of this fact, we shall learn that a Banach space X has the Radon-Nikodým property if and only if for each closed bounded subset A the bounded linear functionals that attain their maximum values on A are dense in X^*. Equally it will be proved that a Banach space has the Radon-Nikodým property if and only if each of its closed bounded sets contains an extreme point of its closed convex hull.

Throughout this section X is a Banach space. If $A \subseteq X$ and $\varepsilon > 0$, $B_\varepsilon(A)$ stands for the set

$$B_\varepsilon(A) = \bigcup_{x \in A} \{y \in X : \|y - x\| \leqq \varepsilon\}.$$

A subset B of X has a finite ε-net if there is a finite subset F of B such that $B \subseteq B_\varepsilon(F)$.

The first theorem shows that the definition of dentable sets can be modified in a way useful later.

THEOREM 1 (HUFF-MORRIS). *Any one of the following statements about a nonempty closed bounded convex subset K of X implies all the others:*

(a) *The set K is not dentable.*

(b) *There exists $\varepsilon > 0$ such that no slice of K has a finite ε-net.*

(c) *There exists $\varepsilon > 0$ such that for each finite set $F \subseteq K$, $K = \overline{\text{co}}(K \backslash B_\varepsilon(F))$.*

PROOF. To prove that (a) implies (b), suppose K is not dentable and that $\|x\| \leqq 1$ for each $x \in K$. Then there exists a $\delta > 0$ such that every slice of K has diameter larger than δ. Let $\varepsilon = \delta/3$, $x^* \in X^*$, $\|x^*\| = 1$ and let $\alpha > 0$. Consider the slice $S(x^*, \sup x^*(K) - \alpha, K) = S$. Suppose there is a finite ε-net $\{x_1, \cdots, x_n\}$ in S. Let

$$H = \{x \in K: x^*(x) = \alpha\} \subsetneq S.$$

Note that H is a closed convex set. By paring the finite ε-net $\{x_1, \cdots, x_n\}$ in S to a "minimal ε-net", we can and do assume that, for some $m \leqq n$,

$$S = \overline{\text{co}}(H \cup (S \cap B_\varepsilon(\{x_1, x_2, \cdots, x_m\})))$$

but

$$S \neq K_1 = \overline{\text{co}}(H \cup (S \cap B_\varepsilon(\{x_2, \cdots, x_m\}))).$$

Let $y_0 \in S \cap B_\varepsilon(x_1)\backslash K_1$. Choose $y^* \in X^*$, $\|y^*\| = 1$ such that

$$a = \sup y^*(K_1) < y^*(y_0) \leqq \sup y^*(S) = c.$$

Choose β with $a < \beta < c$ and $(\beta - a)/(c - a) > 1 - \delta/12$. Let $S' = S(y^*, c - \beta, S)$.

The plan is to show that the diameter of S' is no greater than δ. This fact will then guarantee that S' is not a slice of K. Assuming this has been established, let $\gamma = \sup y^*(K)$ and note that since $S' \subseteq S(y^*, \gamma - \beta, K)$, it follows that there exists $z \in S(y^*, \gamma - \beta, K)$ such that $z \notin S'$. Since $y^*(z) \geqq \beta$ and $z \notin S'$, we see $z \notin S$. Hence $x^*(z) < \alpha$. Now, if $w \in S'$, then $x^*(w) > \alpha$ since $S' \subseteq S$, $H \subseteq K_1$ and $K_1 \cap S' = \varnothing$. It follows that there is $0 < \lambda < 1$ such that

$$x^*(\lambda z + (1 - \lambda)w) = \alpha,$$

i.e., $\lambda z + (1 - \lambda) w \in H$. But $w \in S' = S(y^*, \alpha - \beta, S)$. Hence $y^*(w) \geqq \beta$ and we knew in advance that $y^*(z) \geqq \beta$. Therefore $y^*(\lambda z + (1 - \lambda)w) \geqq \beta$. Accordingly $\lambda z + (1 - \lambda)w \in S'$. This is impossible since this means $\lambda z + (1 - \lambda)w \in H \cap S'$, an empty set. This contradiction means the proof that (a) implies (b) will be complete if we can show that the diameter of S' is no greater than δ.

To this end, let

$$L = \{x \in S \cap B_\varepsilon(x_1): y^*(x) \geqq a\}$$

and

$$K_2 = \{x \in S: y^*(x) \leqq a\}.$$

Note that $K_1 \subseteq K_2$ and $y_0 \in L$. Moreover, one has

$$L \cup K_2 \supseteq H \cup (S \cap B_\varepsilon(\{x_1, \cdots, x_n\})).$$

Hence $S = \overline{\text{co}}(L \cup K_2)$. Now since $\{x \in S: y^*(x) > \beta\}$ is a relatively open dense subset of S' and $\text{co}(L \cup K_2)$ is dense in S, we find that $S' \cap \text{co}(L \cup K_2)$ is also dense in S'. At this point, let $0 \leqq \lambda_i \leqq 1$, $u_i \in L$, $v_i \in K_2$, $i = 1, 2$. Then if $\lambda_i u_i + (1 - \lambda_i)v_i \in S'$, for $i = 1, 2$, then one has

$$\beta \leqq y^*(\lambda_i u_i + (1 - \lambda_i)v_i) \leqq \lambda_i c + (1 - \lambda_i)\, a = \lambda_i\, (c - a) + a.$$

Hence $\lambda_i \geqq (\beta - a)/(c - a) > 1 - \delta/12$ and $(1 - \lambda_i) < \delta/12$ for $i = 1, 2$. From this and the facts that $x \in K$ implies $\|x\| \leqq 1$ and $u_i \in B_\varepsilon(x_1)$ it follows that

$$\begin{aligned}
&\|[\lambda_1 u_1 + (1 - \lambda_1)v_1] - [\lambda_2 u_2 + (1 - \lambda_2)v_2]\| \\
&\qquad \leqq \|\lambda_1 u_1 - \lambda_2 u_2\| + \sum_{i=1}^{2} \|(1 - \lambda_i)v_i\| \\
&\qquad \leqq \|u_1 - (1 - \lambda_1)u_1 - (u_2 - (1 - \lambda_2)u_2)\| + \delta/12 + \delta/12 \\
&\qquad \leqq \|u_1 - u_2\| + \sum_{i=1}^{2} \|(1 - \lambda_i)u_i\| + \delta/12 + \delta/12 \\
&\qquad \leqq 2\varepsilon + \delta/12 + \delta/12 + \delta/12 + \delta/12 \\
&\qquad = 2\varepsilon + \delta/3 = \delta.
\end{aligned}$$

Hence the diameter of S' is at most δ and the proof of (a) implies (b) is complete.

That (b) implies (c) is an easy consequence of the separation theorem, while the truth of (c) implies (a) is apparent.

The next result is a direct consequence of the characterization of Banach spaces with the Radon-Nikodým property in terms of their dentable subsets. Teamed with Theorem 1, this result will be used to reveal some features of nondentable sets that were previously inaccessible to us.

THEOREM 2 (DAVIS-PHELPS). *A Banach space without the Radon-Nikodým property has a bounded open subset whose norm closure is not dentable.*

In fact, a Banach space has the Radon-Nikodým property if and only if each of its equivalent norms has a dentable closed unit ball.

PROOF. Obviously it suffices to prove only the second assertion. The "only if" part is an easy consequence of the facts that the Radon-Nikodým property is unaffected by renormings with equivalent norms and that bounded sets in spaces with the Radon-Nikodým property are dentable.

Conversely, if X lacks the Radon-Nikodým property, there is a bounded subset K of X which is not dentable. By a straightforward argument, $K \cup (-K)$ is not dentable. Therefore $\overline{\text{co}}(K \cup (-K))$ is not dentable by Proposition V.3.2. Another straightforward calculation shows that if B is the closed unit ball of X, then $B_1 = B + \overline{\text{co}}(K \cup -K)$ is a closed bounded absolutely convex nondentable body in X. Evidently B_1 is the nondentable closed unit ball for an equivalent norm on X.

The following corollary of Theorems 1 and 2 places nondentability in a new geometric light. We have already seen in Chapter V that a Banach space X contains a nondentable bounded set if and only if X contains a non-σ-dentable bounded set. Thinking of σ-dentability as dentability with respect to *infinite* convex sums we find that the next result says that X contains a nondentable bounded set if and only if X contains a bounded set which is not dentable with respect to *finite* convex sums. In fact Corollary 3 says more than this.

COROLLARY 3 (HUFF-MORRIS-DAVIS-PHELPS). *A Banach space without the Radon-Nikodým property has a nonempty open bounded convex subset K such that, for some $\varepsilon > 0$,*

$$K = \text{co}(K \backslash B_\varepsilon(\{x_1, \cdots, x_n\}))$$

for every finite subset $\{x_1, \cdots, x_n\}$ *of K. Indeed, K can be chosen to be any open bounded set whose closure is not dentable.*

PROOF. With the help of Theorem 2 choose a nonempty open bounded convex set K such that the closure $\bar{K}$ of K is not dentable. Let $\{x_1, \cdots, x_n\}$ be a finite subset of K. According to Theorem 1, there is an $\varepsilon > 0$ such that

$$\bar{K} = \overline{\text{co}}(\bar{K} \backslash B_\varepsilon(\{x_1, \cdots, x_n\})).$$

Let $J = \bar{K} \backslash B_\varepsilon(\{x_1, \cdots, x_n\})$ and note that the interior J° of J is $J^\circ = K \backslash B_\varepsilon(\{x_1, \cdots, x_n\})$. Now let $y \in J$ and $z \in K$. The half-open line segment $[z, y)$ is in K and points of $[z, y)$ sufficiently close to y are outside the closed set $B_\varepsilon(\{x_1, \cdots, x_n\})$. Thus y is a limit of points in $K \backslash B_\varepsilon(\{x_1, \cdots, x_n\}) = J^\circ$. It follows that $\text{co}(J) \subseteq \overline{\text{co}}(J^\circ)$.

Now recall that K is the interior of $\bar{K} = \overline{\text{co}}(J)$. Since $\text{co}(J) \subseteq \overline{\text{co}}(J^\circ)$, it follows that K is a subset of the interior of $\overline{\text{co}}(J^\circ)$. But if A is a nonempty open convex set, then A is the interior of its closure. Consequently, $K \subseteq \text{co}(J^\circ) = \text{co}(K \backslash B_\varepsilon(\{x_1, \cdots, x_n\}))$. Hence $K = \text{co}(K \backslash B_\varepsilon(\{x_1, \cdots, x_n\}))$, and the proof is finished.

Corollary 3 takes on a particularly significant form when translated to the martingale context. By V.3.1 and V.3.7, the space X lacks the Radon-Nikodým property if and only if there is a martingale (f_n, B_n) of countably-valued functions in $L_1(\mu, X)$, where μ is Lebesgue measure on $[0, 1]$, such that

(i) $\sup_n \|f_n\|_\infty < \infty$ and

(ii) there is $\varepsilon > 0$ such that $\|f_n(t) - f_{n+1}(t)\|_X \geqq \varepsilon$ for all $t \in [0, 1]$ and all n.

With the help of Corollary 3 and the techniques of §V.3, it is easy to prove

COROLLARY 4. *Let* μ *be Lebesgue measure on* $[0, 1]$. *A Banach space X lacks the Radon-Nikodým property if and only if there is a martingale* (f_n, B_n) *of simple functions in* $L_1(\mu, X)$ *such that*

(i) $\sup_n \|f_n\|_\infty < \infty$ *and*

(ii) *there is* $\varepsilon > 0$ *such that* $\|f_n(t) - f_m(s)\|_X \geqq \varepsilon$ *for all* $s, t \in [0, 1]$ *and all positive integers m and n with* $m \neq n$.

A major theorem of James, which is not included in this survey, characterizes weakly compact subsets of a Banach space as weakly closed bounded sets upon which each bounded linear functional attains a maximum value. The next result contrasts dramatically with James's theorem in that it shows that a Banach space lacks the Radon-Nikodým property if and only if there is a norm closed bounded set upon which no nonzero bounded linear functional attains a maximum value. This theorem also contains intuitive evidence that says that in at least some respects, bounded norm closed subsets of a Banach space with the Radon-Nikodým property behave as if they were weakly compact.

THEOREM 5 (HUFF-MORRIS). *Any one of the following conditions is both necessary and sufficient for X to have the Radon-Nikodým property.*

(a) *Every closed bounded subset of X contains an extreme point of its closed convex hull.*

(b) *Every closed bounded subset of X contains an extreme point of its convex hull.*

(c) *For each closed bounded subset A of X there is a nonzero $x^* \in X^*$ and $x_0 \in A$ such that*

$$x^*(x_0) = \sup x^*(A).$$

(d) *For each closed bounded subset A of X the collection of $x^* \in X^*$ that attain their maxima on A is norm-dense in X^*.*

PROOF. The necessity of each of the conditions (a), (b), (c), and (d) follows easily from Theorem 3.3 (and, in the case of (d), its proof) applied to the closed convex hull of A.

The sufficiency of the conditions will be proved simultaneously. The plan is to suppose X lacks the Radon-Nikodým property and to produce an equivalent norm $|||\cdot|||$ such that the closed unit ball $B_{|||\cdot|||}$ for this norm admits a closed subset A in the interior of $B_{|||\cdot|||}$ such that $\overline{\text{co}}(A) = B_{|||\cdot|||}$. Plainly the existence of such a set A violates each of (a), (b), (c), and (d). Thus the proof will be complete upon the construction of A.

To this end, use Theorem III.3.2 to find a separable closed linear subspace Y of X that also lacks the Radon-Nikodým property. By Theorem 2, Y has an equivalent norm whose open unit ball K has a nondentable closure. Let (y_n) be dense in K. With the help of Corollary 3, choose $\varepsilon > 0$ such that $K = \text{co}(K \backslash B_\varepsilon(F))$ for every finite subset F of K. Also observe that this means that, if E_0 and E_1 are finite subsets of K, then there is a finite subset E_2 of K such that $E_2 \cap B_\varepsilon(E_0) = \varnothing$ and $E_1 \subseteq \text{co}(E_2)$. With this in mind, define a sequence (F_n) of finite subsets of K as follows:

Let $F_1 = \{y_1\}$. If $F_2, \cdots, F_{n-1}$ have been defined, choose a finite subset F_n of K subject to the two requirements that $B_\varepsilon(\bigcup_{i=1}^{n-1} F_i) \cap F_n = \varnothing$ and such that $F_{n-1} \cup \{y_n\} \subseteq \text{co}(F_n)$.

Now a pleasant situation has come about. The set $F = \bigcup_{n=1}^\infty F_n$ is a closed set because $\|x-y\| \geqq \varepsilon$ for $x \in F_m$ and $y \in F_n$ for $m \neq n$ and each F_n is finite. Evidently F is a subset of K. Equally evident is the fact that $\overline{\text{co}}(F)$ is the closure of K. This finishes the separable case.

For the nonseparable case, let B denote the closed unit ball of X equipped with its original norm. Write $G_n = F_n + \varepsilon/3(1 - 1/(n+1))B$. Note that each G_n is closed and $\|x-y\| \geqq \varepsilon/3$ for $x \in G_m$ and $y \in G_n$ and $m \neq n$. Consequently $A = \bigcup_{n=1}^\infty G_n$ is norm closed. Next we shall prove that $\text{co}(A)$ is dense in $K + \{x \in X: \|x\| < \varepsilon/3\} = D$. For, if $\eta > 0$ is given and $w \in K$, $x \in X$ and $\|x\| < \varepsilon/3$ there is by the first part of the proof, a $y \in \text{co}(F)$ with $\|w - y\| < \eta$. If n is chosen such that $y \in \text{co}(F_n)$ and $\|x\| \leqq (\varepsilon/3)(1 + 1/(n+1))$ then $y + x \in \text{co}(G_n)$ and $\|(w+x) - (y+x)\| < \eta$. This proves that $\text{co}(A)$ is dense in D.

To finish the proof, note that D is an absolutely convex bounded open subset of X which obviously contains A. The Minkowski functional (or gauge) of D is an equivalent norm $|||\cdot|||$, and its unit ball contains the closed set A. Since $\text{co}(A)$ is dense in D, $\overline{\text{co}}(A)$ is all of the closed unit ball for $|||\cdot|||$. This completes the proof.

A fitting end to this chapter and section is the following thought-provoking corollary of the proof of Theorem 5.

COROLLARY 6 (HUFF-MORRIS). *A Banach space lacks the Radon-Nikodým pro-*

perty if and only if there is a bounded open convex set K in X and a norm closed subset A of K such that the closed convex hull of A coincides with the closure of K.

In this case, K may be selected as the open unit ball for some equivalent renorming of X.

5. Notes and remarks. Some fifty years ago in Latvia, a thirty-year-old mathematician in the School of Railways went to the American Consulate and claimed he had a job waiting for him at Dartmouth University. In his possession was a postcard saying "The weather at Dartmouth is fine." It was by this prearranged signal that J. D. Tamarkin was able to find his way to the United States. In the United States, Tamarkin met J. A. Clarkson and suggested that Clarkson look at differentiability properties of vector-valued functions. This was the beginning of the study of the Radon-Nikodým property and led to Clarkson's **[1936]** fundamental paper. Interestingly enough, this paper which is quite geometric in nature has as its avowed object the isolation of *geometric* conditions on a Banach space X that ensure that X-valued functions of bounded variation are differentiable almost everywhere, a condition which is equivalent to the Radon-Nikodým property. This is how uniformly convex Banach spaces were born. Clarkson **[1936]** introduced the notion of uniform convexity, proved that uniformly convex Banach spaces have the Radon-Nikodým property and established the famous "Clarkson's inequalities" to prove that the L_p spaces $(1 < p < \infty)$ are uniformly convex. In passing, he noted that l_1, although not uniformly convex, has the Radon-Nikodým property while neither c_0 nor $L_1[0, 1]$ have the Radon-Nikodým property. Thus uniform convexity is sufficient but not necessary for the Radon-Nikodým property and neither c_0 nor $L_1[0, 1]$ have equivalent uniformly convex norms. In view of the facts that at the time of Clarkson's paper it was not known that uniformly convex spaces are reflexive (this had to wait until Mil'man **[1938]** and Pettis **[1939a]**) and the Dunford-Pettis theorem was still a few years in the future, we cannot overstate our respect for Clarkson's work. In addition, it is gratifying to us to be able to say that the important concept of uniform convexity owes its origin to the Radon-Nikodým property.

Clarkson **[1936]** also showed that strict convexity does not imply the Radon-Nikodým property by proving that every separable Banach space admits an equivalent strictly convex norm. It is unknown to this day whether each Banach space with the Radon-Nikodým property admits an equivalent strictly convex norm.

As noted in the notes and remarks section of Chapter V, the Radon-Nikodým theorem and the geometry of Banach spaces lapsed into an estrangement of three decades after the Clarkson paper. Then, at Berkeley, California, Rieffel taught a real analysis course in which he opted to present the Bochner integral instead of the classical Lebesgue theory. As rumor has it, all went smoothly until he came to the Radon-Nikodým theorem and its attendant difficulties in infinite dimensional Banach spaces. At this time, Rieffel **[1967]** isolated the notion of dentability and obtained many of the results already discussed in Chapter V. He noted that dentability assumptions are in a sense extremal in character. He observed that extreme points of compact convex sets are denting points[3] and was able to show that for any set Γ each bounded subset of $l_1(\Gamma)$ is dentable. In addition he wondered (a) whether all convex closed bounded sets in l_1 have extreme points, (b) whether weakly compact sets are always dentable, (c) which Banach spaces have only dentable bounded sets, and (d) whether the existence of denting points is in some way related to the existence of strongly exposed points.

During the thirty-year separation of the Radon-Nikodým property and the geometry of Banach spaces, the close relationship between the Radon-Nikodým property and extreme point phenomena was just beneath the surface of some classical work in Banach space theory. For instance, the work of Price **[1940]** and Kreĭn and Mil'man **[1940]** coupled with Alaoglu's **[1940]** theorem established that the closed unit ball of a dual space has no paucity of extreme points. Since the closed unit balls of c_0 and $L_1[0, 1]$ have few, if any, extreme points, neither of these spaces is *isometric* to a separable dual space.

Now let us look at c_0 and $L_1[0, 1]$ from the point of view of the Radon-Nikodým property. Clarkson **[1936]** exhibited c_0-valued and $L_1[0, 1]$-valued functions on $[0, 1]$ of bounded variation that are nowhere differentiable. Gel'fand **[1938]** showed that a function of bounded variation on $[0, 1]$ with values in a separable dual space is differentiable almost everywhere. These two facts team up to lead to the conclusion that *neither* c_0 *nor* $L_1[0,1]$ *is isomorphic to a subspace of a separable dual space*. Somehow it seems that the use of Radon-Nikodým type considerations to prove the latter (stronger) assertion regarding c_0 and $L_1[0, 1]$ came to be universally regarded as being in the nature of a curiosity.

Let us now return to Rieffel's questions. The first of his questions was answered almost as soon as it was asked by Lindenstrauss **[1966a]** who showed that l_1 has the Kreĭn-Mil'man property and in so doing proved Theorem 1.7 and began a chain of elegant papers by various authors on the Kreĭn-Mil'man property. First, Bessaga and Pełczyński **[1966]** showed that all separable dual spaces have the Kreĭn-Mil'man property. Second, Asplund proved that all $l_1(\Gamma)$-spaces have the Kreĭn-Mil'man property and remarked that Lindenstrauss had actually shown that all locally uniformly convex dual spaces have this property. (A Banach space X is locally uniformly convex (Lovaglia **[1955]**) if for each sequence $(x_n)_{n=0}^{\infty}$ in X with $\|x_n\| = 1$ such that $\lim_n \|x_n + x_0\| = 2$, then $\lim_n \|x_n - x_0\| = 0$.) Later extensions of the results mentioned above can be found in John and Zizler **[1974]**, **[1976]** and Troyanski **[1971]**. Scanning the long list of spaces with the Kreĭn-Mil'man property and noting the resemblance to the list of known possessors of the Radon-Nikodým property, Diestel, in 1972, was unable to resist the temptation to ask: Are the Kreĭn-Mil'man and Radon-Nikodým properties equivalent?

The second of Rieffel's questions dealing with the dentability of weakly compact sets took a few years to solve. Before Rieffel's paper appeared, Lindenstrauss **[1965]**, **[1966b]** and Amir and Lindenstrauss **[1968]** had launched a deep analysis of weakly compact sets in Banach spaces. Building on this, Troyanski **[1971]** showed that weakly compact sets live in spaces with equivalent locally uniformly convex norms.

[3] If D is a subset of a Banach space X, a point $d \in D$ is a denting point for D if, for each $\varepsilon > 0$, $d \notin \overline{\text{co}}(D \setminus B_\varepsilon(d))$.

Lindenstrauss **[1963]** had previously shown that weakly compact convex sets in spaces with equivalent locally uniformly convex norms have strongly exposed points. Thus, Troyanski **[1971]** proved that weakly compact sets are dentable in a truly spectacular way. This paper of Troyanski dealt with problems much deeper than the dentability of weakly compact sets and this result is hardly more than a small corollary of the rest of his work. In any case the very nature of his paper makes it relevant to the theory of vector measures.

Even before Troyanski's paper, Lindenstrauss **[1963]** had shown that separable weakly compact sets are dentable. Namioka **[1967]** gave a beautifully elegant proof of this (in this connection the paper of Namioka also contains a proof of the Bessaga-Pełczyński **[1966]** result; this paper really ought to be read by those interested in the extreme point phenomena under discussion). Then Maynard **[1973a]** made a simple, but important, observation; a set is dentable if each of its countable subsets is dentable, a fact which is implicit in the proof of V.3.4. Thus the Maynard-Namioka proof is more economical than the Troyanski proof. Of course, our favorite proof of this fact is the purely measure-theoretic proof given in V.3.10.

The real breakthrough in the study of the Radon-Nikodým property as a geometric property was provided by Maynard **[1973a]** in response to the third of Rieffel's questions. (See the notes and remarks section of Chapter V for more on this.) Though Maynard did not give a complete solution to Rieffel's question the importance of his work cannot be overstated. Using Maynard's work as a basis, Davis and Phelps **[1974]** and Huff **[1974]** completely solved Rieffel's third question. Their answer is given by V.3.4. However, by characterizing the Radon-Nikodým property as a geometric property, Maynard's work allowed the Radon-Nikodým property from a (what some might consider to be heretical) geometric viewpoint. Thus Maynard's paper opened the study of the Radon-Nikodým property to geometers of Banach spaces. Chapter VII is basically a report on *some* of the results obtained by the geometers.

Theorem 1.4 and its preliminary lemmas are from the basic work of Bishop and Phelps **[1961]**, **[1963]**. Our proof of Lemma 1.2 was shown to us by R. E. Huff. Theorem 1.5 was discovered by J. Lindenstrauss soon after the Davis-Huff-Phelps Theorem V.3.4 became known and can be found in Phelps **[1974]**. Lindenstrauss's proof of Theorem 1.5 is geometrically identical to his proof that l_1 has the Kreĭn-Mil'man property. As remarked above, Theorem 1.7 is due to Lindenstrauss **[1966a]**.

Lemmas 2.1 through 2.5 and Theorem 2.6 are due to Stegall **[1975]**. Our presentation differs somewhat from Stegall's and is essentially that of R. E. Huff. Stegall's theorem and the construction leading to it are at the heart of the structure theory of the Radon-Nikodým property for dual spaces. Further it is the key that allows the Radon-Nikodým property to be related to certain other properties of Banach spaces. As a consequence of Stegall's construction any dual space without the Radon-Nikodým property has a bounded infinite δ-tree. At this point no one knows whether a Banach space without the Radon-Nikodým property has a bounded δ-tree. Theorem 2.7 is due to Huff and Morris **[1975]**; Corollary 2.11 is due to Stegall **[1975]** who also showed that *if X is a Banach space with a closed linear subspace Y for which $(X/Y)^*$ and Y^* have the Radon-Nikodým property then X^* has the Radon-Nikodým property*. A recent result along the same lines is the following theorem of G. Edgar.

THEOREM (EDGAR). *If X is a Banach space and Y is a closed linear subspace of X such that X/Y and Y have the Radon-Nikodým property, then X has the Radon-Nikodým property.*

PROOF. It suffices to prove the following: Let (Ω, Σ, μ) be a probability space, $F: \Sigma \to X$ be a vector measure such that $\| F(E) \| \leqq \mu(E)$ for all $E \in \Sigma$; then F has a Radon-Nikodým derivative with respect to μ. Let φ be the quotient map of X onto X/Y. Then φF is an X/Y-valued measure with average range contained in the closed unit ball of X/Y. Hence φF is countably additive, is of bounded variation and is μ-continuous. Since X/Y has the Radon-Nikodým property there is a Bochner μ-integrable function $f: \Omega \to X/Y$ with $f(\Omega)$ contained in the closed unit ball of X/Y such that

$$\varphi F(E) = \int_E f \, d\mu$$

for all $E \in \Sigma$. By Michael's selection theorem (cf. Parthasarathy **[1972**, p. 9]) φ admits a continuous section $\tilde{\varphi}: X/Y \to X$ that maps the closed unit ball of X/Y into the 2-ball of X; naturally $\tilde{\varphi} \cdot f$ is Bochner integrable. Let

$$G(E) = \int_E \tilde{\varphi}(f(\omega)) \, d\mu(\omega) \quad \text{for } E \in \Sigma.$$

Then G is a countably additive, μ-continuous X-valued measure on Σ of bounded variation. Moreover

$$\varphi \cdot G = \varphi \cdot F,$$

that is, $F - G$ has values in $Y = \varphi^{-1}(\{0\})$. In addition $F - G$ has its average range with respect to μ contained in the 3-ball of Y so it has a Radon-Nikodým derivative with respect to μ, say $g: \Omega \to Y$. The function $g + \tilde{\varphi} \cdot f$ is $dF/d\mu$.

The results of §3 are due to Phelps **[1974]**. They provide a striking response to Rieffel's fourth question. Phelps's constructions are global in nature in that they often leave the set in question. Left unanswered therefore is the question: If K is a closed bounded convex subset of a Banach space and each subset of K is dentable does K have a strongly exposed point?[4]

Some of the material of §3 has been executed in the context of locally convex spaces by Gilliam **[1976]** and Saab **[1976]**.

Huff and Morris **[1976]** are responsible for the bulk of §4. To our knowledge, the theorems of this section are the first deep results dealing with the existence of extreme points in nonconvex sets since the time of Mil'man's original theorem that says that, if A is a compact set, then the extreme points of the closed convex hull of A all belong to A. Theorem 4.2 and the idea behind Corollary 4.3 can be found in Davis and Phelps **[1974]**; the rest of §4 is in its pristine form of Huff and Morris **[1976]**.[5]

The work of Chapter VII almost answers Diestel's question (are the Kreĭn-Mil'man and Radon-Nikodým properties equivalent?) and Uhl's question (are

[4]This question has recently been resolved in the positive by Bourgain **[1976]**.

[5]Recently, Bourgain has shown that the Radon-Nikodým property for X is equivalent to each *weakly* closed bounded set having an extreme point.

separable spaces with the Radon-Nikodým property subspaces of separable dual spaces?). Related to both these questions is whether or not separable spaces with the Kreĭn-Mil'man property are subspaces of separable dual spaces. An affirmative answer to this question will solve both Diestel's and Uhl's questions in the affirmative. On the other hand, it is still unknown if the Kreĭn-Mil'man property is separably determined, that is, if each separable closed linear subspace of the Banach space X has the Kreĭn-Mil'man property need X have the Kreĭn-Mil'man property?

Higher duals and the Radon-Nikodým property. Just as smoothness and convexity properties deteriorate badly as we advance from X to X^*, to X^{**}, to X^{***}, etc., when X is nonreflexive, there is a similar deterioration when X^* lacks the Radon-Nikodým property. Perhaps the following table best illustrates the true state of affairs (appropriate definitions follow the table):

	X^* is reflexive if		X^* has the Radon-Nikodým property if
(1a)	$X^{(iv)}$ is strictly convex		
(2a)	X^{***} is smooth	(2b)	X^{***} is strictly convex
(3a)	X^{**} is weakly locally uniformly convex	(3b)	X^{**} is smooth
(4a)	X^* is very smooth	(4b)	X^* is weakly locally uniformly convex
(5a)	X is uniformly convex	(5b)	X is very smooth

Here X is *smooth* if for each $x \in X$ with $\|x\| = 1$, there is a unique $x^* \in X^*$ with $\|x^*\| = 1$ and $x^*(x) = 1$. It is well known that if X is smooth, the mapping $x \to x^*$ above is norm-to-weak* continuous. The space X is *very smooth* if it is smooth and the mapping $x \to x^*$ is norm-to-weak continuous. The space X is weakly locally uniformly convex if for each sequence $(x_n)_{n=0}^{\infty}$ with $\|x_n\| = 1$ and $\lim_n \|x_n + x_0\| = 2$, then $\lim_n x_n = x_0$ weakly.

The results (1a) through (5a) are more or less classical and are due to Mil'man **[1938]**, Pettis **[1939a]**, Smulyan **[1939]**, **[1941]**, Giles **[1974]** and Dixmier **[1948]**. Many of the original proofs can be simplified by use of the Bishop-Phelps theorem; see Day **[1973]** or Diestel **[1975]**. Of related interest is a recent result of Singer **[1975]**: *If X^{***} is strictly convex and X^* contains a proper subspace Y for which the natural map of X into Y^* is an isometry then X is reflexive.*

Theorems (2b) and (3b) are due to Sullivan **[1976]** who obtains a number of related results using the "principle of local reflexivity" of Lindenstrauss and Rosenthal **[1969]**. Theorems (4b) and (5b) can be found in Diestel and Faires **[1974]** although (5b) is essentially due to Bishop and Phelps. Theorem (5b) should be compared to the theorem of Restrepo **[1964]** that states that if X has a Fréchet differentiable norm and X is separable then X^* is separable. Thus, if X is any Banach space with a Fréchet differentiable norm, then X^* has the Radon-Nikodým property by III.3.6. This last fact was approached via IV.1.1 by Leonard and Sundaresan **[1974]** who proved that if X has a Fréchet differentiable norm and $1 < p < \infty$ then the norm of $L_p(\mu, X)$ is Fréchet differentiable and $L_p(\mu, X)^* = L_q(\mu, X^*)$ $(p^{-1} + q^{-1} = 1)$. Turett **[1976]** has extended this result to the context of Orlicz spaces.

Here are three related unanswered questions: If X^* has the Radon-Nikodým property does X admit an equivalent very smooth or Fréchet differentiable norm? Does X admit an equivalent norm that gives X^* a weakly locally uniformly convex *dual* norm? Does X admit an equivalent norm that gives X^{***} a strictly convex *third dual* norm?

Asplund spaces. Following Namioka and Phelps **[1975]**, call a Banach space X an *Asplund space* if every continuous convex real-valued function on an open convex subset of X is Fréchet differentiable at all points of a dense G_δ subset of its domain. Asplund **[1968]** showed that if the word "G_δ" is dropped from this definition the resulting class of spaces is not enlarged. Further, Asplund proved that if X is an Asplund space (Asplund called these spaces "strong differentiability spaces"), then every weak*-closed bounded convex subset of X^* is the weak*-closed convex hull of its weak*-strongly exposed points. Namioka and Phelps **[1975]** proved the converse and went one step farther. They showed that if X is an Asplund space then every closed bounded convex subset of X^* is dentable. According to V.3.7 this proves that *if X is an Asplund space, then X^* has the Radon-Nikodým property.* Phelps asked whether the converse is true. Partial results in this direction were obtained by Morris **[1976]** and John and Zizler **[1976]** who showed that if X is a subspace of a weakly compactly generated Banach space and X^* has the Radon-Nikodým property, then X is an Asplund space. Shortly thereafter, the general case fell victim to Charles Stegall who proved that *if X^* has the Radon-Nikodým property, then X is an Asplund space.* Upon seeing Stegall's proof, I. Namioka was able to give a proof of his own. Since Namioka's argument is "tree-like", we shall include it here.

Namioka's starting point is a lemma from Namioka and Phelps **[1975]**: If X is not an Asplund space, then there is a bounded subset A of X^* and an $\varepsilon > 0$ with the property that diam $(U \cap A) > \varepsilon$ for every weak*-open subset U of X^* such that $U \cap A$ is nonempty. Take such a set A and let U be a nonempty relatively weak*-open subset of A. According to the Namioka-Phelps lemma, there are x_1^* and x_2^* in U and x_1 in X with $\|x_1\| = 1$ such that $x_1^*(x_1) - x_2^*(x_1) > \varepsilon$. Let $U_1 = U$, $U_2 = \{x^* \in U_1\colon x^*(x_1) > x_1^*(x_1) - \varepsilon/3\}$ and $U_3 = \{x^* \in U_1\colon x^*(x_1) < x_2^*(x_1) + \varepsilon/3\}$. By repeated applications of the above argument, one produces a sequence (U_n) of relatively weak*-open subsets of A and a sequence (x_n) in X such that

(a) $\|x_n\| = 1$ for all n,

(b) U_{2n} and U_{2n+1} are subsets of U_n for all n, and

(c) if $x^* \in U_{2n}$ and $y^* \in U_{2n+1}$, then $|x^*(x_n) - y^*(x_n)| \geqq \varepsilon/3$.

Now let Z be the (separable) subspace of X generated by the sequence (x_n). Then Z^* is not separable because it contains a bounded uncountable $(\varepsilon/3)$-net. Indeed for each "branch" $U_1 \supset U_{n_1} \supset U_{n_2} \supset \cdots$, pick $x^* \in \bigcap_j$ weak*-closure (U_{n_j}). If y^* and z^* come from different branches, then there is an n such that y^* is in the weak*-closure of U_{2n} and z^* is in the weak*-closure of U_{2n+1} (or vice-versa). In any case (c) guarantees that $\|y^* - z^*\|_{Z^*} \geqq \varepsilon/3$. Since there are evidently an uncountable number of "branches", Z is a separable subspace of X whose dual is not separable. Hence X^* fails to have the Radon-Nikodým property by Theorem 2.6.

In the notes and remarks section of Chapter III, we noted that if X admits an equivalent Fréchet differentiable norm, then X^* has the Radon-Nikodým property (this follows from Restrepo **[1964]** and III.3.6). This fact combined with Stegall's theorem above gives a short proof of a recent theorem of Ekeland and Lebourg **[1976]** who proved that if a Banach space admits an equivalent Fréchet differentiable norm, then it is an Asplund space. Left open is the following question: Does every Asplund space admit an equivalent Fréchet differentiable norm, or, equivalently, does X^* have the Radon-Nikodým property only if X admits an equivalent Fréchet differentiable norm? As unabashed optimists in Radon-Nikodým matters, we hope this question will be resolved in the affirmative.

The dual of a space with the Radon-Nikodým property. Quite often convexity and smoothness conditions appear in a natural duality. In fact the duality between smoothness conditions on X (in the context of Asplund spaces) and convexity conditions on X^* (in the context of spaces with the Radon-Nikodým property) seem to be in almost a total duality. Collier **[1976]** has studied smoothness conditions on X^* that are equivalent to the Radon-Nikodým property for X.

Let us agree that X^* is *weak*-Asplund* if each weak* lower semicontinuous convex functional is Fréchet differentiable on a norm-dense norm-G_δ subset of its points of norm continuity. Collier **[1976]** has proved that X has the Radon-Nikodým property if and only if X^* is a weak*-Asplund space. Consequently, *X fails to have the Radon-Nikodým property if and only if X admits an equivalent norm whose dual norm is nowhere Fréchet differentiable.* Collier's work is just a beginning; it seems that there must be more recognizable variations of the notion of weak*-Asplund spaces that would make Collier's results all the more valuable.

Hybrid spaces. Clarifying the theme of James **[1960]**, Lindenstrauss **[1971]** has shown that if X is a separable Banach space then there is a separable Banach space Y such that Y^{**} is isomorphic to $X^* \oplus Y$. This fact and its proof leads directly to the construction of the celebrated James space (see James **[1950]**, **[1951]**) and many other fundamental examples and counterexamples. For instance, as James and Lindenstrauss note, if a sequence $(X_n)_{n=0}^{\infty}$ of separable Banach spaces is defined by $X_0 = c_0$ and $X_n^{**} \approx X_{n-1}^* \oplus X_n$ as above, then $X_n^{(n+1)}$ is separable and $X_n^{(n+2)}$ is not separable. Thus by Stegall's Theorem 2.6 and the Dunford-Pettis Theorem III.3.1, $X_n^*, X_n^{**}, \cdots, X_n^{(n+1)}$ all have the Radon-Nikodým property and $X_n^{(n+2)}$ lacks the Radon-Nikodým property.

The James Tree space and the James Hagler spaces. For some time there was hope in the hearts of many that a separable Banach space with nonseparable dual must contain a copy of l_1. In the midst of a spate of fundamental counterexamples, James **[1974]** dashed these hopes by constructing the James Tree space JT. The space JT is a separable dual space, each of whose infinite dimensional subspaces contains a copy of l_2. In addition JT* is nonseparable and $\mathrm{JT}^{(n+2)} = \mathrm{JT}^{(n)} \oplus l_2(\Gamma)$ for a fixed uncountable set Γ and all n. Thus by Theorem 2.6 and Theorem III.3.7 all the even duals of JT have the Radon-Nikodým property and all the odd duals of JT fail the Radon-Nikodým property. If this is not enough, be disheartened by the fact that neither JT nor any of its higher duals contain copies of c_0 or l_1.

Not all of the above facts are due to James; some of them were discovered by Lindenstrauss and Stegall **[1976]** who also showed that there is a weakly measurable function with values in JT* that is not equivalent to any strongly measurable function in spite of the fact that JT* contains no copy of l_∞. Finally, it should be remarked that JT has a boundedly complete basis (x_n) (and so $\mathrm{JT} = [x_n^*]^*$ where $(x_n^*) \subseteq \mathrm{JT}^*$ is the sequence of coefficient functionals of the basis (x_n)) but that $[x_n^*]$ does not have the Kreĭn-Mil'man property.[6]

[6]This fact was proved by R.C. James, R.E. Huff and P.D. Morris.

Instead of cultivating l_2 seeds, Hagler [**1976**] decided to cultivate c_0 seeds and grew the James Hagler space JH. The space JH is a separable Banach space, each of whose infinite dimensional subspaces contains a copy of c_0. The space JH* is nonseparable but has the Schur property (weakly Cauchy sequences are norm convergent). By Stegall's Theorem 2.6, the space JH* lacks the Radon-Nikodým property. Since JH* has the Schur property, every operator from $L_1[0, 1]$ to JH* takes weakly compact sets into norm compact sets but not every such operator is representable. This shows that the converse to III.2.11 is false. The space JH has a host of other properties that make it an important example in the theory of Banach spaces.

An important consequence of the separability of the dual of a Banach space X is the existence of weakly Cauchy subsequences in each bounded sequence in X. The discovery of JT had killed the hopes that separability of dual and noncontainment of l_1 were identical for a separable Banach space, but then Rosenthal [**1974**] established the following striking substitute for real Banach spaces and Dor [**1974**] extended it to complex Banach spaces.

THEOREM (ROSENTHAL). *A Banach space X contains an isomorphic copy of l_1 if and only if there is a bounded sequence in X with no weakly Cauchy subsequence.*

An abundance of corollaries follow from Rosenthal's theorem. Here are some examples from Rosenthal [**1976**] and Odell and Rosenthal [**1975**].

THEOREM (ODELL-ROSENTHAL). *Let X be a separable Banach space. Then each of the following conditions is equivalent to the condition that X contains no copy of l_1.*

(i) *Bounded sequences in X have weakly Cauchy subsequences.*

(ii) *Bounded sequences in X^{**} have weak* convergent subsequences.*

(iii) *The space X is weak* sequentially dense in X^{**}.*

(iv) *Weak*-compact convex sets in X^* are the norm closed convex hulls of their extreme points.*

Haydon [**1976**] has extended some of the Odell-Rosenthal results to general Banach spaces. In particular, Haydon established (i) $\Leftrightarrow$ (iv).

By Corollary 2.8, the dual of a separable Banach space has the Radon-Nikodým property if and only if every closed bounded convex set of the dual is the closed convex hull of its extreme points. Comparing this to part (iv) of the Odell-Rosenthal theorem above, we see that there is an extremely fine dividing line between the statements "X contains no copy of l_1" and "X^* has the Radon-Nikodým property". Of course the James Tree space lives on this dividing line.

The ideas of Rosenthal [**1974**] were used by Johnson [**1976**] to show that *if X^* contains a copy of l_1 but no weak*-null sequence is equivalent to the unit vector basis of l_1, then X contains l_1*. A consequence of this result worth nothing is: *If X is a Grothendieck space whose dual space has the Radon-Nikodým property then X is reflexive; therefore smooth Grothendieck spaces are reflexive.*

One of the outstanding problems in the structure theory of Banach spaces is Rosenthal's problem: Does every infinite dimensional Banach space contain a copy of c_0, l_1 or an infinite dimensional reflexive space? A subsidiary question is whether every infinite dimensional Banach space contains a copy of c_0 or an infinite dimensional subspace with the Radon-Nikodým property. This question can be merged into Rosenthal's problem by proving (if possible) that a Banach space with the Radon-Nikodým property has a weakly sequentially complete subspace.

Trees in Banach spaces. Trees in Banach spaces were first investigated by James [**1974**]. The link between trees in Banach spaces and the Radon-Nikodým property seems to have first appeared implicitly in Maynard [**1973**]. In fact hindsight shows that Example V.1.7 is nothing but an easy special case of Maynard's Theorem V.3.1.

According to Corollary V.2.5 no Banach space with the Radon-Nikodým property contains a bounded infinite δ-tree. Conversely, the proof of Stegall's Theorem 2.6 shows that a dual space without the Radon-Nikodým property contains a bounded infinite δ-tree (the sequence $(T^*(\chi_{C_n})/\mu(C_n))$ is one such tree). It is unknown whether every Banach space without the Radon-Nikodým property has a bounded infinite δ-tree. On the other hand, by Theorem V.3.7 (and V.3.1) every Banach space without the Radon-Nikodým property contains a δ-tree-like structure sometimes called a bounded infinite δ-bush. (A bounded infinite δ-bush can be found inside any set that is not σ-dentable.)

A special type of bounded infinite δ-tree has been studied by Harrell and Karlovitz [**1972**], [**1974**], [**1975**]. Call a Banach space X *flat* if there is a function g: $[0,2] \to X$ such that $\|g(t)\| = 1$ for $0 \leqq t \leqq 2$, $g(0) = -g(2)$ and $\|g(s) - g(t)\| \leqq \|s - t\|$ for all $0 \leqq s, t \leqq 2$. This means that the unit sphere of X has an equator of length 4. The spaces $C[0, 1]$ and $L_1(\mu)$ are flat for nonatomic μ. Harrell and Karlovitz have shown that a Banach space is flat if and only if it contains a bounded infinite δ-tree and an associated system of linear functionals that separate certain parts of the tree from other parts. They call this an infinite supported tree. For more on this see Schäffer [**1976**].

Thus flat spaces lack the Radon-Nikodým property by V.2.5 and hence no flat dual space is separable. Not every dual space without the Radon-Nikodým property contains a subspace isomorphic to a flat space. This fact is due to Karlovitz [**1974**] who proved that if X is flat then so is X^*. Thus no dual of the James Tree space is isomorphic to a flat space since $\mathrm{JT}^{(2n+1)}$ has the Radon-Nikodým property for all $n \geqq 0$. Karlovitz [**1976**] has characterized Banach spaces X that contain a copy of l_1 in terms of a certain type of tree found in X^*. Of course from Lotz [**1976**] it follows that if a dual Banach lattice lacks the Radon-Nikodým property, it contains a subspace that is isomorphic to one of the flat spaces $C[0, 1]$ or $L_1[0, 1]$.

The Bishop-Phelps property. A Banach space X has the *Bishop-Phelps property* if for every Banach space Y and any closed bounded convex subset K of X the collection of all bounded linear operators from X to Y that achieve a maximum norm value on K is dense in the space of all operators from X to Y. In some way, Theorem 4.5 seems to hint of a connection between the Radon-Nikodým property and the Bishop-Phelps property. In addition, another hint of a connection lies in a paper of Lindenstrauss [**1963**]. In that paper he showed that l_1 and the reflexive Banach spaces all have the Bishop-Phelps property. Combining Theorem 3.3 and Theorem 4.2 with Lindenstrauss [**1963**, Theorem 2], one finds that if a Banach space X admits an equivalent locally uniformly convex norm and has the Bishop-Phelps property then it has the Radon-Nikodým property. But aside from Lindenstrauss [**1963**] and Zizler [**1973**] very little was known about the Bishop-Phelps property until the paper of Bourgain [**1976**] appeared very recently. Bourgain has shown that if a

Banach space has the Bishop-Phelps property, then it has the Radon-Nikodým property as well. He also obtained a partial converse by proving that if a Banach space X has the Radon-Nikodým property, then for every Banach space Y and for every closed bounded *absolutely* convex set C in X the operators from X to Y that achieve their maximum norm values on C form a dense G_δ set in the space of all bounded linear operators from X to Y. With his methods, Bourgain was also able to show that a closed bounded convex subset of an *arbitrary* Banach space is the closed convex hull of its strongly exposed points whenever each of its nonempty subsets is dentable. This fact is not a consequence of Theorem 3.3 since the arguments leading to Theorem 3.3 are global arguments. For a localization of Theorem 3.3, see Saab [**1976b**].

Some work of related interest has been done by Uhl [**1976**] who showed that if X is a strictly convex Banach space then the norm attaining operators from $L_1[0, 1]$ to X are dense in the space of all operators if and only if X has the Radon-Nikodým property. In addition, Diestel and Uhl [**1976**] have noted that for many Banach spaces the norm attaining compact operators are dense in the space of compact operators; whether this holds generally is not known.

6. Summary of equivalent formulations of the Radon-Nikodým property. Each of the following conditions is necessary and sufficient for a Banach space X to have the Radon-Nikodým property.

(1) Every closed linear subspace of X has the Radon-Nikodým property.

(2) Every *separable* closed linear subspace of X has the Radon-Nikodým property.

(3) Every function f: $[0, 1] \to X$ of bounded variation is differentiable almost everywhere.

(4) Every function f: $[0, 1] \to X$ of bounded variation is weakly differentiable off a fixed set of measure zero.

(5) Every absolutely continuous function f: $[0, 1] \to X$ is differentiable almost everywhere. In this case we have

$$f(b) - f(a) = \int_a^b f'(t)\,dt$$

for any a and $b \in [0, 1]$.

(6) Every absolutely continuous function f: $[0, 1] \to X$ is weakly differentiable off a fixed set of measure zero. In this case we have

$$x^*(f(b) - f(a)) = \int_a^b x^*f'(t)\,dt$$

for all a and $b \in [0, 1]$ and all $x^* \in X^*$.

(7) Every bounded linear operator from $L_1[0, 1]$ to X factors through l_1.

(7a) For any finite measure space (Ω, Σ, μ) every bounded linear operator from $L_1(\mu)$ to X factors through l_1.

(8) The absolutely summing, integral and nuclear operators from $C[0, 1]$ to X are identical classes.

(8a) For every compact Hausdorff space Ω, the absolutely summing, integral and nuclear operators from $C(\Omega)$ to X are identical classes.

(9) For every Banach space Y the Pietsch integral operators from Y to X coincide (isometrically) with the nuclear linear operators from Y to X.

(10) Every bounded subset of X is dentable.

(10a) Every closed bounded convex subset of X is dentable.

(10b) Every bounded subset of X is σ-dentable.

(11) Every nonempty closed bounded subset of X contains an extreme point of its closed convex hull.

(11a) If D is a nonempty closed bounded subset of X, then some bounded linear functional on X assumes a maximum value on D.

(12) Every nonempty closed bounded convex subset of X is the closed convex hull of its denting points.

(13) Every nonempty closed bounded convex subset of X has a strongly exposed point.

(13a) Every nonempty closed bounded convex subset of X is the closed convex hull of its strongly exposed points.

(14) For each nonempty closed bounded absolutely convex subset B of X and every Banach space Y, the collection of operators that attain their maximum norm on B is dense in the space of operators from X to Y.

(15) For each closed bounded convex subset K of X, the set of elements in X^* that strongly expose some point of K is dense in X^*.

(16) The space X does *not* have the following property: There exists a nonempty closed subset K of the norm-interior of a closed bounded convex set C in X with $\overline{\text{co}}\, K = C$.

(17) For any finite measure space (Ω, Σ, μ) each uniformly integrable $L_1(\mu, X)$-bounded martingale is $L_1(\mu, X)$-convergent.

In case X is isomorphic to the dual of some Banach space Y, then (1)–(17) are equivalent to the following:

(18) Every separable subspace of Y has a separable dual.

(19) Every separable subspace of X is isomorphic to a subspace of a separable dual.

(20) Every nonempty closed bounded convex subset of X has an extreme point.

(20a) Every nonempty closed bounded convex subset of X is the closed convex hull of its extreme points.

(21) For some p with $1 \leqq p < \infty$, we have $L_p([0, 1], Y)^* = L_q([0, 1], X)$ where $(1/p) + (1/q) = 1$.

(21a) For every finite measure space (Ω, Σ, μ) and all p with $1 \leqq p < \infty$, we have $L_p(\mu, Y)^* = L_q(\mu, X)$ where $(1/p) + (1/q) = 1$.

7. The Radon-Nikodým property for specific spaces.

Spaces that have the Radon-Nikodým property.

Reflexive spaces.

Separable duals.

Weakly compactly generated (WCG) duals.

Dual subspaces of WCG spaces.

Locally uniformly convex duals.

Weakly locally uniformly convex duals.

Duals of spaces with Fréchet differentiable norm.

Spaces with a boundedly complete basis.

$l_1(\Gamma)$, Γ any set.

$c_p(H)$, $1 \leqq p < \infty$, the Schatten classes.

$H^p(D)$, D: disk, Hardy classes, $1 \leqq p < \infty$.

$(\Sigma \oplus X_\alpha)_p$, $1 \leqq p < \infty$, if X_α has the Radon-Nikodým property.

The space of unconditionally convergent series in X if X has the Radon-Nikodým property.

N(X), nuclear operators on a reflexive space X with the approximation property.

Quasi-reflexive spaces.

JT, James Tree space.

$\mathscr{L}(l_p, l_q)$, $1 \leqq q < p < \infty$.

X^{**}, X^* when X^{**}/X is separable (Kuo **[1974]**).

X^* if the unit ball of X^{**} is Eberlein compact in the weak* topology (Kuo **[1974]**).

$L_p(\mu, X)$, $1 < p < \infty$, if X has the Radon-Nikodým property.

Spaces that do not have the Radon-Nikodým property.

$L_1[0, 1]$, $BV_0[0, 1]$.

$L_1(\mu)$, μ not purely atomic.

c_0, c, l_∞, $L_\infty[0, 1]$.

$C(\Omega)$, Ω infinite compact Hausdorff.

$K(X)$, compact operators on X, $X = l_p$, L_p or $C(\Omega)$.

$\mathscr{L}(X)$, bounded operators on X, $X = l_p$, L_p or $C(\Omega)$.

JT*, dual of James Tree space.

JH*, dual of James Hagler space.

$H_\infty(D)$, $A(D)$—the disk algebra.

X^* if X contains l_1.

VIII. TENSOR PRODUCTS OF BANACH SPACES

A well-known line of discourse sometimes attributed to R. P. Kaufman runs as follows: "Measures are easy; vectors are easy, and tensors are easy. Therefore vector measures are easy." To a certain point this argument is right—when a vector measure arises as a tensor product of vectors and measures, it is a very well-behaved measure, an "easy" measure. Indeed, any vector measure arising as an indefinite Bochner integral is an "easy" measure. But we have seen that the main problem connected with measures arising as indefinite Bochner integrals is recognizing them. Tensor products are not much help in this. Furthermore, the preceding chapters lay a solid base for a stronger statement: The beautiful theorems about vector measures are not proved by tensor product arguments. Building on this theme, this chapter takes the point of view that many important theorems about tensor products of Banach spaces are proved by vector measure arguments.

The chapter opens with an introduction to the least and greatest reasonable crossnorms on the tensor product of two Banach spaces. §2 deals with some of the duality manipulations so typical in the theory of tensor products. The approximation property and the metric approximation property are the central objects of study in §3. Finally, in the last section, we shall study applications of the theory of vector measures to the theory of tensor products of Banach spaces. In addition we shall see both the theories of vector measures and of tensor products at work in Banach space theory.

1. The least and greatest crossnorms. This section is an introduction to the study of the algebraic tensor product of two Banach spaces equipped with two "reasonable" norms—the "least crossnorm" and the "greatest crossnorm." We will be mainly concerned with terminology, notation and other formalities. Aside from a few examples, most of the action will be postponed to the later sections. Throughout, X and Y are Banach spaces. Also we shall forego any formal definition of the algebraic tensor product $X \otimes Y$ of X and Y.

DEFINITION 1. A norm α on $X \otimes Y$ is called a *reasonable crossnorm* whenever α satisfies the conditions:

(RI) $\alpha(x \otimes y) \leqq \|x\| \, \|y\|$ for all $x \in X$ and $y \in Y$, and

(R2) if $x^* \in X^*$ and $y^* \in Y^*$, then $x^* \otimes y^* \in (X \otimes Y, \alpha)^*$ and has functional norm $\leqq \|x^*\| \, \|y^*\|$.

PROPOSITION 2. *Suppose that α is a reasonable crossnorm on $X \otimes Y$. Then*

(i) $\alpha(x \otimes y) = \|x\| \, \|y\|$ *for all $x \in X$ and $y \in Y$.*

(ii) *If $x^* \in X^*$ and $y^* \in Y^*$ then the norm of $x^* \otimes y^*$ as a member of $(X \otimes Y, \alpha)^*$ is $\|x^*\| \, \|y^*\|$.*

(iii) *If α^* is the norm induced on $X^* \otimes Y^*$ by considering $X^* \otimes Y^*$ as a linear subspace of $(X \otimes Y, \alpha)^*$, then α^* is a reasonable crossnorm on $X^* \otimes Y^*$.*

PROOF. To prove (i), let $x \in X$ and $y \in Y$. Choose $x^* \in X^*$ and $y^* \in Y^*$, each of norm one, such that $x^*(x) = \|x\|$ and $y^*(y) = \|y\|$. Then by (R2) we have $x^* \otimes y^* \in (X \otimes Y, \alpha)^*$ and the functional norm of $x^* \otimes y^* \leqq 1$. Thus we have

$$\|x\| \, \|y\| = |x^*(x)y^*(y)| = |(x^* \otimes y^*)(x \otimes y)| \leqq \alpha(x \otimes y).$$

An appeal to (Rl) provides the reverse inequality.

To prove (ii), let $x^* \in X^*$, $y^* \in Y^*$. Choose (x_n), (y_n) from X, Y, respectively, such that $\|x_n\| = 1 = \|y_n\|$ and $\|x^*\| = \lim_n x^*(x_n)$ and $\|y^*\| = \lim_n y^*(y_n)$. Then we have

$$\begin{aligned}\|x^*\| \, \|y^*\| &= \lim_n |x^*(x_n)| \, |y^*(y_n)| = \lim_n |(x^* \otimes y^*)(x_n \otimes y_n)| \\ &\leqq \overline{\lim_n} \, \alpha(x_n \otimes y_n) \text{norm}(x^* \otimes y^*) \leqq \text{norm}(x^* \otimes y^*).\end{aligned}$$

This proves that the functional norm of $x^* \otimes y^* = \|x^*\| \, \|y^*\|$.

(iii) It is plain that condition (Rl) for α^* is just condition (R2) for α. Hence we need only show that, if $x^{**} \in X^{**}$ and $y^{**} \in Y^{**}$, then $x^{**} \otimes y^{**} \in (X^* \otimes Y^*, \alpha^*)^*$ and the functional norm of $x^{**} \otimes y^{**}$ is no more than $\|x^{**}\| \, \|y^{**}\|$.

Let $x^{**} \in X^{**}$ and $y^{**} \in Y^{**}$. With the help of Goldstine's theorem choose nets (x_β) in X and (y_γ) in Y such that $\|x_\beta\| \leqq \|x^{**}\|$, $\|y_\gamma\| \leqq \|y^{**}\|$, $\lim_\beta x_\beta = x^{**}$ (weak*) and $\lim_\gamma y_\gamma = y^{**}$ (weak*). If $u^* \in X^* \otimes Y^*$, u^* is of the form $u^* = \sum_{i=1}^n x_i^* \otimes y_i^*$ for some $x_1^*, \cdots, x_n^* \in X^*$ and $y_1^*, \cdots, y_n^* \in Y^*$. Thus we have

$$\begin{aligned}|(x^{**} \otimes y^{**})(u^*)| &= \left|\sum_{i=1}^n x^{**}(x_i^*)y^{**}(y_i^*)\right| \\ &= \left|\sum_{i=1}^n \lim_\beta x_i^*(x_\beta) \lim_\gamma y_i^*(y_\gamma)\right| \\ &= \lim_\beta \lim_\gamma \left|\sum_{i=1}^n x_i^*(x_\beta) y_i^*(y_\gamma)\right| \\ &\leqq \overline{\lim_{\beta,\gamma}} \, |(x_\beta \otimes y_\gamma)(u^*)| \\ &\leqq \overline{\lim_{\beta,\gamma}} \, \|x_\beta\| \, \|y_\gamma\| \alpha^*(u^*) \\ &\leqq \|x^{**}\| \, \|y^{**}\| \alpha^*(u^*).\end{aligned}$$

This establishes (R2) for α^*.

In view of (iii), we refer to α^* as the dual crossnorm.

Of particular interest are two reasonable crossnorms: the least reasonable crossnorm and the greatest reasonable crossnorm. A description of each is next.

Let $u \in X \otimes Y$. Define $\lambda(u)$ by

$$\lambda(u) = \sup\{|(x^* \otimes y^*)(u)| : x^* \in X^*, y^* \in Y^*, \|x^*\|, \|y^*\| \leqq 1\}.$$

Obviously λ is a norm on $X \otimes Y$, but in addition λ is a reasonable crossnorm and is the *least* reasonable crossnorm.

PROPOSITION 3. *The norm λ is a reasonable crossnorm on $X \otimes Y$. Moreover if α is any reasonable crossnorm on $X \otimes Y$, then $\lambda(u) \leqq \alpha(u)$ for all $u \in X \otimes Y$.*

PROOF. Let $x \in X$ and $y \in Y$. Then

$$\begin{aligned}\lambda(x \otimes y) &= \sup\{|(x^* \otimes y^*)(x \otimes y)| : x^* \in X^*, y^* \in Y^*, \|x^*\|, \|y^*\| \leqq 1\} \\ &= \sup\{|x^*(x)y^*(y)| : x^* \in X^*, y^* \in Y^*, \|x^*\|, \|y^*\| \leqq 1\} \\ &\leqq \|x\| \, \|y\|.\end{aligned}$$

This shows that λ satisfies (Rl).

If $x^* \in X^*$ and $y^* \in Y^*$, for any $u \in X \otimes Y$, then

$$\begin{aligned}|(x^* \otimes y^*)(u)| &= \|x^*\| \, \|y^*\| \, |((x^*/\|x^*\|) \otimes (y^*/\|y^*\|))(u)| \\ &\leqq \|x^*\| \, \|y^*\| \, \lambda(u)\end{aligned}$$

by definition of λ. Thus $x^* \otimes y^* \in (X \otimes Y, \lambda)^*$ and $\lambda^*(x^* \otimes y^*) \leqq \|x^*\| \, \|y^*\|$. Therefore (R2) is also satisfied by λ. Thus λ is a reasonable crossnorm.

Finally, if α is any reasonable crossnorm on $X \otimes Y$, then for all $u \in X \otimes Y$, $x^* \in X^*$ and $y^* \in Y^*$ with $\|x^*\|, \|y^*\| \leqq 1$, we have $x^* \otimes y^* \in (X \otimes Y, \alpha)^*$ and $|(x^* \otimes y^*)(u)| \leqq \alpha(u)$. Consequently,

$$\lambda(u) = \sup\{|(x^* \otimes y^*)(u)| : x^* \in X^*, y^* \in Y^*, \|x^*\|, \|y^*\| \leqq 1\} \leqq \alpha(u).$$

Usually $X \otimes Y$ equipped with the least reasonable crossnorm is incomplete. Its completion will be denoted by $X \check{\otimes} Y$.

EXAMPLE 4. *The space $L_1(\mu) \check{\otimes} X$ as a space of vector measures.* We shall show that if (Ω, Σ, μ) is a finite measure space, $L_1(\mu) \check{\otimes} X$ is (isometrically isomorphic to) the space $K(\mu, X)$ of all μ-continuous vector measures $G: \Sigma \to X$ whose range is relatively compact, equipped with the semivariation norm.

PROOF. For each $u = \sum_{i=1}^n f_i \otimes x_i \in L_1(\mu) \otimes X$ ($f_i \in L_1(\mu)$, $x_i \in X$), define a vector measure $G_u: \Sigma \to X$ by $G_u(E) = \int_E \sum_{i=1}^n x_i f_i \, d_\mu$. Since G_u has a finite dimensional range, we see that $G_u \in K(\mu, X)$. Moreover

$$\begin{aligned}\|G_u\|(\Omega) &= \sup_{\|x^*\| \leqq 1} |x^* G_u|(\Omega) = \sup_{\|x^*\| \leqq 1} \int_\Omega \left| \sum_{i=1}^n x^*(x_i) f_i \right| d\mu \\ &= \sup_{\|x^*\| \leqq 1} \sup_{g \in L_\infty(\mu), \|g\|_\infty \leqq 1} \int_\Omega \sum_{i=1}^n x^*(x_i) f_i g \, d\mu \\ &= \sup\{(y^* \otimes x^*)(u) : y^* \in L_1(\mu)^*, x^* \in X^*, \|y^*\|, \|x^*\| \leqq 1\} \\ &= \lambda(u).\end{aligned}$$

Thus the mapping $u \to G_u$ takes $L_1(\mu) \check{\otimes} X$ isometrically onto a closed subspace of $K(\mu, X)$.

To complete the discussion, we must show that a dense subset of $K(\mu, X)$ is in the image of $L_1(\mu) \otimes X$ under this mapping. To this end, let $G \in K(\mu, X)$ and, for each partition π, define

$$G_\pi(E) = \sum_{A \in \pi} \frac{G(A)}{\mu(A)} \mu(E \cap A).$$

Then $G_\pi(\Sigma) \subseteq \operatorname{co}(G(\Sigma))$ and $G_\pi \in K(\mu, X)$. Define $T: L_\infty(\mu) \to X$ by $T(f) = \int_\Omega f\, dG$, $f \in L_\infty(\mu)$. According to VI.1, T is compact and by Schauder's theorem $T^*: X^* \to L_\infty(\mu)^*$ is compact. Now, for each $x^* \in X^*$, $x^*G = T^*(x^*)$. Since $G \ll \mu$, the compact operator T^* has its range in $L_1(\mu)$. Since $\lim_\pi E_\pi$ is the identity in the strong operator topology (see III.2) $\lim_\pi E_\pi T^*g = T^*g$ uniformly in $\|g\|_1 \leqq 1$. A simple check similar to that found in III.2.2 reveals that $E_\pi T^* = (TE_\pi)^*$. Consequently $\lim_\pi TE_\pi = T$ in the uniform operator topology. Another simple computation shows $TE_\pi(f) = \int_\Omega f\, dG_\pi$ for each $f \subset L_\infty(\mu)$. Since $\|T - TE_\pi\| = \|G - G_\pi\|(\Omega)$, it follows that $\lim_\pi \|G - G_\pi\|(\Omega) = 0$.

Finally, each measure G_π is obviously in the range of the mapping from $L_1(\mu) \otimes X$ to $K(\mu, X)$ discussed above and thus this mapping carries $L_1(\mu) \otimes X$ onto a *dense* subset of $K(\mu, X)$.

Example 4 can be redirected a bit. Let (Ω, Σ, μ) be a finite measure space and $P_1(\mu, X)$ be the space of all measurable Pettis integrable functions $f: \Omega \to X$ equipped with the norm

$$\|f\|_{P_1} = \sup_{\|x^*\| \leqq 1} \int_\Omega |x^*f|\, d\mu.$$

Simple functions are dense in $P_1(\mu, X)$ because $P_1(\mu, X)$ consists of measurable functions and $\lim_{\mu(E) \to 0} \|f\chi_E\|_{P_1} = 0$ for all $f \in P_1(\mu, X)$. Now for each $f \in P_1(\mu, X)$, define $G_f: \Sigma \to X$ by $G_f(E) = (\text{Pettis})\text{-}\int_E f\, d\mu$. Then $\|G_f\|(\Omega) = \|f\|_{P_1}$ and if f is a simple function, then $G_f \in K(\mu, X)$. Thus the mapping $f \to G_f$ takes $P_1(\mu, X)$ isometrically into $K(\mu, X)$. For reasons similar to those above this mapping takes $P_1(\mu, X)$ onto a dense subset of $K(\mu, X)$. This is summarized as follows.

THEOREM 5. *Let (Ω, Σ, μ) be a finite measure space. The following spaces are identical spaces with identical norms*:

(a) $K(\mu, X)$,

(b) *the completion of* $P_1(\mu, X)$,

(c) $L_1(\mu) \check{\otimes} X$, *and*

(d) *the subspace of* $\mathscr{L}(L_\infty(\mu); X)$ *consisting of compact weak* to weakly continuous operators.*

That (d) is equivalent to the others is a consequence of VI.1.6.

EXAMPLE 6. *Let Ω be a compact Hausdorff space. The space $C(\Omega) \check{\otimes} X$ is isometrically isomorphic to the Banach space $C_X(\Omega)$ of continuous functions $f: \Omega \to X$ equipped with the norm $\|f\| = \sup\{\|f(\omega)\|_X : \omega \in \Omega\}$.*

To see this, define $J: C(\Omega) \otimes X \to C_X(\Omega)$ by $J(\sum_{i=1}^n f_i \otimes x_i)(\omega) = \sum_{i=1}^n f_i(\omega)x_i$. Then one has

$$\begin{aligned}\left\|J\left(\sum_{i=1}^n f_i \otimes x_i\right)\right\| &= \sup\left\{\left\|\sum_{i=1}^n f_i(\omega)x_i\right\| : \omega \in \Omega\right\} \\ &= \sup\left\{\left|x^*\left(\sum_{i=1}^n f_i(\omega)x_i\right)\right| : \omega \in \Omega,\ x^* \in X^*,\ \|x^*\| \leqq 1\right\} \\ &= \sup\left\{\left|\sum_{i=1}^n f_i(\omega)x^*(x_i)\right| : \omega \in \Omega,\ x^* \in X^*,\ \|x^*\| \leqq 1\right\} \\ &= \sup\left\{\sup\left\{\left|\sum_{i=1}^n x^*(x_i) f_i(\omega)\right| : \omega \in \Omega\right\} : x^* \in X^*,\ \|x^*\| \leqq 1\right\} \\ &= \sup\left\{\left\|\sum_{i=1}^n x^*(x_i)f_i\right\|_\infty : x^* \in X^*,\ \|x^*\| \leqq 1\right\} \\ &= \sup\left\{\left|\nu\left(\sum_{i=1}^n x^*(x_i)f_i\right)\right| : \nu \in C(\Omega)^*,\ x^* \in X^*,\ \|\nu\|,\ \|x^*\| \leqq 1\right\} \\ &= \sup\left\{\left|\sum_{i=1}^n x^*(x_i)\nu(f_i)\right| : \nu \in C(\Omega)^*,\ x^* \in X^*,\ \|\nu\|,\ \|x^*\| \leqq 1\right\} \\ &= \lambda\left(\sum_{i=1}^n f_i \otimes x_i\right).\end{aligned}$$

Thus $C(\Omega) \check{\otimes} X$ is isometrically isomorphic with the closed linear subspace of $C_X(\Omega)$ generated by the family of functions of the form $\sum_{i=1}^n f_i(\cdot)x_i$ where $f_1, \cdots, f_n \in C(\Omega)$ and $x_1, \cdots, x_n \in X$; once this family is shown to be dense in $C_X(\Omega)$, the example will be complete.

Let $f: \Omega \to X$ be continuous and $\varepsilon > 0$. Since $f(\Omega)$ is compact there exist $\omega_1, \cdots, \omega_n \in \Omega$ such that, for any $\omega \in \Omega$, $\|f(\omega) - f(\omega_j)\| \leqq \varepsilon/2$ for some j, $1 \leqq j \leqq n$. Let $U_j = \{\omega \in \Omega : \|f(\omega) - f(\omega_j)\| < \varepsilon\}$ for $1 \leqq j \leqq n$. Then $\{U_1, \cdots, U_n\}$ is an open cover of Ω. Let $\{f_1, \cdots, f_n\} \subseteq C(\Omega)$ be a partition of unity subordinate to the open cover $\{U_1, \cdots, U_n\}$, i.e., $\sum_{i=1}^n f_i(\omega) \equiv 1$, $0 \leqq f_j(\omega) \leqq 1$ for all $\omega \in \Omega$ and all j, $1 \leqq j \leqq n$, and $f_j(\omega) = 0$ if $\omega \notin U_j$ for $1 \leqq j \leqq n$. If $g: \Omega \to X$ is defined by $g(\omega) = \sum_{i=1}^n f_i(\omega) f(\omega_i)$, it is straightforward to verify that, for each $\omega \in \Omega$, $\|g(\omega) - f(\omega)\| < \varepsilon$.

The space $X \check{\otimes} Y$ is called the *injective* tensor product of X and Y for reasons justified in

PROPOSITION 7. *Let W be a closed linear subspace of X. Then $W \check{\otimes} Y$ is a closed linear subspace of $X \check{\otimes} Y$ and $Y \check{\otimes} W$ is a closed linear subspace of $Y \check{\otimes} X$.*

PROOF. Since both statements are proved by similar routine computations, only the routine computation proving the first statement will be given here. Let $u = \sum_{i=1}^n w_i \otimes y_i \in W \otimes Y$. Then

$$\begin{aligned}\lambda_{W\otimes Y}(u) &= \sup\left\{\left|\sum_{i=1}^n w^*(w_i)y^*(y_i)\right| : w^* \in W^*,\ y^* \in Y^*,\ \|w^*\|,\ \|y^*\| \leqq 1\right\} \\ &= \sup\left\{\left|w^*\left(\sum_{i=1}^n y^*(y_i)w_i\right)\right| : w^* \in W^*,\ y^* \in Y^*,\ \|w^*\|,\ \|y^*\| \leqq 1\right\} \\ &= \sup\left\{\sup\left\{\left|w^*\left(\sum_{i=1}^n y^*(y_i)w_i\right)\right| : w^* \in W^*,\ \|w^*\| \leqq 1\right\} : y^* \in Y^*, \|y^*\| \leqq 1\right\} \\ &= \sup\left\{\left\|\sum_{i=1}^n y^*(y_i)w_i\right\| : y^* \in Y^*,\ \|y^*\| \leqq 1\right\}\end{aligned}$$

$$= \sup\left\{\sup\left\{\left|x^*\left(\sum_{i=1}^n y^*(y_i)w_i\right)\right| : x^* \in X^*, \|x^*\| \leqq 1\right\} : y^* \in Y^*, \|y^*\| \leqq 1\right\}$$

$$= \sup\left\{\left|\sum_{i=1}^n x^*(w_i)y^*(y_i)\right| : x^* \in X^*, y^* \in Y^*, \|x^*\|, \|y^*\| \leqq 1\right\}$$

$$= \lambda_{X\otimes Y}(u).$$

This completes the proof.

There is another way to describe the least reasonable crossnorm. Let $\mathscr{B}(X, Y)$ be the Banach space of all continuous bilinear functionals on $X \times Y$ under the norm $\|\cdot\|$ defined for $\psi \in \mathscr{B}(X, Y)$ by

$$\|\psi\| = \sup\{|\psi(x, y)| : x \in X, y \in Y, \|x\|, \|y\| \leqq 1\}.$$

Note that each member of $X \otimes Y$ acts naturally as a continuous bilinear functional on $X^* \otimes Y^*$. Moreover, if $u \in X \otimes Y$, then $\lambda(u) = \|u\|$ where u is viewed as a member of $\mathscr{B}(X^*, Y^*)$. Thus there is a natural isometry of $X \check{\otimes} Y$ into $\mathscr{B}(X^*, Y^*)$.

This observation comes in handy in the examination of the greatest reasonable crossnorm.

Define γ on $X \otimes Y$ by $\gamma(u) = \sup\{|\psi(u)| : \psi \in \mathscr{B}(X, Y), \|\psi\| \leqq 1\}$ for $u \in X \otimes Y$. Obviously γ is a seminorm on $X \otimes Y$ but it is also the greatest reasonable crossnorm on $X \otimes Y$, as we shall presently see.

PROPOSITION 8. *The seminorm γ is a reasonable crossnorm on $X \otimes Y$. Moreover if α is any reasonable crossnorm on $X \otimes Y$, then $\alpha(u) \leqq \gamma(u)$ for all $u \in X \otimes Y$.*

PROOF. Since $X^* \check{\otimes} Y^*$ is isometric to a closed subspace of $\mathscr{B}(X^{**}, Y^{**})$, we may write $\|x^* \otimes y^*\|_{\mathscr{B}(X^{**}, Y^{**})} = \|x^*\| \, \|y^*\|$. Consequently the restriction $x^* \otimes y^*|_{\mathscr{B}(X,Y)}$ to $X \otimes Y$ satisfies

$$\|x^* \otimes y^*|_{X\otimes Y}\|_{\mathscr{B}(X,Y)} \leqq \|x^*\| \, \|y^*\|.$$

Thus if $u \in X \otimes Y$, then

$$\lambda(u) = \sup\{|x^* \otimes y^*(u)| : x^* \in X^*, y^* \in Y^*, \|x^*\|, \|y^*\| \leqq 1\}$$
$$\leqq \sup\{|\psi(u)| : \psi \in \mathscr{B}(X, Y), \|\psi\| \leqq 1\} = \gamma(u).$$

This proves that the seminorm γ is a norm on $X \otimes Y$.

Also if $x \in X$ and $y \in Y$, then

$$\gamma(x \otimes y) = \sup\{|\psi(x, y)| : \psi \in \mathscr{B}(X, Y), \|\psi\| \leqq 1\}$$
$$= \sup\{\|x\| \, \|y\| \psi(x/\|x\|, y/\|y\|) : \psi \in \mathscr{B}(X, Y), \|\psi\| \leqq 1\}$$
$$= \|x\| \, \|y\|.$$

Hence γ is a crossnorm on $X \otimes Y$. Since $\lambda(u) \leqq \gamma(u)$ for $u \in X \otimes Y$, a straightforward computation shows γ is reasonable. This proves the first statement.

To prove the second statement, let α be any reasonable crossnorm on $X \otimes Y$ and let $u \in X \otimes Y$. Choose $\psi^* \in (X \otimes Y, \alpha)^*$ such that $\psi^*(u) = \alpha(u)$ and $\|\psi^*\|_{(X\otimes Y,\alpha)^*} = 1$. Define $\psi \in \mathscr{B}(X, Y)$ by $\psi(x, y) = \psi^*(x \otimes y)$. Then

$$|\psi(x, y)| = |\psi^*(x \otimes y)| \leqq \alpha(x \otimes y)\|\psi^*\|_{(X\otimes Y,\alpha)^*}$$
$$= \alpha(x \otimes y) = \|x\| \, \|y\|$$

for each $x \in X$, $y \in Y$. It follows that $\psi \in \mathscr{B}(X, Y)$ and $\|\psi\| \leqq 1$. Hence $\alpha(u) = |\psi^*(u)| = |\psi(u)| \leqq \gamma(u)$ and the proof is complete.

The completion of $X \otimes Y$ under γ will be denoted by $X \hat{\otimes} Y$ and called the *projective* tensor product of X and Y. The completed norm on $X \hat{\otimes} Y$ will still be denoted by γ. For practical work with γ, the following proposition shows that γ is greater than all crossnorms on $X \otimes Y$ and further gives a useful alternative version of γ.

PROPOSITION 9. (a) *If $u \in X \otimes Y$, then*

$$\gamma(u) = \inf\left\{\sum_{i=1}^{n} \|x_i\| \, \|y_i\| : x_i \in X, y_i \in Y, u = \sum_{i=1}^{n} x_i \otimes y_i\right\}.$$

(b) *If $u \in X \hat{\otimes} Y$ and $\varepsilon > 0$, then there exist sequences (x_n) in X and (y_n) in Y such that $\lim_n \|x_n\| = 0 = \lim_n \|y_n\|$, $u = \sum_{n=1}^{\infty} x_n \otimes y_n$ in γ-norm and $\gamma(u) \leqq \sum_{n=1}^{\infty} \|x_n\| \, \|y_n\| \leqq \gamma(u) + \varepsilon$.*

PROOF. To prove (a), let $\alpha(u)$ denote the expression on the right-hand side of the statement. Clearly, α is a seminorm on $X \otimes Y$ that satisfies $\alpha(x \otimes y) \leqq \|x\| \, \|y\|$. Also, if $u = \sum_{i=1}^{n} x_i \otimes y_i$, then

$$\gamma(u) \leqq \sum_{i=1}^{n} \gamma(x_i \otimes y_i) = \sum_{i=1}^{n} \|x_i\| \, \|y_i\|.$$

Accordingly, $\gamma(u) \leqq \alpha(u)$ and α is a reasonable crossnorm on $X \otimes Y$. Since γ is the greatest reasonable crossnorm, $\alpha = \gamma$ and the proof of (a) is finished.

To prove (b), select a sequence $(u_n) \subseteq X \otimes Y$ such that $\gamma(u - u_n) < \varepsilon/2^{n+3}$ for all n. With the help of part (a) write $u_1 = \sum_{i=1}^{i(1)} x_i \otimes y_i$ where $\sum_{i=1}^{i(1)} \|x_i\| \, \|y_i\| \leqq \gamma(u_1) + \varepsilon/2^4 \leqq \gamma(u) + \varepsilon/2^3$. For each $n \geqq 1$, note that

$$\begin{aligned}\gamma(u_{n+1} - u_n) &\leqq \gamma(u - u_{n+1}) + \gamma(u - u_n) \\ &\leqq \varepsilon/2^{n+4} + \varepsilon/2^{n+3} < \varepsilon/2^{n+2}.\end{aligned}$$

This fact and (a) enable us to write

$$u_{n+1} - u_n = \sum_{i(n)+1}^{i(n+1)} x_i \otimes y_i$$

where $\sum_{i=i(n)+1}^{i(n+1)} \|x_i\| \, \|y_i\| < \varepsilon/2^{n+2}$. Obviously the series $u_1 + \sum_{n=1}^{\infty}(u_{n+1} - u_n)$ converges absolutely to u. Moreover straightforward computations based on the above inequalities and the triangle inequality establish that $\gamma(u) \leqq \sum_{i=1}^{\infty} \|x_i\| \, \|y_i\| < \gamma(u) + \varepsilon$. If we can show that the sequences (x_n) and (y_n) can be selected so that $\lim_n \|x_n\| = 0 = \lim_n \|y_n\|$, we shall have finished the proof. To establish this technicality note that if $x_i \otimes y_i$ is written as $n^2 \cdot ((x_i/n) \otimes (y_i/n))$, stringing together sufficiently many parts of $x_i \otimes y_i$ yields another representation of u with the desired additional property.

The term "projective tensor product" derives from the fact that if Z is a closed linear subspace of X then $(X/Z) \hat{\otimes} Y$ is a quotient of $X \hat{\otimes} Y$ and $Y \hat{\otimes} (X/Z)$ is a quotient of $Y \hat{\otimes} X$ for any Banach space Y. The proof of this assertion is given at the end of this section after the idea of tensoring operators has been discussed. The subspace structure of X and Y is not usually preserved when taking projective tensor products. If Z is a closed linear subspace of Y then $\gamma_{X \otimes Y}(u) \leqq \gamma_{X \otimes Z}(u)$ for any $u \in X \otimes Z$. It is not true in general that $\gamma_{X \otimes Y}(u)$ equals $\gamma_{X \otimes Z}(u)$ as the results of the next section will show. Sometimes it is true; for instance, if Z is a complemented subspace of Y by a norm one projection, equality holds. Another situation in which equality holds is discussed next.

EXAMPLE 10. *$L_1(\mu) \hat{\otimes} X$ is isometrically isomorphic to the Banach space $L_1(\mu,X)$.* (Here (Ω, Σ, μ) can be any measure space.) To prove this, first note that the natural inclusion J of $L_1(\mu) \otimes X$ into $L_1(\mu, X)$ is a bounded linear operator of norm $\leqq 1$. Moreover J takes the dense subspace of $L_1(\mu) \hat{\otimes} X$ consisting of elements of the form $\sum_{i=1}^n \chi_{A_i} \otimes x_i$ where $A_1, \cdots, A_n$ are disjoint sets in Σ of finite μ-measure and $x_1, \cdots, x_n \in X$ *onto* the dense subspace of simple functions in $L_1(\mu, X)$. Thus to show that J is an isometry of $L_1(\mu) \hat{\otimes} X$ onto $L_1(\mu, X)$ it suffices to show that

$$\gamma\Big(\sum_{i=1}^n \chi_{A_i} \otimes x_i\Big) \leqq \Big\| J\Big(\sum_{i=1}^n \chi_{A_i} \otimes x_i\Big)\Big\|_{L_1(\mu, X)}$$

where $A_1, \cdots, A_n$ are disjoint sets of finite measure in Σ and $x_1, \cdots, x_n \in X$. This is easy:

$$\begin{aligned}
\gamma\Big(\sum_{i=1}^n \chi_{A_i} \otimes x_i\Big) &\leqq \sum_{i=1}^n \gamma(\chi_{A_i} \otimes x_i) \\
&= \sum_{i=1}^n \|\chi_{A_i}\|_1 \|x_i\| \\
&= \sum_{i=1}^n \|x_i\| \, \mu(A_i) \\
&= \Big\| \sum_{i=1}^n x_i \chi_{A_i} \Big\|_{L_1(\mu,X)} \\
&= \Big\| J\Big(\sum_{i=1}^n \chi_{A_i} \otimes x_i\Big)\Big\|_{L_1(\mu, X)}.
\end{aligned}$$

An immediate consequence is

PROPOSITION 11. *If X is a closed linear subspace of Y, then $L_1(\mu) \hat{\otimes} X$ is a closed linear subspace of $L_1(\mu) \hat{\otimes} Y$.*

This section will be closed with a few words about induced operators. If W, X, Y and Z are Banach spaces and $S: W \to Y$ and $T: X \to Z$ are bounded linear operators, consider $S \otimes T: W \otimes X \to Y \otimes Z$ defined by $(S \otimes T)(\sum_{i=1}^n w_i \otimes x_i) = \sum_{i=1}^n (Sw_i) \otimes (Tx_i)$. Evidently $S \otimes T$ is a well-defined linear operator sending $W \otimes X$ into $Y \otimes Z$. If $W \otimes X$ and $Y \otimes Z$ are both equipped with their respective greatest reasonable crossnorms, then $S \otimes T$ is continuous. For,

$$\begin{aligned}
\gamma\Big((S \otimes T)\Big(\sum_{i=1}^n w_i \otimes x_i\Big)\Big) &= \gamma\Big(\sum_{i=1}^n Sw_i \otimes Tx_i\Big) \\
&\leqq \sum_{i=1}^n \|Sw_i\| \, \|Tx_i\| \\
&\leqq \|S\| \, \|T\| \sum_{i=1}^n \|w_i\| \, \|x_i\|.
\end{aligned}$$

An appeal to Proposition 9 yields

$$\gamma((S \otimes T)(u)) \leqq \|S\| \, \|T\| \, \gamma(u)$$

for each $u \in W \otimes X$. It follows that $S \otimes T$ has a unique bounded linear extension to an operator $S \hat{\otimes} T: W \hat{\otimes} X \to Y \hat{\otimes} Z$ with $\|S \hat{\otimes} T\| \leqq \|S\| \|T\|$.

In a similar fashion, $S \otimes T$ "induces" a bounded linear operator $S \check{\otimes} T: W \check{\otimes} X \to Y \check{\otimes} Z$ for which $\|S \check{\otimes} T\| \leqq \|S\| \|T\|$. The basic calculation this time is the following: If $y^* \in Y^*$ and $z^* \in Z^*$, then

$$\begin{aligned} \left|(y^* \otimes z^*)\left(\sum_{i=1}^n Sw_i \otimes Tx_i\right)\right| &= \left|\sum_{i=1}^n y^*(Sw_i)z^*(Tx_i)\right| \\ &= \left|\sum_{i=1}^n (S^*y^*)(w_i)(T^*z^*)(x_i)\right| \\ &= \|S^*\| \|T^*\| \left|\left(\frac{S^*y^*}{\|S^*\|} \otimes \frac{T^*z^*}{\|T^*\|}\right)\left(\sum_{i=1}^n w_i \otimes x_i\right)\right|. \end{aligned}$$

Thus we have

$$\lambda((S \otimes T)(u)) \leqq \|S^*\| \|T^*\| \lambda(u) = \|S\| \|T\| \lambda(u)$$

for any $u \in W \otimes X$.

With the notion of the tensor product of two operators it is easy to justify the term "projective tensor product". In the next proposition, "id" stands for the identity operator.

PROPOSITION 12. *Let Z be a closed linear subspace of X and let $\psi : X \to X/Z$ be the natural quotient map. Then $\psi \hat{\otimes} \mathrm{id}: X \hat{\otimes} Y \to X/Z \hat{\otimes} Y$ is a quotient map for each Banach space Y. Similarly, $\mathrm{id} \hat{\otimes} \psi : Y \hat{\otimes} X \to Y \hat{\otimes} X/Z$ is a quotient map.*

PROOF. We shall concern ourselves only with $\psi \hat{\otimes} \mathrm{id}$. Let $\varepsilon > 0$. Let $u \in (X/Z) \hat{\otimes} Y$. By Proposition 9(b), there exist sequences $(\hat{x}_n)$ in X/Z and (y_n) in Y such that $\|y_n\| \leqq 1$, $\hat{u} = \sum_{n=1}^\infty \hat{x}_n \otimes y_n$, $\gamma(\hat{u}) \leqq \sum_{n=1}^\infty \|\hat{x}_n\| \|y_n\| \leqq \gamma(\hat{u}) + \varepsilon/2$. Now each $\hat{x}_n$ is in X/Z and so there exists $x_n \in X$ such that $\psi x_n = \hat{x}_n$, $\|\hat{x}_n\| \leqq \|x_n\| \leqq \|\hat{x}_n\| + \varepsilon/2^{n+1}$. Consider $u = \sum_{n=1}^\infty x_n \otimes y_n \in X \hat{\otimes} Y$. It is plain that $(\psi \hat{\otimes} \mathrm{id})(u) = \hat{u}$ and equally plain that $\gamma(\hat{u}) \leqq \gamma(u) + \varepsilon/2 \leqq \gamma(\hat{u}) + \varepsilon$. It follows that $\psi \hat{\otimes}$ id is a quotient map of $X \hat{\otimes} Y$ onto $(X/Z) \hat{\otimes} Y$.

2. The duals of $X \hat{\otimes} Y$ and $X \check{\otimes} Y$. The principal results of this section are the identification of $(X \hat{\otimes} Y)^*$ as the space of continuous bilinear functionals on $X \times Y$ and the isolation of the integral bilinear forms as precisely those bilinear functionals in the dual of $X \check{\otimes} Y$. Throughout X, Y and Z are Banach spaces.

The first identification is mainly a formality. Suppose $\Phi: X \times Y \to Z$ is a continuous bilinear operator. Then $\|\Phi\| = \sup\{\|\Phi(x, y)\| : x \in X, y \in Y, \|x\|, \|y\| \leqq 1\}$. In addition Φ induces a mapping $\Phi': X \otimes Y \to Z$, namely

$$\Phi'\left(\sum_{i=1}^n x_i \otimes y_i\right) = \sum_{i=1}^n \Phi\left(x_i, y_i\right).$$

It is plain that Φ' is a well-defined linear map. Moreover, since

$$\begin{aligned} \left\|\Phi'\left(\sum_{i=1}^n x_i \otimes y_i\right)\right\| &\leqq \sum_{i=1}^n \|\Phi(x_i, y_i)\| \\ &\leqq \|\Phi\| \sum_{i=1}^n \|x_i\| \|y_i\|, \end{aligned}$$

we see that Φ' is a continuous linear operator from $(X \otimes Y, \gamma)$ to Z having operator norm $\leqq \|\Phi\|$.

On the other hand, if $\Psi': X \otimes Y \to Z$ is a γ-continuous linear operator, then Ψ' induces a bilinear map $\Psi: X \times Y \to Z$ defined by $\Psi(x, y) = \Psi'(x \otimes y)$. It is clear that Ψ is well defined and bilinear; since

$$\begin{aligned}\|\Psi(x, y)\| &= \|\Psi'(x \otimes y)\| \leqq \|\Psi'\|\gamma(x \otimes y) \\ &= \|\Psi'\| \, \|x\| \, \|y\|,\end{aligned}$$

Ψ is continuous with bilinear norm $\leqq \|\Psi'\|$.

Summarizing this discussion is a result sometimes called the "universal mapping property" of the projective tensor product.

THEOREM 1. *The correspondence $\Phi \leftrightarrow \Phi'$ is a (natural) isometric isomorphism between the Banach spaces $\mathscr{B}(X, Y; Z)$ of continuous bilinear operators from $X \times Y$ to Z and $\mathscr{L}(X \hat{\otimes} Y; Z)$ of bounded linear operators from $X \hat{\otimes} Y$ to Z.*

In particular, $(X \hat{\otimes} Y)^*$ is the Banach space $\mathscr{B}(X, Y)$ of continuous bilinear functionals on $X \times Y$.

The space $\mathscr{B}(X, Y)$ can be identified in a natural way with $\mathscr{L}(X; Y^*)$. To accomplish this, let $\psi' \in \mathscr{B}(X, Y)$ and define $\psi: X \to Y^*$ by $\psi(x)(y) = \psi'(x, y)$. Evidently ψ is linear and

$$|\psi(x)(y)| = |\psi'(x, y)| \leqq \|\psi'\| \, \|x\| \, \|y\|.$$

Hence ψ is bounded and has operator norm no greater than $\|\psi'\|$. In the other direction, let $\psi \in \mathscr{L}(X; Y^*)$ and define a functional ψ' on $X \times Y$ by $\psi'(x, y) = \psi(x)(y)$. The operation ψ' is bilinear and

$$\begin{aligned}|\psi'(x, y)| &= |\psi(x)(y)| \leqq \|\psi(x)\| \, \|y\| \\ &\leqq \|\psi\| \, \|x\| \, \|y\|.\end{aligned}$$

Therefore ψ' is continuous and has bilinear functional norm no greater than $\|\psi\|$. Summarizing the discussion is

COROLLARY 2. *The Banach space $\mathscr{L}(X; Y^*)$ is isometrically isomorphic to $(X \hat{\otimes} Y)^*$. An element $\psi^* \in (X \hat{\otimes} Y)^*$ corresponds to $\psi' \in \mathscr{B}(X, Y)$ and ψ' corresponds to $\psi \in \mathscr{L}(X; Y^*)$ if and only if*

$$\psi^*(x \otimes y) = \psi'(x, y) = \psi(x)(y)$$

for all $x \in X$ and $y \in Y$.

A particularly striking consequence of Corollary 2 is

COROLLARY 3. *Suppose X has the property that whenever Y is a closed linear subspace of Z then $Y \hat{\otimes} X$ is a closed linear subspace of $Z \hat{\otimes} X$ (under the natural inclusion). Then X^* is an injective Banach space.*

PROOF. Let Y be a closed linear subspace of Z and let $T: Y \to X^*$ be a continuous linear operator. Then T corresponds to $T' \in \mathscr{B}(Y, X)$ via the equation $(Ty)(x) = T'(y, x)$, with $\|T\| = \|T'\|$. But $\mathscr{B}(Y, X) = (Y \hat{\otimes} X)^*$. Hence by the Hahn-Banach theorem and the hypothesis that $Y \hat{\otimes} X$ is a subspace of $Z \hat{\otimes} X$, T' has a continuous linear extension T'' to a member of $(Z \hat{\otimes} X)^* = \mathscr{B}(Z, X)$ with $\|T''\| = \|T'\|$; T'' induces a member T''' of $\mathscr{L}(Z; X^*)$ defined by the equation $(T''' z)(x) = T''(z,x)$. Also $\|T'''\| = \|T''\|$. It is plain that T''' is a norm preserving extension of T from Y to Z. Thus X^* is an injective Banach space.

EXAMPLE 4. *The existence of Banach spaces X for which $Y \hat{\otimes} X$ is not a subspace of $Z \hat{\otimes} X$ even though Y is a subspace of Z.* By Corollary 3, any Banach space X whose dual X^* is not injective has the property that for some Banach space Z there is a closed linear subspace Y of Z for which $Y \hat{\otimes} X$ is not a subspace of $Z \hat{\otimes} X$. By definition, an injective Banach space is complemented in any space in which resides as a closed subspace. Thus if X^* is weakly sequentially complete or weakly compactly generated (and infinite dimensional) then X^* is not injective. In fact, every injective space is contained (complementably) in a space of the form $B(\Sigma)$, where Σ is the σ-field of all subsets of some set. Thus by VI.2, injective spaces must contain l_∞, an obvious impossibility for both weakly sequentially complete and weakly compactly generated Banach spaces.

The isolation of $(X \check{\otimes} Y)^*$ is a bit more complicated but more rewarding than the isolation of $(X \hat{\otimes} Y)^*$, for it is through the space $(X \check{\otimes} Y)^*$ that the theory of vector measures enters the theory of tensor products and secures a leading role in the study of topological tensor products.

From now through the end of this chapter, U_{X^*} will stand for the closed unit ball of X^* in its weak*-topology.

With the help of Alaoglu's theorem consider the compact Hausdorff product space $U_{X^*} \times U_{Y^*}$. Fix $x \in X$ and $y \in Y$ and note that $(x \otimes y)(x^*, y^*) = x^*(x)y^*(y)$ defines (in a natural way) a member of $C(U_{X^*} \times U_{Y^*})$. Extend this correspondence to all of $X \otimes Y$ by writing

$$J\Big(\sum_{i=1}^{n} x_i \otimes y_i\Big)(x^*, y^*) = \sum_{i=1}^{n} x^*(x_i)y^*(y_i).$$

Plainly this defines a linear operator $J: X \otimes Y \to C(U_{X^*} \times U_{Y^*})$. A glance at the definition of the least reasonable crossnorm λ is enough to show that J is an isometry. Thus J extends to an isometric embedding of $X \check{\otimes} Y$ into $C(U_{X^*} \times U_{Y^*})$. This sets up the identification of $(X \check{\otimes} Y)^*$.

THEOREM 5 (GROTHENDIECK). *A continuous bilinear functional ψ on $X \times Y$ defines a member of $(X \check{\otimes} Y)^*$ if and only if there exists a regular Borel measure μ on $U_{X^*} \times U_{Y^*}$ such that, for each $x \in X$ and each $y \in Y$,*

$$\psi(x, y) = \int_{U_{X^*} \times U_{Y^*}} x^*(x)y^*(y)\, d\mu(x^*, y^*).$$

In this case, the norm of ψ as a member of $(X \check{\otimes} Y)^*$ is precisely the variation $|\mu|(U_{X^*} \times U_{Y^*})$ of μ.

PROOF. Adhere to the notation of the discussion preceding the statement of the theorem. Let $\psi \in (X \check{\otimes} Y)^*$ and note that $\psi \circ J^{-1}$ is a bounded linear functional on the closed subspace $J(X \check{\otimes} Y)$ of $C(U_{X^*} \times U_{Y^*})$. With the help of the Hahn-Banach theorem select a norm preserving extension χ of $\psi \circ J^{-1}$ to all of $C(U_{X^*} \times U_{Y^*})$. By the Riesz Representation Theorem, there is a regular Borel measure μ on $U_{X^*} \times U_{Y^*}$ such that

$$\chi(f) = \int_{U_{X^*} \times U_{Y^*}} f(x^*, y^*)\, d\mu(x^*, y^*)$$

for all $f \in C(U_{X^*} \times U_{Y^*})$. In addition, $|\mu|(U_{X^*} \times U_{Y^*}) = \|\chi\| = \|\psi \circ J^{-1}\| \leqq \|\psi\|$. Also for $x \in X$ and $y \in Y$, one has

$$\begin{aligned} \psi(x, y) &= (\psi \circ J^{-1} \circ J)(x \otimes y) = \chi(J(x \otimes y)) \\ &= \int_{U_{X^*} \times U_{Y^*}} x^*(x) y^*(y)\, d\mu(x^*, y^*). \end{aligned}$$

On the other hand, if ψ is representable in the form

$$\psi(x, y) = \int_{U_{X^*} \times U_{Y^*}} x^*(x) y^*(y)\, d\mu(x^*, y^*)$$

for some regular Borel measure μ on $U_{X^*} \times U_{Y^*}$, then the functional ψ^* induced by ψ on $X \otimes Y$ defined by

$$\psi^*\Big(\sum_{i=1}^n x_i \otimes y_i\Big) = \sum_{i=1}^n \psi(x_i, y_i)$$

satisfies

$$\psi^*(u) = \int_{U_{X^*} \times U_{Y^*}} (x^* \otimes y^*)(u)\, d\mu(x^*, y^*)$$

for each $u \in X \otimes Y$. It follows that

$$\begin{aligned} |\psi^*(u)| &= \Big| \int_{U_{X^*} \times U_{Y^*}} (Ju)(x^*, y^*)\, d\mu(x^*, y^*) \Big| \\ &\leqq \int_{U_{X^*} \times U_{Y^*}} |(Ju)(x^*, y^*)|\, d|\mu|(x^*, y^*) \\ &\leqq \|Ju\|_\infty |\mu|(U_{X^*} \times U_{Y^*}) \\ &= \lambda(u) |\mu|(U_{X^*} \times U_{Y^*}). \end{aligned}$$

Therefore ψ^* extends to a continuous linear functional on $X \check{\otimes} Y$ with $\|\psi^*\| \leqq |\mu|(U_{X^*} \times U_{Y^*})$. Combining the two paragraphs completes the proof.

DEFINITION 6. A continuous bilinear functional ψ on $X \times Y$ is *integral* whenever ψ determines a member of $(X \check{\otimes} Y)^*$. The class of integral bilinear functionals on $X \times Y$ is denoted by $\mathscr{B}^\wedge(X, Y)$.

A continuous linear operator $T: X \to Y$ is an *integral operator* (in the sense of Grothendieck) whenever the bilinear functional $\tau \in \mathscr{B}(X, Y^*)$ defined by $\tau(x, y^*) = y^*(Tx)$ belongs to $\mathscr{B}^\wedge(X, Y^*)$.

The integral norm of $\psi \in \mathscr{B}^\wedge(X, Y)$ is just the functional norm of ψ as a member of $(X \check{\otimes} Y)^*$; this norm will be denoted by $\|\psi\|_{\text{int}}$. If $T: X \to Y$ is an integral operator, the integral norm of T is $\|\tau\|_{\text{int}}$, where $\tau \in \mathscr{B}^\wedge(X, Y^*)$ is the bilinear functional on $X \times Y^*$ induced by T.

Basic to much of the theory of integral operators is the next result concerning the "ideal" property of the class of integral operators.

THEOREM 7. *Let W, X, Y and Z be Banach spaces and suppose $T: W \to X$, $S: X \to Y$ and $R: Y \to Z$ are continuous linear operators with S integral. Then $RST: W \to Z$ is integral and $\|RST\|_{\text{int}} \leqq \|R\|\ \|S\|_{\text{int}}\ \|T\|$.*

PROOF. Consider the continuous linear operator $T \check{\otimes} R^*$ induced by $T: W \to X$ and $R^*: Z^* \to Y^*$, $T \check{\otimes} R^*: W \check{\otimes} Z^* \to X \check{\otimes} Y^*$. By Theorem 5, the adjoint of $T \check{\otimes} R^*$ acts as a continuous linear operator $(T \check{\otimes} R^*)^*: \mathscr{B}^\wedge(X, Y^*) \to \mathscr{B}^\wedge(W, Z^*)$. If $\sigma \in \mathscr{B}^\wedge(X, Y^*)$ is given by $\sigma(x, y^*) = y^*Sx$, then $(T \check{\otimes} R^*)^* \sigma \in \mathscr{B}^\wedge(W, Z^*)$ and, as a straightforward calculation shows, for each $w \in W$ and $z^* \in Z^*$,

$$(T \check{\otimes} R^*)(\sigma)(w, z^*) = z^*RSTw.$$

It follows that RST induces the integral bilinear functional $(T \check{\otimes} R^*)^*(\sigma)$; hence RST is integral.

To estimate the integral norm of RST, note

$$\begin{aligned} \|RST\|_{\text{int}} &= \|(T \check{\otimes} R^*)^*(\sigma)\|_{\text{int}} \leqq \|(T \check{\otimes} R^*)^*\| \, \|\sigma\|_{\text{int}} \\ &= \|T \check{\otimes} R^*\| \, \|S\|_{\text{int}} \leqq \|T\| \, \|R^*\| \, \|S\|_{\text{int}} \\ &= \|R\| \, \|S\|_{\text{int}} \, \|T\|. \end{aligned}$$

A frequently useful fact regarding integral operators is contained in

THEOREM 8. *A continuous linear operator $T: X \to Y$ is integral if and only if $JT: X \to Y^{**}$ is integral, where $J: Y \to Y^{**}$ is the natural embedding. In this case,* $\|T\|_{\text{int}} = \|JT\|_{\text{int}}$.

PROOF. If T is integral, Theorem 7 guarantees that JT is integral and $\|JT\|_{\text{int}} \leqq \|T\|_{\text{int}}$.

To conclude to the converse and reverse inequality, suppose $JT: X \to Y^{**}$ is integral and let $\psi \in \mathscr{B}^\wedge(X, Y^{***})$ be the bilinear functional induced by JT, i.e., $\psi(x, y^{***}) = y^{***}JTx$. Let J^* be the natural embedding of Y^* into Y^{***}. Then the continuous linear operator $I_X \check{\otimes} J^*: X \check{\otimes} Y^* \to X \check{\otimes} Y^{***}$ induced by the identity operator I_X on X and J^* has a continuous adjoint. But Theorem 5 guarantees that $(I_X \check{\otimes} J^*)^*$ acts as an operator between $\mathscr{B}^\wedge(X, Y^{***})$ and $\mathscr{B}^\wedge(X, Y^*)$. It is a routine calculation to check that if $x \in X$ and $y^* \in Y^*$ then

$$(I_X \check{\otimes} J^*)^*(\psi)(x, y^*) = y^*Tx.$$

Since the left-hand side of the above equality defines a member of $\mathscr{B}^\wedge(X, Y^*)$, the very definition of integral operator shows that T is an integral operator with integral norm

$$\begin{aligned} \|T\|_{\text{int}} &= \|(I_X \check{\otimes} J^*)^*(\psi)\|_{\text{int}} \\ &\leqq \|(I_X \check{\otimes} J^*)^*\| \, \|\psi\|_{\text{int}} \\ &= \|I_X \check{\otimes} J^*\| \, \|JT\|_{\text{int}} \\ &\leqq \|I_X\| \, \|J^*\| \, \|JT\|_{\text{int}} = \|JT\|_{\text{int}} \end{aligned}$$

and the proof is complete.

The role of the theory of vector measures in the theory of tensor products derives largely from the next theorem. This should be clear to anyone who is familiar with VI.3.10 and its consequences.

THEOREM 9. *A bounded linear operator $T: X \to Y$ is integral if and only if T admits a factorization*

$$\begin{array}{ccccc} X & \xrightarrow{T} & Y & \xrightarrow{J} & Y^{**} \\ S\big\downarrow & & & \nearrow Q & \\ L_\infty(\mu) & \xrightarrow[I]{} & L_1(\mu) & & \end{array}$$

where μ is a finite regular Borel measure on some compact Hausdorff space Ω, J: $Y \to Y^{**}$ is the natural embedding, I: $L_\infty(\mu) \to L_1(\mu)$ is the natural inclusion and S: $X \to L_\infty(\mu)$ and Q: $L_1(\mu) \to Y^{**}$ are bounded linear operators. In this case, Ω, μ, Q and S can be chosen so that $\|Q\|, \|S\| \leqq 1$ and $|\mu|(\Omega) = \|T\|_{\text{int}}$.

PROOF. First we shall show that every integral operator does admit such a factorization. To this end, let T: $X \to Y$ be integral and $\tau \in \mathscr{B}^\frown(X, Y^*)$ be the associated bilinear functional $\tau(x, y^*) = y^*Tx$, where $\|T\|_{\text{int}} = \|\tau\|_{\text{int}}$. Then there is a regular Borel measure μ on $U_{X^*} \times U_{Y^{**}}$ such that

$$y^*Tx = \tau(x, y^*) = \int_{U_{X^*} \times U_{Y^{**}}} x^*(x) y^{**}(y^*) \, d\mu(x^*, y^{**})$$

and $|\mu|(U_{X^*} \times U_{Y^{**}}) = \|\tau\|_{\text{int}} = \|T\|$. Define S: $X \to L_\infty(\mu)$ by $Sx(x^*, y^{**}) = x^*(x)$ and R: $Y^* \to L_\infty(\mu)$ by $Ry^*(x^*, y^{**}) = y^{**}(y^*)$. Evidently S and R are continuous linear operators with $\|S\|, \|R\| \leqq 1$. Moreover, if $x \in X$ and $y^* \in Y^*$, then

$$\begin{aligned} y^*Tx &= \int_{U_{X^*} \times U_{Y^{**}}} x^*(x) y^{**}(y^*) \, d\mu(x^*, y^{**}) \\ &= \int_{U_{X^*} \times U_{Y^{**}}} (Sx)(x^*, y^{**})(Ry^*)(x^*, y^{**}) \, d\mu(x^*, y^{**}) \\ &= (Ry^*)(ISx) = y^*(R^*ISx), \end{aligned}$$

where I: $L_\infty(\mu) \to L_1(\mu)$ is the natural inclusion operator. If J is the natural imbedding of Y into Y^{**}, the above calculation shows $JT = R^*|_{L_1(\mu)} IS$ as desired.

Conversely, suppose JT admits the factorization claimed. To show JT, and hence, by Theorem 8, T is integral it is evidently enough to show that the natural inclusion I is integral and then appeal to Theorem 7. By the Stone Representation Theorem, there exist a compact Hausdorff totally disconnected space Λ and a regular Borel measure λ on Λ such that $L_1(\mu)$ and $L_1(\lambda)$ are isometrically isomorphic as are $L_\infty(\mu)$ and $C(\Lambda)$. It is immediate from Theorem 7 that the inclusions of $L_\infty(\mu)$ in $L_1(\mu)$ and of $C(\Lambda)$ in $L_1(\lambda)$ are simultaneously integral or nonintegral. Thus it is enough to show that the inclusion of $C(\Lambda)$ into $L_1(\lambda)$ is integral. Let $h \in C(\Lambda \times \Lambda)$. Define $l(h)$ by

$$l(h) = \int_\Lambda h(t, t) \, d\lambda(t).$$

Plainly $l \in C(\Lambda \times \Lambda)^*$. By the Riesz Representation Theorem, there is a regular Borel measure ν on $\Lambda \times \Lambda$ such that $\int h \, d\nu = l(h) = \int_\Lambda h(t, t) \, d\lambda(t)$ for each $h \in C(\Lambda \times \Lambda)$. This holds in particular for $h \in C(\Lambda \times \Lambda)$ of the form $h(s, t) = f(s) \, g(t)$, where f and $g \in C(\Lambda)$. But this means that

$$\int_{\Lambda \times \Lambda} f(s) g(t) \, d\nu(s, t) = \int_\Lambda f(t) g(t) \, d\lambda(t)$$

for all f and $g \in C(\Lambda)$. Now note that the right side is just the form of evaluation of a member $g \in L_1(\lambda)^* = C(\Lambda)$ at a member $f \in C(\Lambda)$ viewed, after inclusion, as a member of $L_1(\lambda)$. This completes the proof.

The next corollary unites the work of VI.3 with the current study.

COROLLARY 10. *If $T: X \to Y$ is a Pietsch integral operator then T is integral and $\|T\|_{\text{int}} \leqq \|T\|_{\text{pint}}$. Conversely, if Y is complemented in Y^{**} by a norm one projection, then each integral operator $T: X \to Y$ is Pietsch integral and the integral and Pietsch integral norms of T coincide.*

PROOF. Suppose $T: X \to Y$ is a Pietsch integral operator. By VI.3.10, there exist a compact Hausdorff space Ω, a regular Borel measure μ on Ω and operators $R: X \to L_\infty(\mu)$ and $S: L_1(\mu) \to X$ with $\|S\|, \|R\| \leqq 1$ such that T admits the factorization

$$\begin{array}{ccc} X & \xrightarrow{\;T\;} & Y \\ {\scriptstyle R}\downarrow & & \uparrow{\scriptstyle S} \\ L_\infty(\mu) & \xrightarrow[I]{} & L_1(\mu) \end{array}$$

where $I: L_\infty(\mu) \to L_1(\mu)$ is the natural inclusion map. Moreover for each $\varepsilon > 0$ we can choose Ω and μ to satisfy $|\mu|(\Omega) \leqq \|T\|_{\text{pint}} + \varepsilon$. Now $I: L_\infty(\mu) \to L_1(\mu)$ is integral; this was proved in Theorem 9. Thus by Theorem 7, the operator T is integral. Furthermore, we have

$$\begin{aligned} \|T\|_{\text{int}} &= \|SIR\|_{\text{int}} \leqq \|S\| \; \|I\|_{\text{int}} \, \|R\| \\ &\leqq \|I\|_{\text{int}} = |\mu|(\Omega) \leqq \|T\|_{\text{pint}} + \varepsilon. \end{aligned}$$

Since $\varepsilon > 0$ is arbitrary, $\|T\|_{\text{int}} \leqq \|T\|_{\text{pint}}$ and the first statement is proved.

To prove the second statement suppose $T: X \to Y$ is an integral operator in the sense of Grothendieck. Then Theorem 9 guarantees the existence of a compact Hausdorff space Ω, a regular Borel measure μ on Ω, and operators $R: X \to L_\infty(\mu)$ and $S: L_1(\mu) \to Y^{**}$ such that $\|R\|, \|S\| \leqq 1, |\mu|(\Omega) = \|T\|_{\text{int}}$ and JT admits a factorization

$$\begin{array}{ccccc} X & \xrightarrow{\;T\;} & Y & \xrightarrow{\;J\;} & Y^{**} \quad (\xleftarrow{\;P\;} Y) \\ {\scriptstyle R}\downarrow & & & & \uparrow{\scriptstyle S} \\ L_\infty(\mu) & & \xrightarrow[I]{} & & L_1(\mu) \end{array}$$

where $I: L_\infty(\mu) \to L_1(\mu)$ is the natural inclusion map, $J: Y \to Y^{**}$ is the natural embedding and $P: Y^{**} \to Y$ is the norm one projection whose existence is asserted in the hypothesis of the second statement. Thus T admits to the factorization

$$\begin{array}{ccc} X & \xrightarrow{\;T\;} & Y \\ {\scriptstyle R}\downarrow & & \uparrow{\scriptstyle PS} \\ L_\infty(\mu) & \xrightarrow[I]{} & L_1(\mu) \end{array}$$

By VI.3.10, T is Pietsch integral. Moreover, VI.3.8 and VI.3.9 show that

$$\begin{aligned}\|T\|_{\text{pint}} &= \|PSIR\|_{\text{pint}} \leqq \|P\|\ \|S\|\ \|I\|_{\text{pint}}\|R\| \\ &\leqq \|I\|_{\text{pint}} = |\mu|(\Omega) = \|T\|_{\text{int}}.\end{aligned}$$

This completes the proof.

An important consequence of the factorization theorem is next.

COROLLARY 11. *A bounded linear operator $T: X \to Y$ is integral if and only if the adjoint $T^*: Y^* \to X^*$ is integral. In this case, $\|T\|_{\text{int}} = \|T^*\|_{\text{int}}$.*

PROOF. Suppose $T: X \to Y$ is integral. Let Ω, μ, Q and S be as in the statement of Theorem 9 with $\|Q\|$, $\|S\| \leqq 1$ and $\mu(\Omega) = \|T\|_{\text{int}}$. Take the diagram featured in Theorem 9 and take adjoints to produce the commutative diagram

$$\begin{array}{ccccc} & & Y^{***} & \xrightarrow{J^*} Y^* \xrightarrow{T^*} & X^* \\ & & {\scriptstyle Q^*}\downarrow & & \uparrow{\scriptstyle S^*} \\ L_\infty(\mu) & = & L_1(\mu)^* & \xrightarrow[I^*]{} L_\infty(\mu)^* = & L_1(\mu)^{**} \end{array}$$

Let K be the natural embedding of Y^* into Y^{***} and L be the natural embedding of $L_1(\mu)$ into $L_\infty(\mu)^* = L_1(\mu)^{**}$. Then we have

$$\begin{array}{ccccccc} Y^* \xrightarrow{K} & Y^{***} & \xrightarrow{J^*} & Y^* & \xrightarrow{T^*} & X^* & \\ & {\scriptstyle Q^*}\downarrow & & & & \uparrow{\scriptstyle S^*} & \\ L_\infty(\mu) = & L_1(\mu)^* & \xrightarrow[I]{} & L_1(\mu) & \xrightarrow[L]{} & L_\infty(\mu)^* & = L_1(\mu)^{**} \end{array}$$

Note that $T^* = T^*J^*K = S^*LIQ^*K$. Since I is integral, Theorem 7 guarantees that T^* is integral. Moreover,

$$\begin{aligned}\|T^*\|_{\text{int}} &= \|S^*LIQ^*K\|_{\text{int}} \\ &\leqq \|S^*\|\ \|L\|\ \|I\|_{\text{int}}\ \|Q^*\|\ \|K\| \\ &= \|S\|\ \|L\|\ \|I\|_{\text{int}}\ \|Q\|\ \|K\| \\ &\leqq \|I\|_{\text{int}} = |\mu|(\Omega) = \|T\|_{\text{int}}.\end{aligned}$$

To obtain the converse and reverse inequality, suppose $T: X \to Y$ is a continuous linear operator and $T^*: Y^* \to X^*$ is integral. By the first part of the present proof, $T^{**}: X^{**} \to Y^{**}$ is integral. But the following commutative diagram

$$\begin{array}{ccccc} X & \xrightarrow{T} & Y & \xrightarrow{J} & Y^{**} \\ {\scriptstyle K}\downarrow & & & \nearrow{\scriptstyle T^{**}} & \\ X^{**} & & & & \end{array}$$

where $K: X \to X^{**}$ and $J: Y \to Y^{**}$ are the natural embeddings, shows that JT: $X \to Y^{**}$ is integral and has integral norm $\leqq \|T^{**}K\|_{\text{int}} \leqq \|T^{**}\|_{\text{int}} \leqq \|T^*\|_{\text{int}}$, by Theorem 7. An appeal to Theorem 8 completes the proof.

COROLLARY 12. *A continuous bilinear functional ψ on $X \times Y$ is integral if and only if the continuous linear operator $T_\psi: X \to Y^*$ defined by $(T_\psi x)(y) = \psi(x, y)$ is integral. In this case,* $\|\psi\|_{\text{int}} = \|T_\psi\|_{\text{int}}$.

PROOF. Suppose that T_ψ is integral. Then the bilinear functional τ on $X \times Y^{**}$ given by $\tau(x, y^{**}) = y^{**}(Tx)$ is an integral bilinear functional with $\|\tau\|_{\text{int}} = \|T_\psi\|_{\text{int}}$. Consider the operator $I_X \check{\otimes} J: X \check{\otimes} Y \to X \check{\otimes} Y^{**}$ induced by the identity operator I_X on X and the natural imbedding J of Y into Y^{**}. By Theorem 5 and Definition 6, $(I_X \check{\otimes} J)^*: \mathscr{B}^\wedge(X, Y^{**}) \to \mathscr{B}^\wedge(X, Y)$, i.e., $(I_X \check{\otimes} J)^*$ maps members of $\mathscr{B}^\wedge(X, Y^{**})$ to members of $\mathscr{B}^\wedge(X, Y)$. Now it is straightforward to see that $\psi = (I_X \check{\otimes} J)^*\tau$. Thus ψ is integral. Moreover we have

$$\begin{aligned}\|\psi\|_{\text{int}} &= \|(I_X \check{\otimes} J)^*\tau\|_{\text{int}} \leqq \|(I_X \check{\otimes} J)^*\| \, \|\tau\|_{\text{int}} \\ &= \|I_X \check{\otimes} J\| \, \|T_\psi\|_{\text{int}} = \|I_X\| \, \|J\| \, \|T_\psi\|_{\text{int}} \\ &= \|T_\psi\|_{\text{int}}.\end{aligned}$$

Conversely, if ψ is an integral bilinear functional on $X \times Y$, then there exists a regular Borel measure μ on $U_{X^*} \times U_{Y^*}$ such that

$$\psi(x, y) = \int_{U_{X^*} \times U_{Y^*}} x^*(x) y^*(y) \, d\mu(x^*, y^*)$$

for each $x \in X$ and $y \in Y$. Moreover $\|\psi\|_{\text{int}} = |\mu|(U_{X^*} \times U_{Y^*})$. Define $R: X \to L_\infty(\mu)$ by $(Rx)(x^*, y^*) = x^*(x)$ and $S: Y \to L_\infty(\mu)$ by $(Sy)(x^*, y^*) = y^*(y)$. Then for $x \in X$ and $y \in Y$, we have

$$\begin{aligned}(T_\psi x)(y) &= \psi(x, y) \\ &= \int_{U_{X^*} \times U_{Y^*}} x^*(x) y^*(y) \, d\mu(x^*, y^*) \\ &= \int_{U_{X^*} \times U_{Y^*}} (Rx)(x^*, y^*)(Sy)(x^*, y^*) \, d\mu(x^*, y^*) \\ &= (Rx)(ISy) = ((IS)^*(Rx))(y)\end{aligned}$$

where $I: L_\infty(\mu) \to L_1(\mu)$ is the natural inclusion. Of course, the above calculation shows that $T_\psi = S^*I^*R$. Since I (and hence I^*) is integral, the operator T_ψ is integral and, by Corollary 11,

$$\begin{aligned}\|T_\psi\|_{\text{int}} &= \|S^*I^*R^*\|_{\text{int}} \leqq \|S^*\| \, \|I^*\|_{\text{int}} \, \|R\| \\ &\leqq \|I^*\|_{\text{int}} = \|I\|_{\text{int}} \\ &\leqq |\mu|(U_{X^*} \times U_{Y^*}) = \|\psi\|_{\text{int}}.\end{aligned}$$

This section will be terminated with a few more remarks about natural identifications and an application of the duality theory developed thus far. (We now know how a juggler feels when he is well into his act.) Specifically we shall look at important properties that are hereditarily preserved under the process of taking projective tensor products.

Let $\psi \in \mathscr{B}(X, Y)$. Then ψ can be considered as a member F of $\mathscr{L}(X; Y^*)$. Also $F \in \mathscr{L}(X; Y^*)$ induces in a natural way a $\psi' \in \mathscr{B}(X, Y^{**})$ given by $\psi'(x, y^{**}) = y^{**}(Fx)$. A simple calculation shows $\|\psi\| = \|F\| = \|\psi'\|$. A similar situation happens in $\mathscr{B}^\wedge(X, Y)$. In fact, in this case, if $\psi \in \mathscr{B}^\wedge(X, Y)$, then $F \in \mathscr{L}(X; Y^*)$ is an integral linear operator by Corollary 12 and $\|\psi\|_{\text{int}} = \|F\|_{\text{int}}$. By definition F induces an integral bilinear functional ψ' on $X \times Y^{**}$ with $\|F\|_{\text{int}} = \|\psi'\|_{\text{int}}$. This proves

COROLLARY 13. *The natural inclusion maps of $\mathscr{B}(X, Y)$ into $\mathscr{B}(X, Y^{**})$ and of $\mathscr{B}^\wedge(X, Y)$ into $\mathscr{B}^\wedge(X, Y^{**})$ are isometric for any pair of Banach spaces X and Y. The same applies to the inclusion maps of $\mathscr{B}(X, Y)$ into $\mathscr{B}(X^{**}, Y)$ and of $\mathscr{B}^\wedge(X, Y)$ into $\mathscr{B}^\wedge(X^{**}, Y)$.*

COROLLARY 14. *For any Banach spaces X and Y, the natural inclusion of $X \hat{\otimes} Y$ into $X \hat{\otimes} Y^{**}$ is an isometry. The same is true for the inclusion map of $X \hat{\otimes} Y$ into $X^{**} \hat{\otimes} Y$.*

PROOF. Plainly the inclusion does not increase norm. On the other hand, if $u \in X \hat{\otimes} Y$ then there is $\psi \in \mathscr{B}(X, Y)$ $(= (X \hat{\otimes} Y)^*)$ such that $\|\psi\| = 1$ and $\psi(u) = \gamma(u)$. Let ψ' be ψ viewed as a member of $\mathscr{B}(X, Y^{**})$. As noted in Corollary 13, $\|\psi'\| = \|\psi\|$. Also $\psi'(u') = \psi(u)$ where u' is the image of u under the natural inclusion of $X \hat{\otimes} Y$ into $X \hat{\otimes} Y^{**}$. Thus $\gamma(u) = \psi(u) = \psi'(u') \leqq \gamma(\mu')$. The proof now follows.

3. The approximation and metric approximation properties. At the core of many proofs of properties of the classical Lebesgue spaces is an argument based on the density of simple functions. In spaces of operators, the natural analogue of the class of simple functions is the class of finite rank operators. Therefore a somewhat optimistic way of studying spaces of operators is to establish properties of finite rank operators and then, with the aid of a density theorem, establish the same properties for all members of the space of operators under scrutiny.

Motivated by this heuristic notion is the following

DEFINITION 1. A Banach space X is said to have the *approximation property* if for each compact set $K \subseteq X$ and $\varepsilon > 0$ there is a continuous finite rank operator $T: X \to X$ such that, for all $x \in K$, $\|Tx - x\| \leqq \varepsilon$. If in addition T can always be found with $\|T\| \leqq 1$, then X is said to have the *metric approximation property*.

This section investigates various formulations of the approximation and metric approximation properties. Though, by the famous Enflo example, not all Banach spaces enjoy these properties, many classical spaces do have enough structure to allow good approximation by finite rank operators.

First a few technical lemmas will be needed.

LEMMA 2. *Every compact subset of a Banach space is contained in the closed convex hull of some sequence converging to zero in norm.*

PROOF. Let K be a compact subset of the Banach space X. For each $\varepsilon > 0$ there is a finite set F and a compact set L such that $K \subseteq \operatorname{co}(L \cup F)$ and such that $F \subseteq 2K$ and $\sup_{x \in L}\|x\| \leqq \varepsilon$. Indeed, if $\{y_1, \cdots, y_n\}$ is an $(\varepsilon/2)$-net in K, let $L_i = \{k - y_i: k \in K, \|k - y_i\| \leqq \varepsilon/2\}$, $i = 1, \cdots, n$. Note that $L = 2\bigcup_{i=1}^n \bar{L}_i$ is compact, and $\|x\| \leqq \varepsilon$ for each $x \in L$. Let $F = \{2y_1, \cdots, 2y_n\}$. Then F is finite and $F \subseteq 2K$. In addition, for each $x \in K$ there is an i with $1 \leqq i \leqq n$ such that $\|x - y_i\| < \varepsilon/2$. Thus $2(x - y_i) \in L_i$ and $2y_i \in F$ and

$$x = \tfrac{1}{2}(2(x - y_i) + 2y_i) \in \operatorname{co}(L \cup F).$$

Hence $K \subseteq \operatorname{co}(L \cup F)$.

Now for each n choose a finite set F_n and a compact set L_n such that $\sup_{x\in L_n} \|x\| \leqq 2^{-n}$, $F_n \subseteq 2L_n$ ($L_0 = K$) and $K \subseteq \mathrm{co}(\bigcup_{j=1}^n F_j \cup L_n)$. Then $\bigcup_{n=1}^\infty F_n$ is a countable set which can be listed as a sequence tending to zero in norm.

Let $k \in K$, $\varepsilon > 0$ and choose n such that $2^{-n} < \varepsilon$. Then choose $\lambda_1, \lambda_2, \cdots, \lambda_n, \lambda_{n+1} \geqq 0$ with $\sum_{j=1}^{n+1}\lambda_j = 1$, $f_1, \cdots, f_n \in \bigcup_{j=1}^n F_j$ and $l \in L_n$ such that $k = \sum_{i=1}^n \lambda_i f_i + \lambda_{n+1} l$. But then

$$\left\| k - \sum_{i=1}^n \lambda_i f_i \right\| = \|\lambda_{n+1} l\| \leqq 2^{-n} < \varepsilon.$$

Therefore tacking 0 to the beginning of the sequential listing of $\bigcup_{n=1}^\infty F_n$ allows us to realize K as a member of the closed convex hull of the resulting sequence.

As usual, $\mathscr{L}(X; Y)$ is the Banach space of continuous linear operators from X to Y. We wish to consider $\mathscr{L}(X; Y)$ in the topology of uniform convergence on compact subsets of X, i.e., the locally convex topology generated by the seminorms $\rho_K(T) = \sup\{\|Tk\| : k \in K\}$, where K ranges over the compact subsets of X. This space $\mathscr{L}(X; Y)$ in this topology will be denoted by $\mathscr{L}_c(X; Y)$.

The next lemma describes the dual of $\mathscr{L}_c(X; Y)$.

LEMMA 3. *A linear functional u on $\mathscr{L}(X; Y)$ is in $\mathscr{L}_c(X; Y)^*$ if and only if there exist sequences (y_n^*) in Y^* and (x_n) in X such that $\sum_{n=1}^\infty \|y_n^*\| \|x_n\| < \infty$ and $u(T) = \sum_{n=1}^\infty y_n^*(Tx_n)$ for all $T \in \mathscr{L}(X; Y)$.*

PROOF. A straightforward computation shows that the linear functionals on $\mathscr{L}(X; Y)$ of the above form are in $\mathscr{L}_c(X; Y)^*$. Conversely, if $u \in \mathscr{L}_c(X; Y)^*$, then there exists a compact set K in X such that $|u(T)| \leqq \sup\{\|Tk\| : k \in K\}$ for all $T \in \mathscr{L}(X; Y)$. By Lemma 2, K is contained in the closed convex hull of some sequence (x_n) of members of X that tend to zero in norm. Therefore

$$(*) \qquad |u(T)| \leqq \sup_n \|Tx_n\|.$$

Let $c_0(Y)$ be the Banach space of all sequences in Y that tend to zero in norm. Norm $c_0(Y)$ by the usual sup norm. Let $Z = \{(Tx_n) : T \in \mathscr{L}(X; Y)\}$. Define a linear functional u^* on Z by $u^*((Tx_n)) = uT$. By $(*)$, u^* is a continuous linear functional on the linear subspace Z of $c_0(Y)$. By the Hahn-Banach theorem, u^* has a continuous linear extension, still called u^*, to all of $c_0(Y)$. But $c_0(Y)^*$ is just $l_1(Y^*)$, the space of absolutely convergent series of members of Y^*. Thus there exists a sequence (y_n^*) of members of Y^* for which $\sum_n \|y_n^*\| = \|u^*\|$ and for which $u^*((y_n)) = \sum_{n=1}^\infty y_n^*(y_n)$ for any $(y_n) \in c_0(Y)$. It follows that

$$u(T) = u^*((Tx_n)) = \sum_{n=1}^\infty y_n^*(Tx_n)$$

and the proof is complete.

Now we are in a position to begin to understand why tensor products are of use in examining the approximation property.

THEOREM 4. *Each of the following statements about a Banach space X implies all of the others:*

(i) *The space X has the approximation property.*

(ii) *$X^* \otimes X$ is dense in $\mathscr{L}_c(X; X)$.*

(iii) *For each Banach space Y, $X^* \otimes Y$ is dense in $\mathcal{L}_c(Y; X)$.*

(iv) *For each Banach space Y, $Y^* \otimes X$ is dense in $\mathcal{L}_c(Y; X)$.*

(v) *The natural linear injection of $X^* \hat{\otimes} X$ into $\mathcal{L}(X; X)$ is one-to-one.*

(vi) *If (x_n^*) is a sequence in X^* and (x_n) is a sequence in X such that $\sum_n \|x_n^*\| \, \|x_n\| < \infty$ and $\sum_{n=1}^\infty x_n^*(x)x_n = 0$ for each $x \in X$, then $\sum_{n=1}^\infty x_n^*(x_n) = 0$.*

PROOF. To prove that (i) implies (ii), let $T\colon X \to X$ be a linear continuous operator, K be a compact subset of X and $\varepsilon > 0$. Choose $\delta > 0$ so that $\|Tx\| \leqq \varepsilon$ whenever $\|x\| \leqq \delta$. By (i), there exists $u = \sum_{i=1}^n x_i^* \otimes x_i \in X^* \otimes X$ such that, for each $x \in K$, $\|x - u(x)\| \leqq \delta$. Then we have

$$\left\| Tx - \left(\sum_{i=1}^n x_i^* \otimes Tx_i\right)(x)\right\| = \left\| Tx - \sum_{i=1}^n x_i^*(x)Tx_i \right\|$$
$$= \left\| T\left(x - \sum_{i=1}^n x_i^*(x)x_i\right)\right\| = \|T(x - u(x))\| \leqq \varepsilon,$$

and (ii) follows.

To prove that (ii) implies (iii), suppose $S \in \mathcal{L}(X; Y)$. Define $\phi(S)\colon \mathcal{L}(X; X) \to \mathcal{L}(X; Y)$ by $\phi(S)(T) = S \cdot T$. Evidently ϕ is a linear operator from $\mathcal{L}(X; X)$ to $\mathcal{L}(X; Y)$ which is continuous from $\mathcal{L}_c(X; X)$ to $\mathcal{L}_c(X; Y)$. Moreover, $\phi(S)$ maps $X^* \otimes X$ onto $X^* \otimes S(X)$. If (ii) holds, then $X^* \otimes X$ is dense in $\mathcal{L}_c(X; X)$. Consequently $X^* \otimes X$ is mapped by ϕ onto the subspace $X^* \otimes S(X)$ whose $\mathcal{L}_c(X; Y)$-closure contains $\phi(S)(I_X) = S$ where I_X is the identity operator on X. Clearly (iii) follows.

That (ii) implies (iv) is a consequence of arguments similar to those of the above paragraph involving the operator $\chi(R)\colon \mathcal{L}(X; X) \to \mathcal{L}(Y; X)$ defined by $\chi(R)(T) = T \cdot R$ for $R \in \mathcal{L}(Y; X)$ and $T \in \mathcal{L}(X; X)$. Since the implications "(ii) implies (i)", "(iii) implies (ii)", and "(iv) implies (ii)" are all trivial, the equivalence of (i) through (iv) has been demonstrated.

To prove that (vi) implies (i), note that if the identity operator I_X on X is not in the $\mathcal{L}_c(X; X)$-closure of $X^* \otimes X$, then there exists $u^* \in \mathcal{L}_c(X; X)^*$ such that $u^*(u) = 0$, each $u \in X^* \otimes X$ yet $u^*(I_X) = 1$. By Lemma 3, there exist sequences (x_n^*) and (x_n) of members of X^* and X respectively such that $\sum_n \|x_n^*\| \, \|x_n\| < \infty$ and $u^*(T) = \sum_{n=1}^\infty x_n^*(Tx_n)$ for each $T \in \mathcal{L}(X; X)$. Fixing $x \in X$ and $x^* \in X^*$ and looking at $T = x^* \otimes x$, we get

$$0 = u^*(T) = \sum_{n=1}^\infty x_n^*(Tx_n) = \sum_{n=1}^\infty x_n^*(x)x^*(x_n)$$
$$= x^*\left(\sum_{n=1}^\infty x_n^*(x)x_n\right).$$

Since x^* is arbitrary, we infer $\sum_{n=1}^\infty x_n^*(x)x_n = 0$. By (vi), we have $\sum_{n=1}^\infty x_n^*(x_n) = 0$. On the other hand,

$$1 = u^*(I_X) = \sum_{n=1}^\infty x_n^*(I_X x_n) = \sum_{n=1}^\infty x_n^*(x_n).$$

The resulting contradiction shows that (vi) implies (i). Conversely, if the identity operator on X, I_X, is in the $\mathcal{L}_c(X; X)$-closure of $X^* \otimes X$, then given sequences (x_n^*)

and (x_n) in X^* and X respectively such that $\sum_n \|x_n^*\| \|x_n\| < \infty$ and $\sum_{n=1}^{\infty} x_n^*(x)x_n = 0$ for all $x \in X$, then $\sum_{n=1}^{\infty} x_n^*(x_n) = 0$. In fact, if $x^* \in X^*$, then for each $x \in X$, we have

$$x^*\Big(\sum_{n=1}^{\infty} x_n^*(x)x_n\Big) = 0$$

so that $\sum_{n=1}^{\infty} x_n^* \otimes x_n$, viewed as a member of $\mathscr{L}_c(X; X)^*$, vanishes on elements of $\mathscr{L}(X; X)$ of the form $x^* \otimes x$. Therefore $\sum_{n=1}^{\infty} x_n^* \otimes x_n$ vanishes on the linear span of such elements of $\mathscr{L}(X; X)$, i.e., on $X^* \otimes X$. By continuity, it follows that $\sum_{n=1}^{\infty} x_n^* \otimes x_n$ vanishes on the $\mathscr{L}_c(X; X)$-closure of $X^* \otimes X$; in particular, $\sum_{n=1}^{\infty} x_n^* \otimes x_n$ vanishes on I_X. But

$$\Big(\sum_{n=1}^{\infty} x_n^* \otimes x_n\Big)(I_X) = \sum_{n=1}^{\infty} x_n^*(x_n).$$

Now suppose that the natural inclusion of $X^* \hat{\otimes} X$ into $\mathscr{L}(X; X)$ is one-to-one. Let (x_n^*) and (x_n) be sequences in X^* and X respectively for which $\sum_n \| x_n^* \| \| x_n \| < \infty$ and $\sum_{n=1}^{\infty}(x_n^* \otimes x_n)(x) = \sum_{n=1}^{\infty} x_n^*(x)x_n = 0$ for each $x \in X$. Plainly the operator $\sum_{n=1}^{\infty} x_n^* \otimes x_n$ is in the image of the natural injection of $X^* \hat{\otimes} X$, which vanishes on all of X. Thus $\sum_{n=1}^{\infty} x_n^* \otimes x_n$ is zero as a member of $X^* \hat{\otimes} X$. Therefore

$$\psi^*\Big(\sum_{n=1}^{\infty} x_n^* \otimes x_n\Big) = 0$$

for each $\psi^* \in (X^* \hat{\otimes} X)^*$. One such ψ^* is induced by $\mathrm{tr} \in \mathscr{B}(X^*, X)$ where $\mathrm{tr}(x^*, x) = x^*(x)$. For this ψ^*,

$$0 = \psi^*\Big(\sum_{n=1}^{\infty} x_n^* \otimes x_n\Big) = \sum_{n=1}^{\infty} \mathrm{tr}(x_n^* \otimes x_n) = \sum_{n=1}^{\infty} x_n^*(x_n).$$

This completes the proof that (v) implies (vi).

Suppose (vi) holds. Let $u \in X^* \hat{\otimes} X$. By Proposition 1.9(b) there are sequences (x_n^*) and (x_n) in X^* and X, respectively, such that $u = \sum_{n=1}^{\infty} x_n^* \otimes x_n$, $\sum_{n=1}^{\infty} \|x_n^*\| < \infty$ and $\lim_n \|x_n\| = 0$. Since the natural inclusion of $X^* \hat{\otimes} X$ into $\mathscr{L}(X; X)$ is continuous, $\sum_{n=1}^{\infty} x_n^*(\,\cdot\,)x_n$ defines u as a member of $\mathscr{L}(X; X)$. Suppose that $u(x) = 0$ for each $x \in X$. Then if $\varepsilon > 0$ there is a finite rank continuous linear operator $F : X \to X$ such that $\|Fx_n - x_n\| \leqq \varepsilon$ for all n; this is an easy consequence of the fact that $\|x_n\| \to 0$. Thus

$$\gamma\Big(\sum_{n=1}^{\infty} x_n^* \otimes x_n - \sum_{n=1}^{\infty} x_n^* \otimes Fx_n\Big) \leqq \varepsilon \sum_{n=1}^{\infty} \|x_n^*\|.$$

But for $x \in X$, we have

$$0 = F\Big(\sum_{n=1}^{\infty} x_n^*(x)x_n\Big) = \sum_{n=1}^{\infty} x_n^*(x)F(x_n)$$
$$= \sum_{n=1}^{\infty} (x_n^* \otimes Fx_n)(x).$$

Hence $\sum_{n=1}^{\infty} x_n^* \otimes Fx_n$ is zero as a member of $\mathscr{L}(X; X)$. But $\sum_{n=1}^{\infty} x_n^* \otimes Fx_n = F \cdot (\sum_{n=1}^{\infty} x_n^* \otimes x_n)$ and so is a member of $X^* \otimes X$. Consequently, $\sum_{n=1}^{\infty} x_n^* \otimes Fx_n$ is zero in $X^* \otimes X$ and hence is zero in $X^* \hat{\otimes} X$.

But then we have

$$\gamma\Big(\sum_{n=1}^{\infty} x_n^* \otimes x_n\Big) = \gamma\Big(\sum_{n=1}^{\infty} x_n^* \otimes x_n - \sum_{n=1}^{\infty} x_n^* \otimes Fx_n\Big)$$
$$\leqq \varepsilon \sum_{n=1}^{\infty} \|x_n^*\|.$$

Since $\varepsilon > 0$ was arbitrary, it follows that

$$\gamma(u) = \gamma\Big(\sum_{n=1}^{\infty} x_n^* \otimes x_n\Big) = 0$$

and $u = 0$ as a member of $X^* \hat{\otimes} X$. This completes the proof that (vi) implies (v) and completes the proof of Theorem 4.

A remark at this juncture is in order. A useful equivalent of the approximation property for the Banach space X is the following statement: *For each Banach space Y each compact linear operator T: $Y \to X$ is the limit in operator norm of a sequence of continuous finite rank linear operators F_n: $Y \to X$.* That this property is necessary is an easy consequence of the very definition of the approximation property; one merely needs to approximate the identity operator I_X on X by finite rank operator S_n on $K = \overline{\{Ty: \|y\| \leqq 1\}}$ and look at $F_n = S_n \cdot T$. The converse also holds but takes us too far afield from our present interests to be pursued herein.

COROLLARY 5. *If X^* has the approximation property, then so does X.*

PROOF. Let (x_n^*) and (x_n) be sequences in X^* and X, respectively, such that $\sum_{n=1}^{\infty} \|x_n^*\| \|x_n\| < \infty$ and $\sum_{n=1}^{\infty} x_n^*(x)x_n = 0$ for each $x \in X$. Let J be the natural imbedding of X into X^{**}. Then $\sum_{n=1}^{\infty} \|Jx_n\| \|x_n^*\| = \sum_{n=1}^{\infty} \|x_n\| \|x_n^*\| < \infty$. Also for each $x^* \in X^*$ and $x \in X$, we have

$$\sum_{n=1}^{\infty} (Jx_n)(x^*)x_n^*(x) = \sum_{n=1}^{\infty} x^*(x_n)x_n^*(x)$$
$$= x^*\Big(\sum_{n=1}^{\infty} x_n^*(x)x_n\Big) = 0.$$

Therefore $\sum_{n=1}^{\infty}(Jx_n)(x^*)x_n^* = 0$ for each $x^* \in X^*$. Since X^* has the approximation property, Theorem 4(vi) insures that $\sum_{n=1}^{\infty}(Jx_n)(x_n^*) = 0$. But $\sum_{n=1}^{\infty}(Jx_n)(x_n^*) = \sum_{n=1}^{\infty} x_n^*(x_n)$, so by Theorem 4(vi), we see that X has the approximation property.

THEOREM 6. *Suppose X^* has the approximation property. Then for each Banach space Y and each compact linear operator T: $X \to Y$, there exists a sequence F_n: $X \to Y$ of finite rank continuous linear operators such that $\lim_n \|T - F_n\| = 0$.*

PROOF. Let $\varepsilon > 0$ and suppose $T: X \to Y$ is a compact linear operator. Then T^*: $Y^* \to X^*$ is also compact. Since X^* has the approximation property, there exists $u \in X^{**} \otimes X^*$ such that $\|x^* - ux^*\| < \varepsilon$ for all x^* in the image of the unit ball of Y^* under T^*. Clearly uT^* is a continuous linear operator with finite dimensional range. Thus $T^{**}u^* = (uT^*)^*$ is a continuous linear operator with finite dimensional range. It is easily established that

$$\big\|T^{**}\big|_X u^* - T\big\| < \varepsilon.$$

Since $T^{**}(X^{**}) \subseteq Y$ this finishes the proof.

THEOREM 7. *Suppose X is a Banach space such that X^* has the approximation property. If a continuous linear operator $T: X \to Y$ has a nuclear adjoint, then T is itself nuclear.*

PROOF. Let $u \in Y^{**} \hat{\otimes} X^*$ define the nuclear operator T^*. If we can show that u is actually a member of $Y \hat{\otimes} X^*$, then it is a routine check to see that $u \in X^* \hat{\otimes} Y$, the natural symmetric cousin of u, defines the operator T and therefore that T is nuclear.

Now, by Corollary 2.12, $Y \hat{\otimes} X^*$ is a closed linear subspace of $Y^{**} \hat{\otimes} X^*$. Consequently, to show that u is in $Y \hat{\otimes} X^*$, it suffices to show that any $\psi \in (Y^{**} \hat{\otimes} X^*)^*$ that vanishes on $(Y \hat{\otimes} X^*)$ also vanishes at u.

Let $\psi \in \mathscr{B}(Y^{**}, X^*) = (Y^{**} \hat{\otimes} X^*)^*$. Then ψ corresponds to a continuous linear operator $A: Y^{**} \to X^{**}$. If ψ vanishes on $Y \hat{\otimes} X^*$ then A vanishes on Y. Moreover, if u is represented by $u = \sum_{n=1}^{\infty} y_n^{**} \otimes x_n^*$, where $\sum_{n=1}^{\infty} \|y_n^{**}\| \|x_n^*\| < \infty$, then $\psi(u) = \sum_{n=1}^{\infty} (Ay_n^{**})(x_n^*)$. Consider the member ν of $X^{**} \hat{\otimes} X^*$ given by $\nu = (A \hat{\otimes} \mathrm{id}_{X^*})(u)$. Since X^* has the approximation property, Theorem 4 guarantees that the natural injection of $X^{**} \hat{\otimes} X^*$ into $\mathscr{L}(X^*; X^*)$ is one-to-one. Therefore, if the continuous linear operator $S: X^* \to X^*$ corresponding to $\nu \in X^{**} \hat{\otimes} X^*$ is zero, then it follows that ν is zero as a member of $X^{**} \hat{\otimes} X^*$. This in turn will show that $\chi(\nu) = 0$ for every $\chi \in \mathscr{B}(X^{**}; X^*) = (X^{**} \hat{\otimes} X^*)^*$. Most particularly, if $\chi(x^{**}, x^*) = \chi^{**}(x^*)$, i.e., χ is the trace functional, then $\chi(\nu) = 0$. But it is easily seen that, for this particular χ, $\chi(\nu) = \sum_{n=1}^{\infty} (Ay_n^{**})(x_n^*) = \psi(u)$. Therefore, to show that $\psi(u) = 0$, it suffices to prove that $S = 0$.

To this end, note that $S^* = AT^{**}$; this is a straightforward check. Of course, since $T^*: Y^* \to X^*$ is nuclear, the operator T is compact. Therefore $T^{**}(X^{**}) \subseteq Y$. But A vanishes on Y, so AT^{**} is the zero operator on X^{**}. Hence $S^* = 0$ and so $S = 0$. This completes the proof.

THEOREM 8. *Each of the following statements about a Banach space X implies all of the others*:

(i) *The space X has the metric approximation property.*

(ii) *Each $T \in \mathscr{L}(X; X)$ is in the $\mathscr{L}_c(X; X)$-closure of $\{u \in X^* \otimes X: \|u\| \leqq \|T\|\}$.*

(iii) *For each Banach space Y, each $T \in \mathscr{L}(X; Y)$ is in the $\mathscr{L}_c(X; Y)$-closure of $\{u \in X^* \otimes Y: \|u\| \leqq \|T\|\}$.*

(iv) *For each Banach space Y, each $T \in \mathscr{L}(Y; X)$ is in the $\mathscr{L}_c(Y; X)$-closure of $\{u \in Y^* \otimes X: \|u\| \leqq \|T\|\}$.*

(v) *The natural inclusion of $X \hat{\otimes} X^*$ into $\mathscr{B}^{\wedge}(X^*, X)$ is an isometry.*

(vi) *For each Banach space Y, the natural inclusion of $X \hat{\otimes} Y$ into $\mathscr{B}^{\wedge}(X^*, Y^*)$ is an isometry.*

PROOF. The equivalence of statements (i) through (iv) can be established in a manner similar to that used in the proof of the equivalence of statements (i) through (iv) in Theorem 4. Before completing the proof let us make a few general remarks on the duality between $X \otimes Y$ and various spaces of bilinear functionals.

Let X and Y be Banach spaces. Then the γ-unit ball of $X \otimes Y$ is in duality with the unit ball of $\mathscr{B}(X, Y)$, and the unit ball of $X \otimes Y$ in its relative $\mathscr{B}^{\wedge}(X^*, Y^*)$-norm topology is in duality with the unit ball of $X^* \check{\otimes} Y^*$. Both these statements are the main consequences of the duality established in §2. Thus for these two norms to agree on $X \otimes Y$ (that is, for the natural injection of $X \hat{\otimes} Y$ into $\mathscr{B}^{\wedge}(X^*, Y^*)$ to be an isometry), it is both necessary and sufficient that the unit ball of $X^* \hat{\otimes} Y^*$ is dense in that of $\mathscr{B}(X, Y)$ in the topology of pointwise convergence on $X \times Y$ (the weak*-topology of $\mathscr{B}(X, Y)$ relative to $X \otimes Y$).

Now identifying $\mathscr{B}(X, Y)$ with $\mathscr{L}(X; Y^*)$, we know that if (iii) holds then the unit ball of $X^* \otimes Y^*$ is dense in the unit ball of $\mathscr{L}(X; Y^*)$ relative to the topology of uniform convergence on compact subsets of X. It follows trivially that the unit ball of $X^* \otimes Y^*$ is dense in the topology of pointwise convergence in the unit ball of $\mathscr{L}(X; Y^*)$. Consequently, on $X \otimes Y$, the γ-norm and the relative $\mathscr{B}^{\wedge}(X^*, Y^*)$-norm agree, and this is just the conclusion of (vi).

Now suppose (vi) holds. According to Corollary 2.13 the natural inclusion of $\mathscr{B}^{\wedge}(X^*, X)$ into $\mathscr{B}^{\wedge}(X^*, X^{**})$ is an isometry for any Banach space X. Applying (vi) to $Y = X^*$ shows that the natural inclusion of $X \hat{\otimes} X^*$ into $\mathscr{B}^{\wedge}(X^*, X^{**})$ is an isometry. But the natural inclusion of $X \hat{\otimes} X^*$ into $\mathscr{B}^{\wedge}(X^*, X)$ followed by the natural inclusion of $\mathscr{B}^{\wedge}(X^*, X)$ into $\mathscr{B}^{\wedge}(X^*, X^{**})$ is precisely the natural inclusion of $X \hat{\otimes} X^*$ into $\mathscr{B}^{\wedge}(X^*, X^{**})$. Since the latter two are isometries the former must also be an isometry; thus (v) holds.

Finally (i) follows from (v), since by our discussion preceding the proof of "(iii) implies (vi)", it is readily seen that (v) implies that the unit ball of $\mathscr{L}(X; X)$ is the closure of $\{u \in X^* \otimes X: \|u\| \leqq 1\}$ relative to the topology of pointwise convergence. However, the set $\{u \in X^* \otimes X: \|u\| \leqq 1\}$ is equicontinuous, so its closure taken relative to the topology of pointwise convergence coincides with that taken with respect to uniform convergence on compact subsets of X, and (i) follows.

COROLLARY 9. *If X^* has the metric approximation property, then so does X.*

PROOF. Consider the commutative diagram

$$\begin{array}{ccc} X^* \hat{\otimes} X^{**} & \longrightarrow & \mathscr{B}^{\wedge}(X^{**}, X^*) \\ \uparrow & & \uparrow \\ X^* \hat{\otimes} X & \longrightarrow & \mathscr{B}^{\wedge}(X, X^*) \end{array}$$

where all the maps are natural inclusions. The left vertical map is an isometry by Corollary 2.14, the right vertical map is an isometry of Corollary 2.13. Since X^* has the metric approximation property, the top map is an isometry by Theorem 8(v). Commutativity insures that the bottom horizontal map is an isometry which in turn guarantees that X has the metric approximation property (again, by Theorem 8(v)).

Of special interest to us is the following formulation of the metric approximation property for dual spaces:

THEOREM 10. *If X is a Banach space, then X^* has the metric approximation property if and only if the natural injection of $X^* \hat{\otimes} Y^*$ into $\mathscr{B}^{\wedge}(X, Y)$ is an isometry for each Banach space Y.*

PROOF. By Corollary 2.13, the natural inclusion of $\mathscr{B}^{\wedge}(X; Y)$ into $\mathscr{B}^{\wedge}(X^{**}, Y^{**})$ is an isometry. If X^* has the metric approximation property, then Theorem 8 guarantees that the natural inclusion of $X^* \hat{\otimes} Y^*$ into $\mathscr{B}\hat{}(X^{**}, Y^{**})$ is an isometry. But the diagram

$$X^* \hat{\otimes} Y^* \to \mathscr{B}\hat{}(X, Y) \to \mathscr{B}\hat{}(X^{**}, Y^{**}),$$

where all maps are natural inclusions, commutes, and the right half and the composition are both isometries. Consequently the left half is also an isometry.

Conversely, suppose that for each Banach space Y the natural inclusion of $X^* \hat{\otimes} Y^*$ into $\mathscr{B}\hat{}(X, Y)$ is an isometry. By Corollary 2.14, for each Banach space Z the natural injection of $X^* \hat{\otimes} Z$ into $X^* \hat{\otimes} Z^{**}$ is an isometry. This, in tandem with our hypothesis, shows that for every Banach space Z, the natural inclusion of $X^* \hat{\otimes} Z$ into $\mathscr{B}\hat{}(X, Z^*)$ is an isometry. But by Corollary 2.13, the natural inclusion of $\mathscr{B}\hat{}(X, Z^*)$ into $\mathscr{B}\hat{}(X^{**}, Z^*)$ is an isometry. It follows that the inclusion of $X^* \hat{\otimes} Z$ into $\mathscr{B}\hat{}(X^{**}, Z^*)$ is an isometry since it is the left map of the commutative diagram

$$X^* \hat{\otimes} Z \to \mathscr{B}\hat{}(X, Z^*) \to \mathscr{B}\hat{}(X^{**}, Z^*)$$

where all maps are natural and the right map and composition map are both known to be isometries. Application of Theorem 8(vi) finishes the proof.

Touching firm ground, we shall conclude this section with an example that shows that the classical Lebesgue spaces have the metric approximation property. Here we violate our usual assumption that (Ω, Σ, μ) is a finite measure space.

EXAMPLE 11. Let (Ω, Σ, μ) be any measure space and let $1 \leqq p < \infty$. Then $L_p(\mu)$ has the metric approximation property.

Let $\varepsilon > 0$. We will show that for each $f_1, \cdots, f_n \in L_p(\mu)$ there exists a finite rank continuous linear operator $F\colon L_p(\mu) \to L_p(\mu)$ such that $\|F\| \leqq 1$ and $\|Ff_i - f_i\|_p \leqq \varepsilon$ for $i = 1, \cdots, n$. Condition (ii) of Theorem 9 is easily derived from this.

Let $A \in \Sigma$ be chosen to have finite μ-measure and such that, for $i = 1, \cdots, n$ $\int_{\Omega \setminus A} |f_i|^p \, d\mu \leqq \varepsilon/2$. If $\mu(A) = 0$ then $F = 0$ is the required operator. Otherwise, for each $\beta > 0$ we can decompose A into a finite number of disjoint members of Σ, say $A_1, \cdots, A_{k(\beta)}$, such that on A_j none of the functions of $f_1, \cdots, f_n$ varies by more than $\beta/2$. Define $F\colon L_p(\mu) \to L_p(\mu)$ by

$$Ff = \sum_{i=1}^{k(\beta)} \frac{\int_{A_i} f \, d\mu}{\mu(A_i)} \chi_{A_i}.$$

It is easily verified that F is the desired operator.

Since the dual of any $C(\Omega)$-space is isometric to some $L_1(\mu)$ space, it follows from Corollary 7 and the above that *for any compact Hausdorff space Ω, $C(\Omega)$ has the metric approximation property*. In particular, each $L_\infty(\mu)$ space has the metric approximation property.

4. Applications of tensor products and vector measures to Banach space theory. Tensor products do not enjoy universal popularity among vector measure theorists and some may be wondering why a chapter on tensor products appears in a monograph on vector measures. This section is to assuage the doubting Thomases by using the theory of tensor products together with the theory of vector measures to isolate some important properties of Banach spaces. Most striking of the topics treated in this section is the discovery of the role of the Radon-Nikodým theorem in deciding when a Banach space with the approximation property has the metric approximation property. From this we shall see that although the statement that a reflexive Banach space with the approximation property has the metric approximation property contains not even a hint of measure-theoretic content, its proof is a fairly easy consequence of the Dunford-Pettis-Phillips theorem. Further evidence of the applicability of the theorems of vector measures and tensor products will be found in theorems dealing with conditions ensuring that the space of bounded linear operators $\mathscr{L}(X; Y)$ is reflexive, theorems that relate nuclear operators to integral operators and a theorem of Diestel that says $L_p(\mu, X)$ ($1 \leqq p < \infty$, μ-finite) is weakly compactly generated whenever X is. This last fact carries some extra surprise: for, on the basis of IV.1, we know that $L_p(\mu, X)^*$ is often quite unwieldy. Included in the development of this section is the Davis-Figiel-Johnson-Pełcyzyński Factorization Lemma, a lemma which should find its way into introductory texts quite rapidly.

Further applications of the theory of vector measures of the same genre as those presented in this section that depend on Banach space results outside the assumptions of this text are discussed in some detail in the notes and remarks section following this chapter.

Before starting the proper discussion of this section, think back to the discussion of §2 that centered on the relationship between Pietsch integral and integral operators. The main fact of use here is the following: If Y is complemented in Y^{**} by a norm one projection, then the Grothendieck integral operators from X to Y and the Pietsch integral operators from X to Y coincide for each Banach space X. Moreover, the integral and Pietsch integral norms are the same. This in mind, we shall begin the discussion with a simultaneous generalization of several facts formerly known for reflexive spaces and separable duals.

THEOREM 1. *Let X be a Banach space that is complemented in X^{**} by a norm one projection. If X has the approximation property and the Radon-Nikodým property then X has the metric approximation property.*

PROOF. According to the discussion preceding the statement of Theorem 1, the integral operators from X to X and the Pietsch integral operators from X to X are identical classes with identical norms. Since X has the Radon-Nikodým property, VI.4.8 ensures that the Pietsch integral operators from X to X and the nuclear operators from X to X are identical classes with identical norms. Since X has the approximation property, Theorem 3.4 guarantees that the natural inclusion J of $X^* \hat{\otimes} X$ into $\mathscr{L}(X; X)$ is one-to-one. A quick look at the definition of nuclear operators and the nuclear norm reveals that $T \in \mathscr{L}(X; X)$ is nuclear if and only if T is the range of J and that the nuclear norm of T is given by the quotient norm of $J^{-1}(\{T\})$ as a member of $(X^* \hat{\otimes} X) / J^{-1}(\{0\})$. Since J is one-to-one, $J^{-1}(\{0\}) = \{0\}$ and hence $X^* \hat{\otimes} X$ is isometric to the space of nuclear operators from X to X.

Recapitulating, we have seen that the natural inclusion of $X^* \hat{\otimes} X$ into the integral operators from X to X is an isometric isomorphism. An appeal to Theorem 3.8(v) finishes the proof.

A myriad of corollaries follow. Two of the most notable are given next.

COROLLARY 2 (GROTHENDIECK). *Reflexive Banach spaces with the approximation property enjoy the metric approximation property.*

PROOF. Reflexive Banach spaces are complemented by a norm one projection in their second duals! Furthermore, they have the Radon-Nikodým property.

Noting that all dual spaces are norm one complemented in their second duals by the natural restriction projection, we also have

COROLLARY 3 (GROTHENDIECK). *Separable dual spaces with the approximation property have the metric approximation property.*

PROOF. Separable dual spaces have the Radon-Nikodým property. This combined with the observation above allows Theorem 1 to be applied.

A glance at the list compiled in the notes and remarks section of Chapter VII will help the reader produce any number of corollaries that have the same ring as Corollaries 2 and 3.

The next theorem uses Theorem 1 together with the Radon-Nikodým theorems of Chapter III to look at reflexive spaces of operators.

THEOREM 4. *If X and Y are Banach spaces and one of them has the approximation property, then $\mathscr{L}(X; Y)$ is reflexive if and only if X and Y are reflexive and each member of $\mathscr{L}(X; Y)$ is compact.*

PROOF. To prove the sufficiency, suppose X and Y are reflexive and that one of them has the approximation property. First it will be shown that $K(X; Y)$, the subspace of $\mathscr{L}(X; Y)$ consisting of the compact operators, is isometric to $X^* \check{\otimes} Y$.

To prove this, note that if Y has the approximation property, then this follows from the remark preceding Corollary 3.5. If X has the approximation property, then since X is reflexive Corollary 3.5 guarantees that X^* also has the approximation property. An appeal to Theorem 3.6 reveals that $X^* \otimes Y$ is dense in $K(X; Y)$. Hence, in either case, $K(X; Y)$ is isometric to $X^* \hat{\otimes} Y$.

Now the duality magic of §2 will be brought to bear. Suppose X and Y are reflexive and each member of $\mathscr{L}(X; Y)$ is compact. On the basis of §2, we know that $K(X; Y)^* = (X^* \check{\otimes} Y)^* = \mathscr{B}^\wedge(X^*, Y) = I(X^*; Y^*)$ where $I(X^*; Y^*)$ is the space of all integral operators from X^* to Y^*. Since Y^* is a dual space, the discussion preceding Theorem 1 ensures that $I(X^*; Y^*) = \mathrm{PI}(X^*; Y^*)$. Since Y^* is reflexive, Y^* has the Radon-Nikodým property, and hence $\mathrm{PI}(X^*; Y^*) = \mathrm{N}(X^*; Y^*)$ by VI.4.8. Also since either X^* or Y^* has the approximation property and both are reflexive one of them has the metric approximation property by Theorem 1. In either case, an appeal to Theorem 3.7 or Theorem 3.10 guarantees that $\mathrm{N}(X^*; Y^*) = X^{**} \hat{\otimes} Y^* = X \hat{\otimes} Y^*$. Thus $K(X, Y)^* = X \hat{\otimes} Y^*$. Finally, we have

$$(X \hat{\otimes} Y^*)^* = \mathscr{B}(X, Y^*) = \mathscr{L}(X; Y^{**}) = \mathscr{L}(X; Y).$$

Since $\mathscr{L}(X; Y) = K(X; Y)$ by hypothesis, it follows that $K(X; Y)^{**} = (X \hat{\otimes} Y^*)^* = \mathscr{L}(X; Y) = K(X; Y)$ and that $\mathscr{L}(X; Y) = K(X; Y)$ is reflexive.

Conversely, suppose $\mathscr{L}(X; Y)$ is reflexive. Evidently $\mathscr{L}(X; Y)$ contains isometric copies of Y and X^*. Hence X and Y are both reflexive. Now we are in a position to apply duality arguments as above to conclude $K(X; Y)^{**} = (X \hat{\otimes} Y^*)^* = \mathscr{L}(X; Y)$. Since $\mathscr{L}(X; Y)$ is reflexive, $K(X; Y)$ is also reflexive and it follows immediately that $\mathscr{L}(X; Y) = K(X; Y)$ and this completes the proof.

Recall a classical theorem due to Paley that guarantees that every member of $\mathscr{L}(l_q; l_p)$ is compact provided $1 \leqq p < q < \infty$.

COROLLARY 5. *If* $1 < p < q < \infty$, $\mathscr{L}(l_q; l_p)$ *is reflexive.*

The next theorem should be viewed as a dual of VI.4.8. The proof is completely vector measure-theoretic.

THEOREM 6. *Let X be a Banach space such that X^* has the approximation property. Then X^* has the Radon-Nikodým property if and only if every integral operator from X to Y is nuclear. In this case $I(X; Y)$ and $\mathrm{N}(X; Y)$ are identical classes with identical norms.*

PROOF. First suppose X^* has the Radon-Nikodým property. If $T: X \to Y$ is integral the adjoint $T^* : Y^* \to X^*$ is also integral by Corollary 2.11. Also since X^* is complemented in X^{***} by a norm one projection, it follows that T^* is Pietsch integral. A glance at VI.48. reveals that T^* is nuclear and hence T is nuclear by Theorem 3.7 since X^* has the approximation property. All the norm identities have also been carried along.

Conversely, let Σ be a σ-field of subsets of a point set Ω and $F: \Sigma \to X^*$ be a countably additive vector measure of finite variation μ. Define $T: L_\infty(\mu) \to X^*$ by

$$T(f) = \int f\,dF, \qquad f \in L_\infty(\mu).$$

Now consider the adjoint $T^* : X^{**} \to L_\infty(\mu)^*$. For $x \in X$ and $f \in L_\infty(\mu)$, we have

$$T^*(x)(f) = \int f\,dF(x).$$

Therefore $T^*(x)$ is the derivative f_x of the scalar measure $F(x)$ with respect to μ. Moreover,

$$\left|\int_E f_x\,d\mu\right| \leqq \|F(E)\|\,\|x\| \leqq \mu(E)\|x\|.$$

Hence $\|f_x\| \leqq \|x\|$ μ-almost everywhere. Two important properties emerge. The first is that if $S: X \to L_\infty(\mu)^*$ is the restriction of T^* to X, then S has its range in $L_1(\mu)$ and hence $S^* = T$. The second is that since $Sx = f_x$ and $\|f_x\| \leqq \|x\|$ μ-almost everywhere for each $x \in X$, we see that S admits the following factorization

$$\begin{array}{ccccc} X & \xrightarrow{S} & L_1(\mu) & \xrightarrow{J} & L_\infty(\mu)^* \\ & {\scriptstyle R}\searrow & \nearrow{\scriptstyle I} & & \\ & & L_\infty(\mu) & & \end{array}$$

where I is the natural inclusion map and J is the natural embedding of $L_1(\mu)$ into its second dual. By VI.3.9, the operator S is Pietsch integral and hence S is integral. By hypothesis, S is nuclear. Hence $T = S^*$ is nuclear. Accordingly T has the action

$$T(f) = \sum_{n=1}^{\infty} x_n^* \int_\Omega f\,d\mu_n$$

where each μ_n is a finitely additive measure that vanishes on μ-null sets, $x_n^* \in X^*$ and $\sum_{n=1}^{\infty} |\mu_n|(\Omega)\,\|x_n^*\| < \infty$. Now if we can replace the μ_n's by μ-continuous measures, then we shall be in a good position to find a Radon-Nikodým derivative for F.

To this end, first note that, for each $E \in \Sigma$, $F(E) = \sum_{n=1}^{\infty} x_n^* \mu_n(E)$ and that the series $\sum_{n=1}^{\infty} x_n^* \mu_n$ converges to F in the variation norm of $bv(\Sigma; X^*)$. Also note that by I.5.9, the operation of associating to each $G \in bv(\Sigma; X^*)$ the μ-continuous part of G is a norm one projection P on $bv(\Sigma; X^*)$. Hence

$$F = PF = \sum_{n=1}^{\infty} P(x_n^* \mu_n) = \sum_{n=1}^{\infty} x_n^* \hat{\mu}_n$$

where μ_n is the μ-continuous part of the scalar measure μ_n and the convergence is in the variation norm. Now let g_n be the Radon-Nikodým derivative of $\hat{\mu}_n$ with respect to μ. Since $\sum_{n=1}^{\infty} \|x_n^*\| \, |\mu_n|(\Omega) < \infty$ and $|\hat{\mu}_n|(\Omega) \leq |\mu_n|(\Omega)$, we see that $\sum_{n=1}^{\infty} x_n^* g_n$ is absolutely convergent in $L_1(\mu, X)$ norm to a function g which evidently satisfies

$$F(E) = \sum_{n=1}^{\infty} x_n^* \hat{\mu}_n(E) = \int_E g \, d\mu$$

for all $E \in \Sigma$. Hence X^* has the Radon-Nikodým property. (Note that in this last part the approximation assumption was unnecessary.)

The linear topological structure of the projective tensor product of two Banach spaces is quite complicated. As an illustration, note that if $X = l_p = Y$ for some $1 < p < \infty$ then $X \hat{\otimes} Y$ contains a complemented subspace isometric to l_1 and is therefore nonreflexive. Worse than this is the realization that if $X = l_p(\Gamma) = Y$ for an uncountable set Γ then $X \hat{\otimes} Y$ contains a complemented subspace isometric to $l_1(\Gamma)$ and therefore is not even weakly compactly generated. In spite of these rather grim facts, the operation of taking projective tensor products does not destroy all the good properties of Banach spaces.

THEOREM 7. *Suppose that X and Y are Banach spaces and that X^* and Y^* have the Radon-Nikodým property and at least one of them has the approximation property. Then* $\mathrm{N}(X; Y^*) = X^* \hat{\otimes} Y^*$ *has the Radon-Nikodým property.*

PROOF. First note that the reasoning employed in the proof of Theorem 4 shows that $\mathrm{N}(X; Y^*) = X^* \hat{\otimes} Y^*$ is the dual space of $X \check{\otimes} Y$. Therefore to show $X^* \hat{\otimes} Y^*$ has the Radon-Nikodým property it suffices by III.3.6 to show that each separable subspace of $X \check{\otimes} Y$ has a separable dual.

Let S be a separable subspace of $X \check{\otimes} Y$. Then it is easily seen that there exist separable closed linear subspaces X_0 and Y_0 of X and Y, respectively, such that S is isometric to a subspace of $X_0 \check{\otimes} Y_0$. By a corollary of Stegall's theorem in VII.2, X_0^* and Y_0^* are separable. Therefore X_0^* and Y_0^* have the Radon-Nikodým property. Therefore $(X_0 \check{\otimes} Y_0)^* = \mathscr{B}^\wedge(X_0, Y_0) = I(X_0; Y_0^*) = \mathrm{PI}(X_0; Y_0^*) = \mathrm{N}(X_0; Y_0^*)$, by the results of §2, the remarks preceding Theorem 1 and Theorem VII.4.8. But an obvious calculation shows that if X_0^* and Y_0^* are separable, then the space $\mathrm{N}(X_0; Y_0^*)$ of nuclear linear operators from X_0 to Y_0^* is also separable. Therefore $(X_0 \check{\otimes} Y_0)^*$ is separable. But S is a closed linear subspace of $(X_0 \check{\otimes} Y_0)$ so S^* is isomorphic to a quotient of the separable space $(X_0 \check{\otimes} Y_0)^*$ and is therefore itself separable. This completes the proof.

The next lemma provides a way of making reflexive Banach spaces from weakly compact subsets of arbitrary Banach spaces. This lemma has (at least) two important virtues. A number of basic facts about Banach spaces are easy consequences of it and its proof is strikingly elementary.

LEMMA 8 (DAVIS, FIGIEL, JOHNSON, PEŁCZYŃSKI). *Let X be a Banach space with closed unit ball $B = \{x \in X: \|x\| \leqq 1\}$ and let W be a convex symmetric bounded subset of X. For each positive integer n, let $U_n = 2^n W + 2^{-n}B$. Denote by $\| \ \|_n$ the Minkowski functional or gauge of U_n, i.e.,*

$$\|x\|_n = \inf\{\alpha > 0: x \in \alpha U_n\}.$$

For $x \in X$, let $|||x|||$ be given by $|||x||| = (\sum_{n=1}^{\infty} \|x\|_n^2)^{1/2}$. Let $Y = \{x \in X: |||x||| < \infty\}$ and denote by C the $|||\cdot|||$-closed unit ball of Y. Let $J: Y \to X$ be the natural inclusion. Then

(i) *$W \subseteq C$,*

(ii) *$(Y, |||\cdot|||)$ is a Banach space and J is continuous,*

(iii) *$J^{**}: Y^{**} \to X^{**}$ is one-to-one and $Y = (J^{**})^{-1}(X)$, and*

(iv) *$(Y, |||\cdot|||)$ is reflexive if and only if W is relatively weakly compact in X.*

PROOF. (i) If $w \in W$, then $\|w\|_n \leqq 2^{-n}$ for all n. Thus,

$$|||w||| = \left(\sum_{n=1}^{\infty} \|w\|_n^2\right)^{1/2} \leqq \left(\sum_{n=1}^{\infty} 2^{-2n}\right)^{1/2} \leqq 1.$$

Hence $w \in C$ whenever $w \in W$.

(ii) Let X_n be the linear space X equipped with the norm $\| \ \|_n$. Since W is bounded, $\| \ \|_n$ is equivalent to $\| \ \|$ and X_n is a Banach space. Let Z be the Banach space of all sequences (x_n) in X such that $\sum_{n=1}^{\infty} \|x_n\|_n^2 < \infty$, under the norm $\|(x_n)\| = (\sum_{n=1}^{\infty} \|x_n\|_n^2)^{1/2}$. Define $\psi: Y \to Z$ by $\psi(y) = (Jy, Jy, \cdots)$. The operator ψ is readily seen to be a linear isometry and $\psi(Y)$ is precisely the set $\{z \in Z: z = (x_n), x_n = x_1$ for all $n\}$ which is a closed linear subspace of Z. Consequently $(Y, |||\cdot|||)$ is a Banach space. Also $J: Y \to X$ can be viewed as the composition of ψ with the natural projection of Z onto its first coordinate. Since each of these maps is continuous and linear and since X_1 is isomorphic to X it follows that J is bounded.

(iii) It is not difficult to establish that Z^* is the Banach space of sequences (x_n^*) of members of X^* such that $\sum_{n=1}^{\infty} \|x_n^*\|_n^2 < \infty$ equipped with the norm $\|(x_n^*)\| = (\sum_{n=1}^{\infty} \|x_n^*\|_n^2)^{1/2}$ and that Z^{**} is the Banach space of sequences (x_n^{**}) of members of X^{**} such that $\sum_{n=1}^{\infty} \|x_n^{**}\|_n^2 < \infty$ equipped with the norm $\|(x_n^{**})\| = (\sum_{n=1}^{\infty} \|x_n^{**}\|_n^2)^{1/2}$. All the pairings are natural.

If we consider $\Psi: Y \to Z$ as in (ii), then it is easily verified that $\psi^{**}: Y^{**} \to Z^{**}$ has the form $\psi^{**}(y^{**}) = (J^{**}y^{**}, J^{**}y^{**}, \cdots)$. Since ψ is an isometry, ψ^{**} is also an isometry. Evidently this means $J^{**}: Y^{**} \to X^{**}$ is a one-to-one map. Also $\psi^{**-1}(\psi(Y)) = Y$ which means (in terms of J) that $(J^{**})^{-1}(X) = Y$. The proof of (iii) is now complete.

(iv) First note that the weak*-closure of JC in X^{**} is precisely the image under J^{**} of the closed unit ball of Y^{**}. In fact, by Alaoglu's theorem the closed unit ball in Y^{**} is weak*-compact. By Goldstine's theorem, C (the closed unit ball in Y) is weak*-dense in the closed unit ball of Y^{**}. Thus, since J^{**} is weak*-continuous, $JC = J^{**}C$ is weak*-dense in the image of the closed unit ball of Y^{**} under J^{**}.

Now if W is relatively weakly compact in X, then the weak closure $\bar{W}$ of W is weakly compact in X. Thus the sets

$$K_n = 2^n \bar{W} + 2^{-n}\{x^{**} \in X^{**}: \|x^{**}\| \leqq 1\}$$

all contain JC and are weak*-compact in X^{**}. Thus each K_n contains the weak*-closure of JC which is precisely the image of the closed unit ball of Y^{**} under J^{**}. Hence we have

$$\begin{aligned} J^{**}(\text{closed unit ball of } Y^{**}) &\subseteq \bigcap_{n=1}^{\infty} K_n \\ &\subseteq \bigcap_{n=1}^{\infty} X + 2^{-n}\{x^{**} \in X^{**}: \|x^{**}\| \leqq 1\} = X. \end{aligned}$$

Therefore $Y^{**} \subseteq J^{**-1}(X) \subseteq Y$. This shows that Y is reflexive when W is relatively weakly compact. The converse is obvious and the proof of Lemma 8 is complete.

Flowing from this lemma are a number of basic facts about weakly compact operators and weakly compact sets in Banach spaces. Only a few will be noted here. For a few more, see the notes and remarks section. The first fact should find its way into basic courses very quickly for it reduces much of the study of weakly compact operators to operators on or into reflexive Banach spaces.

COROLLARY 9. *Let $T: Z \to X$ be a weakly compact linear operator. Then there exist a reflexive Banach space Y and continuous linear operators $S: Z \to Y$ and R: $Y \to X$ such that $T = RS$.*

PROOF. In the notation of Lemma 8, let W be the image under T of the closed unit ball of Z. Then W is relatively weakly compact so that the Banach space Y of Lemma 8 is reflexive. Let $R = J$ and $S = J^{-1}T$.

An efficiency expert could use Lemma 8 to reorganize much of III.2 and III.3 in a very tight way. One possible rearrangement of III.2 and III.3 proceeds as follows. First prove that separable dual spaces have the Radon-Nikodým property. Then prove that a Banach space has the Radon-Nikodým property if and only if each of its separable subspaces has this property. Finally, factorization proves that all weakly compact operators on $L_1(\mu)$ factor through separable reflexive spaces (hence separable duals). Preferring a presentation that is closer to the genetic approach than the approach suggested above, we decided against using factorization in Chapter III.

Another immediate consequence of the construction on which Lemma 8 is based is the following fundamental corollary.

COROLLARY 10. *If K is a weakly compact subset of a Banach space X, then there exist a reflexive Banach space R and a continuous linear one-to-one operator T: $R \to X$ such that $T(R)$ contains K. Consequently a Banach space X is weakly compactly generated if and only if there is a reflexive Banach space R and a one-to-one bounded linear operator T: $R \to X$ such that $T(R)$ is dense in X.*

At this writing, the general structure of weakly compact subsets of $L_p(\mu, X)$ is unknown. In fact the next result which deals with this problem is not tractable by the methods of IV.1. It is an easy consequence of factorization.

COROLLARY 11 (DIESTEL). *Let X be a Banach space, let (Ω, Σ, μ) be a finite measure space and let $1 \leqq p < \infty$. If X is weakly compactly generated, then $L_p(\mu, X)$ is also weakly compactly generated.*

PROOF. According to Corollary 10, there is a reflexive Banach space R and a bounded linear operator $T: R \to X$ whose range is dense in X.

Now suppose $1 < p < \infty$. By IV.1, $L_p(\mu, R)$ is reflexive. Moreover, the operator $\tilde{T}: L_p(\mu, R) \to L_p(\mu, X)$ defined by $(\tilde{T}_p f)(\cdot) = T(f(\cdot))$ is a bounded linear operator with dense range. Hence $L_p(\mu, X)$ is weakly compactly generated.

If $p = 1$, let J be the natural inclusion of $L_2(\mu, X)$ into $L_1(\mu, X)$ and define S: $L_2(\mu, R) \to L_1(\mu, X)$ by $S = J\tilde{T}_2$ where $\tilde{T}_2$ is as above. Then S is a bounded linear operator with dense range and the proof is complete.

Next factorization is used to obtain any easy proof of a theorem of Grothendieck.

THEOREM 12 (GROTHENDIECK). *Let X, Y and Z be Banach spaces and suppose T: $X \to Y$ and $S: Y \to Z$ are continuous linear operators.*

If T is integral and S is weakly compact, then ST is nuclear.

If T is weakly compact and S is integral, then ST is nuclear into Z^{**}. If one assumes in this case that X^* has the approximation property, then ST is nuclear into Z.

PROOF. If T is integral and S is weakly compact, then by Corollary 9, there exists a reflexive Banach space W and continuous linear operators $A: Y \to W$ and B: $W \to Z$ such that $S = BA$. Now $AT: X \to W$ is an integral operator into the reflexive Banach space W. Consequently AT is Pietsch integral into a Banach space with the Radon-Nikodým property. Theorem VI.4.8 tells us that AT is nuclear so that $ST = BAT$ is also nuclear.

Now suppose that T is weakly compact and S is integral. Then $T^*: Y^* \to X^*$ is weakly compact by Gantmacher's theorem and $S^*: Z^* \to Y^*$ is integral by Corollary 2.11. By the first part of the proof, T^*S^* is nuclear. But $T^*S^* = (ST)^*$. Generally, we know that $ST = (ST)^{**}|_X$ so that ST is nuclear into Z^{**} since $(ST)^{**} = (T^*S^*)^*$ is the adjoint of a nuclear linear operator and therefore is nuclear. In case X^* possesses the approximation property then $ST: X \to Z$ has a nuclear adjoint and, by Theorem 3.8, is nuclear into Z.

Finally we close this section with a striking theorem of Rosenthal. Though this result is not dependent upon the methods of tensor products, it is included here because it follows (easily) from the factorization lemma (through Corollary 10) and serves as an excellent illustration of how factorization in tandem with the Dunford-Pettis-Phillips theorem can be used in Banach space theory.

THEOREM 13 (ROSENTHAL). *Let (Ω, Σ, μ) be a finite measure space. Then every weakly compact subset of $L_\infty(\mu)$ is norm separable.*

PROOF (FIGIEL). Let K be a weakly compact subset of $L_\infty(\mu)$. By Corollary 10 there is a reflexive Banach space R and a bounded linear operator $T: R \to L_\infty(\mu)$ such that $T(R)$ contains K. Clearly T is the adjoint of some bounded linear operator $S: L_1(\mu) \to R^*$. Since R is reflexive, R^* is also reflexive. Hence S is a weakly compact linear operator on $L_1(\mu)$. By III.2.12, the operator S has separable range, i.e., there exists a separable reflexive Banach space F through which S factors. Taking adjoints, one easily sees that T factors through a separable space and therefore has separable range.

5. Notes and remarks. The systematic study of tensor products of Banach spaces was initiated in the 1940's by Dunford and Schatten **[1946]**, Schatten **[1943a]**, **[1943b]**, **[1946]**, **[1947]** and Schatten and von Neumann **[1946]**, **[1948]**. Schatten **[1950]** organized and extended the basic results of these early papers. Also in this paper is the first appearance of the greatest and least reasonable crossnorms. Propositions 1.2, 1.3, 1.7, 1.8, and 1.9(a) can be found by leafing through the pages of Schatten **[1950]**.

It is clear from these early works that tensor products were developed as a tool for studying spaces of operators. Ironically, progress in the study of tensor products was impeded by the very single-mindedness of this objective. Though interpretations of tensor products as spaces of operators were known, little time was spent on other interpretations. Thus it was natural that those who studied tensor products of Banach spaces followed the path that was so successfully cleared in the study of tensor products of Hilbert spaces. The real breakthrough came with Grothendieck **[1955a]**, **[1955b]**, **[1956a]**, **[1956b]**. Some of his important contributions are surveyed in this chapter.

As we noted above, much of the first section is in Schatten **[1950]**. Proposition 1.9(b), however, makes its first appearance in Grothendieck **[1955a]** as does Proposition 1.12. Example 1.10 and Proposition 1.11 are due to Grothendieck. They are clear indications that vector measures may be of use in studying tensor products. Example 1.6 is also due to Grothendieck **[1955a]**. The identification of $L_1(\mu) \check{\otimes} X$ with the completion of the space of Pettis integrable strongly measurable functions is due to Gil de la Madrid **[1965]**, **[1966]**.

Theorem 2.1 was proved by Schatten and von Neumann **[1946]**, **[1948]** but curiously the derivation of Corollary 2.3 is due to Grothendieck **[1955b]**. Actually Grothendieck proved a considerably stronger theorem than Corollary 2.3:

THEOREM (GROTHENDIECK). *Let X be a Banach space such that whenever Y and Z are Banach spaces and Y is a closed linear subspace of Z, then $Y \hat{\otimes} X$ is a closed subspace of $Z \hat{\otimes} X$. Then X is isometrically isomorphic to $L_1(\mu)$ for some measure μ.*

Dixmier **[1953]**, Grothendieck **[1957b]** and Sakai **[1971]** have established relatives of this result in the context of C^* and W^* algebras. Stegall and Rutherford **[1972]** have generalized the above theorem to obtain an isomorphic classification of $\mathscr{L}_1$ spaces.

Call a norm on a tensor product $X_1 \otimes Y_1$ a $\otimes$-norm (Grothendieck **[1956a]**) if it arises as a special case of a method for norming $X \otimes Y$ for every pair of Banach spaces X and Y. The greatest and least reasonable crossnorms are excellent examples. Evidently the norm on $L_1(\mu, X)$ comes from a $\otimes$-norm on $L_1(\mu) \otimes X$. Kwapien **[1972a]** has shown that if $1 < p < \infty$ then there is no $\otimes$-norm such that for all μ and X the norm on $L_p(\mu, X)$ comes from a $\otimes$-norm. (In particular the $L_p(\mu, X)$ norm is not the completion of $L_p(\mu, X)$ $(1 < p < \infty)$ under either the greatest or least reasonable crossnorms, a fact which is also very easy to see directly.) Further investigation of the role of $L_p(\mu, X)$ in the study of operator ideals and tensor products can be found in Gordon and Saphar **[1976]**, Nielson **[1972]**, Persson **[1969]**, Persson and Pietsch **[1969]** and Saphar **[1972]**.

All the results in §2 dealing with integral operators are due to Grothendieck **[1955a]**. They obviously provide another point of entry for vector measures into Banach space theory.

A few remarks must be made at this juncture. The study of integral operators conducted in §2 gives good evidence of the importance of a basic advance due to Grothendieck: the method of factorization. The *idea* of factoring an operator through classical Banach spaces is simple and possibly occurred to others before Grothendieck. What sets the work of Grothendieck apart is that he not only thought about factoring operators; he did it! Integral operators factor through classical spaces and to Grothendieck this meant something. He unravelled its meaning and in so doing opened new vistas to functional analysts, the ramifications of which are only now being felt and understood with proper appreciation.

To further illustrate the power of Grothendieck's factorization methods (especially in the light of the theory of vector measures) consider the classes of 2-absolutely summing and absolutely summing operators. Each of these classes enjoys the "ideal" property established for nuclear operators in VI.4.2(ii). A basic fact about 2-summing operators (see Grothendieck **[1955a]**, Pietsch **[1967]**) is that if $T: X \to Y$ is 2-summing then there is a regular Borel measure on U_{X^*} such that for all $x \in X$

$$(*) \qquad \|Tx\|^2 \leq \int_{U_{X^*}} |x^*(x)|^2 \, d\mu(x^*).$$

It follows that T admits a factorization in the form

$$\begin{array}{ccc} X & \xrightarrow{\;T\;} & Y \\ {\scriptstyle A}\downarrow & & \uparrow{\scriptstyle B} \\ C(U_{X^*}) & \xrightarrow[\;I\;]{} & L_2(\mu) \end{array}$$

where $A: X \to C(U_{X^*})$, $B: L_2(\mu) \to Y$ are bounded linear operators and $I: C(U_{X^*}) \to L_2(\mu)$ is the natural inclusion. The noteworthy and simple fact here is that I is 2-summing. A direct consequence of $(*)$ and Schwarz's inequality is that the composition of two 2-summing operators is absolutely summing (i.e., 1-summing). What is not so clear is the following theorem of Grothendieck **[1955a]** which ultimately hinges on the fact that $L_2(\mu)$ has the Radon-Nikodým property.

THEOREM (GROTHENDIECK). *The composition of two 2-summing operators is nuclear.*

PROOF. Let $T: X \to Y$ and $S: Y \to Z$ be 2-summing operators. Then T and S admit the factorization

$$\begin{array}{ccccccc} X & \xrightarrow{\quad T \quad} & & Y & & \xrightarrow{\quad S \quad} & Z \\ {\scriptstyle A}\downarrow & & {\scriptstyle B}\nearrow & & \searrow{\scriptstyle C} & & \uparrow{\scriptstyle D} \\ C(U_{X^*}) & \xrightarrow[\;I_1\;]{} & L_2(\mu) & & C(U_{Y^*}) & \xrightarrow[\;I_2\;]{} & L_2(\nu) \end{array}$$

where A, B, C and D are bounded linear operators and I_1 and I_2 are the 2-summing natural inclusions. Now $CBI_1: C(U_{X^*}) \to C(U_{Y^*})$ is 2-summing by the ideal structure of 2-summing operators. Therefore $I_2CBI_1: C(U_{X^*}) \to L_2(\nu)$ is absolutely summing. *Since* $L_2(\nu)$ *has the Radon-Nikodým property,* I_2CBI_1 is nuclear by VI.4.7. Therefore $D(I_2CBI_1)A = ST$ is nuclear by VI.4.2(ii).

An immediate consequence of the above is the famous

DVORETSKY-ROGERS THEOREM. *If each unconditionally convergent series in the Banach space* X *is absolutely convergent then* X *is finite dimensional.*

PROOF. A Banach space X in which every unconditionally convergent series is absolutely convergent has an absolutely summing (hence 2-summing) identity operator. But the identity operator is idempotent! Hence it is nuclear (and, in particular, compact) by the last theorem.

Though Grothendieck had many spectacular successes in functional analysis, his most profound theorem was probably the slowest in gaining any recognition. It awaited the fundamental paper of Lindenstrauss and Pełczyński **[1968]** to be properly exposed. Reformulating Grothendieck's own work, Lindenstrauss and Pełczynski gave a proof of the

FUNDAMENTAL THEOREM OF THE METRIC THEORY OF TENSOR PRODUCTS. *Let* (a_{ij}) *be an* $n \times n$ *real matrix such that* $|\sum_{i,j=1}^{n} a_{i,j} t_i s_j| \leqq 1$ *whenever* $|t_i|, |s_j| \leqq 1$ *for* $i, j = 1, \cdots, n$. *Then there is a* $K > 0$ *such that if* H *is a Hilbert space with inner product* $(\ ,\ )$ *and* $x_1, \cdots, x_n, y_1, \cdots, y_n \in H$ *have norm one, then*

$$\left| \sum_{i,j=1}^{n} a_{i,j}(x_i, y_j) \right| \leqq k.$$

This inequality (sometimes known as Grothendieck's inequality) subsumes a number of classical inequalities of Hardy, Littlewood and Orlicz. Using it Grothendieck **[1956a]** and Lindenstrauss and Pełczyński **[1968]** were able to derive the following

THEOREM (GROTHENDIECK, LINDENSTRAUSS, PEŁCZYŃSKI). (1) *Every operator from an* $L_1(\mu)$ *space to a Hilbert space is absolutely summing.*

(2) *Every operator from a* $C(\Omega)$ *space to an* $L_p(\mu)$ *space* $(1 \leqq p \leqq 2)$ *is* 2-*summing.*

An amusing and simple consequence of the above results and the vector measure theory developed in Chapters VI and VIII is the

SIX LEMMA. *Let* X *and* Y *be Banach spaces and* $T: X \to Y$ *be a bounded linear operator that admits a factorization*

$$\begin{array}{ccccccccccc} X & & & & \xrightarrow{\quad\quad T \quad\quad} & & & & & & Y \\ {\scriptstyle S}\downarrow & & & & & & & & & & \uparrow{\scriptstyle D} \\ Z_1 & \xrightarrow[I]{} & Z_2 & \xrightarrow[W]{} & Z_3 & \xrightarrow[E]{} & Z_4 & \xrightarrow[L]{} & Z_5 & \xrightarrow[R]{} & Z_6 \end{array}$$

where D, R, L, E, W, I, *and* S *are all bounded linear operators and* Z_i, $i = 1, 2, 3, 4, 5, 6$, *are all Banach spaces of type* $L_1(\mu)$, $L_2(\mu)$ *or* $C(\Omega)$ *such that exactly two of the* Z_i'*s are of each type and no type appears consecutively in the above factorization. Then* T *is nuclear!*

Of course the converse holds.

Although it is clear that Banach and his coworkers knew many of the equivalent formulations of the approximation property (see Pełczyński [**1972**]) their first formal treatment is found in Grothendieck [**1955a**]. It was our good fortune to come into possession of some unpublished lecture notes of H. P. Rosenthal on the approximation property. These notes greatly influenced our presentation particularly since they delocally convexified many of the previously known proofs. We have not copied Rosenthal's notes; so please do not blame him for any criticisms that come to mind. The spirit of §3 is still that of Grothendieck [**1955a**].

Theorem 4.1 is implicit in Grothendieck [**1955a**]. Both Corollary 4.2 and Corollary 4.3 appear explicitly in Grothendieck [**1955a**]. In his notes, Rosenthal observes that these results do not depend on tensor products. In fact the reader may find some entertainment in stripping the tensor products from our proofs. Johnson, Rosenthal and Zippin [**1971**] have applied these results to some delicate structures in Banach spaces. Also appearing more or less implicitly in Grothendieck [**1955a**] are Theorems 4.6 and 4.7.

Theorem 4.1 is not true for arbitrary Banach spaces. Indeed Figiel and Johnson [**1973**] have produced a Banach space with the approximation property without the metric approximation property.

Reflexivity of spaces of operators. Theorem 4.4 is due to Ruckle [**1972**] and Holub [**1973**]. Kalton [**1974a**] removed the approximation assumptions by showing that if X and Y are reflexive and each bounded linear operator from X to Y is compact then $\mathscr{L}(X;\ Y)$ is reflexive. Feder and Saphar [**1975**] have shown that the space of compact operators between reflexive Banach spaces is either reflexive or is not a dual space; moreover Feder and Saphar demonstrate that the second possibility does indeed arise in nature.

That reflexive spaces have the Radon-Nikodým property plays a crucial part in practically all known general results regarding the reflexivity of spaces of operators. Possibly the most spectacular such result is due to Gordon, D. R. Lewis and Retherford [**1973**]:

THEOREM (GORDON-LEWIS-RETHERFORD). *Let X and Y be reflexive Banach spaces and suppose X has the approximation property. Then the space* AS$(X;\ Y)$ *of absolutely summing operators from X to Y is reflexive.*

The proof makes essential use of Corollary 4.2, Theorem 4.6 and the criteria for weak compactness in $L_1(\mu, X)$ presented in IV.2.

Recently D.R. Lewis [**1976**] has conducted an intensive study of $\otimes$ norms with the Radon-Nikodým property and in so doing has given a number of examples of reflexive spaces of operators ((p, q)-summing operators, quasi-p-integral operators, etc.).

Weak sequential completeness of $X \check{\otimes} Y$, $X \hat{\otimes} Y$. The most penetrating study of weak convergence in $X \check{\otimes} Y$ is that of D. R. Lewis [**1973**]. An easy consequence of the identification of $(X \check{\otimes} Y)^*$ with the integral bilinear functionals is that a sequence (u_n) in $X \check{\otimes} Y$ is weakly null if and only if, for each $x^* \in X^*$ and $y^* \in Y^*$, $\lim_n (x^* \otimes y^*)(u_n) = 0$. Armed with this Lewis derived a sufficient condition for conditional weak compactness and necessary and sufficient conditions for relative weak compactness in $X \check{\otimes} Y$. Unfortunately, one of his conditions for relative weak compactness requires information regarding the closure of a set of operators in the weak operator topology and is therefore likely to be difficult to apply. On the other hand, it is the first criterion available and so is bound to be the basis for future improvements.

In the direction of weak sequential completeness, D.R. Lewis **[1973]** has obtained the definitive result.

THEOREM (D. R. LEWIS). *Suppose X and Y are Banach spaces and one of them has the metric approximation property. The space $X \check{\otimes} Y$ is weakly sequentially complete if and only if* (i) *both X and Y are weakly sequentially complete and* (ii) *every weak*-to-weakly continuous linear operator $T: X^* \to Y$ is compact.*

Suppose (i) and (ii) hold and let $(u_n) \subseteq X \check{\otimes} Y$ be a weakly Cauchy sequence. For each $x^* \in X^*$, $(u_n(x^*))$ is weakly Cauchy in Y and, each $y^* \in Y^*$, $(u_n^*(y^*))$ is weakly Cauchy in X. By (i), $u_0 x^* = \text{weak} \lim_n u_n x^*$ and $v_0 y^* = \text{weak} \lim_n u_n^* y^*$ exist for each $x^* \in X^*$ and $y^* \in Y^*$ and both u_0 and v_0 are bounded linear operators. Moreover, if $x^* \in X^*$ and $y^* \in Y^*$,

$$\begin{aligned} y^*(v_0^* x^*) &= (v_0 y^*)(x^*) = \lim_n (u_n^* y^*)(x^*) \\ &= \lim_n y^*(u_n x^*) = y^*(u_0 x^*). \end{aligned}$$

So $v_0^* = u_0$ and u_0 is weak*-weakly continuous. By (ii), u_0 is compact. Since either X or Y has the approximation property, $u_0 \in X \check{\otimes} Y$. But if $x^* \in X^*$ and $y^* \in Y^*$ then

$$\begin{aligned} \lim_n (x^* \otimes y^*)(u_n) &= \lim_n y^*(u_n x^*) \\ &= y^* u_0(x^*) = (x^* \otimes y^*)(u_0) \end{aligned}$$

and so $u_n \to u_0$ weakly.

For the converse we first note that (ii) is easily replaced by: (ii′) every weak*-weakly continuous linear operator from Y to X is compact.

A moment's reflection on the symmetry of the injective tensor product and the above reformulation of (ii) allows us to assume that it is Y that has the metric approximation property and that $X \check{\otimes} Y$ is weakly sequentially complete. Both X and Y are isometric to a subspace of $X \otimes Y$ so both are weakly sequentially complete.

To prove that (ii) is also necessary let $w: X^* \to Y$ be a weak*-weakly continuous operator, i.e., $w^* Y^* \subseteq X$ and let (x_n^*) be a bounded sequence in X^*. To prove w is compact we will show that there exists $v \in X \check{\otimes} Y$ such that $wx_n^* = vx_n^*$ for all n!

Let Z be the closed linear space generated by the set $\{wx_1^*, wx_2^*, \cdots\}$. Then Z is a separable subspace of the closed linear span W of $\{wx^*: \|x^*\| \leq 1\}$, a weakly compactly generated subspace of Y. But Amir and Lindenstrauss **[1968]** have proved that for each separable subspace Z of a weakly compactly generated Banach space W there exists a separable subspace Z_0 of W containing Z and a norm one linear projection P from W onto Z_0. Keep in mind that the injectivity of $\check{\otimes}$ insures that

$$X \check{\otimes} Z \subseteq X \check{\otimes} Z_0 \subseteq X \check{\otimes} W \subseteq X \check{\otimes} Y,$$

where "$\subseteq$" denotes containment as a linear topological subspace under the natural identification.

Let (y_n) be a dense sequence in Z_0. Since Y has the metric approximation property, for each n there is a $u_n \in \mathscr{L}(X; Y)$ such that u_n has finite dimensional range, $\|u_n\| \leqq 1$ and $\|u_n y - y\| \leqq \|y\|/n$ for all $y \in$ linear span$\{y_1, \cdots, y_n\}$. Define $v_n = u_n \circ P \circ w$. It is plain that each v_n is a finite rank continuous linear operator whose adjoint takes Y^* into X. Therefore each $v_n \in X \check{\otimes} Y$. Moreover an easy calculation shows that if $x^* \in X^*$ and $y^* \in Y^*$ then $\lim_n y^* v_n x^* = y^* P w x^*$ so (v_n) is a weakly Cauchy sequence in $X \check{\otimes} Y$. Since $X \check{\otimes} Y$ is weakly sequentially complete, there is a $v \in X \check{\otimes} Y$ for which weak $\lim_n v_n = v$. But a quick check shows that $vx_n^* = Pwx_n^*$ for each n. This completes the proof.

COROLLARY. *If $1 \leqq p < 2$ then the completion $L_1(\mu) \check{\otimes} l_p$ of the space of measurable Pettis integrable functions into l_p is weakly sequentially complete.*

For $2 \leqq p$ the space $L_1(\mu) \check{\otimes} l_p$ is ***not*** *weakly sequentially complete.*

A basic open problem in the study of the projective tensor product of two Banach space is whether $X \hat{\otimes} Y$ is weakly sequentially complete whenever X and Y are. Other than the progress reported in the notes and remarks section of Chapter IV there is but one additional bit of information known: If $1 < p < \infty$ then $L_p(\mu) \hat{\otimes} Y$ is weakly sequentially complete whenever Y is. This is due to D. R. Lewis and Wojtazczyk [**1976**] and uses in an essential way the fact that $L_p[0, 1]$ has an unconditional basis.

The Radon-Nikodým property for spaces of operators. We cite three *typical* problems that remain open.

(1) If X and Y have the Radon-Nikodým property does $X \hat{\otimes} Y$? Alternatively, if X^* and Y have the Radon-Nikodým property does the space N$(X; Y)$ of nuclear operators from X to Y also have the Radon-Nikodým property?

(Diestel and Uhl [**1976**] note that the answer is affirmative whenever X and Y are dual spaces one of which has the approximation property.)

(2) If X^* and Y have the Radon-Nikodým property need AS$(X; Y)$ also have the property?

(Again if Y is a dual space with the approximation property then the response is known to be affirmative; the proof is identical to that of the Diestel-Uhl observation cited above with the Persson-Pietsch [**1969**] duality being used instead of the duality between injective and projective tensor products.)

(3) For $\mathscr{L}(X; Y)$ to have the Radon- Nikodým property is it necessary that every operator from X to Y be compact?

(Diestel and Morrison [**1976**] have some results which indicate the response should be affirmative.)

Integral and nuclear operators into L_1. Among the more beautiful results of Grothendieck [**1955a**] are those characterizing the integral and nuclear operators into $L_1(\mu)$. A subset K of $L_1(\mu)$ is *lattice bounded* if there is a $g \in L_1(\mu)$ such that $|f| \leqq g$ μ-almost everywhere for all $f \in K$. The set K is equimeasurable whenever given $\varepsilon > 0$ there is a set Ω_ε such that $\mu(\Omega \backslash \Omega_\varepsilon) \leqq \varepsilon$ and $\{f \chi_{\Omega_\varepsilon} : f \in K\}$ is a relatively compact subset of $L_\infty(\mu)$.

THEOREM (GROTHENDIECK). *Let X be any Banach space with unit ball B, (Ω, Σ, μ) be any measure space and $T: X \to L_1(\mu)$ be a bounded linear operator. Then*

(1) *T is integral if and only if $T(B)$ is lattice bounded.*

(2) *T is nuclear if and only if $T(B)$ is lattice bounded and equimeasurable.*

On the basis of Chapter VI, this result has a striking vector measure-theoretic interpretation. To wit: *a vector measure with values in an L_1 space is of bounded variation if and only if its range is lattice bounded in which case the measure has an approximate Radon-Nikodým derivative with respect to its variation if and only if its range is also equimeasurable.*

A few words on the proof of Grothendieck's result are in order.

If $T: X \to L_1(\mu)$ is an integral operator, then $T^*: L_\infty(\mu) \to X^*$ is also integral and therefore corresponds to a countably additive vector measure $F: \Sigma \to X^*$ whose variation $|F|$ is bounded. It is easy to check that $d|F|/d\mu = g$ is a bound for $T(B)$. On the other hand, if $T(B)$ is lattice bounded, then there is a $g \in L_1(\mu)$ so that $|Tx| \leqq g$ μ-almost everywhere for all $x \in B$. Let $\nu: \Sigma \to \boldsymbol{R}$ be the measure $\nu(E) = \int_E g\, d\mu$. The condition $|Tx| \leqq g$ μ-almost everywhere ensures that $T(B) \subseteq$ unit ball of $L_\infty(\nu)$. It follows that T admits the factorization

$$\begin{array}{ccc} X & \xrightarrow{\quad T \quad} & L_1(\mu) \\ {\scriptstyle A}\downarrow & & \uparrow{\scriptstyle C} \\ L_\infty(\nu) & \xrightarrow[\quad I \quad]{} & L_1(\nu) \end{array}$$

where $A: X \to L_\infty(\nu)$ is formally T, $I: L_\infty(\nu) \to L_1(\nu)$ is the natural (integral) inclusion operator and $C(h) = (h/g)\chi_{\{\omega\in\Omega: g(\omega)\neq 0\}}$. Since I is integral, the operator T is also integral.

If $T: X \to L_1(\mu)$ is nuclear, then T is integral; so $T(B)$ is lattice bounded. Also $T^*: L_\infty(\mu) \to X^*$ is nuclear and therefore from the results of VI.4, T^* is representable, i.e., there is a $g \in L_1(\mu, X^*)$ such that $T^*h = \int hg\, d\mu$ for all $h \in L_\infty(\mu)$. It follows that, for each $x \in X$, $(Tx)(\cdot) = g(\cdot)(x)$ holds μ-almost everywhere. It is easy to deduce from this that $T(B)$ is equimeasurable.

If $T(B)$ is lattice bounded, then there is a $g \in L_1(\mu)$ such that $|Tx| \leqq g$ μ-almost everywhere for all $x \in B$. Thus the range of T is supported by a σ-finite restriction of μ. Let (A_n) be an increasing sequence of sets of finite μ-measure such that $g = 0$ μ-almost everywhere outside of $\bigcup_n A_n$. By equimeasurability we can assume $T(B)|_{A_n}$ is relatively compact in $L_\infty(\mu|_{A_n})$ for each n. For each n let $T_n: X \to L_\infty(\mu|_{A_n})$ be given by $T_n x = Tx|_{A_n}$. For $\omega \in A_n$ define $f_n(\omega) \in X^*$ by $f_n(\omega)(x) = T_n(x)(\omega)$. Then each $f_n: A_n \to X^*$ satisfies $\|f_n\| \leqq g$ μ-almost everywhere and f_n is measurable (T_n is compact). Let $f: \Omega \to X^*$ be the map that on A_n coincides with f_n and is 0 outside $U_n A_n$. It is easy to see that $T^*h = \int hf\, d\mu$ for all $h \in L_\infty(\mu)$ and T^* is nuclear. This proves that T is nuclear.

Factorization of weakly compact and compact operators. As remarked in the text, we believe the factorization scheme of W. J. Davis, Figiel, Johnson and Pełczyński **[1974]** to be so basic and elegant that it is certain to be standard fare in elementary functional analysis texts quite soon. It provides easy proofs of a number of known results (witness Theorems 4.12, 4.13; also Gantmacher's theorem and much of Chapter III) and allows for new, previously inaccessible, results to be obtained (Corollary 4.11 is a simple example). Diestel and Faires **[1976]** have used factorization to show that if X^* has the Dunford-Pettis property then the injective tensor product of any weakly compact operator with a weakly compact operator whose domain is X is weakly compact. W. B. Johnson and Lindenstrauss **[1974]** use factorization to show every subspace of X is weakly compactly generated whenever both X and X^* are weakly compactly generated. The Davis-Figiel-Johnson-Pełczyński paper itself contains many applications obtaining new results as well as elegant proofs of some previously very difficult theorems. For instance, the fact that the unit ball of the dual of a weakly compactly generated space is weak* sequentially compact is given a much shorter proof than that of Amir-Lindenstrauss **[1968]** while the *method* of factorization shows that separable dual spaces imbed in spaces with boundedly complete bases. It is clear that the Davis-Figiel-Johnson-Pełczyński paper is a basic contribution to the study of weakly compact subsets of Banach spaces; as such it is recommended reading to vector measure theorists.

A related result of independent interest is due to W. B. Johnson **[1971]** and Figiel **[1973]** and ought to be mentioned.

THEOREM (FIGIEL, W. B. JOHNSON). *If T is a compact linear operator from X to Y, then there exist a reflexive Banach space R and compact linear operators $A: X \to R$ and $B: R \to Y$ such that T admits the factorization*

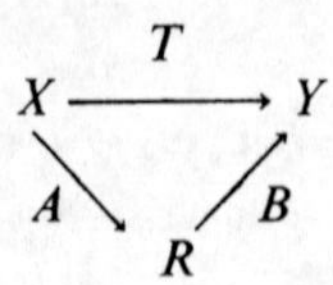

Both W. B. Johnson **[1971]** and Figiel **[1973]** give an in-depth analysis of this factorization scheme obtaining a number of interesting results, among them the conclusion that precisely one of Grothendieck's conjectures: "all reflexive Banach spaces have the approximation property" and "not all Banach spaces have the approximation property" is correct. In the pre-Enflo period it was not known which one!

Unanswered at present is the question: Does every unconditionally converging operator factor through a Banach space not containing c_0? An affirmative response in tandem with Pełczyński's elegant **[1960]** proof that every operator from a $C(\Omega)$ to a space not containing c_0 is weakly compact would allow a more compact presentation of §2 in Chapter VI. It could also have a number of consequences in the study of unconditionally converging operators on tensor products.

The theory of tensor products of Banach spaces is intimately related to the study of Banach operator ideals. There has been rapid progress in this subject during the past decade. Already there exist several excellent surveys of this field of investigation: the Pietsch **[1972]** "pre-book" and Retherford **[1975]** are highly recommended sources of information. The forthcoming monograph of Pietsch and the Springer-Verlag lecture notes of Lotz should also be consulted. Our comments on operator ideals have therefore been limited to those aspects in which vector measure theory has obviously (or could have) played a substantial role. It can be taken as an article of our faith that vector measure techniques will play an ever increasing role in the study of operator ideals to the mutual benefit of each area of investigation.

IX. THE RANGE OF A VECTOR MEASURE

The intriguing connection between the geometry of subsets of Banach spaces and vector measure theory is not confined to Radon-Nikodým considerations. The range of a vector measure has some vivid geometric properties, and we shall study some of these properties in this chapter.

Famous for its subtlety and utility, the Liapounoff convexity theorem is one of the central classical theorems of the theory of vector measures. In §1 we shall see that this theorem is false in infinite dimensional Banach spaces and is true in finite dimensional spaces for roughly the same reasons.

The second section deals with the celebrated theorem of Rybakov that says that if F is a strongly additive X-valued measure then $F \ll |x^*F|$ for some $x^* \in X^*$.

Extreme point phenomena of the range of a vector measure are studied in the third section. Here the main work centers around extreme points, denting points, exposed points and strongly exposed points of the range of a vector measure. We shall see how Rybakov's theorem is linked with the search for exposed and strongly exposed points in the range of a vector measure.

1. The Liapounoff Convexity Theorem. One of the most beautiful and best-loved theorems of the theory of vector measures is the Liapounoff Convexity Theorem which states that the range of a nonatomic vector measure with values in a finite dimensional space is compact and convex. Subtle enough to have intrigued many mathematicians, Liapounoff's theorem has been proved and reproved by a variety of methods over the years since Liapounoff's original proof in 1940. As interesting as the finite dimensional case is, the infinite dimensional version is even more subtle since it is apparently false. In fact, as we shall see later, Liapounoff's Convexity Theorem fails in every infinite dimensional Banach space.

EXAMPLE 1 (UHL). *A nonatomic vector measure of bounded variation whose range is closed but is blatantly nonconvex and noncompact.* Let Σ be the Borel sets in $[0, 1]$ and μ be Lebesgue measure. Define $G\colon \Sigma \to L_1(\mu)$ by $G(E) = \chi_E$. If $\pi \subseteq \Sigma$ is a partition of $[0, 1]$, it is evident that $\sum_{E\in\pi} \|G(E)\|_1 = \sum_{E\in\pi} \mu(E) = 1$. By an easy application of the Dominated Convergence Theorem, $G(\Sigma)$ is closed in $L_1(\mu)$. To see that $G(\Sigma)$ is not a convex set, note that $\frac{1}{2}\chi[0, 1] \in \operatorname{co}G(\Sigma)$ while if, $E \in \Sigma$,

$$\|G(E) - \tfrac{1}{2}\chi_{[0,1]}\|_1 = [\mu([0, 1]\backslash E) + \mu(E)]/2 = \tfrac{1}{2}.$$

To see that $G(\Sigma)$ is not compact, let $E_n = \{t \in [0, 1]: \sin(2^n \pi t) > 0\}$ for each positive integer n. A brief computation shows that $\| G(E_m) - G(E_n) \| = \frac{1}{4}$ for $m \neq n$. Hence $G(\Sigma)$ is not compact.

The range of a vector measure can fail to be convex for reasons far more subtle than those used above.

EXAMPLE 2 (LIAPOUNOFF). *An l_2-valued nonatomic vector measure of bounded variation whose range is not convex.* Let Σ be the Borel sets in $[0, 2\pi]$ and μ be Lebesgue measure. Select a complete orthogonal system $(w_n)_{n=0}^{\infty}$ in $L_2(\mu)$ such that each χ_n assumes only the values ± 1 and such that $w_0 = \chi_{[0,2\pi]}$ while $\int_0^{2\pi} w_n \, d\mu = 0$ for $n \geqq 1$ (the Walsh functions will work). For each n, define λ_n on Σ by

$$\lambda_n(E) = 2^{-n} \int_E [(1 + w_n(t))/2] \, d\mu(t),$$

$E \in \Sigma$. Define $G: \Sigma \to l_2$ by

$$G(E) = (\lambda_0(E), \lambda_1(E), \cdots, \lambda_n(E), \cdots).$$

Then $\|G(E)\|_{l_2} \leqq 2\mu(E)$ for each $E \in \Sigma$. Hence G is a countably additive vector measure of bounded variation which is evidently nonatomic. Now consider $G([0, 2\pi]) = (2\pi, \pi/2, \pi/4, \cdots)$ and suppose there is $E \in \Sigma$ such that $G(E) = G([0, 2\pi))/2$. Then we have

$$\pi = \lambda_0(E) = \int_E d\mu = \mu(E)$$

and for $n \geqq 1$ we have

$$2^{-n-1}\pi = \lambda_n(E) = 2^{-n} \int_E [(1 + w_n(t))/2] \, d\mu(t) = 2^{-n}\mu(E \cap U_n),$$

where $U_n = \{s \in [0, 2\pi] : w_n(s) = +1\}$. From this and the identities $\mu(U_n) = \mu(E) = \pi$, it follows that

$$\mu(E \cap U_n) = \mu(E \backslash U_n) = \mu(U_n \backslash E) = \mu([0, 2\pi] \backslash (E \cup U_n)) = \pi/2 \quad \text{for } n > 0.$$

Define $f = \chi_E - \chi_{[0,2\pi]\backslash E}$. Then we have $\int_0^{2\pi} f w_0 \, d\mu = \pi - \pi = 0$ and, for $n > 0$, we have

$$\begin{aligned} \int_0^{2\pi} f w_n \, d\mu &= \mu(U_n \cap E) + \mu([0, 2\pi] \backslash (E \cup U_n)) \\ &= \mu(E \backslash U_n) - \mu(U_n \backslash E) \\ &= 0. \end{aligned}$$

Since $f \in L_2([0, 2\pi])$ and $f \neq 0$ this contradicts the completeness of $(w_n)_{n=0}^{\infty}$ and shows that $G(\Sigma)$ is not convex.

After the devastation of Examples 1 and 2, let us see what can be salvaged in the infinite dimensional situation. Examples 1 and 2 suggest that nonatomicity may not be a particularly strong property of vector measures, particularly from the point of view of the Liapounoff theorem in the infinite dimensional context. Let us look at the finite dimensional situation in an attempt to understand what nonatomicity means.

Let $G: \Sigma \to \boldsymbol{R}^n$ have the form

$$G(E) = (\mu_1(E),\cdots,\mu_n(E)), \qquad E \in \Sigma,$$

where each μ_i is a countably additive finite signed scalar measure on Σ. Set $\mu(E) = \sum_{k=1}^n |\mu_k|(E)$. Then $\|G\|(E) \to 0$ if and only if $\mu(E) \to 0$. Now if G is nonatomic, then μ is nonatomic. Consequently, if $E \in \Sigma$ and $\mu(E) > 0$ the mapping on the infinite dimensional subspace $\{f\chi_E : f \in L_\infty(\mu)\}$ that takes f into $\int_E f\,dG$ is never one-to-one. It turns out that in the infinite dimensional case, this latter condition is precisely what is needed to make Liapounoff's theorem work. Before we see why this is true, let us collect a few facts from Chapter I. Throughout, Σ is a σ-field of subsets of Ω and X is a Banach space.

LEMMA 3. (a) *Let G: $\Sigma \to X$ be a countable additive vector measure. Then there exists a finite nonnegative countably additive (scalar) measure μ on Σ such that $\mu(E) = 0$ if and only if $G(E \cap F) = 0$ for all $F \in \Sigma$.*

(b) *If μ is such a measure, then the operator $f \to \int_\Omega f\,dG$, $f \in L_\infty(\mu)$, is continuous for the weak*-topology on $L_\infty(\mu)$ and the weak topology on X.*

(c) $\overline{\text{co}}\, G(\Sigma) = \{\int_\Omega f\,dG : 0 \leqq f \leqq 1, f \in L_\infty(\mu)\}$.

PROOF. Assertion (a) is Corollary I.2.6, while (b) is contained in Corollary I.2.7. To prove (c), let

$$U = \{f \in L_\infty(\mu) : 0 \leqq f \leqq 1\} \quad \text{and} \quad V = \left\{\int_\Omega f\,dG : f \in U\right\}.$$

Since U is weak*-compact, a glance at (b) proves that V is a weakly compact set which is plainly convex. In addition, we have

$$G(\Sigma) = \left\{\int_\Omega \chi_E\,dG : E \in \Sigma\right\} \subseteq V.$$

Hence $\overline{\text{co}}(G(\Sigma)) \subseteq V$. To prove the reverse inclusion, note that sums of the form $\sum_{i=1}^n \alpha_i G(E_i)$, $0 \leqq \alpha_1 \leqq \alpha_2 \leqq \cdots \leqq \alpha_n \leqq 1$, $E_i \cap E_j = \emptyset$ for $i \neq j$, are dense in V. Now we will sum the last summation by Abel partial summation; i.e.,

$$\sum_{i=1}^n \alpha_i G(E_i) = \beta_1 G\left(\bigcup_{i=1}^n E_i\right) + \sum_{j=2}^n \beta_j G\left(\bigcup_{i=j}^n E_i\right),$$

where $\beta_1 = \alpha_1$ and $\beta_j = \alpha_j - \alpha_{j-1}$ for $j \geqq 2$. Since $\sum_{j=1}^n \beta_j \leqq 1$ and $0 \in G(\Sigma)$, it follows that $\sum_{i=1}^n \alpha_i G(E_i) \in \text{co}(G(\Sigma))$. Hence $\text{co}(G(\Sigma))$ is dense in V and so $\overline{\text{co}}(G(\Sigma)) = V$.

With these formalities out of the way, the main theorem is ready for attack.

THEOREM 4 (KNOWLES; LIAPOUNOFF CONVEXITY THEOREM IN THE WEAK TOPOLOGY). *Let G and μ be as in the statement of Lemma 3. Any one of the following statements about G implies all the others.*

(i) *If $E \in \Sigma$ and $\mu(E) > 0$, then the operator $f \to \int_E f\,dG$ on $L_\infty(\mu)$ is not one-to-one on the subspace of functions in $L_\infty(\mu)$ vanishing off E.*

(ii) *For each $E \in \Sigma$, $\{G(A \cap E) : A \in \Sigma\}$ is a weakly compact convex set in X.*

(iii) *If $0 \neq f \in L_\infty(\mu)$, there exists a function $g \in L_\infty(\mu)$ such that $\|fg\|_\infty > 0$ but $\int_\Omega fg\,dG = 0$.*

PROOF. Note that (iii) implies (i). To show that (i) implies (iii), let $f \in L_\infty(\mu)$, $\|f\|_\infty$

> 0. Then there is an $\varepsilon > 0$ and $E \in \Sigma$ such that $|f\chi_E| > \varepsilon$ and $\mu(E) > 0$. According to (i), there is an $h \in L_\infty(\mu)$ such that $\|h\chi_E\|_\infty > 0$ and $\int_E h\, dG = 0$. Set $g = h/f$ on E and $g = 0$ off E. Then $fg = h$ on E and so $\|fg\chi_E\|_\infty > 0$. Also $\int_\Omega fg\, dG = \int_E h\, dG = 0$. Hence (iii) holds.

To check that (ii) implies (i), suppose (i) is false. Without loss of generality, we can and do assume that $E = \Omega$, so that the operator $f \to \int_\Omega f\, dG$ $(f \in L_\infty(\mu))$ is one-to-one. Evidently this means that $G(\Sigma) = \{\int_\Omega \chi_E\, dG : E \in \Sigma\}$ is a proper subset of $V = \{\int_\Omega f\, dG : f \in L_\infty(\mu), 0 \leqq f \leqq 1\}$. But by Lemma 3, $V = \overline{\text{co}}(G(\Sigma))$; thus $G(\Sigma)$ cannot be both closed and convex. Hence (ii) is false.

To complete the proof, we shall verify that (iii) implies (ii). Again it is enough to show that $G(\Sigma)$ is convex and weakly compact since the same argument can be applied to $\{G(A \cap E) : A \in \Sigma\}$ when $E \in \Sigma, E \neq \Omega$. Let $f \in L_\infty(\mu)$ be such that $0 \leqq f \leqq 1$. Since by Lemma 3, the operator $\int_\Omega(\cdot)\, dG$ is continuous for the weak*-topology on $L_\infty(\mu)$ and the weak topology on X, the set

$$H = \left\{g \in L_\infty(\mu) : 0 \leqq g \leqq 1, \int_\Omega f\, dG = \int_\Omega g\, dG\right\}$$

is a weak*-compact convex set in $L_\infty(\mu)$ and therefore has extreme points. If we can show that the extreme points of H, denoted by ext(H), are all contained in $\{\chi_E : E \in \Sigma\}$ it will then follow immediately that there exists $E \in \Sigma$ such that $G(E) = \int_\Omega f\, dG$. Then an appeal to Lemma 3(c) will yield the equalities

$$\overline{\text{co}}(G(\Sigma)) = \left\{\int_\Omega f\, dG : f \in L_\infty(\mu), 0 \leqq f \leqq 1\right\} = G(\Sigma)$$

and prove that $G(\Sigma)$ is weakly compact and convex.

To this end, suppose $f_0 \in$ ext H but $\|f_0 - \chi_E\|_\infty > 0$ for each $E \in \Sigma$. A simple calculation shows that there exists $f_1 \in L_\infty(\mu)$ with $\|f_1\|_\infty > 0$ such that $0 \leqq f_0 \pm f_1 \leqq 1$. An appeal to (iii) gives us a $g_1 \in L_\infty(\mu)$, that may be selected with $\|g_1\|_\infty \leqq 1$, such that $\|f_1 g_1\|_\infty > 0$ but $\int_\Omega f_1 g_1\, dG = 0$. Then $f_0 \pm f_1 g_1 \in H$; thus f_0 is *not* an extreme point of H, a contradiction. This completes the proof.

Of course, in light of the remarks before Lemma 3, the classical Liapounoff theorem follows as an immediate corollary.

COROLLARY 5 (LIAPOUNOFF). *Let Σ be a σ-field of subsets of Ω, X be a finite dimensional Banach space and $G: \Sigma \to X$ be a countably additive vector measure. If G is nonatomic, then the range of G is a compact convex subset of X.*

In addition, the finite dimensional version of Liapounoff's theorem can be used to show that the weak closure of the range of a nonatomic vector measure with values in a Banach space is weakly compact and convex. For this suppose $F: \Sigma \to X$ is countably additive and has no atoms. Then x^*F is a nonatomic signed measure for each $x^* \in X^*$. Hence if $x_1^*, x_2^*, \cdots, x_n^* \in X^*$, then $G = (x_1^*F, x_2^*F, \cdots, x_n^*F)$ is a measure whose range is compact and convex. Now if x is in the closed convex hull of $F(\Sigma)$, then it is easy to see that $(x_1^*(x), \cdots, x_n^*(x))$ belongs to the closed convex hull of $G(\Sigma)$. Hence $(x_1^*(x), x_2^*(x), \cdots, x_n^*(x)) \in G(\Sigma)$. In other words, there exists $E \in \Sigma$ such that $x_i^*(x) = x_i^*F(E)$ for all $1 \leqq i \leqq n$. This means x belongs to the weak closure of $F(\Sigma)$. So the weak closure of $F(\Sigma)$ is weakly compact (I.2.7) and convex.

Liapounoff's Example 2 is a concrete version of the proof of the next corollary which shows that Corollary 5 characterizes finite dimensional Banach spaces.

COROLLARY 6. *Let Σ be the σ-field of Borel subsets of* [0, 1]. *If X is an infinite dimensional Banach space, then there is a countably additive vector measure of bounded variation $G: \Sigma \to X$ and a set $E \in \Sigma$ such that $\{G(A \cap E): A \in \Sigma\}$ is not a weakly compact convex set in X.*

PROOF. Let μ be Lebesgue measure on Σ. Select a sequence (f_n) in $L_1(\mu)$ such that $\|f_n\|_1 = 1$ and such that the only $g \in L_\infty(\mu)$ with $\int_{[0,1]} f_n g \, d\mu = 0$ for all n is $g = 0$. Choose a sequence of pairs (x_n^*, x_n) such that $x_n^* \in X^*$, $x_n \in X$, $x_m^*(x_n) = 0$ if $m \neq n$ and $x_n^*(x_n) = 1$ for all n (the existence of such a biorthogonal system is not hard to establish). Define $T: L_\infty(\mu) \to X$ by

$$T(g) = \sum_{n=1}^{\infty} x_n (2^n \|x_n\|)^{-1} \int_{[0,1]} f_n g \, d\mu.$$

If $T(g) = 0$, then $x_n^* T(g) = (2^n \|x_n\|)^{-1} \int_{[0,1]} f_n g \, d\mu = 0$ for all n. Hence T is one-to-one on $L_\infty(\mu)$. To produce the advertised measure, define, for $E \in \Sigma$, $G(E) = T(\chi_E)$. It is not hard to see that G is countably additive and that $T(\cdot) = \int_{[0,1]} (\cdot) \, dG$. Hence G violates (i) of Theorem 4. By Theorem 4, G is as advertised.

The next corollary gives an internal characterization of vector measures that obey Liapounoff's theorem.

COROLLARY 7. *Let Σ be a σ-field of subsets of Ω and $G: \Sigma \to X$ be a countably additive vector measure. For each $A \in \Sigma$ the set $\{G(A \cap E): E \in \Sigma\}$ is a weakly compact convex subset of X if and only if for each $B \in \Sigma$ there is a set $E \in \Sigma$ such that $G(E \cap B) = G(B)/2$.*

PROOF. The necessity is obvious. To prove the sufficiency, suppose there is a set $A \in \Sigma$ such that $G(\Sigma \cap A)$ is not a weakly compact convex subset of X. Let μ be a finite measure related to G as in Lemma 3. Then necessarily $\mu(A) > 0$. According to Theorem 4, there is a set $B \in \Sigma$ with $B \subseteq A$ and $\mu(B) > 0$ such that the operator $f \to \int_B f \, dG$ is one-to-one on the subspace of functions f in $L_\infty(\mu)$ vanishing off B. Now if there is a set $E \in \Sigma$ with $E \subseteq B$ and $G(E) = G(B)/2$, then $G(B \backslash E) = G(B)/2$ as well. Hence

$$\int_B (\chi_E - \chi_{B \backslash E}) \, dG = (G(B) - G(B))/2 = 0,$$

a contradiction which completes the proof.

Now we shall use Theorem 4 for some positive examples.

EXAMPLE 8. *An $L_1(\mu)$-valued measure whose range is weakly compact and convex.* Let (Ω, Σ, μ) be a finite measure space and λ be Lebesgue measure on the Borel σ-field B of subsets of [0, 1]. Let Ω_1 be $\Omega \times [0, 1]$ and Σ_1 be the product σ-field $\Sigma \times B$. Define a vector measure $G: \Sigma_1 \to L_1(\mu)$ by

$$G(E)(\omega) = \int_{[0,1]} \chi_E(\omega, t) \, d\lambda(t),$$

for $E \in \Sigma_1$ and $\omega \in \Omega$.

Suppose $E \in \Sigma_1$ and $\|G\|(E) \neq 0$. For $(\omega, t) \in E$, set

$$\alpha(\omega) = \left(\int_{[0,1]} (t\chi_E(\omega, t)\, d\lambda(t) \Big/ \int_{[0,1]} \chi_E(\omega,t)\, d\lambda(t)\right),$$

observing the convention $0/0 = 0$. Now if $(\omega, t) \in \Omega_1$ set

$$f(\omega, t) = [t - \alpha(\omega)]\, \chi_E(\omega, t).$$

An easy computation shows that f is a bounded Σ_1-measurable function on Ω_1. Moreover, $|f|$ is positive on a set A with $\|G\|(A) \neq 0$. On the other hand,

$$\left(\int_E f\, dG\right)(\omega) = \int_{[0,1]} f(\omega, t)\, \chi_E(\omega, t)\, d\lambda(t) = 0$$

for all $\omega \in \Omega$. Thus Theorem 4 guarantees that $G(\Sigma)$ is weakly compact and convex.

Building on Example 8 is

EXAMPLE 9. *A c_0-valued measure whose range is weakly compact and convex.* Let Ω be $[0, 1] \times [0, 1]$ and Σ be the σ-field of Borel subsets of Ω. Let μ be Lebesgue measure on $[0, 1]$ and $\mu \times \mu$ be Lebesgue product measure on Σ. Define $\lambda_n\colon \Sigma \to R$ by

$$\lambda_n(E) = \int_E t^n\, d\mu \times d\mu(s, t)$$

and set

$$G(E) = (\lambda_1(E), \lambda_2(E), \cdots, \lambda_n(E), \cdots).$$

Then $G\colon \Sigma \to c_0$ is a countably additive vector measure. We shall show that G has a weakly compact and convex range.

For $E \in \Sigma$ and $t \in [0, 1]$ set

$$F(E)(t) = \int_{[0,1]} \chi_E(s, t)\, d\mu(s).$$

Then $F(E) \in L_1([0, 1])$ and, by Example 8, F has a weakly compact and convex range. Define $T\colon L_1([0, 1]) \to c_0$ by $T(g) = \left(\int_{[0,1]} t^n g(t)\, d\mu(t)\right)_{n=1}^{\infty}$ for $g \in L_1([0,1])$. The operator T is bounded and linear. Also $TF(E) = G(E)$. Since F has weakly compact and convex range and since T is weakly continuous and linear, it follows that G has a weakly compact and convex range.

In tandem, Examples 2 and 9 reveal another negative aspect of the Liapounoff theorem: Knowing that a vector measure arises as an indefinite Bochner integral seems to be of little importance in determining whether the vector measure in question has a weakly compact and convex range. Indeed, the measures of Examples 2 and 9 both arise as indefinite Bochner integrals. In any case, the next result tells us that the *closure* of the range of the measure of Example 2 is both convex and *norm* compact, while the range of the measure of Example 9 has a convex and norm compact range.

THEOREM 10. (UHL; WEAK LIAPOUNOFF THEOREM FOR THE STRONG TOPOLOGY). *Let Σ be a σ-field of subsets of Ω and suppose X has the Radon-Nikodým property. If $G\colon \Sigma \to X$ is of bounded variation, nonatomic and countably additive, then the norm closure of $G(\Sigma)$ is convex and norm compact.*

PROOF. Let μ be the variation of G. Since X has the Radon-Nikodým property, there is a function $f \in L_1(\mu, X)$ such that

$$G(E) = \int_E f\,d\mu, \quad E \in \Sigma.$$

For each partition π of Ω into members of Σ, define the operator T_π: $L_\infty(\mu) \to X$ by $T_\pi(g) = \int_\Omega E_\pi(f) g\,d\mu$, where $E_\pi(f) = \sum_{E \in \pi} (\int_E f\,d\mu/\mu(E))\chi_E$ and the usual $0/0 = 0$ convention is in force. Since π contains only finitely many sets, each T_π is a compact linear operator. If T: $L_\infty(\mu) \to X$ is defined by $T(g) = \int_\Omega fg\,d\mu$ then

$$\| T(g) - T_\pi(g) \| \leqq \| E_\pi(f) - f \|_1 \, \| g \|_\infty.$$

An appeal to Lemma III.2.1 shows that $\lim_\pi \| T - T_\pi \| = 0$ in the uniform operator topology. Hence T is compact. Thus $G(\Sigma) = \{T(\chi_E): E \in \Sigma\}$ is relatively compact in X.

To prove that the closure of $G(\Sigma)$ is convex, let x_1 and x_2 be in the closure of $G(\Sigma)$. Let $\varepsilon > 0$ and choose E_1, E_2 such that $\| x_i - G(E_i) \| < \varepsilon/2$ for $i = 1, 2$. Choose a partition π that refines the trivial partitions $\{E_1, \Omega\backslash E_1\}$, $\{E_2, \Omega\backslash E_2\}$ and satisfies $\| E_\pi(f) - f \|_1 < \varepsilon/2$. Then $\int_{E_i} E_\pi(f)\,d\mu = G(E_i)$ for $i = 1, 2$. Moreover, the measure $\int_{(\cdot)} E_\pi(f)\,d\mu$ has a convex range since G (and therefore μ) is nonatomic and has finite dimensional range.[7] Thus if $0 \leqq \alpha \leqq 1$, there is a set $E_0 \in \Sigma$ such that $\int_{E_0} E_\pi(f)\,d\mu = \alpha G(E_1) + (1 - \alpha) G(E_2)$. Accordingly, we have

$$\begin{aligned} \| \alpha x_1 + (1-\alpha) x_2 - G(E_0) \| &\leqq \alpha \| x_1 - G(E_1) \| + (1-\alpha) \| x_2 - G(E_2) \| \\ &\quad + \left\| \int_{E_0} E_\pi(f)\,d\mu - \int_{E_0} f\,d\mu \right\| \\ &< \alpha\varepsilon/2 + (1-\alpha)\,\varepsilon/2 + \varepsilon/2 = \varepsilon. \end{aligned}$$

This completes the proof.

After understanding Theorem 10, we begin to understand just how subtle Liapounoff's Example 2 is, for the range of the measure in Liapounoff's Example 2 has a convex and norm compact norm closure. Thus we see that, even though Examples 1 and 2 purport to be examples of the same phenomenon, they are very different examples.

2. Rybakov's theorem. Throughout this section let Σ be a σ-field of subsets of the point set Ω and X be a Banach space. If F: $\Sigma \to X$ is a countably additive vector measure then the Bartle, Dunford and Schwartz Theorem I.2.6 produces a finite nonnegative real-valued measure μ on Σ such that $F \ll \mu$. What this theorem does not say directly is that μ may be taken to be of the form $|x^*F|$ for certain x^*'s in X^*. This is Rybakov's theorem and is the central topic of this section.

LEMMA 1. *Let μ_1 and μ_2 be countably additive finite measures on Σ. Then $\mu_i \ll |t\mu_1 + (1-t)\mu_2|$ obtains for all but countably many real numbers t.*

PROOF. Let $\lambda = |\mu_1| + |\mu_2|$ and f_i be the Radon-Nikodým derivative of μ_i with respect to λ, $i = 1, 2$. For each real number α, let

$$E_\alpha = \{\omega \in \Omega: f_1(\omega) + \alpha f_2(\omega) = 0\} \cap \{\omega \in \Omega: f_1(\omega) \text{ or } f_2(\omega) \neq 0\}.$$

[7]This can be seen directly without recourse to the Liapounoff theorem.

If $\alpha \neq \beta$, then $E_\alpha \cap E_\beta = \emptyset$. Hence for all α outside a countable set J, $f_1(\omega) + \alpha f_2(\omega)$ is nonzero for λ-almost all $\omega \in \{\omega \in \Omega : f_1(\omega) \text{ or } f_2(\omega) \neq 0\}$. It follows that $\mu_i \ll |\mu_1 + \alpha\mu_2|$, $i = 1, 2$, for all α outside J. Thus $\mu_i \ll |t\mu_1 + (1-t)\mu_2|$ ($i = 1, 2$) for all t in the set $\{(\alpha + 1)^{-1} : \alpha \in J, \alpha \neq -1\}$. This completes the proof.

THEOREM 2 (RYBAKOV). *Let $F: \Sigma \to X$ be a countably additive vector measure. Then there is $x^* \in X^*$ such that $F \ll |x^*F|$.*

PROOF. The collection of scalar measures $\{x^*F : \|x^*\| \leqq 1\}$ is convex and uniformly countably additive. By the proof of I.2.4, there is a sequence (x_n^*) in the unit ball of X^* such that

$$F \ll \sum_{n=1}^{\infty} \beta_n |x_n^* F| = \lambda$$

where $\beta_n \geqq 0$ and can be selected such that $\sum_{n=1}^{\infty} \beta_n = 1$. Let $y_1^* = x_1^*$ and, proceeding by induction, use the lemma to find y_k^* in the unit ball of X^* such that $y_{k-1}^* F \ll |y_k^* F|$ and $x_k^* F \ll |y_k^* F|$. Evidently $|x_k^* F| \ll |y_k^* F|$ and $|y_{k-1}^* F| \ll |y_k^* F|$ for all k. For each k, let f_k be the Radon-Nikodým derivative of $y_k^* F$ with respect to λ. Let $E_k = \{\omega \in \Omega : f_k(\omega) = 0\}$. Since $|y_k^* F| \ll |y_{k+1}^* F|$ for all k, there is no loss of generality in choosing the f_k's such that $E_{k+1} \subseteq E_k$ for all k. Let $E = \bigcap_{k=1}^{\infty} E_k$. Then $\lambda(E) = 0$ since $|x_k^* F|(E) = 0$ for all k. Thus $\lim_k \lambda(E_k) = 0$. Pass to a subsequence, if necessary, to have $\lambda(E_k) < 2^{-(k+1)}$ for all k and note that this does not ruin anything.

Next, two sequences (α_n) and (δ_n) of positive real numbers will be defined. Let $\alpha_1 = 1$ and choose $\delta_1 > 0$ such that

$$\lambda(\{\omega \in \Omega : |f_1(\omega)| < \delta_1\}) < \tfrac{1}{2}.$$

Let $g_1 = \alpha_1 f_1$ $(= f_1)$. Suppose $\alpha_1, \cdots, \alpha_{n-1}$ and $\delta_1, \cdots, \delta_{n-1}$, have been defined. Let $g_{n-1} = \sum_{k=1}^{n-1} \alpha_k f_k$. By the proof of Lemma 1, select a number α_n such that

(a) $0 < \alpha_n \leqq 2^{-n+1}$,

(b) $g_n = g_{n-1} + \alpha_n f_n$ is nonzero λ-almost everywhere on $\{\omega : g_{n-1}(\omega) \text{ or } f_n(\omega) \neq 0\}$, and

(c) $\lambda(\{\omega : |\alpha_n f_n(\omega)| \geqq \delta_{n-1}/4\}) \leqq 2^{-n}$.

By the fact that $\lambda(\{\omega \in \Omega : f_n(\omega) = 0\}) < 2^{-n+1}$ and (b), select δ_n such that $0 < \delta_n < \delta_{n-1}/4$ and $\lambda(\{\omega : |g_n(\omega)| \leqq \delta_n\}) < 2^{-n}$.

Now since $0 < \alpha_n \leqq 2^{-n+1}$ and (f_n) is $L_1(\lambda)$-bounded, it follows that $\sum_{n=1}^{\infty} |\alpha_n f_n|$ converges in $L_1(\lambda)$-norm and hence λ-almost everywhere since it is a series of nonnegative functions. Hence

$$g = \sum_n \alpha_n f_n = \lim_n g_n$$

converges in $L_1(\lambda)$-norm and λ-almost everywhere.

Now let $A = \{\omega \in \Omega : g(\omega) = 0\}$. Suppose $\lambda(A) = a > 0$ and let $A_n = \{\omega : |g_n(\omega)| > \delta_n\}$ and $B_n = \{\omega : |\alpha_n f_n(\omega)| \geqq (\delta_{n-1})/4\}$. Since $\lambda(\{\omega : |g_n(\omega)| \leqq \delta_n\}) < 2^{-n}$, we see that $\lim_n \lambda(A \cap A_n) = \lambda(A) = a$. By (c), we have $\lambda(B_n) < 2^{-n}$ for all n. Choose n_0 so large that $\lambda(A \cap A_{n_0}) > a/2$ and $\sum_{n=n_0+1}^{\infty} 2^{-n} < a/3$. Then

$$\lambda\Big((A \cap A_{n_0}) \setminus \bigcup_{n=n_0+1}^{\infty} B_n\Big) > a/6,$$

and in particular there is a point $\omega \in (A \cap A_{n_0}) \setminus \bigcup_{n=n_0+1}^{\infty} B_n$ such that $g(\omega) = \sum_{n=1}^{\infty} \alpha_n f_n(\omega)$. But then

$$\begin{aligned} |g(\omega)| &= \left| g_{n_0}(\omega) - \sum_{n=n_0+1}^{\infty} (-\alpha_n f_n(\omega)) \right| \\ &\geqq |g_{n_0}(\omega)| - \sum_{n=n_0+1}^{\infty} |\alpha_n f_n(\omega)| \\ &> \delta_{n_0} - \sum_{i=1}^{\infty} \delta_{n_0} 4^{-i} \\ &= (2/3)\, \delta_{n_0} > 0, \end{aligned}$$

which contradicts the fact that $\omega \in A$ and proves that $\lambda(A) = 0$.

Now let $x^* = \sum_{n=1}^{\infty} \alpha_n y_n^* / \sum_{n=1}^{\infty} \alpha_n$. Note that $\|x^*\| \leqq 1$. Also note that, by the Radon-Nikodým theorem, $(\sum_{i=1}^{\infty} \alpha_i)^{-1} g$ is the Radon-Nikodým derivative of x^*F with respect to λ. Since $\lambda(\{\omega: g(\omega) = 0\}) = 0$, we see that $\lambda \ll |x^*F|$. Thus $F \ll \lambda \ll |x^*F|$. This completes the proof.

COROLLARY 3 (WALSH). *If $F: \Sigma \to X$ is a countably additive measure the collection of $x^* \in X^*$ such that $F \ll |x^*F|$ is dense in X^*.*

PROOF. Apply Lemma 1.

COROLLARY 4. *Corollary 3 remains true for strongly additive vector measures on fields of sets.*

PROOF. Apply the Stone space techniques of I.5.

The final corollary is a consequence of the proof of Theorem 2 and Lemma 1.

COROLLARY 5. *Let K be a bounded convex collection of countably additive scalar measures on Σ. If K is uniformly countably additive, the measures $\mu \in K$ such that $\lim_{|\mu|(E)\to 0} |\lambda|(E) = 0$ uniformly in $\lambda \in K$ is dense in K.*

3. Extreme point phenomena. Extreme point phenomena prosper in the theory of vector measures. For convex subsets of spaces with the Radon-Nikodým property this fact was graphically illustrated in Chapter VII. In this section, we shall see that the closed convex hull of the range of a vector measure has its own vivid extremal structure.

Throughout this section Σ is a σ-field of subsets of a point set Ω and X is a Banach space.

THEOREM 1. *Let $F: \Sigma \to X$ be a countably additive vector measure. Every extreme point of $\overline{\text{co}}(F(\Sigma))$ belongs to $F(\Sigma)$.*

PROOF. Select μ as in the statement of Lemma 1.3 and observe that if $A \in \Sigma$ and $\mu(A) > 0$, then there is a subset $B \subseteq A$, $B \in \Sigma$, such that $F(B) \neq 0$. Now suppose x is an extreme point of $\overline{\text{co}}(F(\Sigma))$. According to Lemma 1.3, there is an $f \in L_\infty(\mu)$ with $0 \leqq f \leqq 1$ such that $x = \int_\Omega f\, dF$. Now if f is not μ-almost everywhere equal to the characteristic function χ_E of some set $E \in \Sigma$, then there is an $\varepsilon > 0$ and a set $A \in \Sigma$ of positive μ-measure such that $\varepsilon\chi_A \leqq f\chi_A \leqq (1 - \varepsilon)\chi_A$. In addition A can be selected such that $F(A) \neq 0$. Let $f_n = f + (-1)^n \varepsilon \chi_A$, $n = 1, 2$. Then $0 \leqq f_n \leqq 1$ and

$$x = \left(\int_\Omega f_1\,dF + \int_\Omega f_2\,dF\right)\Big/2.$$

Moreover, $\int_\Omega f_1\,dF \neq \int_\Omega f_2\,dF$ and, by Lemma 1.3, both of these integrals are in $\overline{\text{co}}(F(\Sigma))$. Thus x is not an extreme point, a contradiction which shows f is of the form χ_E and that $x \in F(\Sigma)$. This completes the proof.

The following technical corollary of the proof will be useful later.

COROLLARY 2. *In the notation of the proof of Theorem 1, if $f \in L_\infty(\mu)$, $0 \leqq f \leqq 1$, and $\int_\Omega f\,dF$ is an extreme point of $\overline{\text{co}}\,F(\Sigma)$, then $f = \chi_E$ μ-almost everywhere for some set $E \in \Sigma$ that is uniquely determined within a set of μ-measure zero.*

PROOF. Only the uniqueness assertion needs to be checked. This follows directly from the special property of μ.

COROLLARY 3 (ANANTHARAMAN). *Let $F: \Sigma \to X$ be a countably additive vector measure. A point of $\overline{\text{co}}(F(\Sigma))$ is an exposed point of $\overline{\text{co}}(F(\Sigma))$ if and only if it belongs to $F(\Sigma)$ and is an exposed point of $F(\Sigma)$.*

PROOF. All exposed points of $\overline{\text{co}}(F(\Sigma))$ belong to $F(\Sigma)$ because they are all extreme points of $\overline{\text{co}}(F(\Sigma))$.

For the rest of the proof, let K be the weak closure of $F(\Sigma)$. If $x^* \in X^*$, the scalar measure x^*F has a closed and bounded range. Hence

$$\sup x^*F(\Sigma) = \sup x^*(K) = \sup x^*\,\overline{\text{co}}(F(\Sigma)).$$

It follows that an exposed point of $\overline{\text{co}}(F(\Sigma))$ is an exposed point of K and that an exposed point of K is an exposed point of $F(\Sigma)$.

On the other hand, if $x \in F(\Sigma)$ is exposed by $x^* \in X^*$, then x^* supports $\overline{\text{co}}(F(\Sigma))$ in a weakly compact convex face K_1 of $\overline{\text{co}}(F(\Sigma))$. All the extreme points of K_1 are also extreme points of $\overline{\text{co}}(F(\Sigma))$ and therefore belong to $F(\Sigma)$. It follows from Corollary 2 that $K_1 = \{x\}$. Hence x^* exposes $\overline{\text{co}}(F(\Sigma))$ at x. This completes the proof.

If D is a subset of a Banach space X let us agree that $x \in X$ is a *denting point* of D if there is no $\varepsilon > 0$ such that $x \in \overline{\text{co}}(D\backslash B_\varepsilon(x))$.

THEOREM 4 (ANANTHARAMAN). *Let $F: \Sigma \to X$ be a countably additive vector measure. The extreme points of $\overline{\text{co}}(F(\Sigma))$ are all denting points of $\overline{\text{co}}(F(\Sigma))$.*

PROOF. Proceeding by contradiction, suppose x_0 is an extreme point of $\overline{\text{co}}(F(\Sigma))$ that is not a denting point of $\overline{\text{co}}(F(\Sigma))$. It follows easily that there exists an $\varepsilon > 0$ and a sequence (x_n) in the convex hull of $\overline{\text{co}}(F(\Sigma)\backslash B_\varepsilon(x_0))$ such that $\lim_n x_n = x_0$. Now write $x_n = \sum_{i=1}^{k(n)} \alpha_i(n) y_i(n)$ where $\alpha_i(n) \geqq 0$, $\sum_{i=1}^{k(n)} \alpha_i(n) = 1$, $y_i(n) \in \overline{\text{co}}(F(\Sigma))$ and $\|y_i(n) - x_0\| \geqq \varepsilon$ for all n and i. Let μ be as in the statement of Lemma 1.3. According to Lemma 1.3 there is for each n and i a function $f_i(n) \in L_\infty(\mu)$ with $0 \leqq f_i(n) \leqq 1$ μ-almost everywhere such that $y_i(n) = \int_\Omega f_i(n)\,dF$. Write $f_n = \sum_{i=1}^{k(n)} \alpha_i(n) f_i(n)$ and note that $0 \leqq f_n \leqq 1$. Now (f_n) has a subnet (g_α) converging in the weak*-topology of $L_\infty(\mu)$ to a function $f_0 \in L_\infty(\mu)$. Evidently $0 \leqq f_0 \leqq 1$ μ-almost everywhere. But the operator $f \to \int_\Omega f\,dF$ from $L_\infty(\mu)$ to X is weak*-to-weakly continuous by Lemma 1.3. Hence $x_0 = \lim_n x_n = \text{weak}\lim_\alpha \int_\Omega g\hat{\alpha}\,dF$. Now by Corollary 2, $f_0 = \chi_{E_0}$ for some $E_0 \in \Sigma$. If the operator $f \to \int_\Omega f\,dF$ from $L_\infty(\mu)$ to X is denoted by T, we have shown that χ_{E_0} belongs to the weak*-closure in $L_\infty(\mu)$ of $\text{co}(T^{-1}(\overline{\text{co}}(F(\Sigma)\backslash B_\varepsilon(x_0))))$. But $\{f \in L_\infty(\mu): 0 \leqq f \leqq 1\}$ and $\{f \in L_1(\mu): 0 \leqq f \leqq 1\}$ are set-theoretically identical and the identity mapping between them establishes a homeomorphism from the former in its weak*-topology to the latter in its weak topology. It follows that χ_{E_0} is in the weak closure of the convex set

$$\text{co}\,(\{f \in L_1(\mu): 0 \leqq f \leqq 1\}\backslash T^{-1}\,(B_\varepsilon(x_0))).$$

Since T is $L_1(\mu)$-norm continuous on $\{f \in L_1(\mu): 0 \leqq f \leqq 1\}$ to X, $T^{-1}(B_\varepsilon(x_0))$ contains an $L_1(\mu)$-δ-ball about χ_{E_0}; hence by Mazur's theorem

$$\chi_{E_0} \in \overline{\text{co}}(\{f \in L_1(\mu): 0 \leqq f \leqq 1\}\backslash B_\delta(x_0)).$$

Thus χ_{E_0} is not a denting point of $\{f \in L_1(\mu): 0 \leqq f \leqq 1\}$. But χ_{E_0} is strongly exposed by χ_{E_0} if $\mu(E_0) > 0$ or is strongly exposed by χ_Ω if $\mu(E_0) = 0$. Hence by V.3.10, the element χ_{E_0} is a denting point of $\{f \in L_1(\mu): 0 \leqq f \leqq 1\}$, a contradiction which completes the proof.

The next theorem establishes a concrete relationship between Rybakov's theorem and exposing linear functionals. It is proved by repeated applications of the Hahn Decomposition Theorem.

THEOREM 5 (ANANTHARAMAN). *Let $F\colon \Sigma \to X$ be a countably additive vector measure and $x^* \in X^*$. Then $F \ll |x^*F|$ if and only if x^* exposes $F(\Sigma)$.*

In particular, any one of the following statements about $E \in \Sigma$ and $x^* \in X^*$ imply all the others:

(1) *$F(\Sigma)$ is exposed at $F(E)$ by x^*.*

(2) *$F \ll |x^*F|$ and x^*F achieves its maximum value at E.*

(3) *If μ is as in Lemma 1.3 and $A \in \Sigma$ satisfies $\mu(A \,\Delta\, E) > 0$, then $x^*F(A) < x^*F(E)$.*

PROOF. Plainly the first assertion follows directly from the equivalence of (1) and (2).

To prove that (1) implies (2), let $F(E)$ be exposed by x^*. Suppose there is a set $A \in \Sigma$ with $|x^*F|(A) = 0$ but $F(A) \neq 0$. Let P be a positive set for x^*F (and $\Omega\backslash P$ be a negative set) in the sense of the Hahn decomposition. Either $F(A \cap P)$ or $F(A\backslash P)$ is not zero. Suppose $F(A \cap P)$ is not zero. Since $|x^*F|(A \cap P) = 0$, we have

$$x^*F(P\backslash A \cap P) = x^*F(P) = \max x^*F(\Sigma).$$

Since x^* exposes $F(\Sigma)$ at $F(E)$, it follows that $F(P\backslash(A \cap P)) = F(P)$. But

$$F(P\backslash(A \cap P)) = F(P) - F(A \cap P) \neq F(P),$$

a contradiction. If $F(A\backslash P)$ is not zero replace $A \cap P$ by $A\backslash P$ above to obtain a contradiction.

To prove that (2) implies (3), suppose $F \ll x^*F$ and x^*F achieves its maximum at E. Let P and $\Omega\backslash P$ be the Hahn decomposition of x^*F into its positive and negative sets respectively. Then $x^*F(E) = x^*F(P)$. Accordingly,

$$\begin{aligned} x^*F(E\backslash P) &= x^*F(E) - x^*F(E \cap P) \\ &= x^*F(P) - x^*F(E \cap P) \\ &= x^*F(P\backslash E). \end{aligned}$$

But if $x^*F(P\backslash E) > 0$, then $x^*F((P\backslash E) \cup E) > x^*F(E)$. On the other hand,

$x^*F(P\backslash E) \geqq 0$. Hence $x^*F(P\backslash E) = 0$ and therefore $|x^*F|(P\backslash E) = 0$. Similarly $|x^*F|(E\backslash P) = 0$. This proves that (2) implies (3) since $|x^*F| \ll \mu$ and $\mu \ll |x^*F|$.

For the proof that (3) implies (1), let P and $\Omega\backslash P$ again be the Hahn decomposition of Ω into positive and negative sets with respect to x^*F respectively. Evidently $x^*F(E) = x^*F(P)$ and by hypothesis every subset of $E \Delta P$ in Σ has F measure zero. Thus

$$\begin{aligned} F(E) &= F(E \cap P) + F(E\backslash P) \\ &= F(E \cap P) \\ &= F(E \cap P) + F(P\backslash E) \\ &= F(P). \end{aligned}$$

This completes the proof.

Corollary 6 (Anantharaman). *Let $F: \Sigma \to X$ be a countably additive vector measure. The exposed points of $F(\Sigma)$ are strongly exposed in $\overline{\text{co}}(F(\Sigma))$.*

Proof. Let $x^* \in X^*$ expose a point $F(E)$ in $\overline{\text{co}}(F(\Sigma))$. By the last theorem, $F \ll |x^*F|$, E is a positive set and $\Omega\backslash E$ is a negative set for the Hahn decomposition of Ω. By Lemma 1.3, we have

$$\overline{\text{co}}(F(\Sigma)) = \left\{\int_\Omega f\,dF : 0 \leqq f \leqq 1, f \in L_\infty(|x^*F|)\right\}.$$

Now if (f_n) is a sequence in $L_\infty(|x^*F|)$ with $0 \leqq f_n \leqq 1$ for all n such that

$$\lim_n x^* \textstyle\int f_n\,dF = x^*F(E),$$

we must show

$$\lim_n \left\| \int_\Omega f_n\,dF - F(E)\right\| = 0.$$

Now we have

$$\begin{aligned} \lim_n \int_\Omega |\chi_E - f_n|\,d|x^*F| &= \lim_n \left[\int_E |1 - f_n|\,d|x^*F| + \int_{\Omega\backslash E} |f_n|\,d|x^*F|\right] \\ &= \lim_n \left[\int_E (1 - f_n)\,dx^*F - \int_{\Omega\backslash E} f_n\,dx^*F\right] \\ &= \lim_n \left[x^*F(E) - x^*\int_\Omega f_n\,dF\right] = 0. \end{aligned}$$

Therefore $\lim_n f_n = \chi_E$ in $L_1(|x^*F|)$-norm. But by II.4.1 integration with respect to F is $L_1(|x^*F|)$-norm continuous on $\{f \in L_\infty(\mu): 0 \leqq f \leqq 1\}$. Hence $\lim_n \|\int_\Omega f_n\,dF - F(E)\| = 0$. This completes the proof.

Corollary 7. *Let $F: \Sigma \to X$ be a countably additive vector measure. The collection of members of X^* that strongly expose $\overline{\text{co}}\, F(\Sigma)$ is dense in X^*.*

Proof. This is a direct consequence of Corollary 2.3, Theorem 5 and Corollary 6.

4. Notes and remarks. Liapounoff **[1940]** proved the finite dimensional version of Theorem 1.4. As one can easily see from the proof of Theorem 1.10, the real issue in proving Liapounoff's theorem is proving that the range is closed. Convexity comes along as an automatic bonus. Several authors have treated the finite dimensional version of Liapounoff's theorem. See Sierpinski **[1922]**, Halmos **[1948]**, Blackwell **[1951a]**, **[1951b]**, Dvoretsky, Wald and Wolfowitz **[1951a]**, Olech **[1966]**, Lindenstrauss **[1966c]** and Hermes and LaSalle **[1969]**.

The Lindenstrauss **[1966c]** proof is the most incisive and elegant proof of Liapounoff's theorem in the finite dimensional case. Many of the ideas on which Lindenstrauss based his proof can be found in Karlin **[1953]** but the Lindenstrauss proof is in no sense an immediate extension of Karlin's work. The infinite dimensional version of Liapounoff's theorem remained resistant to analysis for a long time. Kingman and Robertson **[1968]** pierced its armor in a special case and the general case fell victim to Knowles **[1974]** who combined some of the ideas of Lindenstrauss, Kingman and Robertson with his own and thus proved Theorem 1.4.

It is well known that Liapounoff's theorem in the finite dimensional case is intimately related to the "bang-bang" principle of optimal control theory (cf. Hermes and LaSalle **[1969]**, LaSalle **[1960]** and Olech **[1966]**). Thanks to Theorem 1.4, it is now available for use in infinite dimensional optimal control theory. For more on this see Kluvanek and Knowles **[1974a]**, **[1975]** and Knowles **[1976]**.

Corollary 1.7 is from Kluvanek and Knowles **[1975]**; it has its finite dimensional origins in Halmos **[1948]**. Examples 1.8 and 1.9 are from Kluvanek and Knowles **[1975]**. Example 1.8 is the essential content of a similar theorem of Romanovskii and Sudakov **[1965]**. Example 1.2 is from Liapounoff **[1946]**; see also Pełczyński **[1959]**. Example 1.1 and Theorem 1.10 are from Uhl **[1969e]**.

Those interested in the locally convex version of Liapounoff's theorem should consult Kluvanek and Knowles **[1975]** for an excellent discussion of this topic.

Recently the finite dimensional version of the Liapounoff Convexity Theorem has been used to advantage in Banach space theory. See Hagler and Stegall **[1973]**, Maurey **[1975]** and Dor **[1975b]**.

The Bartle, Dunford and Schwartz Theorem I.2.6. was proved nearly twenty years before Rybakov **[1970]** proved Theorem 2.2. There is an alternative proof of Rybakov's theorem that is highly Banach space theoretic. According to Lindenstrauss **[1963]**, a weakly compact subset of a Banach space with an equivalent strictly convex norm has an exposed point. Amir and Lindenstrauss **[1968]** showed that every weakly compact set in a Banach space lives in a strictly convexifiable Banach space. Hence weakly compact sets have exposed points. This fact together with Theorem 3.5 provides a proof of Rybakov's theorem that might be favored by some.

A strengthened form of Corollary 2.3 was proved by Walsh **[1971]** who showed that $\{x^* \in X^*: F \ll |x^*F|\}$ is a dense G_δ subset of X^*. Walsh also obtained a result related to Corollary 2.5. Related results have been obtained for group-valued measures by Drewnowski **[1973b]**, **[1974b]** and Musial **[1973a]**, **[1973b]**.

Our treatment of Rybakov's theorem is taken from some unpublished notes written by R. Huff and P. Morris **[1973]**.

Knowles **[1976]** has isolated a puzzlingly interesting class of vector measures F such that $F \ll |x^*F|$ for all nonzero $x^* \in X^*$. These measures are closely related to the concept of normality in time optimal control theory.

Theorem 3.1 is due to Tweddle **[1972]** and Kluvanek **[1973]**. The rest of §3 is from the beautiful paper of Anantharaman **[1973]** who also treats the locally convex case. Traces of §3 can be found in Liapounoff **[1940]** and in Halmos **[1948]**.

Most of the existing knowledge about the range of a vector measure rests on the Bartle, Dunford and Schwartz Theorem I.2.7 and Liapounoff's Theorem 1.5. These theorems highlight the special nature of the range of a vector measure but by no means exhaust the expanding body of information concerning these sets.

One interesting fact about the range of a vector measure is due to Bolker **[1969]** and Kluvanek and Knowles **[1975]**. (Ryll-Nardzewski obtained this next result but never published it.)

THEOREM. *If X is a Banach space, Σ is a σ-field and $F\colon \Sigma \to X$ is a countably additive vector measure, then the closed convex hull of $F(\Sigma)$ is the range of a vector measure that obeys Liapounoff's Theorem* 1.5.

PROOF. Let $\mathscr{B}$ be the Borel σ-field of subsets of $[0, 1]$ and $\lambda\colon \mathscr{B} \to [0, 1]$ be Lebesgue measure. Let $\Omega' = \Omega \times [0, 1]$ and Σ' be the σ-field generated by $\Sigma \times \mathscr{B}$. Define $F'\colon \Sigma' \to X$ by

$$F'(E) = \int_\Omega \lambda(\{t \in [0,1] : (\omega, t) \in E\})\, dF(\omega), \qquad E \in \Sigma'.$$

Evidently F' is a $\mu \times \lambda$ continuous X-valued vector measure, where $\mu\colon \Sigma \to [0, 1]$ is countably additive and F is μ-continuous.

We shall show first that $F'(\Sigma') = \overline{\mathrm{co}}(F(\Sigma)) = \{\int f\, dF\colon 0 \leqq f \leqq 1,\ f \in L_\infty(\mu)\}$. It is clear that $F'(\Sigma') \subseteq \overline{\mathrm{co}}(F(\Sigma))$. To check the reverse inclusion, let $f\colon \Omega \to [0, 1]$ be μ-measurable and look at $E' = \{(\omega, t) \in \Omega' : 0 \leqq t \leqq f(\omega)\} \in \Sigma'$. Plainly $F'(E') = \int_\Omega f\, dF$. Hence $F'(\Sigma') = \overline{\mathrm{co}}\,(F(\Sigma))$.

Now we shall show that F' obeys Theorem 1.5. Let f be a $\mu \times \lambda$ bounded measurable function. By Fubini's theorem if $x^* \in X^*$, then

$$\begin{aligned} x^*\left(\int_{\Omega'} f(\omega, t)\, dF'(\omega, t)\right) &= \int_\Omega \int_0^1 f(\omega, t)\, dx^*F'(\omega, t) \\ &= \int_\Omega \int_0^1 f(\omega, t)\, d\lambda(t)\, dx^*F(\omega) \\ &= x^* \int_\Omega \int_0^1 f(\omega, t)\, d\lambda(t)\, dF(\omega). \end{aligned}$$

Therefore we have

$$\int_{\Omega'} f(\omega, t)\, dF'(\omega, t) = \int_\Omega \int_0^1 f(\omega, t)\, d\lambda(t)\, dF(\omega).$$

Now, if $E \in \Sigma'$ is not F'-null, then

$$f(\omega, t) = \left[t - \frac{\int_0^1 t\chi_E(\omega, t)\, d\lambda(t)}{\int_0^1 \chi_E(\omega, t)\, d\lambda(t)}\right]\chi_E$$

is a bounded $\mu \times \lambda$ measurable function on Ω' which is not F'-null. Proceeding as in Example 1.9, we find $\int_E f\, dF' = 0$. An appeal to Theorem 1.4 finishes the proof.

The range of a finite dimensional vector measure. Considerable attention has been paid to the problem of characterizing those sets in a finite dimensional space that are ranges of vector measures. Not every closed bounded convex subset of a finite dimensional l_p space is the range of a vector measure. Lindenstrauss **[1964d]** showed that the unit ball of any two dimensional Banach space is the range of a vector measure but the unit ball of three dimensional l_1 is not. Bolker **[1969]** showed that for $2 \leqq p \leqq \infty$, the unit ball of an n-dimensional l_p space is the range of a vector measure and conjectured that this is not the case for $1 < p < 2$. This conjecture has been proven true by Dor **[1976]**. For more information on the range of a finite dimensional vector measure see Dvoretsky, Wold, and Wolfowitz **[1951a]**, **[1951b]**, Blackwell **[1951a]**, **[1951b]**, Bregtagnolle, Dacunha-Castelle and Krivine **[1966]**, Bolker **[1966]**, Halmos **[1948]**, Herz **[1963]**, Kaufman and Rickert **[1966]**, Rickert **[1967a]**, **[1967b]**, Schneider **[1975]** Schwarz **[1967]**, Witsenhausen **[1973]**, as well as the papers mentioned above.

The range of an infinite dimensional vector measure. The characterization of those sets in an infinite dimensional Banach space that are the ranges of vector measures is a difficult problem. The known results seem to be either very general or very specific and there is not yet a satisfactory relationship between the general results and the specific results.

Definitive general results have been obtained by Kluvanek **[1975]**, **[1976]**. Bolker **[1969]** has shown that a set K in a finite dimensional space is the closed convex hull of the range of a vector measure if and only if it is a zonoid. There is a highly nontrivial extension of Bolker's theorem to the infinite dimensional situation. This is the subject of Kluvanek **[1975]**, **[1976]**. Here the properties of conical measures and closed vector measures and the Daniell integral are brought to bear in a crucial way. The work of Choquet **[1969]** is essential to Kluvanek's treatment.

Kluvanek **[1975]** also established a relation between negative definite functions and the range of a vector measure. Let us agree that if X is a Banach space a real-valued function ϕ on X^* is *negative definite* if for any collection $x_1^*, x_2^*, \cdots, x_n^*$ in X^* and real numbers $\alpha_1, \alpha_2, \cdots, \alpha_n$ with $\sum_{j=1}^n \alpha_j = 0$ the inequality

$$\sum_{j=1}^{n} \sum_{i=1}^{n} \alpha_i \alpha_j \phi(x_i^* - x_j^*) \leqq 0$$

holds. Kluvanek **[1975]**, **[1976]** proved that a weakly compact convex set K in a Banach space that is symmetric about zero is the closed convex hull of the range of a countably additive vector measure on a σ-field if and only if the function $x^* \to \sup|x^*(K)|$ is negative definite.

Anantharaman and Garg **[1976]** have obtained alternate characterizations of the range of a vector measure.

Banach (Kaczmarz and Steinhaus **[1951**, p. 250]) gave a crisp proof of the fact that the unit ball of l_2 is the range of a countably additive vector measure. More generally, Bregtagnolle, Dacunha-Castelle and Krivine **[1966]** and Rosenthal **[1973]** showed that the unit ball of $L_p[0, 1]$ and of l_p for $2 \leqq p \leqq \infty$ is the range of a vector measure. A consequence of Grothendieck's inequality (Grothendieck **[1956a]**, Lindenstrauss and Pełczyński **[1968]**) is the fact that if $1 < p < 2$ then the ball of $L_p[0, 1]$ and of l_p is not the range of a vector measure (see Diestel and Seifert **[1976]**).

Diestel and Seifert **[1976]** have shown that a weakly compact order interval in a Banach lattice is the range of a countably additive vector measure on a σ-field. In the same paper they show that the range of a strongly additive vector measure on a field has the Banach-Saks property, i.e., every sequence of values of a strongly additive vector measure has a subsequence whose arithmetic means converge in norm.

Landers **[1973]** has shown that the range of a nonatomic vector measure is arcwise connected.

I. Aharoni,
[1974] *Every separable metric space is Lipschitz equivalent to a subset of c_0*, Israel J. Math. **19**, 284–291.

C. A. Akemann,
[1967] *The dual space of an operator algebra*, Trans. Amer. Math. Soc. **126**, 286–302. MR **34** #6549.
[1968] *Sequential convergence in the dual of a W^*-algebra*, Comm. Math. Phys. **7**, 222–224. MR **37** #3363.

C. A. Akemann, P. G. Dodds and J. L. B. Gamelen,
[1972] *Weak compactness in the dual space of a C^*-algebra*, J. Functional Analysis **10**, 446–450. MR **49** #9637.

L. Alaoglu,
[1940] *Weak topologies of normed linear spaces*, Ann. of Math. (2) **41**, 252–267. MR **1**, 241.

G. D. Alexander,
[1972] *Banach spaces possessing a Radon-Nikodým property*, Rev. Roumaine Math. Pures Appl. **17**, 485–487. MR **46** #5991.
[1976] *Nuclear operators on the space of continuous functions* (to appear).

G. D. Alexander and C. Swartz,
[1973] *Linear operators on C_X*, Czechoslovak Math. J. **23 (98)**, 231–234. MR **47** #3949.

A. Alexiewicz,
[1949] *On sequences of operations*. I, Studia Math. **11**, 1–30. MR **12**, 418.

I. Amemiya and Kôji Shiga,
[1957] *On tensor products of Banach spaces*, Kōdai Math. Sem. Rep. **9**, 161–178. MR **21** #3766.

D. Amir,
[1962a] *Continuous functions' spaces with the bounded extension property*, Bull. Res. Council Israel Sect. F **10F**, 133–138. MR **26** #592.
[1962b] *Continuous functions' spaces with the separable projection property*, Bull. Res. Council Israel Sect. F **10F**, 163–164. MR **27** #566.
[1964] *Projections onto continuous function spaces*, Proc. Amer. Math. Soc. **15**, 396–402. MR **29** #2634; **30**, p. 1205.

D. Amir and J. Lindenstrauss,
[1968] *The structure of weakly compact sets in Banach spaces*, Ann. of Math. (2) **88**, 35–46. MR **37** #4562.

R. Anantharaman,
[1973] *On exposed points of the range of a vector measure*, Vector and Operator Valued Measures and Applications (Proc. Sympos., Snowbird Resort, Alta, Utah, 1972), Academic Press, New York, pp. 9–22. MR **48** #11436.

R. Anantharaman and K. M. Garg,
[1976a] *On the range of a vector measure* (to appear).
[1976b] *Openness and monotonicity of a vector measure* (to appear).

N. J. M. Anderson and J. P. R. Christensen,
[1973] *Some results on Borel structures with applications to subseries convergence in Abelian topological groups*, Israel J. Math. **15**, 414–420.

T. Andô,
[1961] *Convergent sequences of finitely additive measures*, Pacific J. Math. **11**, 395–404. MR **25** #1255.

P. Antosik,
[1971] *On the Mikusiński diagonal theorem*, Bull. Acad. Polon. Sci. Sér. Sci. Math. Astronom. Phys. **19**, 305–310. MR **45** #5714.

[1972] *A generalization of the diagonal theorem*, Bull. Acad. Polon. Sci. Sér. Sci. Math. Astronom. Phys. **20**, 373–377. MR **48** #2162.

[1973a,b] *Mappings from L-groups into topological groups*. I, II, Bull. Acad. Polon. Sci. Sér. Sci. Math. Astronom. Phys. **21**, 145–153; ibid. **21**, 155–160. MR **47** #7425.

[1976] *The uniform approach to Nikodym and Vitali-Hahn-Saks type theorems* (to appear).

V. I. Arkin and V. L. Levin,

[1972] *Convexity of values of vector integrals, theorems of measurable sampling and variational problems*, Uspehi Mat. Nauk **27**, no. 3 (165), 21–77 = Russian Math. Surveys **27**, no. 3, 21–85.

N. Aronszajn and P. Panitchpakdi,

[1956] *Extension of uniformly continuous transformations and hyperconvex metric spaces*, Pacific J. Math. **6**, 405–439. MR **18**, 917.

G. Arsene and Ş. Strătilă,

[1965] *Prolongement des mesures vectorielles*, Rev. Roumaine Math. Pures Appl. **10**, 333–338. MR **32** #180.

E. Asplund,

[1968] *Fréchet differentiability of convex functions*, Acta Math. **121**, 31–47. MR **37** #6754.

[1976] *Boundedly Krein-compact Banach spaces*, Aarhus preprint.

G. F. Bachelis and H. P. Rosenthal,

[1971] *On unconditionally converging series and biorthogonal systems in a Banach space*, Pacific J. Math **37**, 1–5. MR **47** #778.

W. G. Bade and P. C. Curtis, Jr.,

[1960] *The Wedderburn decomposition of commutative Banach algebras*, Amer. J. Math. **82**, 851–866. MR **23** #A529.

J. W. Baker,

[1973] *Projection constants for C(S) spaces with the separable projection property*, Proc. Amer. Math. Soc. **41**, 201–204. MR **47** #9242.

J. W. Baker and J. Wolfe,

[1976] *Averaging operators and C(S)-spaces with the separable projection property*, Canad. J. Math. (to appear).

S. Banach,

[1932] *Théorie des operations linéaires*, Monografie Mat., Vol. 1, Warzawa.

[1937] *Appendix to S. Saks's book, "Theory of the integral"*, Monografie Mat. **7**, Warsaw.

R. G. Bartle,

[1956] *A general bilinear vector integral*, Studia Math. **15**, 337–352. MR **18**, 289.

[1966] *The elements of integration*, Wiley, New York. MR **34** #293.

R. G. Bartle, N. Dunford and J. T. Schwartz,

[1955] *Weak compactness and vector measures*, Canad. J. Math. **7**, 289–305. MR **16**, 1123.

J. Batt,

[1967] *Integraldarstellungen linearer Transformationen und scwache Kompaktheit*, Math. Ann. **174**, 291–304. MR **36** #6960.

[1969] *Applications of the Orlicz-Pettis theorem to operator-valued measures and compact and weakly compact transformations on the space of continuous functions*, Rev. Roumaine Math. Pures Appl. **14** 907–935.

[1970] *Nonlinear compact mappings and their adjoints*, Math. Ann. **189**, 5–25. MR **42** #8329.

[1972] *Strongly additive transformations and integral representations with measures of nonlinear operators*, Bull. Amer. Math. Soc. **78**, 474–478. MR #**45** 5831.

[1973a] *Die Verallgemeinerungen des Darstellungssatzes von F. Riesz und ihre Anwendungen*, Jber. Deutsch. Math. Verein. **74**, 147–181.

[1973b] *Nonlinear integral operators on $C(S, E)$*, Studia Math. **48**, 145–177. MR **49** #1110.

[1974] *On weak compactness in spaces of vector-valued measures and Bochner integrable functions in connection with the Radon-Nikodym property of Banach spaces*, Rev. Roumaine Math. Pures Appl. **19**, 285–304. MR **49** #5831.

J. Batt and E. J. Berg,

[1969] *Linear bounded transformations on the space of continuous functions*, J. Functional Analysis **4**, 215–239. MR **40** #1798.

J. Batt and H. Konig,

[1959] *Darstellung linearer transformationen durch vektorwertige Riemann-Stieltjes integrale*, Arch. Math. (Basel) **19**, 273–287.

B. Beauzamy,

[1976] *Opérateurs uniformement convexifiants* (to appear).

A. Beck,

[1961] *On the strong law of large numbers*, Ergodic Theory (Proc. Internat. Sympos., Tulane Univ., New Orleans, La., 1961), Academic Press, New York, pp. 21–53. MR **28** #3470.

[1962] *A convexity condition in Banach spaces and the strong law of large numbers*, Proc. Amer. Math. Soc. **13**, 329–334. MR **24** #A3681.

G. Bennett and N. J. Kalton,

[1973] *Inclusion theorems for K spaces*, Canad. J. Math. **25**, 511–524. MR **48** #836.

E. J. Berg (see J. Batt).

C. Bessaga,

[1972] *Topological equivalence of non-separable Banach spaces*, Sympos. on Infinite-Dimensional Topology, Ann. of Math. Studies, no. 69, Princeton Univ. Press, Princeton, N. J; Univ. of Tokyo Press, Tokyo, pp. 3–14. MR **49** #11514.

C. Bessaga and A. Pełczyński,

[1958] *On bases and unconditional convergence of series in Banach spaces*, Studia Math. **17**, 151–164. MR **22** #5872.

[1966] *On extreme points in separable conjugate spaces*, Israel J. Math **4**, 262–264. MR **35** #2126.

[1975] *Topics in infinite dimensional topology*, Monografie Mat., Warsaw.

R. Bilyeu and P. Lewis,

[1976] *Some mapping properties of representing measures* (to appear).

G. Birkhoff,

[1935] *Integration of functions with values in a Banach space*, Trans. Amer. Math. Soc. **38**, 357–378.

E. Bishop and K. de Leeuw,

[1959] *The representations of linear functionals by measures on sets of extreme points*, Ann. Inst. Fourier (Grenoble) **9**, 305–331. MR **22** #4945.

E. Bishop and R. R. Phelps,

[1961] *A proof that every Banach space is subreflexive*, Bull. Amer. Math. Soc. **67**, 97–98. MR **23** #A503.

[1963] *The support functionals of a convex set*, Proc. Sympos. Pure Math., vol. 7, Amer. Math. Soc., Providence, R. I., pp. 27–35. MR **27** #4051.

D. Blackwell,

[1951a] *The range of certain vector integrals*, Proc. Amer. Math. Soc. **2**, 390–395. MR **12**, 810.

[1951b] *On a theorem of Lyapunov*, Ann. Math. Statist. **22**, 112–114. MR **12**, 486.

S. Bochner,

[1933a] *Integration von Funktionen deren Werte die Elemente eines Vectorraumes sind*, Fund. Math. **20**, 262–276.

[1933b] *Absolut-additive abstrakte Mengenfunktionen*, Fund. Math. **21**, 211–213.

[1933c] *Abstrakte fastperiodische Funktionen*, Acta Math. **61**, 149–184.

[1939] *Additive set functions on groups*, Ann. of Math. (2) **40**, 769–799. MR **1**, 110.

[1940] *Finitely additive integral*, Ann. of Math. (2) **41**, 495–504. MR **1**, 336.

S. Bochner and R. S. Phillips,

[1941] *Additive set functions and vector lattices*, Ann. of Math. (2) **42**, 316–324. MR **2**, 315.

S. Bochner and A. E. Taylor,

[1938] *Linear functionals on certain spaces of abstractly-valued functions*, Ann. of Math. (2) **39**, 913–944.

W. M. Bogdanowicz,

[1965a] *A generalization of the Lebesgue-Bochner-Stieltjes integral and a new approach to the theory of integration*, Proc. Nat. Acad. Sci. U.S.A. **53**, 492–498. MR **31** #306.

[1965b] *Integral representation of linear continuous operators from the space of Lebesgue-Bochner summable functions into any Banach space*, Proc. Nat. Acad. Sci. U.S.A. **54**, 351–354. MR **32** #4538.

[1966] *An approach to the theory of Lebesgue-Bochner measurable functions and to the theory of measure*, Math. Ann. **164**, 251–269. MR **34** #1483.

W. M. Bogdanowicz and B. Kritt,

[1967] *Radon-Nikodym differentiation of one vector-valued volume with respect to another*, Bull. Acad. Polon. Sci. Sér. Sci. Math. Astronom. Phys. **15**, 479–486. MR **36** #6578.

E. D. Bolker,

[1966] *Functions resembling quotients of measures*, Trans. Amer. Math. Soc. **124**, 292–312. MR **33** #5834.

[1969] *A class of convex bodies*, Trans. Amer. Math. Soc. **145**, 323–345. MR **41** #921.

[1971] *The zonoid problem*, Amer. Math. Monthly **78**, 529–531.

J. Bourgain,

[1976] *Dentability and the Bishop-Phelps property*, Israel J. Math. (to appear).

R. D. Bourgin,

[1971] *Barycenters of measures on certain noncompact convex sets*, Trans. Amer. Math. Soc. **154**, 323–340. MR **42** #6582.

R. D. Bourgin and G. A. Edgar,

[1976] *Noncompact simplexes in Banach spaces with the Radon-Nikodym property*, J. Functional. Anal. **23**, 162–176.

J. W. Brace,

[1953] *Transformations on Banach spaces*, Dissertation, Cornell Univ., Ithaca, N.Y.

[1954] *On weakly compact transformations* (unpublished).

J. Bretagnolle, D. Dacunha-Castelle and J. L. Krivine,

[1966] *Lois stables et espaces L^p*. Ann. Inst. H. Poincaré Sect. B (N. S.) **2**(1965/66), 231–259. MR **34** #3605.

J. K. Brooks,

[1969a] *On the Vitali-Hahn-Saks and Nikodým theorems*, Proc. Nat. Acad. Sci. U.S.A. **64**, 468–471. MR **42** #3241.

[1969b] *Decomposition theorems for vector measures*, Proc. Amer. Math. Soc. **21**, 27–29. MR **38** #6024.

[1971] *On the existence of a control measure for strongly bounded vector measures*, Bull. Amer. Math. Soc. **77**, 999–1001. MR **44** #4178.

[1972] *Weak compactness in the space of vector measures*, Bull. Amer. Math. Soc. **78**, 284–287. MR **48** #2760.

[1973] *Equicontinuous sets of measures and applications to Vitali's integral convergence theorem and control measures*, Advances in Math. **10**, 165–171. MR **47** #8807.

J. K. Brooks and N. Dinculeanu,

[1974] *Strong additivity, absolute continuity and compactness in spaces of measures*, J. Math. Anal. Appl. **45**, 156–175. MR **49** #11230.

J. K. Brooks and R. S. Jewett,

[1970] *On finitely additive vector measures*, Proc. Nat. Acad. Sci. U.S.A. **67**, 1294–1298. MR **42** #4697.

J. K. Brooks and P. W. Lewis,

[1974] *Linear operators and vector measures*, Trans. Amer. Math. Soc. **192**, 139–162. MR **49** #3585.

[1975] *Linear operators and vector measures*. II, Math. Z. **144**, 45–53. MR **51** #13749.

D. Burgess (see Burgess Davis).

D. L. Burkholder,

[1966] *Martingale transforms*, Ann. Math. Statist. **37**, 1494–1504. MR **34** #8456.

[1973] *Distribution function inequalities for martingales*, Ann. Probability **1**, 19–42. MR **51** #1944.

D. L. Burkholder, B. J. Davis and R. F. Gundy,

[1972] *Integral inequalities for convex functions of operators on martingales*, Proc. Sixth Berkeley Sympos. Math. Statist. and Probability 2, pp. 223–240.

C. L. Bryne,

[1976] *Some sufficient conditions for the weak compactness of subsets of $L_1(m, X)$* (to appear).

M. Cambern,

[1974] *The isometries of $L^p(X, K)$*, Pacific J. Math. **55**, 9–17. MR **51** #6401.

C. Caratheodory,

[1927] *Vorlesungen uber reele Funktionen*, 2nd ed., Teubner, Leipzig; reprint, Chelsea, New York 1960.

D. Cartwright,

[1974] *The order completeness of some spaces of vector-valued functions*, Bull. Austral. Math. Soc. **11**, 57–61. MR **50** #14207.

R. V. Chacon and N. A. Friedman,

[1965] *Additive functionals*, Arch. Rational Mech. Anal. **18**, 230–240. MR **30** #2329.

R. V. Chacon and L. Sucheston,

[1975] *On convergence of vector-valued asymptotic martingales*, Z. Wahrscheinlichkeitstheorie und Verw. Gebiete **33**, 55–59.

J. Chaney,

[1972] *Banach lattices of compact maps*, Math. Z. **129**, 1–19. MR **47** #891.

S. D. Chatterji,

[1960] *Martingales of Banach-valued random variables*, Bull. Amer. Math. Soc. **66**, 395–398. MR **22** #10008.

[1963] *Weak convergence in certain special Banach spaces*, MRC Report #443.

[1964] *A note on the convergence of Banach-space valued martingales*, Math. Ann. **153**, 142–149. MR **28** #4583.

[1968] *Martingale convergence and the Radon-Nikodým theorem in Banach spaces*, Math. Scand. **22**, 21–41. MR **39** #7645.

G. Y. H. Chi,

[1973] *The Radon-Nikodým theorem for vector measures with values in the duals of some nuclear spaces*, Vector and Operator Valued Measures and Applications (Proc. Sympos., Snowbird Resort, Alta, Utah, 1972), Academic Press, New York, pp. 85–95. MR **49** #5833.

T. Cho and A. Tong,

[1973] *A note on the Radon-Nikodym theorem*, Proc. Amer. Math. Soc. **39**, 530–534. MR **47** #3634.

G. Choquet,

[1969] *Measures coniques, affines et cylindriques*, Symposia Mathematica, Vol. II (INDAM, Rome, 1968), Academic Press, London, pp. 145–182. MR **41** #5921.

J. P. R. Christensen (see also N. J. M. Anderson),

[1971] *Borel structures and a topological zero-one law*, Math. Scand. **29**, 245–255 (1972). MR **47** #2021.

J. A. Clarkson,

[1936] *Uniformly convex spaces*, Trans. Amer. Math. Soc. **40**, 396–414.

C. E. Cleaver,

[1969] *Representations of linear functionals in reflexive Banach function spaces*, Nederl. Akad. Wetensch Proc. Ser. A **72** = Indag. Math. **31**, 462–464. MR **40** #4749.

H. B. Cohen, M. A. Labbe and J. Wolfe,

[1972] *Norm reduction of averaging operators*, Proc. Amer. Math. Soc. **35**, 519–523. MR **48** #2741.

J. S. Cohen,

[1973] *Absolutely p-summing, p-nuclear operators and their conjugates*, Math. Ann. **201**, 177–200. MR **51** #6472.

J. B. Collier,

[1975] *A class of strong differentiability spaces*, Proc. Amer. Math. Soc. **53**, 420–422.

[1976] *The dual of a space with the Radon-Nikodym property*, Pacific J. Math. **64**, 103–106.

J. B. Collier and M. Edelstein,

[1974] *On strongly exposed points and Frechet differentiability*, Israel J. Math. **17**, 66–68.

A. Costé,

[1971] *Convergence des séries dans les espaces F-normés de fonctions mesurables*, Bull. Acad. Polon. Sci. Sér. Sci. Math. Astronom, Phys., 131–134. MR **44** #5425.

[1976] *An example of an operator in L^1 [0, 1]*, Private communication.

R. Cristescu,

[1959] *Intégrales de Hellinger dans les espaces de Banach*, Bull. Math. Soc. Sci. Math. Phys. R. P. Roumaine (N. S.) **3 (51)**, 277–282. MR **24** #A423.

[1965] *Sur la représentation des opérateurs lineaires par des intégrales vectorielles de Hellinger*, Rév. Roumaine Math. Pures Appl. **10**, 437–441. MR **33** #6389.

[1971] *Intégrales de Hellinger dans les espaces linéaires ordonnés*, Rev. Roumaine Math. Pures Appl. **16**, 483–486. MR **47** #5223.

P. Curtis (see W. Bade).

D. Dacunha-Castelle (see J. Bretagnolle).

G. B. Dantzig and A. Wald,

[1951] *On the fundamental lemma of Neyman and Pearson*, Ann. Math. Statist. **22**, 87–93. MR **12**, 622.

R. B. Darst,

[1961] *A note on abstract integration*, Trans. Amer. Math. Soc. **99**, 292–297. MR **22** #11092.

[1962a] *A decomposition for complete normed abelian groups with applications to spaces of additive set functions*, Trans. Amer. Math. Soc. **103**, 549–559. MR **25** #1256.

[1962b] *A decomposition of finitely additive set functions*, J. Reine Angew. Math. **210**, 31–37. MR **25** #1257.

[1962c] *A note on integration of vector valued functions*, Proc. Amer. Math. Soc. **13**, 858–863. MR **26** #284.

[1962d] *A continuity property for vector valued measurable functions*, Pacific J. Math. **12**, 469–479. MR **26** #6827.

[1963] *The Lebesgue decomposition*, Duke Math. J. **30**, 553–556. MR **28** #176.

[1966] *A direct proof of Porcelli's condition for weak convergence*, Proc. Amer. Math. Soc. **17**, 1094–1096. MR **34** #6505.

[1967] *On a theorem of Nikodym with applications to weak convergence and von Neumann algebras*, Pacific J. Math. **23**, 473–477. MR **38** #6360.

[1969] *Properties of vector valued finitely additive set functions*, Proc. Amer. Math. Soc. **23**, 528–535. MR **40** #294.

[1970a] *The Vitali-Hahn-Saks and Nikodym theorems for additive set functions*, Bull. Amer. Math. Soc. **76**, 1297–1298. MR **41** #8625.

[1970b] *The Lebesgue decomposition, Radon-Nikodym derivative, conditional expectation and martingale convergence for lattices of sets*, Pacific J. Math. **35**, 581–600. MR **45** #491.

R. B. Darst and G. A. DeBoth,

[1971] *Two approximation properties and a Radon-Nikodym derivative for lattices for sets*, Indiana Univ. Math. J. **21** (1971/72), 355–362. MR **45** #5302.

R. B. Darst and E. Green,

[1968] *On a Radon-Nikodym theorem for finitely additive set functions*, Pacific J. Math. **27**, 255–259. MR **38** #4635.

F. K. Dashiell,

[1976] *A theorem on weak compactness of measures with applications to the Baire classes* (to appear).

F. K. Dashiell and J. Lindenstrauss,

[1973] *Some examples concerning strictly convex norms on C(K) spaces*, Israel J. Math. **16**, 329–342. MR **50** #964.

Burgess Davis (see D. L. Burkholder).

W. J. Davis,

[1973] *The Radon-Nikodym property*, Séminaire Maurey-Schwartz 1973–1974, Exposé 69.

W. J. Davis, D. W. Dean and I. Singer,

[1971] *Multipliers and unconditional convergence of biorthogonal expansions*, Pacific J. Math. **37**, 35–39. MR **47** #2316.

W. J. Davis, T. Figiel, W. B. Johnson and A. Pełczyński,

[1974] *Factoring weakly compact operators*, J. Functional Analysis **17**, 311–327. MR **50** #8010.

W. J. Davis and R. P. Phelps,

[1974] *The Radon-Nikodym property and dentable sets in Banach spaces*, Proc. Amer. Math. Soc. **45**, 119–122. MR **49** #9591.

M. M. Day,

[1942] *Operators in Banach spaces*, Trans. Amer. Math. Soc. **51**, 583–608. MR **4**, 14.

[1972] *A geometric proof of Asplund's differentiability theorem*, Israel J. Math. **13**, 277–280.

[1973] *Normed linear spaces*, 3rd ed., Springer-Verlag, Berlin and New York. MR **49** #9588.

P. W. Day (see K. Sundaresan).

D. Dean (see W. J. Davis).

C. Debieve,

[1973] *Intégration par rapport à une mesure vectorielle*, Ann. Soc. Sci. Bruxelles Sér. I **87**, 165–185. MR **47** #6987.

G. A. DeBoth (see R. B. Darst).

M. De Wilde and M. T. De Wilde-Nibes,

[1968] *Intégration par rapport à des measures à valeurs vectorielles*, Rev. Roumaine Math. Pures Appl. **13**, 1529–1537. MR **41** #423.

J. Diestel,

[1972] *The Radon-Nikodym property and the coincidence of integral and nuclear operators*, Rev. Roumaine Math. Pures Appl. **17**, 1611–1620. MR **48** #12052.

[1973a] *Applications of weak compactness and bases to vector measures and vectorial integration*, Rev. Roumaine Math. Pures Appl. **18**, 211–224. MR **47** #5590.

[1973b] *On the essential uniqueness of an example of C. E. Rickart*, Comment. Math. Prace Mat. **17**, 263–264. MR **48** #494.

[1973c] *Grothendieck spaces and vector measures*, Vector and Operator Valued Measures and Applications (Proc. Sympos., Snowbird Resort, Alta, Utah, 1972), Academic Press, New York, pp. 97–108. MR **49** #3538.

[1975a] L_X^1 *is weakly compactly generated if X is*, Proc. Amer. Math. Soc. **48**, 508–510. MR **51** #3893.

[1975b] *Geometry of Banach spaces—Selected topics*, Lecture Notes in Math., vol. 485, Springer-Verlag, Berlin and New York.

[1976a] *Remarks on weak compactness in* $L_1(\mu, X)$, Glasgow Math. J. (to appear).

[1976b] *The Radon-Nikodym property and spaces of operators*, Measure Theory, Springer Lecture Notes in Math., vol. 538.

J. Diestel and B. Faires,

[1974] *On vector measures*, Trans. Amer. Math. Soc. **198**, 253–271. MR **50** #2912.

[1976] *Remarks on the classical Banach operator ideals*, Proc. Amer. Math. Soc. **58**, 189–196.

J. Diestel and T. J. Morrison,

[1976] *The Radon-Nikodym property for the space of operators* (to appear).

J. Diestel and C. J. Seifert,

[1976] *An averaging property of the range of a vector measure*, Bull. Amer. Math. Soc. **82**, 907–909.

J. Diestel and J. J. Uhl, Jr.,

[1976] *The Radon-Nikodym theorem for Banach space valued measures*, Rocky Mountain J. Math. **6**, 1–46.

J. Dieudonné,

Sur le theorem de Lebesgue-Nikodym.

[1941] I, Ann. of Math. (2) **42**, 547–555. MR **3**, 50.

[1944] II, Bull. Soc. Math. France **72**, 193–239. MR **7**, 305.

[1948] III, Ann. Univ. Grenoble Sect. Sci. Math. Phys. (N. S.) **23**, 25–53. MR **10**, 519.

[1951a] IV, J. Indian Math. Soc. (N. S.) **15**, 77–86. MR **13**, 447.

[1951b] V, Canad. J. Math. **3**, 129–139. MR **13**, 448.

[1951c] *Sur la convergence des suites de measures de Radon*, Anais. Acad. Brasil. Ci. **23**, 21–38, 277–282. MR **13**, 121; **13**, 634.

N. Dinculeanu (see also J. K. Brooks),

[1957] *Sur la représentation intégrale des certaines opérations linéaires.* I, C. R. Acad. Sci. Paris **245**, 1203–1205. MR **19**, 870.

[1958] *Mesures vectorielles et opérations linéaires*, C. R. Acad. Sci. Paris **246**, 2328–2331. MR **20** #3254.

[1959a] *Sur la représentation intégrale des certaines opérations linéaires.* II, Compositio Math. **14**, 1–22. MR **21** #2908.

[1959b] *Sur la représentation intégrale des certaines opérations linéaries.* III, Proc. Amer. Math. Soc. **10**, 59–68. MR **21** #2909.

[1962] *Remarks on the integral representation of vector measures and linear operations on* L_0^p, Rev. Roumaine Math. Pures Appl. 7, 287–300. MR **26** #5124.

[1965] *Linear operations on spaces of totally measurable functions*, Rev. Roumaine Math. Pures Appl. **10**, 1493–1524. MR **34** #6506.

[1966a] *Integral representation of linear operators.* I, II, Stud. Cerc. Mat. **18**, 349–385, 483–536. MR **35** #4366a,b.

[1966b] *Integral representation of vector measures and linear operations*, Studia Math. **25** (1964/65), 181–205. MR **33** #243.

[1967a] *Vector measures*, Pergamon Press, New York. MR **34** #6011b.

[1967b] *Integral representation of dominated operations on spaces of continuous vector fields*, Math. Ann. **173**, 147–180. MR **37** #6726.

[1973] *Linear operations on L^p-spaces*, Vector and Operator Valued Mesaures and Applications (Proc. Sympos., Snowbird Resort, Alta, Utah, 1972), Academic Press, New York, pp. 109–124. MR **49** #5816.

N. Dinculeanu and C. Foiaş,

[1959] *Measures vectorielles et opérations linéaires sur L^p_E*, C. R. Acad. Sci. Paris **248**, 1759–1762. MR **21** #3753.

[1961] *Sur la représentation intégrale des certains opérations linéaires*. IV, Canad. J. Math. **13**, 529–556. MR **26** #596.

N. Dinculeanu and I. Kluvánek,

[1967] *On vector measures*, Proc. London Math. Soc. (3) **17**, 505–512. MR **35** #5571.

N. Dinculeanu and P. W. Lewis,

[1970] *Regularity of Baire measures*, Proc. Amer. Math. Soc. **26**, 92–94. MR **41** #5588.

N. Dinculeanu and J. J. Uhl, Jr.,

[1973] *A unifying Radon-Nikodym theorem for vector measures*, J. Multivariate Anal. **3**, 184–203. MR **48** #6361.

J. Dixmier,

[1948] *Sur un théorème de Banach*, Duke Math. J. **15**, 1057–1071. MR **10**, 306.

[1953] *Formes linéaires sur un anneau d'opérateurs*, Bull. Soc. Math. France **81**, 9–39. MR **16**, 539.

I. Dobrakov,

[1970a] *On integration in Banach spaces*. I, Czechoslovak. Math. J. **20** (**95**), 511–536. MR **51** #1391.

[1970b] *On integration in Banach spaces*. II, Czechoslovak Math. J. **20** (**95**), 680–695. MR **51** #1392.

[1971] *On representation of linear operators on $C_0(T, X)$*, Czechoslovak. Math. J. **21**, 13–30.

[1972] *On subadditive operators on $C_0(T)$*, Bull. Acad. Polon. Sci. Sér. Sci. Math. Astronom. Phys. **20**, 561–562. MR **47** #7402.

P. G. Dodds (see also C. A. Akemann),

[1975a] *Sequential convergence in the order duals of certain classes of Riesz spaces*, Trans. Amer. Math. Soc. **203**, 391–403. MR **50** #10748.

[1975b] *o-weakly compact mappings of Riesz spaces*, Trans. Amer. Math. Soc. **214**, 389–402.

J. L. Doob,

[1950] *Stochastic processes*, Wiley, New York; Chapman and Hall, London, 1953, MR **15**, 445.

L. E. Dor,

[1975a] *On sequences spanning a complex l_1 space*, Proc. Amer. Math. Soc. **47**, 515–516. MR **50** #10774.

[1975b] *On projections in L_1*, Ann. of Math. (2) **102**, 463–474.

[1976] *Potentials and isometric embeddings in L_1*, Israel J. Math. **24**, 260–268.

V. M. Doubrovsky (Dubrovskii),

[1947a] *On some properties of completely additive set functions and their application to a generalization of a theorem of Lebesgue*, Mat. Sb. **20** (**62**), 317–329. MR **9**, 19.

[1947b] *On the basis of a family of completely additive functions of sets and on the properties of uniform additivity and equi-continuity*, Dokl. Akad. Nauk. SSSR (N.S.) **58**, 737–740. (Russian) MR **9**, 275.

[1948] *On properties of absolute continuity and equi-continuity*, Dokl. Akad. Nauk SSSR (N.S.) **63**, 483–486. MR **10**, 361.

L. Drewnowski,

Topological rings of sets, continuous set functions, integration.

[1972a] I, Bull. Acad. Polon. Sci. Sér. Sci. Math Astronom. Phys. **20**, 269–276. MR **46** #5558.

[1972b] II, ibid. **20**, 277–286. MR **46** #5558.

[1972c] III, ibid. **20**, 439–445. MR **47** #5200.

[1972d] *Equivalence of Brooks-Jewett, Vitali-Hahn-Saks and Nikodym theorems*, Bull. Acad. Polon. Sci. Sér. Sci. Math. Astronom. Phys. **20**, 725–731. MR **47** #431.

[1973a] *On the Orlicz-Pettis type theorems of Kalton*, Bull. Acad. Polon. Sci. Sér. Sci. Math. Astronom. Phys. **21**, 515–518. MR **48** #4260.

[1973b] *Decompositions of set functions*, Studia Math. **48**, 23–48. MR **49** #518.

[1974a] *On subseries convergence in some function spaces*, Bull. Acad. Polon. Sci. Sér. Sci. Math. Astronom. Phys. **22**, 797–803. MR **51** #1366.

[1974b] *On control submeasures and measures*, Studia Math. **50**, 203–224. MR **50** #4881.

[1974c] *Some remarks on rearrangements of series*, Comment. Math. Prace. Mat. **18** (1974/75), 7–9. MR **51** #3851.

[1975] *Another note on Kalton's theorems*, Studia Math. **52** (1974/75), 233–237. MR **50** #13433.

[1976a] *An extension of a theorem of Rosenthal on operators act from $l_\infty(\Gamma)$*, Studia Math. (to appear).

[1976b] *Un théorème sur les opérateurs de $l_\infty(\Gamma)$*, C. B. Acad. Sci. Paris Ser. A-B **281,** A967–A969.

[1976c] *On complete submeasures*, Comment. Math. (to appear).

L. Drewnowski and W. Orlicz,

[1968a] *A note on modular spaces.* X, Bull. Acad. Polon. Sci. Sér. Sci. Math. Astronom. Phys. **16,** 809–814. MR **39** #3279.

[1968b] *A note on modular spaces.* XI, Bull. Acad. Polon. Sci. Sér. Sci. Math. Astronom. Phys. **16**, 877–882. MR **39** #6068.

[1968c] *On orthogonally additive functionals*, Bull. Acad. Polon. Sci. Sér. Sci. Math. Astronom. Phys. **16**, 883–888. MR **39** #6069.

[1969a] *On representation of orthogonally additive functionals*, Bull. Acad. Polon. Sci. Sér. Sci. Math. Astronom. Phys. **17**, 167–173. MR **40** #3296.

[1969b] *Continuity and representation of orthogonally additive functionals*, Bull. Acad. Polon. Sci. Sér. Sci. Math. Astronom. Phys. **17**, 647–653. MR **41** #806.

L. E. Dubins,

[1957] *Generalized random variables*, Trans. Amer. Math. Soc. **84**, 273–309. MR **19,** 21.

E. Dubinsky, A. Pełczyński and H. P. Rosenthal,

[1972] *On Banach spaces X for which $\Pi_2(\mathscr{L}_\infty, X) = B(\mathscr{L}_\infty, X)$*, Studia Math. **44**, 617–648. MR **51** #1350.

N. Dunford (see also R.G. Bartle),

[1935] *Integration in general analysis*, Trans. Amer. Math. Soc. **37**, 441–453.

[1936a] *Integration and linear operations*, Trans. Amer. Math. Soc. **40**, 474–494.

[1936b] *A particular sequence of step functions*, Duke Math. J. **2**, 166–170.

[1937] *Integration of vector-valued functions*, Bull. Amer. Math. Soc. **43**, 24 (abstract).

[1938] *Uniformity in linear spaces*, Trans. Amer. Math. Soc. **44**, 305–356.

[1939] *A mean ergodic theorem*, Duke Math. J. **5**, 635–646. MR **1**, 18.

N. Dunford and A. P. Morse,

[1936] *Remarks on the preceding paper of James A. Clarkson*, Trans. Amer. Math. Soc. **40**, 415–420.

N. Dunford and B. J. Pettis,

[1940] *Linear operations on summable functions*, Trans. Amer. Math. Soc. **47**, 323–392. MR **1**, 338.

N. Dunford and R. Schatten,

[1946] *On the associate and conjugate space for the direct product*, Trans. Amer. Math. Soc. **59**, 430–436. MR **7**, 455.

N. Dunford and J. T. Schwartz,

[1958] *Linear operators.* Part I, Interscience, New York and London. MR **22** #8302.

A. Dvoretsky, A. Wald and J. Wolfowitz,

[1951a] *Elimination of randomization in certain statistical decision procedures and zero-sum two-person games*, Ann. Math. Statist. **22**, 1–21, MR **12**, 515.

[1951b] *Relations among certain ranges of vector measures*, Pacific J. Math. **1**, 59–74. MR **13**, 331.

M. Edelstein (see also J. Collier),

[1970] *Lines and hyperplanes associated with families of closed and bounded sets in conjugate Banach spaces*, Canad. J. Math. **22**, 933–938. MR **42** #6584.

[1973] *Smoothability versus dentability*, Comment. Math. Univ. Carolinae **14**, 127–133. MR **47** #9243.

[1976a] *Weakly proximinal sets*, J. Approximation Theory (to appear).

[1976b] *Concerning dentability* (to appear).

[1976c] *Convex hulls of strongly exposed points* (to appear).

M. Edelstein and A. C. Thompson,

[1972] *Some results on nearest points and support properties of convex sets in c_0*, Pacific J. Math. **40**, 553–560. MR **46** #7855.

G. A. Edgar (see also R. D. Bourgin),

[1975a] *Disintegration of measures and the vector-valued Radon-Nikodym theorem*, Duke Math. J. **42**, 447–450.

[1975b] *A noncompact Choquet theorem*, Proc. Amer. Math. Soc. **49**, 354–358. MR **51** #8793.

[1976] *Extremal integral representations*, J. Functional Anal. **23**, 145–161.

R. E. Edwards,

[1955] *Vector-valued measures and bounded variation in Hilbert space*, Math. Scand. **3**, 90–96.

[1965] *Functional analysis*, Holt, Rinehart and Winston, New York. MR **36** #4308.

P. van Eldik and J. J. Grobler,

[1973] *Strictly singular and nearly weakly compact operators in Banach function spaces*, Nederl. Akad. Wetensch Proc. Ser. A **76** = Indag. Math. **35**, 92–97. MR **47** #7509.

P. Enflo,

[1970a] *Uniform structures and square roots in topological groups.* I, Israel J. Math. **8**, 230–252. MR **41** #8568.

[1970b] *Uniform structures and square roots in topological groups.* II, Israel J. Math. **8**, 253–272. MR **41** #8568.

[1970c] *On the nonexistence of uniform homeomorphisms between L_p-spaces*, Ark. Mat. **8** (1969), 103–105. MR **42** #6600.

[1972] *Banach spaces which can be given an equivalent uniformly convex norm*, Israel J. Math. **13**, 281–288 (1973). MR **49** #1073.

[1973] *A counterexample to the approximation problem*, Acta Math. **130**, 309–317.

B. Faires (see also J. Diestel),

[1974a] *On Vitali-Hahn-Saks type theorems*, Bull. Amer. Math. Soc. **80**, 670–674. MR **51** #13681.

[1974b] *Grothendieck spaces and vector measures*, Ph.D. Dissertation, Kent State Univ., August 1974.

[1976a] *The Pettis property for non-σ-complete Boolean algebras* (to appear).

[1976b] *Varieties and vector measures*, Math. Nach. (to appear).

B. Faires and T. J. Morrison,

[1976] *The Jordan decomposition of vector-valued measures*, Proc. Amer. Math. Soc. **60**, 139–143.

M. Feder and P. Saphar,

[1976] *Spaces of compact operators and their dual spaces*, Israel J. Math., **21**, no.1, 38–49. MR **51** #13762.

C. Fefferman,

[1967] *A Radon-Nikodym theorem for finitely additive set functions*, Pacific J. Math. **23**, 35–45. MR **35** #6791.

[1968] *L_p spaces over finitely additive measures*, Pacific J. Math. **26**, 265–271. MR **38** #1515.

G. Fichtenholz and L. Kantorovitch,

[1935] *Sur les opérations linéaries dans l'espace des fonctions*, Studia Math. **5**, 70–98.

T. Figiel (see also W. J. Davis),

[1973] *Factorization of compact operators and applications to the approximation problem*, Studia Math. **45**, 191–210. MR **49** #1070.

T. Figiel and W. B. Johnson,

[1973] *The approximation property does not imply the bounded approximation property*, Proc. Amer. Math. Soc. **41**, 197–200. MR **49** #5782.

T. Figiel and G. Pisier,

[1976] *Rademacher averages in uniformly convex spaces* (to appear).

C. A. Fischer,

[1921] *Necessary and sufficient conditions that a linear transformation be completely continuous*, Bull. Amer. Math. Soc. **27**, 10–17.

W. Fischer and U. Scholer,

[1976a] *The range of vector measures into Orlicz spaces* (to appear).

[1976b] *On derivatives of vector measures into* $l_p(X)$, $0 < p < 1$ (to appear).

[1976c] *On boundedness of vector measures* (to appear).

[1976d] *Integral representations of additive operators with respect to a vector measure and its applications* (to appear).

C. Foiaş and I. Singer (see also N. Dinculeanu),

[1960] *Some remarks on the representation of linear operators in spaces of vector-valued continuous functions*, Rév. Roumaine Math. Pures Appl. **5**, 729–752. MR **24** #A1618.

[1965] *Points of diffusion of linear operators and almost diffuse operators in spaces of continuous functions*, Math. Z. **87**, 434–450. MR **31** #5093.

R. M. Fortet and E. Mourier,

[1955] *Les fonctions aléatoires comme éléments aleatoires dans les espaces de Banach*, Studia Math. **15**, 62–79. MR **19**, 1202.

[1954] *Résultats complémentaires sur les éléments aléatoires prenant leurs valeurs dans un espace de Banach*, Bull. Sci. Math. (2) **78**, 14–30. MR **16**, 149.

G. Fox,

[1967] *Extension of a bounded vector measure with values in a reflexive Banach space*, Canad. Math. Bull. **10**, 525–529. MR **37** #1552.

M. Fréchet,

[1907a] *Sur les ensembles des fonctions et les opérations linéaires*, C. R. Acad. Sci. Paris **144**, 1414–1416.

[1907b] *Sur les opérations linéaires*. III, Trans. Amer. Math. Soc. **8**, 433–446.

D. Friedland,

[1976] *On closed subspaces of weakly compactly generated Banach spaces*, Israel J. Math. (to appear).

N. A. Friedman and M. Katz (see also R. V. Chacon),

[1965] *A representation theorem for additive functionals*, Arch. Rational Mech. Anal. **21**, 49–57. MR **33** #245.

[1969] *On additive functionals*, Proc. Amer. Math. Soc. **21**, 557–561. MR **39** #4653.

N. A. Friedman and A. E. Tong,

[1971] *On additive operators*, Canad. J. Math. **23**, 468–480. MR **44** #785.

S. Găină,

[1963] *Extension of vector measures*, Rev. Roumaine Math. Pures Appl. **8**, 151–154. MR **29** #1297.

J. L. B. Gamlen (see also C. A. Akemann),

[1974] *On a theorem of A. Pełczyński*, Proc. Amer. Math. Soc. **44**, 283–285. MR **49** #5786.

P. Gänssler,

[1971] *Compactness and sequential compactness in spaces of measures*, Z. Wahrscheinlichkeitstheorie und Verw. Gebiete **17**, 124–146. MR **44** #793.

K. M. Garg (see R. Anantharaman).

H. G. Garnir and J. Schmets,

[1964] *Théorie de l'intégration par rapport à une mesure dans un espace linéaire à semi-normes*, Bull. Soc. Roy. Sci. Liège **33**, 381–410. MR. **31** #1351.

I. M. Gel'fand,

[1936] *Sur un lemme de la théorie des espaces linéaires*, Comm. Inst. Sci. Math. Méc. Univ. de Kharkoff et Soc. Math. Kharkoff (4) **13**, 35–40.

[1938] *Abstrakte Funktionen und lineare Operatoren*, Mat. Sb. (N. S.) **4(46)**, 235–286.

J. Gil de Lamadrid,

Measures and tensors,

[1965] I, Trans. Amer. Math. Soc. **114**, 98–121. MR **31** #3851.

[1966] II, Canad. J. Math. **18**, 762–793. MR **35** #4367.

J. R. Giles,

[1974] *A non-reflexive Banach space has non-smooth third conjugate space*, Canad. Math. Bull. **17**, 117–119. MR **50** #5429.

D. Gilliam,

[1976a] *Geometry and the Radon-Nikodym theorem in strict Mackey convergence spaces*, Pacific J. Math. (to appear).

[1976b] *On integration and the Radon-Nikodym theorem in quasicomplete locally convex topological vector spaces* (to appear).

D. B. Goodner,

[1950] *Projections in normed linear spaces*, Trans. Amer. Math. Soc. **69**, 89–108. MR **12**, 266.

R. K. Goodrich,

[1968] *A Riesz representation theorem in the setting of locally convex spaces*, Trans. Amer. Math. Soc. **131**, 246–258. MR **36** #5731.

[1970] *A Riesz representation theorem*, Proc. Amer. Math. Soc. **24**, 629–636.

Y. Gordon and D. R. Lewis,

[1974] *Absolutely summing operators and local unconditional structures*, Acta Math. **133**, 27–48.

Y. Gordon, D. R. Lewis and J. R. Retherford,

[1973] *Banach ideals of operators with applications*, J. Functional Analysis **14**, 85–129. MR **52** #1388.

Y. Gordon and P. Saphar,

[1975] *Ideal norms on $E \otimes L_p$*, Technion preprint series MT-264.

G. G. Gould,

[1965] *Integration over vector-valued measures*, Proc. London Math. Soc. (3) **15**, 193–225. MR **30** #4894.

[1966] *The Daniell-Bourbaki integral for finitely additive measures*, Proc. London Math. Soc. (3) **16**, 297–320. MR **34** #4445.

M. Gowurin,

[1936] *Uber die Stieltjesche Integration abstrakter Funktionen*, Fund. Math. **27**, 254–268.

W. H. Graves,

[1976] *A topological linearization of vector measures* (to appear).

E. Green (see R. B. Darst).

N. E. Gretsky,

[1968] *Representation theorems on Banach function spaces*, Mem. Amer. Math. Soc. No. 84, 56 pp.

N. E. Gretsky and J. J. Uhl. Jr.,

[1972] *Bounded linear operators on Banach function spaces of vector-valued functions*, Trans. Amer. Math. Soc. **167**, 263–277. MR **45** #4178.

J. J. Grobler (see also P. van Eldik),

[1970] *Compactness conditions for integral operators in Banach function spaces*, Nederl. Akad. Wetensch Proc. Ser. A **73** = Indag. Math. **32**, 287–294. MR **42** #5105.

[1973] *On Uhl's characterization of operators of finite double norm*, Math. Colloq. U.C.T. **8**, 101–120.

A. Grothendieck,

[1953] *Sur les applications linéaires faiblement compactes d'espaces du type $C(K)$*, Canad. J. Math. **5**, 129–173. MR **15**, 438.

[1954] *Sur certains sous-espaces vectoriels de L^p*, Canad. J. Math. **6**, 158–160. MR **15**, 439.

[1955a] *Produits tensoriels topologiques et espaces nucléaires*, Mem. Amer. Math. Soc. No. **16**, MR **17**, 763.

[1955b] *Une caractérisation vectorielle-métrique des espaces L^1*, Canad. J. Math. **7**, 552–561. MR **17**, 877.

[1953a] *Résumé de la théorie métrique des produits tensoriels topologiques*, Bol. Soc. Mat. São Paolo **8**, 1–79. MR **20** #1194.

[1953b] *Sur certaines classes de suites dans les espaces de Banach, et le théorème de Dvoretsky-Rogers*, Bol. Soc. Mat. São Paulo **8**, 81–110. MR **20** #1195.

[1954] *Espaces vectorielles topologiques*, Inst. Mat. Pura Appl., Univ. de São Paulo, São Paulo, 240 pp. MR **17**, 1110.

[1957b] *Un résultat sur le dual d'une C*-algébre*, J. Math. Pures Appl. (a) **36**, 97–108. MR **19**, 665.

R. L. Gundy (see D. L. Burkholder).

W. Hackenbroch,

[1973] *Zum Radon-Nikodymschen Satz für positive Vektormaße*, Math. Ann. **206**, 63–65. MR **48** #2340.

J. Hagler,

[1973] *Some more Banach spaces which contain L^1*, Studia Math. **46**, 35–42. MR **48** #11995.

[1976a] *Nonseparable "James tree" analogues of the continuous functions on the Cantor set* (to appear).

[1976b] *A counterexample to several questions about Banach spaces* (to appear).

J. Hagler and C. Stegall,

[1973] *Banach spaces whose duals contain complemented subspaces isomorphic to $C[0, 1]$*, J. Functional Analysis **13**, 233–251. MR **50** #2874.

H. Hahn,

[1922] *Uber Folgen linearer Operationen*, Monatsh. Math. Phys. **32**, 3–88.

[1932] *Reelle Funktionen*, Akad. Verlag., Leipzig; reprint, Chelsea, New York.

P. R. Halmos,

[1948] *The range of a vector measure*, Bull. Amer. Math. Soc. **54**, 416–421. MR **9**, 574.

[1950] *Measure theory*, Van Nostrand, Princeton, N. J. MR **11**, 504.

I. Halperin,

[1954] *Uniform convexity in function spaces*, Duke Math. J. **21**(1954), 195–204. MR **15**, 880.

J. C. Hankins and R. M. Rakestraw,

[1976] *Extremal structure of convex cones in Banach spaces* (to appear).

R. E. Harrell and L. A. Karlovitz,

[1970] *Girths and flat Banach spaces*, Bull. Amer. Math. Soc. **76**, 1288–1291. MR **42** #2285.

[1972] *Nonreflexivity and the girth of spheres*, Inequalities, III (Proc. Third Sympos., Univ. California, Los Angeles, 1969), Academic Press, New York, 121–127. MR **49** #7734.

[1974] *The geometry of flat Banach spaces*, Trans. Amer. Math. Soc. **192**, 209–218. MR **49** #3501.

[1975] *On tree structures in Banach spaces*, Pacific J. Math. **59**, 85–91.

M. Hasumi,

[1958] *The extension property of complex Banach spaces*, Tôhoku Math. J. (2) **10**, 135–142. MR **20** #7209.

R. Haydon,

[1976] *Some more characterizations of Banach spaces containing l_1* (to appear).

E. Helly,

[1912] *Uber lineare Funktional operationen*, S.-B. K. Akad Wiss. Wien Math.-Naturwiss, Kl. **121**, (IIa) 265–297.

L. L. Helms,

[1958] *Mean convergence of martingales*, Trans. Amer. Math. Soc. **87**, 439–446. MR **20** #1350.

H. Hermes and J. P. LaSalle,

[1969] *Functional analysis and time optimal control*, Academic Press, New York.

C. S. Herz,

[1963] *A class of negative-definite functions*, Proc. Amer. Math. Soc. **14**, 670–676. MR **28** #1477

E. Hewitt (see K. Yosida).

T. H. Hildebrandt,

[1934] *On bounded functional operations*, Trans. Amer. Math. Soc. **36**, 868–875.

[1953] *Integration in abstract spaces*, Bull. Amer. Math. Soc. **59**, 111–139. MR **14**, 735.

E. Hille and R. S. Phillips,

[1957] *Functional analysis and semi-groups*, Amer. Math. Soc. Colloq. Publ., vol. 31, Amer. Math. Soc., rev. ed., Providence, R. I, 1957. MR **19**, 664.

E. Hille and J. D. Tamarkin,

[1934] *On the theory of linear integral equations*. II, Ann. of Math. (2) **35**, 445–455.

K. Hoffman,

[1962] *Banach spaces of analytic functions*, Prentice-Hall, Englewood Cliffs, N. J. MR **24** #A2844.

J. Hoffman-Jørgensen,

[1971] *Vector measures*, Math. Scand. **28**, 5–32. MR **46** #5564.

[1974] *Sums of independent Banach space valued random variables*, Studia Math. **52**, 159–186. MR **50** #8626.

J. R. Holub,

[1973] *Reflexivity of L(E, F)*, Proc. Amer. Math. Soc. **39**, 175–177. MR **47** #3956.

J. E. Huneycutt, Jr.,

[1969] *Extensions of abstract valued set functions*, Trans. Amer. Math. Soc. **141**, 505–513, MR **42** #6182.

R. E. Huff,

[1973a] *The Yosida-Hewitt decomposition as an ergodic theorem*, Vector and Operator Valued Measures and Applications (Proc. Sympos., Snowbird Resort, Alta, Utah, 1972), Academic Press, New York, pp. 133–139. MR **48** #8743.

[1973b] *Vector measures*, Pennsylvania State Univ. (mimeographed notes).

[1974] *Dentability and the Radon-Nikodým property*, Duke Math. J. **41**, 111–114. MR **49** #5783.

[1976] *The Radon-Nikodým property for Banach spaces*, Measure Theory, Springer Lecture Notes in Math., vol. 538.

R. E. Huff and P. D. Morris,

[1973] *On control measures for sets of measures*.

[1975] *Dual spaces with the Kreĭn-Mil'man property have the Radon-Nikodým property*, Proc. Amer. Math. Soc. **49**, 104–108. MR **50** #14220.

[1976] *Geometric characterizations of the Radon-Nikodym property in Banach spaces*, Studia Math. (to appear).

A. Ionescu Tulcea,

[1974] *On measurability, pointwise convergence and compactness*, Bull. Amer. Math. Soc. **80**, 231–236. MR **50** #4870.

A. and C. Ionescu Tulcea,

[1962] *On the lifting property*. II, J. Math. Mech. **11**, 773–795. MR **26** #2881.

[1963] *Abstract ergodic theorems*, Trans. Amer. Math. Soc. **107**, 107–124. MR **27** #606.

[1969] *Topics in the theory of lifting*, Ergebnisse Math. Grenzgebiete, Band 48, Springer-Verlag, New York. MR **43** #2185.

C. Ionescu Tulcea (see A. Ionescu Tulcea).

J. R. Isbell and Z. Semadeni,

[1963] *Projection constants and spaces of continuous functions*, Trans. Amer. Math. Soc. **107**, 38–48. MR **26** #4169.

R. C. James,

[1950] *Bases and reflexivity of Banach spaces*, Ann. of Math. (2) **52**, 518–527. MR **12**, 616.

[1951] *A non-reflexive Banach space isometric with its second conjugate space*, Proc. Nat. Acad. Sci. U.S.A. **37**, 174–177. MR **13**, 356.

[1972a] *Some self-dual properties of normed linear spaces*, Sympos. on Infinite Dimensional Topology, Ann. of Math. Studies, No. 69, pp. 159–175.

[1972b] *Super-reflexive Banach spaces*, Canad. J. Math. **24**, 896–904. MR **47** #9248.

[1974] *A separable somewhat reflexive Banach space with nonseparable dual*, Bull. Amer. Math. Soc. **80**, 738–743.

R. C. James and J. J. Schäffer,

[1972] *Super-reflexivity and the girth of spheres*, Israel J. Math. **11**, 398–404. MR **46** #4175.

R. S. Jewett (see J. K. Brooks).

K. John and V. Zizler,

[1973] *A note on renorming of dual spaces*, Bull. Acad. Polon. Sci. Sér. Sci. Math. Astronom. Phys, **21**, 47–50. MR **47** #9246.

[1974a] *Smoothness and its equivalents in weakly compactly generated Banach spaces*, J. Functional Analysis **15**, 1–11.

[1974b] *Duals of Banach spaces which admit nontrivial smooth functions*, Bull. Austral. Math. Soc. **11**, 161–166.

[1976] *A note on strong differentiability spaces*, Comment. Math. Univ. Carolinae **17**, 127–134.

J. A. Johnson,

[1974] *Extreme measurable selections*, Proc. Amer. Math. Soc. **44**, 107–112. MR **49** #5818.

W. B. Johnson (see also W. J. Davis and T. Figiel),

[1971] *Factoring compact operators*, Israel J. Math. **9**, 337–345. MR **44** #7318.

W. B. Johnson and J. Lindenstrauss,

[1974] *Some remarks on weakly compactly generated spaces*, Israel J. Math. **17**, 219–230.

W. B. Johnson and H. P. Rosenthal,

[1972] *On ω^*-basic sequences and their applications to the study of Banach spaces*, Studia Math. **43**, 77–92. MR **46** #9696.

W. B. Johnson, H. P. Rosenthal and M. Zippin,

[1971] *On bases, finite dimensional decompositions and weaker structures in Banach spaces*, Israel J. Math. **9**, 488–506. MR **43** #6702.

S. Kaczmarz and H. Steinhaus,

[1951] *Theorie der Orthogonalreihen*; reprint, Chelsea, New York.

M. I. Kadec,

[1967] *A proof of the topological equivalence of all separable infinite-dimensional Banach spaces*, Funkcional. Anal. i Priložen **1**, 61–70 = Functional Anal. Appl. **1**, 53–62. MR **35** #700.

[1969] *On the integration of almost periodic functions with values in a Banach space*, Funkcional Anal. i Priložen **3**, 71–74 = Functional Anal. Appl. **3**, 228–230. MR **40** #4680.

M. I. Kadec and A. Pełczyński,

[1962] *Bases, lacunary sequences and complemented subspaces in the spaces L_p*, Studia Math. **21**, 161–176. MR **27** #2851.

J. P. Kahane,

[1968] *Some random series of functions* Heath, Lexington, Mass. MR **40** #8095.

S. Kakutani,

[1941a] *Concrete representation of abstract (L)-spaces and the mean ergodic theorem*, Ann. of Math. (2) **42**, 523–537. MR **2**, 318.

[1941b] *Concrete representation of abstract (M)-spaces. (A characterization of the space of continuous functions)*, Ann. of Math. (2) **42**, 994–1024. MR **3**, 205.

[1948] *On equivalence of infinite product measures*, Ann. of Math. (2) **49**, 214–224. MR **9**, 340

N. J. **Kalton** (see also G. Bennett),

[1971] *Subseries convergence in topological groups and vector spaces*, Israel J. Math. **10**, 402–412. MR **45** #3628.

[1974a] *Spaces of compact operators*, Math. Ann. **208**, 267–278. MR **49** #5904.

[1974b] *Topologies on Riesz groups and applications to measure theory*, Proc. London Math. Soc. (3) **28**, 253–273. MR **51** #10577.

[1975] *Exhaustive operators and vector measures*, Proc. Edinburgh Math. Soc. **19**, 291–300.

L. Kantorovitch (see also G. Fichtenholz),

[1940] *Linear operations in semi-ordered spaces*. I, Mat. Sb. **7** (49), 209–284. MR **2**, 317.

S. Kaplan,

[1970] *On weak compactness in the space of Radon measures*, J. Functional Analysis **5**, 259–298. MR **41** #5932.

S. Karlin,

[1953] *Extreme points of vector functions*, Proc. Amer. Math. Soc. **4**, 603–610. MR **15**, 109.

L. A. Karlovitz (see also R. E. Harrell),

[1973] *On the duals of flat Banach spaces*, Math. Ann. **202**, 245–250.

M. Katz (see N. A. Friedman).

R. P. Kaufman and N. W. Rickert,

[1966] *An inequality concerning measures*, Bull. Amer. Math. Soc. **72**, 672–676. MR **33** #1430.

J. L. Kelley,

[1952] *Banach spaces with the extension property*, Trans. Amer. Math. Soc. **72**, 323–326. MR **13**, 659.

D. C. Kemp,

[1976] *A note on smoothability in Banach spaces*, Math. Ann. **218**, 211–218.

S. S. Khurana,

Measures and barycenters of measures on convex sets in locally convex spaces.

[1969a] I, J. Math. Anal. Appl. **27**, 103–115. MR **39** #4631.

[1969b] II, ibid. **28**, 222–229. MR **39** #6042.

[1972a] *Measures having barycenters*, J. Math. Anal. Appl. **40**, 622–624. MR **47** #5543.

[1972b] *Characterization of extreme points*, J. London Math. Soc. (2) **5**, 102–104. MR **47** #761.

[1972c] *Barycenters, extreme points, and strongly extreme points*, Math. Ann. **198**, 81–84. MR **48** #9347.

[1973] *Barycenters, pinnacle points, and denting points*, Trans. Amer. Math. Soc. **180**, 497–503. MR **47** #6985.

J. F. C. Kingman and A. P. Robertson,

[1968] *On a theorem of Lyapunov*, J. London Math. Soc. **43**, 347–351. MR **37** #367.

I. Kluvánek (see also N. Dinculeanu),

On the theory of vector measures.

[1961] I, Mat.-Fyz. Časopis. Sloven. Akad. Vied **11**, 173–191. MR **27** #1558a.

[1966] II, ibid., **16**, 76–81. MR **34** #1475.

[1965] *Intégrale vectorielle de Daniell*, Mat.-Fyz Časopis Sloven. Akad. Vied **15**, 146–161. MR **38** #295.

[1967] *Completion of vector measure spaces*, Rev. Roumaine Math. Pures Appl. **12**, 1483–1488. MR **37** #4223.

[1973a] *The extension and closure of vector measure*, Vector and Operator Valued Measures and Applications (Proc. Sympos., Snowbird Resort, Alta, Utah, 1972), Academic Press, New York, 175–190. MR **49** #521.

[1973b] *The range of a vector-valued measure*, Math. Systems Theory **7**, 44–54. MR **48** #495.

[1975] *The range of a vector measure*, Bull. Amer. Math. Soc. **81**, 609–611. MR **51** #13182.

[1976] *Characterization of the closed convex hull of the range of a vector measure*, J. Functional. Anal. **21**, 316–329.

I. Kluvánek and G. Knowles,

[1974a] *Attainable sets in infinite dimensional spaces*, Math. Systems Theory **7**, 344–351.

[1974b] *Liapunov decomposition of a vector measure*, Math. Ann. **210**, 123–127. MR **49** #10853.

[1975] *Vector measures and control systems*, North-Holland, Amsterdam.

G. Knowles (see also I. Kluvánek),

[1974] *Liapunov vector measures*, SIAM J. Control **13**, 294–303.

[1976] *Time optimal control of infinite dimensional systems* (to appear).

S. Koji (see J. Ameniya).

H. Konig (see J. Batt).

M. Krein and D. Mil'man,

[1940] *On extreme points of regular convex sets*, Studia Math. **9**, 133–138. MR **3**, 90.

B. Kritt (see also W.M. Bogdanowicz),

[1966] *An approach to the theory of Radon-Nikodým differentiation of vector-valued volumes*, M.A. Thesis, Catholic Univ. of America.

J. L. Krivine (see J. Bregtagnolle).

K. Kunisawa,

[1940] *Some theorems on abstractly-valued functions in an abstract space*, Proc. Imp. Acad. Tokyo **16**, 68–72. MR **1**, 337.

T. Kuo,

[1974] *Grothendieck spaces and dual spaces possessing the Radon-Nikodým property*, Ph. D. Thesis, Carnegie-Mellon Univ.

[1975] *Weak convergence of vector measures on F-spaces*, Math. Z. **143**, Heft 2, 175–180. MR **51** #11106.

[1975a] *On conjugate Banach spaces with the Radon-Nikodým property*, Pacific J. Math. **59**, 497–503.

[1976] *Grothendieck spaces and linear operators defined on them* (to appear).

J. Kupka,

[1972] *Radon-Nikodým theorems for vector valued measures*, Trans. Amer. Math. Soc. **169**, 197–217. MR **47** #433.

[1974] *A short proof and generalization of a measure theoretic disjointization lemma*, Proc. Amer. Math. Soc. **45**, 70–72, MR **49** #7411.

S. Kwapień,

[1970] *On a theorem of L. Schwartz and its applications to absolutely summing operators*, Studia Math. **38**, 193–201. MR **43** #3822.

[1972a] *On operators factorizable through L_p-spaces*, Bull. Soc. Math. France **31–32**, 215–225.

[1972b] *Isomorphic characterizations of inner product spaces by orthogonal series with vector valued coefficients*, Studia Math. **44**, 583–595. MR **49** #5789.

[1974] *On Banach spaces containing* c_0. A supplement to the paper of J. Hoffman Jorgensen: "Sums of independent Banach space valued random variables", Studia Math. **52**, 187–188. MR **50** #8627.

M. A. Labbe (see H. B. Cohen).

I. Labuda,

[1972a] *Sur quelques généralisations des théorèmes de Nikodym et de Vitali-Hahn-Saks*, Bull. Acad. Polon. Sci. Sér. Sci. Math. Astronom. Phys. **20**, 447–456. MR **49** #5290.

[1972b] *Sur le théorème de Bartle-Dunford-Schwartz*, Bull. Acad. Polon. Sci. Sér. Sci. Math. Astronom. Phys. **20**, 549–553. MR **47** #5218.

Sur quelques théorèmes du type d'Orlicz-Pettis.

[1973a] I, Bull. Acad. Polon. Sci. Sér. Sci Math. Astronom. Phys. **21**, 127–132. MR **47** #5219.

[1973b] II, (and L. Drewnowski) ibid. **21**, 119–126. MR **47** #5220.

[1973c] III, ibid. **21**, 599–605. MR **49** #9150.

[1973c] *A generalization of Kalton's theorem*, Bull. Acad. Polon. Sci. Sér. Sci. Math. Astronom. Phys. **21**, 509–510. MR **48** #2342.

[1974] *Denumerability conditions and Orlicz-Pettis type theorems*, Comment Math. Prace Mat. **18**, 45–49. MR **50** #13436.

[1975] *Sur les measures exhaustives et certaines classes d'espaces vectoriels topologiques considérés par W. Orlicz et L. Schwartz*, C. R. Acad. Sci. Paris Sér A-B **280**, A997–A999. MR **51** #6419.

[1976a] *A note on exhaustive measures*, Comm. Math. (to appear).

[1976b] *Exhaustive measures in arbitrary topological vector spaces*, Studia Math. (to appear).

D. Landers,

[1973] *Connectedness properties of the range of vector and semi-measures*, Manuscripta Math. **9**, 105–112. MR **51** #10578.

D. Landers and L. Rogge,

[1971a] *Equicontinuity and convergence of measures*, Manuscripta Math. **5**, 123–131. MR **45** #7012.

[1971b] *The Hahn-Vitali-Saks and the uniform boundedness theorem in topological groups*, Manuscripta Math. **4**, 351–359. MR **44** #402.

[1972] *Cauchy convergent sequences of regular measures with values in a topological group*, Z. Wahrscheinlichkeitstheorie und Verw. Gebiete **21**, 188–196. MR **46** #9272.

J. P. LaSalle (see also H. Hermes),

[1960] *The time optimal control problem*, Contributions to the Theory of Nonlinear Oscillations, Vol. V, Princeton Univ. Press, Princeton, N. J., pp. 1–24. MR **26** #2704.

K. Lau,

[1976] *On almost Chebyshev subspaces*, J. Approximation Theory (to appear).

E. B. Leach and J. H. M. Whitfield,

[1972] *Differentiable functions and rough norms on Banach spaces*, Proc. Amer. Math. Soc. **33**, 120–126. MR **45** #2471.

S. Leader,

[1953] *The theory of* L^p*-spaces for finitely additive set functions*, Ann. of Math. (2) **58**, 528–543. MR **15**, 326.

H. Lebesgue,

[1909] *Sur les intégrales singulières*, Ann. de Toulouse **1**, 27–117.

K. deLeeuw (see E. Bishop).

I. E. Leonard and K. Sundaresan,

[1974a] *Smoothness and duality in* $L_p(E, \mu)$, J. Math. Anal. Appl. **46**, 513–522. MR **49** #9608.

[1974b] *Geometry of Lebesgue-Bochner function spaces—smoothness*, Trans. Amer. Math. Soc. **198**, 229–251.

V. L. Levin (see V. I. Arkin).

J. S. Lew,
[1971] *The range of a vector measure with values in a Montel space*, Math. Systems Theory **5**, 145–147. MR **46** #1999.

D. R. Lewis (see also Y. Gordon),
[1968] *On the Radon-Nikodým theorem* (unpublished).
[1970] *Integration with respect to vector measures*, Pacific. J. Math. **33**, 157–165. MR **41** #3206.
[1970b] *Weak integrability in L_1 spaces* (unpublished).
[1972a] *On integrability and summability in vector spaces*, Illinois J. Math. **16**, 294–307. MR **45** #502.
[1972b] *A vector measure with no derivative*, Proc. Amer. Math. Soc **32**, 535–536. MR **45** #5309.
[1972c] *Spaces on which each absolutely summing map is nuclear*, Proc. Amer. Math. Soc **31**, 195–198. MR **47** #775.
[1973] *Conditional weak compactness in certain inductive tensor products*, Math. Ann. **201**, 201–209. MR **48** #4761.
[1976] *Banach ideas of operators and the Radon-Nikodým property*, Notices Amer. Math. Soc. **23**, A-106. Abstract #731-31-1.

D. R. Lewis and C. Stegall,
[1973] *Banach spaces whose duals are isomorphic to $l_1(\Gamma)$*, J. Functional Analysis **12**, 177–187. MR **49** #7731.

P. W. Lewis (see also R. Bilyeu, J. K. Brooks and N. Dinculeanu),
[1969] *Extension of operator valued set functions with finite semivariation*, Proc. Amer. Math. Soc. **22**, 563–569. MR **39** #7061.
[1970] *Some regularity conditions on vector measures with finite semi-variation*, Rev. Roumaine Math. Pures Appl. **15**, 375–384. MR **41** #8626.
[1971a] *Vector measures and topology*, Rev. Roumaine Math. Pures Appl. **16**, 1201–1209. MR **46** #7472.
[1971b] *Addendum to: "Vector measures and topology,"* Rev. Roumaine Math. Pures Appl. **16**, 1211–1213. MR **46** #7472.
[1973a] *Permanence properties of absolute continuity conditions*, Vector and Operator Valued Measures and Applications (Proc. Sympos., Snowbird Resort, Alta, Utah, 1972), Academic Press, New York, pp. 197–206. MR **49** #7770.
[1973b] *Variational semi-regularity and norm convergence*, J. Reine Angew. Math. **260**, 23–30. MR **48** #870.
[1974] *Strongly bounded operators*, Pacific J. Math. **53**, 207–209. MR **51** #3966.

A. Lyapunov (Liapounoff),
[1940] *Sur les fonctions-vecteurs complètement additives*, Izv. Akad. Nauk SSSR Ser. Mat. **4**, 465–478. (Russian) MR **2**, 315.
[1946] *Sur les fonctions-vecteurs complètement additives*, Izv. Akad. Nauk SSSR Ser. Mat. **10**, 277–279. (Russian) MR **8**, 157.

W. Linde,
[1976a] *Die Radon-Nikodým Eigenschaft von Banach Räumen und der Zusammerhang zwischen absolut 1-summierenden und 1-radonisierenden Operatoren* (to appear).
[1976b] *An Operator ideal in connection with the Radon-Nikodým property of Banach spaces*, Math. Nach. **71**, 65–73.

L. Lindenstrauss (see also D. Amir, F. K. Dashiell and W. B. Johnson),
[1963] *On operators which attain their norm*, Israel J. Math. **1**, 139–148. MR **28** #3308.
[1964a] *Extension of compact operators*, Mem. Amer. Math. Soc. No. **48**. MR **31** #3828.

[1964b] *On nonlinear projections in Banach spaces*, Michigan Math. J. **11**, 263–287. MR **29** #5088.

[1964c] *On a certain subspace of l_1*, Bull. Acad. Polon. Sci. Sér. Sci. Math. Astronom. Phys. **12**, 539–542. MR **30** #5153.

[1964d] *On the extension of operators with a finite-dimensional range*, Illinois J. Math. **8**, 488–499. MR **29** #6317.

[1965] *On reflexive spaces having the metric approximation property*, Israel J. Math **3**, 199–204. MR **34** #4874.

[1966a] *On extreme points in l_1*, Israel J. Math. **4**, 59–61. MR **34** #589.

[1966b] *On nonseparable reflexive Banach spaces*, Bull. Amer. Math. Soc. **72**, 967–970. MR **34** #4875.

[1966c] *A short proof of Liapounoff's convexity theorem*, J. Math. Mech. **15**, 971–972. MR **34** #7754.

[1972] *Weakly compact sets—their topological properties and the Banach spaces they generate*, Proc. Sympos. Infinite-Dimensional Topology (Louisiana, 1967), Ann. of Math. Studies, no. 69, Princeton Univ. Press, Princeton, N.J.; Univ. of Tokyo Press, Tokyo, 1972, pp. 235–293. MR **49** #11514.

J. Lindenstrauss and A. Pełczyński,

[1968] *Absolutely summing operators in L_p-spaces and their applications*, Studia Math. **29**, 275–326. MR **37** #6743.

J. Lindenstrauss and H. P. Rosenthal,

[1969] *The $\mathscr{L}_p$ spaces*, Israel J. Math. **7**, 325–349. MR **42** #5012.

J. Lindenstrauss and C. Stegall,

[1975] *Examples of separable spaces which do not contain l_1 and whose duals are non-separable*, Studia Math. **56**, 81–105.

J. Lindenstrauss and L. Tzafriri,

[1973] *Classical Banach spaces*, Springer-Verlag Lecture Notes in Math., vol. 338, Berlin-Heidelberg-New York.

G. G. Lorentz and D. G. Wertheim,

[1953] *Representation of linear functionals on Köthe spaces*, Canad. J. Math. **5**, 568–575. MR **15**, 324.

H. P. Lotz,

[1974] *Minimal and reflexive Banach lattices*, Math. Ann. **209**, 117–126. MR **50** #10751.

[1976] *The Radon-Nikodým property in Banach lattices* (to appear).

A. R. Lovaglia,

[1955] *Locally uniformly convex Banach spaces*, Trans. Amer. Math. Soc. **78**, 225–238. MR **16**, 596.

W. A. J. Luxemburg and A. C. Zaanen,

[1963] *Compactness of integral operators in Banach function spaces*, Math. Ann. **149** (1962/63), 150–180. MR **26** #2905.

P. Mankiewicz,

[1972] *On Lipschitz mappings between Fréchet spaces*, Studia Math. **41**, 225–241. MR **46** #7838.

[1973] *On the differentiability of Lipschitz mappings in Fréchet spaces*, Studia Math. **45**, 15–29. MR **48** #9390.

[1974] *On topological, Lipschitz, and uniform classification of LF-spaces*, Studia Math. **52**, 109–142.

B. Maurey,

[1972] *Intégration dans les espaces p-normés*, Ann. Scuola Norm. Sup. Pisa (4) **26**, 911–931.

[1974a] *Théorèmes de factorization pour les operateurs linéaires a valeurs dans les espaces L^p*, Asterisque, no. 11, Soc. Math. France, Paris, 163 pp. MR **49** #9670.

[1974b] *Sous-espaces complémentés de L^p, d'apres P. Enflo*, Seminaire Maurey-Schwartz 1974–75, Expose 3, École Polytechnique.

B. Maurey and G. Pisier,

[1973a] *Un théorème d'extrapolation et ses conséquences*, C. R. Acad. Sci. Paris Sér. A-B **277**, A39–A42. MR **49** #7747.

[1973b] *Caractérisation d'une classe d'espaces de Banach par des propriétés de séries aléatoires vectorielles*, C. R. Acad. Sci. Paris Ser. A-B **277**, A687–A690. MR **48** #9352.

H. B. Maynard,

[1970] *A Radon-Nikodým theorem for vector-valued measures*, Ph. D. Thesis, Univ. of Colorado.

[1971] *A Radon-Nikodým theorem for finitely additive bounded measures* (unpublished).

[1973] *A geometrical characterization of Banach spaces having the Radon-Nikodym property*, Trans. Amer. Math. Soc. **185**, 493–500.

[1973b] *A general Radon-Nikodým theorem*, Vector and Operator Valued Measures and Applications (Proc. Sympos., Snowbird Resort, Alta, Utah, 1972), Academic Press, New York, pp. 233–246. MR **48** #4264.

S. Mazur and W. Orlicz,

[1948] *Sur les espaces métriques linéaires*. I, Studia Math. **10**, 184–208. MR **10**, 611.

[1953] *Sur les espaces métriques linéaires*. II, Studia Math. **13** ,137–179. MR **16**, 932.

C. W. McArthur,

[1967] *On a theorem of Orlicz and Pettis*, Pacific J. Math. **22**, 297–302. MR **35** #4702.

E. J. McShane,

[1950] *Linear functionals on certain Banach spaces*, Proc. Amer. Math. Soc. **1**, 402–408. MR **12**, 110.

M. Métivier,

[1967] *Martingales à valeurs vectorielles. Applications à la dérivation des mesures vectorielles*, Ann. Inst. Fourier (Grenoble) **17**, fasc. 2, 175–208. MR **40** #926.

[1969] *On strong measurability of Banach valued functions*, Proc. Amer. Math. Soc. **21**, 747–748. MR **39** #6373.

J. G. Mikusiński,

[1970] *A theorem on vector matrices and its applications in measure theory and functional analysis*, Bull. Acad. Polon. Sci. Sér. Sci. Math. Astronom. Phys. **18**, 193–196. MR **42** #7849.

[1971] *On a theorem of Nikodým on bounded measures*, Bull. Acad. Polon. Sci. Sér. Math. Astronom. Phys. **19**, 441–444. MR **48** #496.

D. P. Mil'man (see also M. Kreĭn),

[1938] *On some criteria for the regularity of spaces of the type* (B), Dokl. Akad. Nauk SSSR **20**, 243–246. (Russian)

V. D. Mil'man,

[1970] *The geometric theory of Banach spaces*. I, Uspehi Mat. Nauk **25**, no. 3 (153), 113–174 = Russian Math Surveys **25**, no. 3, 111–170.

[1971] *The geometric theory of Banach spaces*. II, Uspehi Mat. Nauk **26**, no. 6 (162), 73–149 = Russian Math. Surveys **26**, no. 6, 79–163.

G. Mokobodzki,

[1971] *Noyaux absolument mesurables on basiques et opérateurs nucléaires*, Seminar Goulaouic-Schwartz 1971–1972, Exposé VI.

V. J. Mizel,

[1969] *Representation of nonlinear transformations on L^p spaces*, Bull. Amer. Math. Soc. **75**, 164–168. MR **38** #4983.

V. J. Mizel and K. Sundaresan,

[1971] *Representation of vector valued nonlinear functions*, Trans. Amer. Math. Soc. **159**, 111–127. MR **43** #5368.

S. Moedomo and J. J. Uhl, Jr.

[1971] *Radon-Nikodým theorems for the Bochner and Pettis integrals*, Pacific J. Math. **38**, 531–536. MR **46** #7473.

P. D. Morris (see R. E. Huff).

T. J. Morrison (see J. Diestel and B. Faires).

A. P. Morse (see N. Dunford).

E. Mourier (see also M. Fortet),

[1953] *Eléments aléatoires dans un espace de Banach*, Ann. Inst. H. Poincaré **13**, 161–244. MR **16**, 268.

M. E. Munroe,

[1946a] *A note on weak differentiability of Pettis integrals*, Bull. Amer. Math. Soc. **52**, 167–174. MR **7**, 307.

[1946b] *A second note on weak differentiability of Pettis integrals*, Bull. Amer. Math. Soc. **52**, 668–670. MR **8**, 33.

K. Musiał,

[1973a] *Absolute continuity of vector measures*, Colloq. Math. **27**, 319–321. MR **48** #6362.

[1973b] *Absolute continuity and the range of group valued measure*, Bull. Acad. Polon. Sci. Sér. Sci. Math. Astronom. Phys. **21**, 105–113. MR **50** #2441.

L. Nachbin,

[1950] *A theorem of the Hahn-Banach type for linear transformations*, Trans. Amer. Math. Soc. **68**, 28–46. MR **11**, 369.

[1961] *Some problems in extending and lifting continuous linear transformations*, Proc. Internat. Sympos. Linear Spaces (Jerusalem, 1960), Jerusalem Academic Press, Jerusalem; Pergamon, Oxford, pp. 340–350. MR **24** #A2826.

M. Nakamura,

[1949] *Notes on Banach space.* IX: *Vitali-Hahn-Saks' theorem and K-spaces*, Tôhoku Math. J. (2) **1**, 106–108. MR **11**, 184.

M. Nakamura and G. Sunouchi,

[1942] *Note on Banach spaces* IV: *On a decomposition of additive set functions*, Proc. Imp. Acad. Tokyo **18**, 333–335. MR **8**, 256.

H. Nakano,

[1941] *Über das System aller stetigen Funktionen auf einem topologischen Raum*, Proc. Imp. Acad. Tokyo **17**, 308–310. MR **7**, 249.

I. Namioka,

[1967] *Neighborhoods of extreme points*, Israel J. Math. **5**, 145–152. MR **36** #4323.

[1974] *Separate continuity and joint continuity*, Pacific J. Math. **51**, 515–531. MR **51** #6693.

I. Namioka and R. R. Phelps,

[1975] *Banach spaces which are Asplund spaces*, Duke Math. J. **42**, 735–750.

J. von Neumann (see also R. Schatten),

[1929] *Zur allgemeinen Theorie des Masses*, Fund. Math. **13**, 73–116.

[1932] *Zur Operatorenmethode in der klassischen Mechanik*, Ann. of Math (2) **33**, 587–642, 789–791.

[1949] *On rings of operators. Reduction theory*, Ann. of Math. (2) **50**, 401–485. MR **10**, 548.

J. Neveu,

[1965] *Relations entre les théorie des martingales et la théorie ergodique*, Colloq. Internat. La Théorie du Potentiel (Orsay, 22–26 juin, 1964), Éditions du Centre National de la Recherche Scientifiques, Paris; Ann. Inst. Fourier (Grenoble) **15**, fasc. 1, 31–42. MR **36** #2778.

N. J. Nielsen,

[1973] *On Banach ideals determined by Banach lattices and their applications*, Dissertationes Math. **109**, 1–62.

O. M. Nikodým,

[1930] *Sur une généralisation des intégrales de M. J. Radon*, Fund. Math. **15**, 131–179.

[1931] *Contribution à la théorie des fonctionnelles linéaries en connexion avec la théorie de la mesure des ensembles abstraits*, Mathematica (Cluj) **5**, 130–141.

[1933a] *Sur les familles bornées de fonctions parfaitement additives d'ensemble abstrait*, Monatsh. Math. Phys. **40**, 418–426.

[1933b] *Sur les suites convergentes de fonctions parfaitement additive d'ensemble abstrait*, Monatsh. Math. Phys. **40**, 427–432.

E. Odell and H. P. Rosenthal,

[1976] *A double-dual characterization of separable Banach spaces containing l_1* (to appear).

S. Ohba,

[1971] *The decomposition theorems for vector measures*, Yokohama Math. J. **19**, 23–28. MR **47** #2028.

[1973] *Extensions of vector measures*, Yokohama Math. J. **21**, 61–66. MR **49** #527.

[1974] *σ-S-bounded vector measures*, Kanagawa Univ. Reports **12**, 1–5.

C. Olech,

[1966] *Extremal solutions of a control system*, J. Differential Equations **2**, 74–101. MR **32** #7284.

W. Orlicz (see also L. Drewnowski and S. Mazur),

[1929] *Beitrage zur Theorie der Orthogonalent wicklungen*. II, Studia Math. **1**, 241–255.

[1948] *Sur les opérations linéaires dans l'espace des fonctions bornées*, Studia Math. **10**, 60–89. MR **9**, #595.

[1968] *Absolute continuity of vector-valued finitely additive set functions*. I, Studia Math. **30**, 121–133. MR **37** #2929.

T. Parthasarathy,

[1972] *Selection theorems and their applications*, Lecture Notes in Math., vol. 263, Springer-Verlag, Berlin and New York, 1972.

A. Pełczyński (see also W. J. Davis, E. Dubinsky, M. Kadec and J. Lindenstrauss),

[1959] *Remarks on the vector measures*, Prace Mat. **3**, 69–72. (Polish) MR **23** #A3820.

[1960] *Projections in certain Banach spaces*, Studia Math. **19**, 209–228. MR **23** #A3441.

[1961] *On the impossibility of embedding of the space L in certain Banach spaces*, Colloq. Math. **8**, 199–203. MR **24** #A1008.

[1962] *Banach spaces on which every unconditionally converging operator is weakly compact*, Bull. Acad. Polon. Sci. Sér. Sci. Math. Astronom. Phys. **10**, 641–648. MR **26** #6785.

[1968] *On Banach spaces containing $L_1(\mu)$*, Studia Math. **30**, 231–246. MR **38** #521.

[1973] *On some problems of Banach*, Uspehi Mat. Nauk **28**, 67–75 = Russian Math. Surveys **28**, 67–76.

[1975] *On Banach space properties of uniform algebras*, Proc. Rie Conf. on Functional Analysis and Function Theory, Kristiansand, June 9–14.

A. Pełczyński and Z. Semadini,

[1959] *Spaces of continuous functions*. III: *Spaces $C(\Omega)$ for Ω without perfect subsets*, Studia Math. **18**, 211–222. MR **21** #6528.

J. Pellaumail,

[1970] *Sur la dérivation d'une mesure vectorielle*, Bull Soc. Math. France **98**, 359–368. MR **43** #2178.

A. Persson,

[1969] *On some properties of p-nuclear and p-integral operators*, Studia Math. **33**, 213–222. MR **40** #769.

A. Persson and A. Pietsch,

[1969] *p-nukleare und p-integrale Abbildungen in Banachräumen*, Studia Math. **33**, 19–62. MR **39** #4645.

B. J. Pettis (see also N. Dunford),

[1938a] *On integration in vector spaces*, Trans. Amer. Math. Soc. **44**, 277–304.

[1938b] *Linear functionals and completely additive set functions*, Duke Math. J. **4**, 552–565.

[1939a] *A proof that every uniformly convex space is reflexive*, Duke Math. J. **5**, 249–253.

[1939b] *Differentiation in Banach spaces*, Duke Math. J. **5**, 254–269.

[1939c] *Absolutely continuous functions in vector spaces*, Bull. Amer. Math. Soc. **45**, 677.

[1951] *Uniform Cauchy points and points of equicontinuity*, Amer. J. Math. **73**, 602–614. MR **13**, 217.

J. Pfanzagl,

[1971] *Convergent sequences of regular measures*, Manuscripta Math. **4**, 91–98. MR **44** #2901.

R. R. Phelps (see also E. Bishop, W. J. Davis and I. Namioka),

[1964] *Weak* support points of convex sets in E^**, Israel J. Math. **2**, 177–182. MR **31** #3832.

[1966] *Lectures on Choquet's theorem*, Van Nostrand Math. Studies, No. 7, Van Nostrand, Princeton, N. J. MR **33** #1690.

[1974] *Dentability and extreme points in Banach spaces*, J. Functional Analysis **16**, 78–90. MR **50** #5427.

R. S. Phillips (see also S. Bochner and E. Hille),

[1940a] *On linear transformations*, Trans. Amer. Math. Soc. **48**, 516–541. MR **2**, #318.

[1940b] *Integration in a convex linear topological space*, Trans. Amer. Math. Soc. **47**, 114–145. MR **2**, 103.

[1943] *On weakly compact subsets of a Banach space*, Amer. J. Math. **65**, 108–136. MR **4**, 218.

A. Pietsch (see also A. Persson),

[1965] *Abbildungen von abstrakten Massen*, Wiss. Z. Friedrich-Schiller-Univ. Jenal Thüringen **14**, 281–286. MR **37** #5355.

[1967] *Absolut p-summierende Abbildungen in normierten Räumen*, Studia Math. **28** (1966/67), 333–353. MR **35** #7162.

[1968] *p-majorisierbare vektorwertige Masse*, Wiss. Z. Friedrich-Schiller-Univ. Jena/Thüringen **18**, Hebt **2**, 243–247. MR **41** #8627.

[1972] *Theorie der Operatorenideale*, Friedrich-Schiller-Univ., Jena. MR **50** #14267.

G. Pisier (see also T. Figiel and B. Maurey),

[1975] *Martingales with values in uniformly convex spaces*, Israel J. Math. **20**, 326–350.

G. B. Price,

[1940] *The theory of integration*, Trans. Amer. Math. Soc. **47**, 1–50. MR **1**, 239.

D. Przeworska-Rolewicz and S. Rolewicz,

[1966] *On integrals of functions with values in a complete linear metric space*, Studia Math. **26**, 121–131. MR **33** #564.

H. Rademacher,

[1919] *Über partielle und totale Differentiziebarkeit von Funkionen mehrerer Variebeln und über die Transformation der Doppel integrale*, Math. Ann. **79**, 340–359.

J. Radon,

[1919] *Über lineare Funktionaltransformationen und Funktional gleichungen*, S.-B. Akad. Wiss. Wien **128**, 1083–1121.

R. M. Rakestraw (see J. C. Hankins).

M. M. Rao,

[1964] *Decomposition of vector measures*, Proc. Nat. Acad. Sci. U. S. A. **51**, 771–774; erratum, ibid. **52**, 864. MR **30** #219a, b.

[1967] *Abstract Lebesgue-Radon-Nikodým theorems*, Ann. Mat. Pures Appl. (4) **76**, 107–131. MR **36** #6579.

[1973] *Remarks on a Radon-Nikodým theorem for vector measures*, Vector and Operator Valued Measures and Applications (Proc. Sympos., Snowbird Resort, Alta, Utah, 1972), Academic Press, New York, pp. 303–317. MR **49** #3074.

G. Restrepo,

[1964] *Differentiable norms in Banach spaces*, Bull. Amer. Math. Soc. **70**, 413–414. MR **28** #4338.

J. R. Retherford (see also Y. Gordon and C. Stegall),

[1975] *Applications of Banach ideals of operators*, Bull. Amer. Math. Soc. **81**, 978–1012.

C. E. Rickart,

[1942] *Integration in a convex linear topological space*, Trans. Amer. Math. Soc. **52**, 498–521. MR **4**, 162.

[1943] *Decomposition of additive set functions*, Duke Math. J. **10**, 653–665. MR **5**, 232.

[1944] *An abstract Radon-Nikodým theorem*, Trans. Amer. Math. Soc. **56**, 50–66. MR **6**, 70.

N. W. Rickert (see also R. P. Kaufman),

[1967a] *The range of a measure*, Bull. Amer. Math. Soc. **73**, 560–563. MR **35** #333.

[1967b] *Measures whose range is a ball*, Pacific J. Math. **23**, 361–371. MR **36** #5296.

M. A. Rieffel,

[1967] *Dentable subsets of Banach spaces, with application to a Radon-Nikodým theorem*, Functional analysis (Proc. Conf., Irvine, Calif., 1966), (B. R. Gelbaum, editor), Academic Press, London; Thompson, Washington, D. C., pp. 71–77. MR **36** #5668.

[1968] *The Radon-Nikodym theorem for the Bochner integral*, Trans. Amer. Math. Soc. **131**, 466–487. MR **36** #5297.

F. Riesz

[1907] *Sur une espèce de géométrie analytiques des systèmes de fonctions sommables*, C. R. Acad. Sci. Paris **144**, 1409–1411.

[1910] *Untersuchungen über System integrierbarer Funktionen*, Math. Ann. **69**, 449–497.

A. P. Robertson (see also J. F. C. Kingman),

[1969] *On unconditional convergence in topological vector spaces*, Proc. Roy. Soc. Edinburgh Sect. A **68**, 145–157. MR **39** #7387.

L. Rogge (see D. Landers).

S. Rolewicz (see D. Przeworska-Rolewicz).

I. V. Romanovskiĭ and. N. Sudakov,

[1965] *On the existence of independent partitions*, Trudy Mat. Inst. Steklov **79**, 5–10 = Proc. Steklov Inst. Math. No. **79**, (1966), 1–7. MR **36** #3386.

U. Rønnow,

[1967] *On integral representation of vector valued measures*, Math. Scand. **21**, 45–53 (1968). MR **39** #4354.

H. P. Rosenthal (see also G. F. Bachelis, E. Dubinsky, W. B. Johnson, J. Lindenstrauss and E. Odell),

[1968] *On complemented and quasi-complemented subspaces of quotients of $C(S)$ for Stonian S*, Proc. Nat. Acad. Sci. U.S.A. **60**, 1165–1169. MR **37** #6740.

[1970a] *On relatively disjoint families of measures, with some applications to Banach space theory*, Studia Math. **37**, 13–36. MR **42** #5015.

[1970b] *On injective Banach spaces and the spaces $L^\infty(\mu)$ for finite measures μ*, Acta Math. **124**, 205–248. MR **41** #2370.

[1973] *On subspaces of L^p*, Ann. of Math. (2) **97**, 344–373. MR **47** #784.

[1974a] *A characterization of Banach spaces containing l^1*, Proc. Nat. Acad. Sci. U.S.A. **71**, 2411–2413. MR **50** #10773.

[1974b] *The hereditary problem for weakly compactly generated Banach spaces*, Compositio Math. **28**, 83–111.
[1975] *The Banach spaces $C(K)$ and $L^p(\mu)$*, Bull. Amer. Math. Soc. **81**, 763–781.
[1976a] *Pointwise compact subsets of the first Baire class* (to appear).
[1976b] *Some applications of p-summing operators to Banach space theory*, Studia Math. (to appear).

G. Rothenberger,
[1973] *Nichtlineare schwach kompakte Transformationen aus Räumen total-mess bares Funktionen*, Diplomarbeit, Munich.

W. H. Ruckle,
[1972] *Reflexivity of $L(E, F)$*, Proc. Amer. Math. Soc. **34**, 171–174. MR **45** #868.

V. I. Rybakov,
[1970] *Theorem of Bartle, Dunford and Schwartz on vector-valued measures*, Mat. Zametik **7**, 247–254 = Math. Notes **7**, 147–151. MR **41** #5591.
[1968] *The Radon-Nikodým theorem and integral representation of vector measures*, Dokl. Akad. Nauk SSSR **180**, 282–285 = Soviet Math. Dokl. **9**, 620–623, MR **37** #5356.

E. Saab,
[1973a] *Dentabilité et points extrémaux dans les espaces localement convexes*, Séminaire Choquet, 1973/74, exposé 13, Secrétariat mathématique, Paris.
[1973b] *Points extrémaux et propriété de Radon-Nikodým dans les espaces de Fréchet dentables*, Séminaire Choquet, 1973/74, exposé 19, Secrétariat mathématique, Paris.
[1974] *Familles absolument sommables et propriété de Radon-Nikodým*, Séminaire Choquet, 1974/75, exposé C7, Secrétariat mathématique, Paris.
[1975] *Dentabilité, points extrémaux et propriété de Radon-Nikodým*, Bull. Sci. Math. (2) **99**, 129–134.
[1976a] *On the Radon-Nikodým property in locally convex spaces of type (BM)* (to appear).
[1976b] *Strongly extreme points in subsets of absolutely convex dentable sets* (to appear).

J. Saint Raymond,
[1976] *Representation integrale dans certains convexes*, Séminaire Choquet, Scerétariat mathématique, Paris.

S. Sakai,
[1964] *Weakly compact operators on operator algebras*, Pacific J. Math. **14**, 659–664. MR **29** #488.
[1971] *C^*-algebras and W^*-algebras*, Ergebnisse der Mathematik und ihrer Grenzgebiete, Band 60, Springer-Verlag, New York and Berlin.

S. Saks,
[1933] *Addition to the note on some functionals*, Trans. Amer. Math. Soc. **35**, 965–970.
[1938] *Integration in abstract metric spaces*, Duke Math. J. **4**, 408–411.

R. Salem,
[1942] *On sets of multiplicity for trigonometrical series*, Amer. J. Math. **64**, 531–538. MR **4**, 38.

P. Saphar (see also M. Feder and Y. Gordon),
[1972] *Applications p-décomposantes et p-absolument sommantes*, Israel J. Math. **11**, 164–179. MR **45** #5779.

F. S. Scalora,
[1961] *Abstract martingale convergence theorems*, Pacific J. Math. **11**, 347–374. MR **23** #A684.

H. H. Schaefer,
[1972] *Normed tensor products of Banach lattices*, Israel J. Math. **13**, 400–415 (1973). MR **48** #12078.
[1974] *Banach lattices and positive operators*, Grundlehren der Math. Wiss., vol. 215, Springer-Verlag, Berlin and New York.

H. H. Schaefer and U. Schlotterbeck,

[1975] *On the approximation of kernel operators by operators of finite rank*, J. Approximation Theory **13**, 33–39. MR **50** #14326.

J. J. Schäffer (see also R. C. James),

[1967a] *Inner diameter, perimeter and girth of spheres*, Math. Ann. **173**, 59–79. MR **36** #1959.

[1967b] *Addendum: Inner diameter, perimeter, and girth of spheres*, Math. Ann. **173**, 79–82. MR **36** #1959.

[1970] *Minimum girth of spheres*, Math. Ann. **184** (1969/70), 169–171. MR **41** #4210; erratum, MR **42**, p. 1824.

[1971a] *Spheres with maximum inner diameter*, Math. Ann. **190** (1970/71), 242–247. MR **42** #8254.

[1971b] *On the geometry of spheres in L-spaces*, Israel J. Math. **10**, 114–120. MR **45** #5720.

[1976] *Girth, perimeter, radius and diameter of spheres in normed spaces* (to appear).

J. J. Schäffer and K. Sundaresan,

[1970] *Reflexivity and the girth of spheres*, Math. Ann. **184** (1969/70), 163–168. MR **41** #4209.

R. Schatten (see also N. Dunford),

[1943a] *On the direct product of Banach spaces*, Trans. Amer. Math. Soc. **53**, 195–217. MR **4**, 161.

[1943b] *On reflexive norms for the direct product of Banach spaces*, Trans. Amer. Math. Soc. **54**, 498–506. MR **5**, 99.

[1946] *The cross-space of linear transformations*, Ann. of Math. (2) **47**, 73–84. MR **7**, 455.

[1947] *On projections with bound* 1, Ann. of Math. (2) **48**, 321–325. MR **10**, 128.

[1950] *A theory of cross-spaces*, Ann. of Math. Studies, no. 26, Princeton Univ. Press, Princeton, N. J. MR **12**, 186.

R. Schatten and J. von Neumann,

[1946] *The cross-space of linear transformations*. II, Ann. of Math. (2) **47**, 608–630. MR **8**, 31.

[1948] *The cross-space of linear transformations*. III, Ann. of Math. (2) **49**, 557–582. MR **10**, 256.

J. Schmets (see also H. G. Garnirer),

[1966] *Sur une généralisation d'un théorème de Lyapounov*, Bull. Soc. Roy. Sci. Liège **35**, 185–194. MR **34** #296.

R. Schneider,

[1975] *Zonoids whose polars are zonoids*, Proc. Amer. Math. Soc. **50**, 365–368.

U. Scholer (see W. Fischer).

J. T. Schwartz (see R. G. Bartle and N. Dunford).

U. Schlotterbeck (see H. H. Schaefer).

L. Schwartz,

[1969a] *Un théorème de convergence dans les* L^p, $0 \leqq p < +\infty$, C. R. Acad. Sci. Paris Sér. A-B **268**, A704–A706. MR **39** #1958.

[1969b] *Un théorème de dualité pour les applications radonifiantes*, C. R. Acad. Sci. Paris Sér A-B **268**, A1410–A1413.

[1969c] *Séminaire*, Ecole Polytechnique, 1969–1970.

G. Schwarz,

[1967] *Variations on vector measures*, Pacific J. Math. **23**, 373–375. MR **36** #3946.

G. L. Seever,

[1968] *Measures on F-spaces*, Trans. Amer. Math. Soc. **133**, 267–280. MR **37** #1976.

C. Seifert (see J. Diestel).

Z. Semadeni (see also J. R. Isbell and A. Pełczyński),

[1964] *On weak convergence of measures and σ-complete Boolean algebras*, Colloq. Math. **12**, 229–233. MR **30** #4889.

[1971] *Banach spaces of continuous functions*. Vol. 1, Monografie Mat., Tom 55, PWN, Warsaw. MR **45** #5730.

A. H. Shuchat,

[1972a] *Approximation of vector-valued continuous functions*, Proc. Amer. Math. Soc. **31**, 97–103. MR **44** #7267.

[1972b] *Integral representation theorems in topological vector spaces*, Trans. Amer. Math. Soc. **172**, 373–397. MR **47** #826.

W. Sierpinski,

[1922] *Sur les fonctions d'ensemble additives et continues*, Fund. Math. **3**, 240–246.

[1938] *Fonctions additives non complètement additives et fonctions non mesurables*, Fund. Math. **30**, 96–99.

I. Singer (see also W. J. Davis and C. Foiaş),

[1957] *Linear functionals on the space of continuous mappings of a compact Hausdorff space into a Banach space*, Rev. Math. Pures Appl. **2**, 301–315. (Russian) MR **20** #3445.

[1958] *Les duals de certaines espaces de Banach de champs de vecteurs*. I, Bull. Sci. Math. (2) **82**, 29–40. MR **20** #3447.

[1959a] *Les duals de certaines espaces de Banach de champs de vecteurs*. II, Bull. Sci. Math. (2) **83**, 73–96. MR **22** #9840.

[1959b] *Sur les applications linéaires intégrales des espaces de fonctions continues*. I, Rev. Roumaine Math. Pures Appl. **4**, 391–401. MR **22** #5883.

[1959c] *Sur les applications linéaires majorées des espaces de fonctions continues*, Att. Accad. Naz. Lincei Rend. Cl. Sci. Fis. Mat. Nat. (8) **27**, 35–41. MR **22** #5885.

[1960a] *Sur la représentation intégral des applications linéaires continues des espaces* L_F^p $(1 \leqq p < +\infty)$, Atti. Accad. Naz. Lincei Rend. Cl. Sci. Fis. Mat. Nat. (8) **29**, 28–32. MR **23** #A2751.

[1960b] *Sur une class d'applications linéaires continues des espaces* L_F^p $(1 \leqq p < +\infty)$, Ann. Sci. École Norm Sup. (3) **77**, 235–256. MR **25** #3351.

[1960c] *Sur les applications linéaires intégrales des espaces de fonctions continues à valeurs vectorielles*, Acta. Math. Acad. Sci. Hungar. **11**, 3–13. MR **22** #5884.

[1975] *On the problem of non-smoothness of non-reflexive second conjugate spaces*, Bull. Austral. Math. Soc. **12**, 407–416.

M. Sion,

[1969] *Outer measures with values in a topological group*, Proc. London Math. Soc. (3) **19**, 89–106. MR **39** #398.

[1973] *A theory of semigroup valued measures*, Lecture Notes in Math, vol. 355, Springer-Verlag, Berlin and New York.

G. Sirvant,

[1950] *Weak compactness in Banach spaces*, Studia Math. **11**, 71–94. MR **14**, 183.

V. L. Šmul'yan,

[1939] *On some geometrical properties of the unit sphere in the space of the type* (B), Mat. Sbornik **6**, 77–94. (Russian).

[1941] *Sur la structure de la sphère unitaire dans l'espace de Banach*, Mat. Sb. **9 (51)**, 545–561.

A. Sobczyk,

[1941] *Projection of the space* (m) *on its subspace* (c_0), Bull. Amer. Math. Soc. **47**, 938–947. MR **3**, 205.

T. W. Starbird,

[1976] *Subspaces of* L^1 *containing* L^1, Ph. D. Thesis, Univ. of California, Berkeley, 1976.

C. Stegall (see also J. Hagler, D. R. Lewis and J. Lindenstrauss),

[1972] *Duals of certain spaces with the Dunford-Pettis property*, Notices Amer. Math. Soc. **19** (7) (1972), A-799. Abstract #699–B31.

[1973] *Banach spaces whose duals contain* $l_1(L)$ *with applications to the study of dual* $L_1(\mu)$ *spaces*, Trans. Amer. Math. Soc. **176**, 463–477. MR **47** #3953.

[1975] *The Radon-Nikodým property in conjugate Banach spaces*, Trans. Amer. Math. Soc. **206**, 213–223. MR **51** #10581.

C. Stegall and J. R. Retherford,

[1972] *Fully nuclear and completely nuclear operators with applications to $\mathscr{L}_1$- and $\mathscr{L}_\infty$-spaces*, Trans. Amer. Math. Soc. **163**, 457–492.

H. Steinhaus (see also S. Kaczmarz),

[1919] *Additive und stetige Funktional operationen*, Math. Z. **5**, 186–221.

W. J. Stiles,

[1970] *On subseries convergence in F-spaces*, Israel J. Math. **8**, 53–56. MR **41** #7405.

M. H. Stone,

[1937] *Applications of the theory of Boolean rings to general topology*, Trans. Amer. Math. Soc. **41**, 375–481.

[1949] *Boundedness properties in function-lattices*, Canad. J. Math. **1**, 176–186. MR **10**, 546.

S. Strătilă (see G. Arsene).

L. Sucheston (see R. V. Chacon).

V. N. Sudakov (see I. V. Romanovskii).

F. E. Sullivan,

[1976] *Geometrical properties determined by the higher duals of a Banach space* (to appear).

K. Sundaresan (see also I. E. Leonard, V. J. Mizel and J. J. Schäffer),

[1969] *Extreme points of the unit cell in Lebesgue-Bochner function spaces.* I, Proc. Amer. Math. Soc. **23**, 179–184.

[1970] *Extreme points of the unit cell in Lebesgue-Bochner function spaces*, Colloq. Math **22**, 111–119.

[1971] *Some geometric properties of the unit cell in spaces $C(X; B)$*, Bull. Acad. Polon. Sci. Sér. Sci. Math. Astronom. Phys. **19**, 1007–1012. MR **46** #2411.

[1974] *Banach lattices of Lebesgue-Bochner function spaces and conditioned expectation operators.* I, Bull. Inst. Math. Acad. Sinica **2**, 165–184. MR **50** #8058.

[1975] *Banach lattices of Lebesgue-Bochner function spaces and conditioned expectation operators*, II, Bull. Inst. Math. Acad. Sinica **3**, 249–257.

[1976] *The Radon-Nikodým theorem for Lebesgue-Bochner function spaces* (to appear).

K. Sundaresan and P. W. Day,

[1972] *Regularity of group valued Baire and Borel measures*, Proc. Amer. Math. Soc. **36**, 609–612. MR **46** #5567.

G. Sunochi (see M. Nakamura).

C. Swartz (see also G. Alexander),

[1972a] *Unconditionally converging operators on the space of continuous functions*, Rev. Roumaine Math. Pures Appl. **17**, 1695–1702. MR **48** #12137.

[1972b] *Vector measures of bounded variation*, Rev. Roumaine Math. Pures Appl. **17**, 1703–1704. MR **48** #12053.

[1973a] *An operator characterization of vector measures which have Radon-Nikodým derivatives*, Math. Ann. **202**, 77–84. MR **48** #4263.

[1973b] *The product of vector-valued measures*, Bull. Austral. Math. Soc. **8**, 359–366. MR **47** #6988.

[1973c] *Vector measures and nuclear spaces*, Rev. Roumaine Math. Pures Appl. **18**, 1261–1268. MR **49** #7772.

[1973d] *Absolutely summing and dominated operators on spaces of vector-valued continuous functions*, Trans. Amer. Math. Soc. **179**, 123–131. MR **47** #9330.

[1973e] *Weakly compact operators from a B-space into the space of Bochner integrable functions*, Colloq. Math. **28**, 69–76. MR **50** #14342.

[1974] *Products of vector measures*, Mat. Časopis Sloven. Akad. Vied. **24**, 289–299.

[1976] *Unconditionally converging and Dunford-Pettis operators on $C_X(S)$*, Studia Math. **57**, 85–90.

K. Swong,

[1964] *A representation theory of continuous linear maps*, Math. Ann. **155**, 270–291; errata, ibid. **157**, 178. MR **29** #2642.

J. Szulga and W. A. Woyczynski,

[1976] *Convergence of submartingales in Banach lattices*, Ann. of Probability (to appear).

J. D. Tamarkin (see E. Hille).

A. E. Taylor (see S. Bochner).

G. Erik F. Thomas,

[1968] *Sur le théorème d'Orlicz et un problème de M. Laurent Schwartz*, C. R. Acad. Sci. Paris Sér. A-B **267**, A7–A10. MR **37** #6432.

[1970a] *L'intégration par rapport a une mesure de Radon vectorielle*, Ann. Inst. Fourier (Grenoble) **20**(2), 55–191.

[1970b] *Le théorème de Lebesgue-Nikodým pour les mesures vectorielles et les applications* 1-*sommantes*, C. R. Acad. Sci. Paris **271**, 872–875.

[1970c] *Sur les applications linéaires* 1-*sommantes et nikodymisantes*, C. R. Acad. Sci. Paris Sér. A **278**, 253–255. MR **49** #9672.

[1974] *The Lebesgue-Nikodým theorem for vector valued Radon measures*, Mem. Amer. Math. Soc. no. 139.

[1975] *Integration of functions with values in locally convex Suslin spaces*, Trans. Amer. Math. Soc. **212**, 61–81.

[1976] *On Radon maps with values in arbitrary topological vector spaces and their integral extensions* (to appear).

A. C. Thompson (see M. Edelstein).

A. E. Tong (see also T. Cho and N. A. Friedman),

[1971] *Nuclear mappings on* $C(X)$, Math. Ann. **194**, 213–224. MR **45** #4204.

T. Traynor,

[1972a] *Decomposition of group-valued additive set functions*, Ann. Inst. Fourier (Grenoble) **22**, fasc. 3, 131–140. MR **48** #11439.

[1972b] *A general Hewitt-Yosida decomposition*, Canad. J. Math. **24**, 1164–1169. MR **47** #3636.

[1973] *S-bounded additive set functions*, Vector and Operator-Valued Measures and Applications (Proc. Sympos., Snowbird Resort, Alta, Utah, 1972), Academic Press, New York, 355–365. MR **48** #11440.

S. L. Troyanski,

[1971] *On locally uniformly convex and differentiable norms in certain non-separable Banach spaces*, Studia Math. **37** (1970/71), 173–180. MR **46** #5995.

[1972] *On equivalent norms and minimal systems in nonseparable Banach spaces*, Studia Math. **43**, 125–138. MR **48** #2734.

Ju. B. Tumarkin,

[1970] *On locally convex spaces with basis*, Dokl. Akad. Nauk SSSR **195**, 1278–1281 = Soviet Math. Dokl. **11**, 1672–1675. MR **42** #6575.

J. B. Turett,

[1976] *Fenchel-Orlicz spaces*, Ph.D. Dissertation, Univ. of Illinois, Urbana.

J. B. Turett and J. J. Uhl, Jr.,

[1976] $L_p(\mu, X)$ $(1 < p < \infty)$ *has the Radon-Nikodým property if X does by martingales*, Proc. Amer. Math. Soc. **61**, 347–350.

P. Turpin,

[1971a] *Généralisations d'un théorème de S. Mazur et W. Orlicz*, C. R. Acad. Sci. Paris Sér. A–B **273**, A457–A460. MR **45** #4106.

[1971b] *Variantes de résultats de S. Mazur et W. Orlicz et convexités généralisées*, C. R. Acad. Sci. Paris Sér. A–B **273**, A506–A509. MR **45** #4107.

[1972] *Mesures vectorielles pathologiques*, C. R. Acad. Sci. Paris Sér. A–B **275**, A981–A984. MR **47** #6989.

[1975a] *Suites sommables dans certains espaces de fonctions mesurables*, C. R. Acad. Sci. Paris **280**, 349–352.

[1975b] *Une mesure vectorielle non bornée*, C. R. Acad. Sci. Paris **280**, 509–511.

[1976] *Convexités dans les espaces vectoriels topologiques généraux* (to appear).

P. Turpin and L. Waelbroeck,

[1968] *Intégration et fonctions holomorphes dans les espaces localement pseudo-convexes*, C. R. Acad. Sci. Paris Sér. A–B **267**, A160–A162. MR **38** #2598.

I. Tweddle,

[1968] *Weak compactness in locally convex spaces*, Glasgow Math. J. **9**, 123–127. MR **39** #752.

[1970] *Unconditional convergence and vector-valued measures*, J. London Math. Soc. (2) **2**, 603–610. MR **42** #5006.

[1970a] *Vector-valued measures*, Proc. London Math. Soc. (3) **20**, 469–489. MR **41** #3707.

[1972a] *The extremal points of the range of a vector-valued measure*, Glasgow Math. J. **13**, 61–63. MR **46** #9290.

[1972b] *The range of a vector-valued measure*, Glasgow Math. J. **13**, 64–68. MR **46** #9291.

L. Tzafriri (see J. Lindenstrauss).

D. J. Uherka,

[1969] *Generalized Stieltjes integrals and a strong representation theorem for continuous linear maps on a function space*, Math. Ann. **182**, 60–66. MR **40** #705.

J. J. Uhl, Jr. (see also J. Diestel, N. Dinculeanu, N. E. Gretsky, S. Moedomo and J. B. Turett),

[1967] *Orlicz spaces of finitely additive set functions*, Studia Math. **29**, 19–58. MR **37** #1985.

[1969a] *The compactness of certain dominated operators*, Rev. Roumaine Math. Pures Appl. **14**, 1635–1637. MR **41** #2452.

[1969b] *Applications of Radon-Nikodým theorems to martingale convergence*, Trans. Amer. Math. Soc. **145**, 271–285. MR **40** #4983.

[1969c] *The Radon-Nikodým theorem and the mean convergence of Banach space valued martingales*, Proc. Amer. Math. Soc. **21**, 139–144. MR **38** #6680.

[1969d] *Martingales of vector valued set functions*, Pacific J. Math. **30**, 533–548. MR **40** #1767.

[1969e] *The range of a vector-valued measure*, Proc. Amer. Math. Soc. **23**, 158–163. MR **41** #8628.

[1970] *Vector integral operators*, Nederl. Akad. Wetensch. Proc. Ser. A **73** = Indag. Math. **32**, 463–478. MR **44** #2103.

[1971a] *Extensions and decompositions of vector measures*, J. London Math. Soc. (2) **3**, 672–676. MR **44** #4181.

[1971b] *On a class of operators on Orlicz spaces*, Studia Math. **40**, 17–22. MR **46** #2479.

[1971c] *Abstract martingales in Banach spaces*, Proc. Amer. Math. Soc. **28**, 191–194. MR **44** #826.

[1972a] *A note on the Radon-Nikodým property for Banach spaces*, Rev. Roumaine Math. Pures Appl. **17**, 113–115.

[1972b] *Martingales of strongly measurable Pettis integrable functions*, Trans. Amer. Math. Soc. **167**, 369–378; erratum, ibid. **181** (1973), 507. MR **45** #2785; **47** #7807.

[1972c] *A characterization of strongly measurable Pettis integrable functions*, Proc. Amer. Math. Soc. **34**, 425–427. MR **47** #5222.

[1973a] *A survey of mean convergence of martingales of Pettis integrable functions*, Vector and Operator-Valued Measures and Applications (Proc. Sympos., Snowbird Resort, Alta, Utah, 1972), Academic Press, New York, pp. 379–385. MR **49** #528.

[1973b] *Applications of a lemma of Rosenthal to vector measures and series in Banach spaces* (unpublished).

[1975] *Norm attaining operators on $L^1[0, 1]$ and the Radon-Nikodým property*, Pacific J. Math **63**, 293–300.

[1976] *Pettis mean convergence of vector-valued asymptotic martingales* Z. Wahrscheinlichkeits-theory und Verw. Gebiete (in press).

N. Th. Varopoulos,

[1966] *Sets of multiplicity in locally compact abelian groups*, Ann. Inst. Fourier (Grenoble) **16**, fasc. 2, 123–158. MR **35** #3379.

W. A. Veech,

[1971] *A short proof of Sobczyk's theorem*, Proc. Amer. Math. Soc. **28**, 627–628. MR **43** #879.

W. P. Veith,

[1971] *Bounded Banach-valued functions with almost periodic differences*, Boll. Un. Mat. Ital. (4) **4**, 220–224. MR **45** #9059.

[1974] *A bounded difference property for classes of Banach-valued functions*, Trans. Amer. Math. Soc. **190**, 49–56.

G. Vitali,

[1905] *Sulle funzioni integrali*, Atti. R. Accad. delle Sci. di Torino **40**, 753–766.

[1907] *Sull' integrazione per serie*, Rend. del Circ. Mat. de Palermo **23**, 137–155.

D. Vogt,

[1967] *Integrationstheorie in p-normierten Räumen*, Math. Ann. **173**, 219–232. MR **36** #345.

A. Wald (see G. B. Dantzig and A. Dvoretsky).

L. Waelbroeck (see P. Turpin).

B. J. Walsh,

[1971] *Mutual absolute continuity of sets of measures*, Proc. Amer. Math. Soc. **29**, 506–510. MR **43** #4998.

R. Wegmann,

[1970] *Der Wertebereich von Vektorintegralen*, Z. Wahrscheinlichkeitstheorie und Verw. Gebiete **14** (1969/70), 203–238. MR **41** #7064,

B. B. Wells, Jr.,

[1969] *Weak compactness of measures*, Proc. Amer. Math. Soc. **20**, 124–130. MR **38** #6343,

D. G. Wertheim (see G. G. Lorentz).

J. H. M. Whitfield (see E. B. Leach).

H. Witsenhausen,

[1973] *Metric inequalities and the zonoid problem*, Proc. Amer. Math. Soc. **40**, 517–520.

J. Wolfe (see also J. W. Baker and H. B. Cohen),

[1971] *Injective Banach spaces of type $C(T)$*, Ph. D. Thesis, Univ. of California, Berkeley.

[1974] *Injective Banach spaces of type $C(T)$*, Israel J. Math. **18**, 133–140. MR **50** #5445.

J. Wolfowitz (see A. Dvoretsky).

T. K. Wong,

[1971] *On a class of absolutely p-summing operators*, Studia Math. **39**, 181–189. MR **46** #7948.

W. A. Woyczyński,

[1969] *Sur la convergence des séries dans les espaces de type* (L), C. R. Acad. Sci. Paris Sér A-B **268**, A1254-A1257. MR **39** #6049.

[1975] *Geometry and martingales in Banach spaces*, Proc. Fourth Winter School on Probability (Karpacz, 1975), Lecture Notes in Math., vol. 472, Springer-Verlag, Berlin and New York, pp. 229–275. MR **51** #14179.

J. D. M. Wright,

[1968] *Applications to averaging operators of the theory of Stone algebra valued modular measures*,

Quart. J. Math. Oxford Ser. (2) **19**, 321–331. MR **40** #1576.

[1969a] *Stone-algebra-valued measures and integrals*, Proc. London Math. Soc. (3) **19**, 107–122. MR **39** #1625.

[1969b] *A Radon-Nikodým theorem for Stone algebra valued measures*, Trans. Amer. Math. Soc. **139**, 75–94. MR **40** #1577.

[1970] *A decomposition theorem*, Math. Z. **115**, 387–391. MR **42** #4691.

[1971] *Vector lattice measures on locally compact spaces*, Math. Z. **120**, 193–203. MR **45** #2450.

K. Yosida and E. Hewitt,

[1952] *Finitely additive measures*, Trans. Amer. Math. Soc. **72**, 46–66. MR **13**, 543.

A. C. Zaanen (see also W. A. J. Luxemburg),

[1953] *Linear analysis*, Interscience, New York; North-Holland, Amsterdam; Noordhoff Groningen. MR **15**, 878.

M. Zippin (see W. B. Johnson).

V. Zizler (see also K. John),

[1973] *On some extremal problems in Banach spaces*, Math. Scand. **32**, 214–224 (1974). MR **49** #11217.

SUBJECT INDEX

AUTHOR INDEX

Numbers refer to pages in the Notes and Remarks section of each chapter where reference is made to an author or work of an author.